Micro- and Nanostructured Multiphase Polymer Blend Systems

Phase Morphology and Interfaces

Micro- and Nanostructured Multiphase Polymer Blend Systems

Phase Morphology and Interfaces

Edited by
Charef Harrats • Sabu Thomas
Gabriel Groeninckx

CRC Press
Taylor & Francis Group
Boca Raton London New York

CRC Press is an imprint of the
Taylor & Francis Group, an **informa** business

A TAYLOR & FRANCIS BOOK

First published 2006 by Taylor & Francis

Published 2019 by CRC Press
Taylor & Francis Group
6000 Broken Sound Parkway NW, Suite 300
Boca Raton, FL 33487-2742

© 2006 by Taylor & Francis Group, LLC
CRC Press is an imprint of Taylor & Francis Group, an Informa business

First issued in paperback 2019

No claim to original U.S. Government works

ISBN-13: 978-0-367-45403-6 (pbk)
ISBN-13: 978-0-8493-3734-5 (hbk)

Library of Congress Card Number 2005046252

Library of Congress Cataloging-in-Publication Data

Micro- and nanostructured multiphase polymer blend systems : phase morphology and interfaces / edited by Charef Harrats, Sabu Thomas, Gabriel Groeninckx.
    p. cm.
Includes bibliographical references and index.
ISBN 0-8493-3734-8
    1. Polymers--Microstructure. 2. Nanostructured materials. 3. Phase transformations (Statistical physics) I. Harrats, Charef. II. Thomas, Sabu. III. Groeninckx, Gabriel.

QC173.4.P65M53 2005
668.9--dc22   2005046252

Visit the Taylor & Francis Web site at
http://www.taylorandfrancis.com

and the CRC Press Web site at
http://www.crcpress.com

Dedication

The editors Charef Harrats and Sabu Thomas have the great pleasure of dedicating this book to Professor Gabriel Groeninckx for the occasion of his retirement (October 2005), for the great and valuable contribution he made during his scientific career in the field of polymer blends and related domains.

Preface

Many new multicomponent polymeric materials have been developed during the past two decades. The large number of scientific papers, industrial patents, scientific meetings, and exhibitions devoted to this class of materials is a sufficient witness to their strategic importance.

The present book is mainly centered around the phase morphologies and interface multiphase polymer blend systems. Since phase morphologies and interface properties are two of the main parameters that determine the thermal, physical, chemical, and mechanical properties of polymer blends, the book covers a very sensitive area, one that relates to the performance of polymer blend–based materials.

This book is intended to be a reference for basic and practical knowledge about phase morphology in multiphase polymer blend systems for students, engineers, and researchers. The way the topics are gathered, the selection of the contributors, and the survey of the phase morphology area (from its theoretical to its practical aspects) make this book an outstanding scientific reference for those involved in the field of polymer materials design. The book is an easy-to-consult volume for teachers giving courses in the field of polymer materials science.

The state-of-the-art challenges and future prospects of micro- and nanostructured polymer blends are discussed in Chapter 1. Emphasis is put on the role that phase morphology plays in controlling the mechanical properties of polymer blends. Details about the theory, the experimental aspects, and the design of blends exhibiting droplet-in-matrix and two-phase cocontinuous phase morphologies are highlighted in Chapter 2 and Chapter 3, respectively. Chapter 4 considers the processing and experimental aspects of the phase morphology development in polymer blends.

The role of the interface and the phase morphology with respect to the mechanical properties of multiphase copolymer systems are discussed in Chapter 5. The theory, experiments, and adhesion aspects of polymer-polymer interfaces in immiscible polymer blends are clarified in Chapter 6. The behavior of the phase morphology of polymer blends when they are subjected to solidification and shear conditions are systematically elucidated in Chapter 7.

Chapter 8 and Chapter 11 are devoted to the study of thermoset/thermoset systems. The former considers simultaneous interpenetrated thermosets by emphasizing recent developments in vinylester/epoxy hybrid systems, whereas the latter highlights the different ways in which nanostructures may be generated in thermosetting polymers. A general classification is proposed based on whether phase segregation occurs before, during, or after the polymerization process.

The most important features of Chapter 9 are related to dynamically vulcanized multicomponent polymer systems. This chapter reports on the peculiarities of morphology formation in the dynamic vulcanization of polymer blends. The mechanism of phase morphology formation under conditions of selective cross-linking of a major rubber phase in a minor thermoplastic matrix is described in detail.

New strategies for controlling the (nano)morphology and ultimately the mechanical properties of multiphase polymeric materials based on polyamide-12 (PA12) are discussed in Chapter 10. The phase morphology controlled by the self-assembly of linear polystyrene-b-polyisoprene-b-polyamide12 triblock copolymers, formed *in situ* during processing, is described. Design of a nanostructured PA12 matrix using reactive PS-PIP diblock copolymers is studied as a function of the relative weight fraction of the PS block within the copolymer.

In Chapter 12, an overview of the crystallization phenomena occurring in polymers and polymer blends with a confined morphology is presented. It is shown that the crystallization behavior of such systems can be drastically affected compared to the bulk state and that crystallization takes place in different steps at much higher supercoolings. The interrelation between phase morphology and rheology of polymer blends is clearly highlighted in Chapter 13. More specifically, the focus is put on three main topics, including the structure-rheology relationship in compatibilized blends, the effects of elasticity on structure development, and the studies that relate the rheological response to the underlying morphology in concentrated blends.

Charef Harrats
Sabu Thomas
Gabriel Groeninckx

Contributors

Philippe Cassagnau
Laboratoire des Matériaux Polymères et
des Biomatériaux
Ingénierie des Matériaux Polymères
Villeurbanne, France

Yves Deyrail
Laboratoire des Matériaux Polymères et
des Biomatériaux
Ingénierie des Matériaux Polymères
Villeurbanne, France

Ivan Fortelný
Academy of Sciences of the Czech
Republic
Institute of Macromolecular Chemistry
Prague, Czech Republic

René Fulchiron
Laboratoire des Matériaux Polymères et
des Biomatériaux
Ingénierie des Matériaux Polymères
Villeurbanne, France

Johannes G.P. Goossens
Eindhoven University of Technology
Department of Polymer Technology
Eindhoven, The Netherlands

Gabriel Groeninckx
Katholieke Universiteit Leuven
Department of Chemistry
Leuven, Belgium

Charef Harrats
Katholieke Universiteit Leuven
Department of Chemistry
Heverlee, Belgium

Robert Jérôme
University of Liège
Center for Education and Research on
Macromolecules
Liège, Belgium

J. Karger-Kocsis
Institut für Verbundwerkstoffe GmbH
Kaiserslautern University of
Technology
Kaiserslautern, Germany

Christian Koulic
Total Feluy S.A.
Polyethylene Department
Feluy, Belgium

V.B.F. Mathot
Katholieke Universiteit Leuven
Department of Chemistry
Leuven, Belgium

Nafaa Mekhilef
Arkema Inc.
King of Prussia Technical Center
King of Prussia, Pennsylvania

Paula Moldenaers
Katholieke Universiteit Leuven
Department of Chemical Engineering
Leuven, Belgium

Jean-Pierre Pascault
Institut National des Sciences
Appliqués de Lyon
Laboratoire des Matériaux
Macromoléculaires
Villeurbanne, France

Peter Van Puyvelde
Katholieke Universiteit Leuven
Department of Chemical Engineering
Leuven, Belgium

Hans-Joachim Radusch
Martin Luther Universität
Halle Wittenberg Institut für
  Werkstoffwissenschaft LS
  Kunststofftechnik
Merseburg, Germany

H. Reynaers
Katholieke Universiteit Leuven
Department of Chemistry
Leuven, Belgium

Manfred Stamm
Leibniz Institute of Polymer Research
Dresden, Germany

Uttandaraman Sundararaj
University of Alberta
Department of Chemical & Materials
  Engineering
Edmonton, Alberta, Canada

Sabu Thomas
Mahatma Gandhi University
School of Chemical Sciences
Kottayam, Kerala, India

R.T. Tol
Katholieke Universiteit Leuven
Department of Chemistry
Leuven, Belgium

Roland Weidisch
Leibniz Institute of Polymer Research
Dresden, Germany

Roberto J.J. Williams
University of Mar del Plata and
  National Research Council
Institute of Materials Science and
  Technology
Mar del Plata, Argentina

Hideaki Yokoyama
Nanotechnology Research Institute
National Institute of Advanced
  Industrial Science and Technology
Ibaraki, Japan

Contents

1 Micro- and Nanostructured Polymer Blends: State of the Art, Challenges, and Future Prospects

Sabu Thomas, Charef Harrats, and Gabriel Groeninckx

CONTENTS

1.1 INTRODUCTION

Polymer blending has been identified as the most versatile and economical method to produce new multiphase polymeric materials that are able to satisfy the complex demands for performance. Over the past few decades the number of polymer blends has grown tremendously. In fact the design and development of these multiphase polymer blend materials are strongly dependent on two major parameters: the control of the interface and the control of the morphology. In general, the term *morphology*

1

refers to the shape and organization on a scale above the atomic level (e.g., the arrangement of polymer molecules into amorphous or crystalline regions) and the manner in which they are organized into more complex units. On the other hand, morphology of a polymer blend indicates the size, shape, and spatial distribution of the component phases with respect to each other. Since it is well established that most of the properties — mechanical, optical, rheological, dielectrical, and barrier properties — of polymer blends are strongly influenced by the type and fineness of the phase structure, the study of the control of the morphology of polymer blends has emerged as an area of continuous interest to polymer material scientists in the last few decades (1–5).

When two immiscible polymers are mixed, the size, shape, and distribution of one phase into the other depends on material parameters (i.e., blend composition, viscosity ratio, elasticity ratio, and interfacial tension) as well as on processing conditions (i.e., temperature, time, and intensity of mixing, and the nature of the flow). Therefore, the greatest challenge in the field of multiphase polymer blend research is the manipulation of the phase structure via a judicious control of the melt flow during processing and the interfacial interactions between the components. The mechanism of morphology development from pellet-sized or powder-sized particles in polymer blends is directly derived from the complex interplay of material parameters and processing conditions. As a result of this, for a given blend, various types of useful morphologies (Figure 1.1) for different end properties such as high strength and toughness, toughness coupled with stiffness, good barrier properties, and high flow can be obtained by a judicious mixing process (2). However, from the point of view of a broader classification, multiphase polymer blends may be divided into two major categories:

1. Blends with a discrete phase structure (i.e., droplets in matrix)
2. Blends with a bicontinuous phase structure (i.e., cocontinuous)

Other types of morphologies include fibrillar (Figure 1.2), core shell (Figure 1.3), and onion ring-like (Figure 1.4) morphologies.

During the last few years, nanostructured polymer blend systems have become increasingly important (6). In nanoblends, the scale of dispersion of one polymer into another is in general below 100 nm. Nanostructured polymer blends very often exhibit unique properties that are directly attributed to the presence of structural entities having dimensions in the nanometer range. The idealized morphology of these polymer blend systems is characterized by the molecular level dispersion of the phases that leads to a considerable enhancement in the mechanical properties, especially the modulus. Nanostructured self-assembled polymer blends are a novel class of materials with enhanced electronic and optical properties. By using advanced processing methods, polymer blends self assemble with nanoscale microstructures and with significantly increased interfacial contact. The latter is responsible for these enhanced electronic and optical properties (6).

This chapter deals with the development of the phase morphology in polymer blends with an emphasis on micro- and nanophase morphologies. The discussion focuses on the relationships between phase morphology and ultimate mechanical

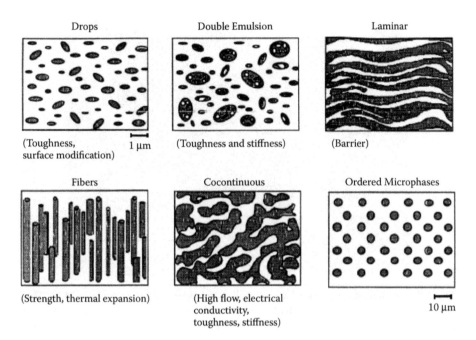

Drops	Double Emulsion	Laminar
(Toughness, surface modification) 1 μm	(Toughness and stiffness)	(Barrier)
Fibers	Cocontinuous	Ordered Microphases
(Strength, thermal expansion)	(High flow, electrical conductivity, toughness, stiffness)	10 μm

FIGURE 1.1 Schematic of useful morphologies of polymer blends. (Reproduced from Macosko, C.W., *Macromol. Symp.*, 149, 171–184, 2000. With permission.)

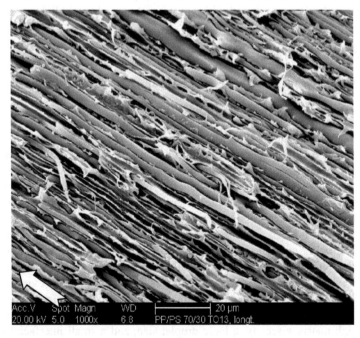

FIGURE 1.2 Scanning electron microscopy (SEM) micrograph of a fibrillar phase morphology obtained from extruded 70:30 wt% polypropylene-polystyrene blend.

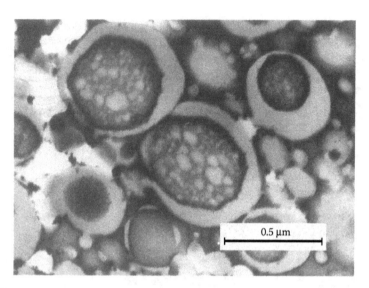

FIGURE 1.3 SEM micrograph of a core shell morphology. (Reproduced from Mark, H.F., Bikales, M., Overberger, C.G., Menges, G., and Kroschwitz, J.I., Eds., Encyclopedia of Polymer Science and Engineering, 2nd Ed., John Wiley & Sons, Inc., New York, 1987. With permission.)

properties. The recent advances in nanostructured polymer blend systems as well as the future challenges and opportunities in this field are also highlighted.

1.2   TYPES OF PHASE MORPHOLOGIES

1.2.1   DISPERSED PHASE AND MATRIX MORPHOLOGY

1.2.1.1   Droplet and Matrix Morphology

The morphology formation during melt-mixing of immiscible polymers involves processes such as liquid drops stretching into threads, breakup of the threads into smaller droplets, and coalescence of the droplets into larger ones (7). The balance of these competing processes determines the final particle size of the blends that results upon solidification of the blends. There are a variety of parameters that control the droplet and matrix morphology in polymer blends (Figure 1.5). These include the viscosity ratio, blend composition, elasticity ratio, shear stress, and interfacial tension. The minor component will be finely dispersed in the matrix if its viscosity is lower than that of the matrix (8). A coarse dispersion is expected in the opposite case. However, the most used parameter remains the viscosity ratio. A viscosity ratio close to unity allows for the generation of the smallest particle size. According to Taylor's analysis (9,10), the deformation of a droplet is enhanced by large shear rates, a high matrix viscosity, a large droplet size, and a small interfacial tension. The deformation is retarded by a large interfacial tension, high dispersed phase viscosity, and a small droplet size. Taylor noted that if a lower or equal viscosity

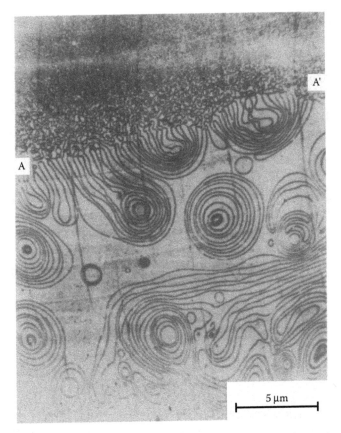

FIGURE 1.4 Transmission electron microscopy (TEM) micrograph of an onion ring morphology. (Reproduced from Mark, H.F., Bikales, M., Overberger, C.G., Menges, G., and Kroschwitz, J.I., Eds., Encyclopedia of Polymer Science and Engineering, 2nd Ed., John Wiley & Sons, Inc., New York, 1987. With permission.)

droplet is placed in a high viscosity matrix, it is readily drawn into a long filamentous ligament that eventually breaks up. This was not the case for high viscosity drops placed in a low viscosity matrix; these tend to retain their spherical shape. Interestingly, several studies on Newtonian systems in both shear and extensional flows were in agreement with Taylor's analysis. According to Grace (11) and Rumscheidt and Mason [12], the droplets deform into ellipsoids under the influence of shear flow but do not break up at a viscosity ratio $p > 3.7$. However, it has been reported that drop breakup was possible over a wide range of viscosity ratios in extensional flow for newtonian systems (13,14). In the case of viscoelastic systems like polymer blends, since the individual components exhibit a large normal stress in flow, the extension of Taylor's analysis to such systems has some limitations. It is also important to mention that the final phase morphology of polymer blends is the result of the equilibrium between domain breakup and coalescence at the end of the mixing process. However, most of the studies on polymer blends regarding domain size, expressed in terms of capillary number and viscosity ratio, do not account for the

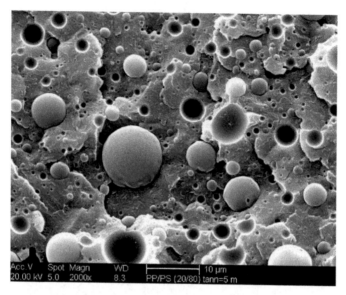

FIGURE 1.5 SEM micrograph of droplet and matrix morphology of 20PS/80PP blends.

coalescence process. An empirical relation that fit the capillary master curve was proposed by Wu (15):

$$\eta_m D \gamma / \Gamma = 4(p)^{\pm 0.84} \tag{1.1}$$

where η_m is the matrix viscosity, D is the shear rate, γ is the particle diameter, Γ is the interfacial tension, and p is the particle to matrix viscosity ratio. The value of $\eta_m D\gamma/\Gamma$ is positive when $p > 1$ and negative when $p < 1$. The formula suggests the existence of a minimum particle size when the viscosities of the two phases are closely matched. As the viscosity moves away form unity in either direction, the dispersed particles become larger. The results obtained by Serpe et al. (16) are in good agreement with Wu's relation. In the case of ethylene-propylene-diene-mono-mer/polypropylene (EPDM/PP) blends, Karger-Kocsis et al. (17) reported that a fine dispersion can be achieved when the value of viscosity ratio is near unity. Avgeropolous et al. (18) have shown that in the case of ethylene-propylene-diene-monomer/butadiene rubber (EPDM/BR) blends, the particle size of the dispersed phase increases with the ratio of the mixing viscosities of the components. In the case of polycarbonate/polypropylene blends, Favis and Chalifoux (19) have shown that major particle disintegration occurs even at a torque ratio of 13 (torque of the dispersed phase relative to the major phase). The dispersed phase dimension increased by a factor of 3 to 4 from a torque ratio of 2 to 13. A minimum particle size was obtained at a torque ratio of approximately 0.25. From experiments and Taylor's predictions, it is shown that a lower viscosity ratio will yield a finer dispersed particle size under the conditions of steady breakup in shear flow. It is important to note that in the case where capillary instabilities play an important role, the mechanism controlling the influence of the viscosity ratio of the components can be very different.

The dispersed phase morphology of polymer blends is also strongly affected by the composition of the blends. Increasing the fraction of the dispersed phase results in an increase in the size of the coalescing particles. Indeed, by increasing the concentration of the second phase, the number of particles in the system increases, leading to an increased number of particle-particle collisions. First, the particles collide and undergo deformation, resulting in the expulsion of the matrix material from the interparticle region. The particles then rupture, and a neck is formed. Finally, the drops coalesce into one particle. The major factors influencing the coalescence include the volume fraction of the dispersed phase, the diameter of the particles, the viscosity of each phase, the interfacial tension, and the mobility of the interface. Two types of coalescence mechanisms exist: the first being determined by the equilibrium thermodynamics and the second caused by the flow. Coalescence occurs in flowing systems as well as in quiescent systems. The latter type follows the Oswald ripening process, which is characterized by a linear increase of the drop volume with time. The process involves diffusion from the smaller drops to the larger ones. Flow modifies the rate of the process, which depends on the drop dynamic cross-section. According to Tokita (20), the final drop diameter in polymer blends results from two competitive processes: continuous break up and coalescence of the dispersed particles. The equilibrium drop diameter should increase with concentration of the dispersed phase, number of drops, and the interfacial tension coefficient; the equilibrium drop diameter should decrease with shear stress. According to Tokita, at equilibrium when breakdown and coalescence are balanced, the following expression holds for the particle size at equilibrium, d_e:

$$d_e \approx 24 P_r \Gamma \pi / \tau_{12} (\Phi_d + 4 P_r E_{DK} / \pi \tau_{12} \Phi^2_d) \qquad (1.2)$$

where τ_{12} is the shear stress, Γ is again the interfacial tension, E_{DK} is the bulk breaking energy, Φ_d is the volume fraction of the dispersed phase, and P_r is the probability that collision will result in coalescence. The theory clearly predicts that the particle size at equilibrium diminishes as the magnitude of the stress field increases, the interfacial tension decreases, and the volume fraction of the dispersed phase decreases. This is in agreement with the pioneering work of Taylor (9,10). The experimental results reported by Favis and Willis (21) are in good agreement with the model of Tokita. For a mobile interface, Elmendorp and Van der Vegt (22) developed an expression to describe the shear induced coalescence of spherical particles. The critical coalescence time, t_c, is

$$t_c = (3 \eta_m R / 2 \Gamma) \ln(R / 2 h_c) \qquad (1.3)$$

where h_c is the critical separation time and R is the radius of the particle. However, according to Utracki and Shi (23), owing to the large size of the polymer blend droplets and the very low diffusion coefficient of polymer molecules in the melt, it is hard to interpret the coalescence phenomena based on brownian motion.

The influence of the elasticity ratio on the dispersed phase morphology in binary polymer-polymer blends is still not well understood. According to Van Oene (14), there are two modes of dispersion in capillary flow: stratification and droplet-fiber formation. The formation of these microstructures is controlled by the particle size, the interfacial tension, and the difference in viscoelastic properties between the two phases. For blends of polymethylmethacrylate (PMMA) and polystyrene (PS) exhibiting large second normal stress, particles of PMMA were formed in the PS matrix. Addition of low molecular weight PMMA to the same blend resulted in stratification. When the dispersed particle size is smaller than 1 μm, the difference in morphology (droplets vs. stratification) vanished, showing that the elastic contribution to the interfacial tension was no longer dominant. The elastic contribution to the interfacial tension can lead to the encapsulation of the less elastic component by the more elastic one. Van Oene used the following equation to relate the effective interfacial tension to the droplet diameter and the normal stress functions:

$$\Gamma_{eff} = \Gamma + d/12 \ (N_{2d} - N_{2m}) \tag{1.4}$$

where Γ_{eff} is the effective interfacial tension under dynamic conditions, Γ is the static interfacial tension, d is the droplet diameter, and N_{2d} and N_{2m} are the second normal stress functions for the dispersed phase and matrix, respectively. Levitt et al. (24) have shown that in simple shear flow, polypropylene (PP) drops stretch perpendicularly to the flow direction in a more highly elastic matrix. The effect was found to be proportional to the second normal stress differences between the two phases. The recent studies of Migler (25) and Hobbie and Migler (26) clearly demonstrated that because of the high droplet elasticity, the droplet can align in the vorticity direction rather than in the flow direction. This behavior was related to the high normal forces in the droplets and the presence of closed strain lines that form in the flow gradient plane. The role of the molecular weight on the coalescence process has been examined recently by Park et al. (27). The droplet trajectories during glancing collisions were carefully monitored. By using matrix fluids of varying molecular weights and keeping the viscosity ratio constant, it was observed that coalescence is facilitated as the molecular weight of the matrix fluid was increased. Of course, one might naturally expect a lower coalescence at high matrix viscosities due to steric hindrance and the consequent slowing down of the drainage process. The existence of the so-called slip layer around the polymer interfaces might be contributing to the increased coalescence.

The influence of shear on the phase morphology has also been investigated. Based on Taylor's analysis, the dispersed phase size should be inversely proportional to the applied shear stress. Min et al. (28) found that for polyethylene/polystyrene (PE/PS) blend system, affine morphology is created by the application of a high shear stress. In this blend system, the shear stress appears to predominate over the viscosity ratio. However, other authors (29–31) have shown that varying the shear stress by a factor of two to three has little effect on the particle size. Favis (8) has further shown that the dispersed phase morphology is not highly sensitive to changes in shear stress and shear rate in an internal mixer. These results suggest that the Taylor theory overestimates particle size. The studies performed by Sundararaj and Macosko (32), and Cigana et al. (33) came to the same conclusion, and they related this to the viscoelastic nature of the droplet.

The dispersed phase morphology can be controlled by interface modification via the incorporation of compatibilizers. The compatibilizer can be added into the blend as a third component (physical compatibilization) (34–36), or it can be generated in situ during processing (reactive compatibilization) (37,38). The major effects of compatibilization are to reduce the particle size of the dispersed phase, to narrow the particle size distribution, to increase the interfacial adhesion between the dispersed phase and the matrix, and to stabilize the phase morphology against coalescence. In fact, the decreased interfacial tension and reduced coalescence account for the decrease in dispersed phase particle size. Recently, the role of block copolymers on the suppression of droplet coalescence has been reported by Lyu and coworkers (39). The effect of the Marangoni stress on the deformation and coalescence in compatibilized polymer blends has been investigated by Van Puyvelde et al. (40). In well compatibilized blends where there is strong interaction between the dispersed phase and matrix, the average particle size is independent of composition right up to the region of dual phase continuity. This was found to be true for a large number of systems such as SMA (styrene-co-maleic anhydride copolymer)/bromobutyl blends compatibilized with dimethyl amino ethanol and ethylene-propylene rubber (EPR/PS) blends compatibilized using tapered styrene-ethylene-butylene rubber diblock copolymer (22,33,41). In all these cases, the uncompatibilized blends showed measurable coalescence at concentrations above 5% of the dispersed phase. Recently also a number of investigations have been performed on the rheology and morphology of compatibilized polymer blends (42–46).

1.2.1.2 Fibrillar Morphology

Fiberlike morphologies (Figure 1.2) can be generated by the deformation of the dispersed phase. Several parameters such as capillary instabilities, deformation behavior, coalescence, and the state of the interface influence the formation of fiberlike morphologies. Elongational and orientating flow fields, as encountered in injection molding processes, are more effective than shear flow fields in transforming droplets into fibrous domains (47–51). The fiber formation was observed at the entrance of the capillary where the contraction generates an elongational flow. The role of the viscosity on the droplet deformation was also elucidated (49). Elmendorp and Maalcke (51) have noticed important effects of the normal forces generated by a viscoelastic droplet on the minor phase deformation and breakup. Favis [8] reported that the elongational flow generated during unidirectional melt-drawing is very effective for the formation of dispersed fibers. According to Tomotika's theory, fiber formation will be stabilized if the interfacial tension is low and the matrix viscosity and thread radius are high. The role of the length of the capillary in the formation of fiberlike morphologies has been studied by LaMantia et al. (52). The liquid crystalline polymer (LCP) fibers that are formed at the converging region of the capillary could retain their fibrillar structure when extruded with a capillary having small length to diameter (L/D) ratio. However, when the L/D ratio was increased, for example, to 40, the LCP phase lost its fibrillar structure. This was attributed to the high shear that was generated, which causes breakup of the fibers, and also to the orientation relaxation time of the LCP, which is of shorter duration than the average time it takes for the polymer to flow through the capillary. The role of coalescence in fiber formation has also been reported. The coalescence of the dispersed particles will lead to the modification of the spherical droplets. The studies of Tsebrenko et al. (53,54) have shown that at the entrance region of the capillary rheometer, the dispersed domains undergo coalescence to develop fiberlike morphologies.

Recently, *in situ* reinforced polymer-polymer composites, so-called microfibrillar composites (MFCs), have been developed by fibrillation of the dispersed phase. Melt blending of two crystallisable immiscible thermoplastic polymers have been performed followed by cold or hot drawing and the subsequent annealing of the drawn blend. Upon drawing, the blends are oriented and microfibrils are formed in the discontinuous (minor) phase. The structure of the material is further developed by subsequent annealing heat treatment. The temperature and duration of this processing step have been shown to significantly affect the structure and properties of the blend. If the annealing temperature is set below the melting points of both components, the microfibrillar structure imparted by drawing is preserved and further improved as a result of physical processes, such as additional crystallisation, minimization of defects in the crystalline regions, and relaxation of residual stress in the amorphous regions. Details of the process and control of the MFC technique is reported elsewhere (55).

1.2.1.3 Droplet in Droplet Morphology

The formation of a droplet in droplet morphology or a composite droplet morphology has been reported in polymer blends (8). The terms subinclusion, core shell structure, and salamilike structure have also been used to describe this type of morphology (Figure 1.6). It is important to emphasize that the origin and the mechanisms for the formation of a particle in particle morphology are not yet fully understood. For uncompatibilized blend systems, several researchers have reported on the particle in particle morphology (56–58). In the case of a binary blend, particle in particle morphology can be spontaneously generated when blending polymers are near the phase inversion region or by selectively imposing phase inversion and subsequently controlling the time of mixing. Hobbs et al. (58) reported on the spontaneous development of composite droplet morphology for a series of ternary blend systems. They have suggested that the subinclusion formation can be predicted by the Harkins equation,

$$\lambda_{31} = \Gamma_{12}\ \Gamma_{32}\ \Gamma_{13} \qquad (1.5)$$

where λ_{31} is the spreading coefficient for component 3 to encapsulate component 1, and Γ_{12}, Γ_{32}, and Γ_{13} are the interfacial tensions between the respective polymer pairs in the blend. When λ_{31} is positive, component 3 will be encapsulated by component 1. The work of Guo and coworkers (59) indicated that the interfacial tension plays the major role in establishing the phase structure. A less significant but nevertheless important role is played by the surface area of the dispersed phase. Pagnoulle and Jérôme (60) have reported that depending on the mixing sequence and on the grafting kinetics, particle in particle morphology can be forced or occurs spontaneously during the compatiblization of modified styrene-acrylonitrile (SAN) with modified EPDM. According to these authors, subinclusions can be spontaneously generated by coa-

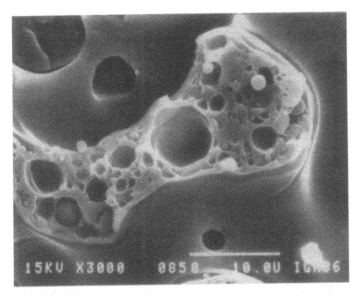

FIGURE 1.6 SEM micrograph of droplet in droplet (composite droplet) morphology. (Reproduced from Favis, B.D., "Factors influencing the morphology of immiscible blends in melt processing" in Polymer Blends, Vol. 1, Formulation, Paul, D.R. and Bucknall, C.B., Eds., John Wiley & Sons, New York, 2000, pp. 501–537. With permission.)

lescence of poorly stabilized dispersed particles during the later stage of the mixing, i.e., when coalescence dominates over the melting or softening process. Very recently, the spontaneous or forced formation of particle in particle morphology during reactive processing of polybutylene terephthalate (PBT) with ethyl-methacrylate-glycidyl methacrylate (E-MA-GMA) random terpolymer has been investigated by Martin et al. (61). Using a one-step compounding process, a composite droplet morphology — consisting of a PBT matrix, a E-MA-GMA dispersed phase, and PBT — subinclusions are generated as a result of the coalescence of poorly stabilized E-MA-GMA particles during the melt processing.

1.2.2 COCONTINUOUS PHASE MORPHOLOGY

In conventional terms, one can define a cocontinuous phase structure as the coexistence of at least two continuous structures within the same volume in which each component is a polymer phase with its own internal networklike structure from which its properties result (62–71). A typical cocontinuous phase morphology is shown in Figure 1.7. In fact, at different compositions above the percolation threshold, various levels of continuity exist. The percentage of continuity can be defined as the weight ratio of the minor phase involved in a continuous path divided by the total weight of the minor phase. Cocontinuous polymer blends have a number of advantages, making them ideal for a wide range of applications. Some of the useful properties of cocontinuous polymer blends include synergistic mechanical properties, controlled electrical conductivity, and selective permeability (62,63).

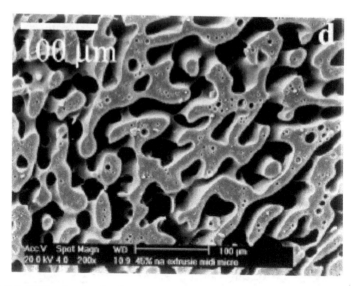

FIGURE 1.7 SEM micrograph of a cocontinuous morphology of PS/PA6 blends. (Reproduced from Tol, R.T., Groeninckx, G., Vinckier, I., Moldenaers, P., and Mewis, J., *Polymer*, 45, 2587–2601, 2004. With permission.)

The concept of cocontinuity was first introduced in experiments that studied a narrow range of compositions where phase inversion occurred. Most of the studies focused on the prediction of the composition of this phase inversion. An important challenge in the study of cocontinuous polymer blends is the accurate determination of their morphology. A variety of methods have been used for detecting cocontinuity, including solvent extraction, microscopy with image analysis, electrical conductivity measurements, and rheological measurements. Although a large number of techniques exist, solvent extraction has been the most common choice for the characterization of cocontinuity.

A number of models based on the viscosity ratio have been applied to predict the composition of the phase inversion region. These models are considered in Chapter 3, which is devoted to cocontinuous phase morphologies.

Li and Favis (72) have reported on the role of the blend interface type on the cocontinuous morphology. A classification of the blend interfaces that provides a general framework for their role in cocontinuous morphology development has been proposed. The blend systems were classified into three types. Type 1 systems are described as immiscible but compatible, demonstrating strong interactions at the interface, i.e., low interfacial tension and a stable threadlike dispersed phase even at low concentrations. Consequently, the droplet lifetime during mixing is lower than the thread lifetime. Such systems attain cocontinuity through thread-thread coalescence. A low percolation threshold and a broad cocontinuous region are the main features for type 1 system continuity development. Type 2 systems are immiscible and incompatible, and have a high interfacial tension. They often show a droplet dispersed phase structure at lower compositions, i.e., the droplet lifetime is greater than the thread lifetime during melt mixing. Such systems attain cocontinuity through droplet-droplet coalescence. The main features of the type 2 systems include a higher percolation threshold than type 1, a narrower cocontinuous region than type 1, and the dependence of dispersed phase size on composition. Type 3 systems are ternary compatibilized blends. Such systems attain cocontinuity through reduced droplet coalescence. These systems are characterized by a higher percolation threshold than type 2, a narrow cocontinuous region, and nondependence of the dispersed phase size with composition. More detailed information on the effect of thermal annealing and compatibilization on the development and stability of cocontinuous phase morphologies is discussed in Chapter 3.

1.3 MORPHOLOGY AND ULTIMATE MECHANICAL PROPERTIES

1.3.1 MORPHOLOGY AND IMPACT PROPERTIES

The relationship between morphology and impact properties of multiphase polymer blend systems has been studied extensively over the past several decades. Both rubbers and engineering thermoplastics are incorporated into brittle plastics to improve their impact properties. The impact modification by rubber toughening involves the incorporation of small amounts of rubber (mostly between 3 to 20 vol%) in rigid polymeric materials such as glassy thermoplastics, semicrystalline thermoplastics, and thermosets in order to enhance their fracture resistance. In fact, acrylonitrile-butadiene-styrene (ABS) and high impact polystyrene (HIPS) were commercialized in the late 1940s in order to develop high impact materials. The failure characteristics of rubber toughened polymers are extremely complex and are affected by composition, morphology, and testing conditions. The optimum rubber particle size varies depending on the chemical structure of the matrix. However, the value is typically in the range of 0.1 to 10 μm (73,74).

In recent years, it has become clear that optimization of rubber toughening requires the rubber particle to undergo strain softening and strain hardening in turn. At first, the rubber particle cavitation, i.e., formation of holes in the rubber phase, occurs. This helps to weaken the rubber particle's resistance to deformation, thereby initiating yielding in the matrix at reduced stress and allowing the particles to cold draw. At the later stages of deformation, the stretching of the rubber fibrils within the cavitated particles introduces a significant degree of strain hardening. The combination of response to the applied stress makes the rubber particle cavitation an effective response to the triaxial tensile stress generated at a crack tip because it allows the surrounding matrix to deform by all available mechanisms. The matrix may respond by shear yielding, multiple crazing, or both depending upon its properties, and the strained rubber particles then stabilize the most highly strained region of the polymer. The phenomenon of rubber cavitation was first noted in the 1970s. Rubber cavitation can be studied by electron microscopy, light scattering, measurements of thermal contraction and expansion, and dynamic mechanical spectroscopy. The stretching of the rubber phase and the subsequent fibrillation are the major contribution to strain hardening. Once the rubber particles have cavitated, the surrounding matrix polymer is able to yield and stretch in a way that was previously impossible. The shell of rigid polymer enclosing the rubber phase expands through biaxial extension, thereby increasing the dimension of the cavitated particle. Shear yielding is the process by which ductile polymer materials extend to high strain in standard tests. The chain segments slip past each other in response to shear test, with the result that a small element of the material in the yield zone changes shape while remaining close to constant volume. This plastic deformation process generates a significant rise in temperature. In recent years, it has been shown that stress whitening in many rubber modified systems is caused by cavitation or debonding of the rubber phase. In fact, the role of the rubber particles is to cavitate internally or to debond from the matrix, thereby relieving hydrostatic tension and initiating the ductile shear yielding mechanism. In the case of rubber modified polyvinyl chloride (PVC), polycarbonate (PC), PBT, nylon 6, nylon66, and epoxy resin, rubber cavitation and associated matrix shear yielding have been reported (74). For core shell rubber modified PC/copolyester, PC/PE, PVC/MBS, and carboxyl terminated butadiene-acrylonitrile (CTBN)/epoxy systems, particle debonding from the matrix has been demonstrated. In these blend materials, the crazing mechanism is suppressed due to the high entanglement density of the matrix. The competition between rubber particle cavitation or debonding at the interface and crazing of the matrix will be determined by the entanglement density of the matrix. PMMA with moderate entanglement density is reported to deform by shear yielding. ABS deforms both by crazing and rubber cavitation followed by shear yielding. However, the relative contribution of both mechanisms to the total deformation in ABS depends on the size of the rubber particles. Bubeck et al. (75) have reported that besides crazing, rubber cavitation can also take place in HIPS.

The particle size has a strong influence on the toughening efficiency. It has been shown that particles below 200 nm are no longer effective in increasing the toughness of nylon6/rubber blends. This has been explained by the fact that very small particles do not cavitate and are not able to initiate shear yielding in the matrix (73). It has also been shown that the brittle to tough transition is influenced by the concentration of the rubber phase. Wu (76) has shown that the brittle to tough transition in nylon/rubber blends takes place at a critical interparticle distance (IPDc), also called critical matrix ligament thickness. The IPD is related to the size of the rubber particles and the rubber volume fraction, $_r$, by the following relation:

$$IPD = d[k(/6\Phi_r)^{1/3} \, 1] \tag{1.6}$$

where the parameter k is a measure of the packing arrangement of a particular lattice and d is the average diameter of dispersed rubber particles.

The role of particle size distribution has been studied by Wu (77). According to Wu, a wide particle size distribution is disadvantageous for the impact behavior of rubber modified blends with pseudoductile matrices because the interparticle distance increases with increasing size of polydispersity at a given rubber fraction. However, Okamoto et al. (78) reported that a bimodal distribution can lead to an enhancement of the toughness. With respect to the role of the interfacial adhesion, a distinction has been made between matrices deforming by multiple shear yielding and those deforming by multiple crazing. For matrices deforming by multiple crazes, good adhesion between the rubber particles and the matrix is required because the rubber particles must act as effective craze stoppers. For matrices deforming by shear yielding as a result of internal rubber cavitation or interface debonding, a good interfacial strength is not required. Dompas and coworkers (79–82) developed a model for the internal cavitation of rubber modified pseudoductile matrices. Their model describes the relation between cavitation for a given relative deformation with elastic properties, the molecular characteristics of the rubber, and the rubber particle size. According to these authors, the particle size required for cavitation, d, is given by the following equation:

$$d = 12(\gamma_r + \Gamma_{sc})/K_r\Delta^{4/3} \tag{1.7}$$

where Δ is the volume strain, K_r is the rubber bulk modulus, Γ_{sc} is the surface energy per unit area, and γ_r is a contribution from the van der Waals surface tension. A similar energy balance model was also developed by Lazzeri and Bucknall (83). Later, Ayre and Bucknall (84) made use of model experiments to verify the energy balance model. Detection of incipient rubber particle cavitation in toughened PMMA using dynamic mechanical tests was reported by Bucknall et al. (85). In their work, rubber toughened PMMA (RTPMMA) was subjected to a range of compressive and tensile axial stresses. In an interesting study on the toughening of SAN with acrylic core shell particles, Steenbrick and coworkers (86) have reported that a high toughness could be achieved if the particle core has a low cross-linking density (and a corresponding low modulus and low cavitation resistance). Easily cavitating particles produced higher toughness. The authors suggested that the mechanical properties of the rubber particle core play a major role in the toughening phenomena.

Multiple crazing is found to be the most prominent toughening mechanism for HIPS, ABS, and RTPMMA. For all three polymers, the matrix is a brittle thermoplastic that generates crazes at strains between 0.3 and 1.0% (73). In fact, multiple crazes were first observed in HIPS. Both optical and transmission electron microscopy (TEM) showed numerous small crazes running between the rubber particles. According to Bucknall (87), crazes are initiated because of stress concentration at the equator of the particles. Donald and Kramer (88) reported that the ease of craze initiation depends to a large extent on the particle size and that crazes are rarely initiated by particles smaller than 1 µm. They suggested that small particles are unable to nucleate crazes and to terminate them before growing to catastrophic sizes. In cases where the energy absorption in the rubber modified system arises entirely from multiple crazing in the matrix, the optimum rubber particle size should depend on the intrinsic capability of the matrix to initiate crazes, i.e., on the entanglement density of the polymer. The optimum rubber particle size ($d_{optimum}$) is related to the entanglement density v_e by the following relation (74):

$$\text{Log}(d_{optimum}) = 1.19 - 14.1v_e \tag{1.8}$$

However, it has been shown by Bubeck et al. (89) and Magalhaes and Borggreve (90) that the occurrence of crazing accounts for about half of the plastic strain in

HIPS and the strain due to the noncrazing mechanism occurs before that due to the crazing mechanism. In fact, in HIPS the cavitation process is very complex. It might be a combination of the development of small voids within the rubber membranes of the salami particles or interface failure, both external and internal, i.e., between the rubber membranes and the polystyrene matrix occlusions. The cavitation process in HIPS is associated with the formation of shear bands in the PS matrix. Rios-Guerrero et al. (91) studied the deformed region around the arrested crack tip in HIPS by electron microscopy. Extensive crazing is observed in most of the studied sections, but several types of cavitation along with apparent narrow shear and dilatation bands have been observed just ahead of the crack tip. Their observations were in agreement with those of Lazzeri and Bucknall (83,92) and Schirrer et al. (93) who suggested that particle cavitation is a prerequisite for shear yielding of the matrix under severe conditions of the notched impact. The effect of rubber particle size on the toughening behavior of rubber modified PMMA with different test methods have been studied by Cho and coworkers (94). In the case of impact tests, maximum impact strength was obtained around a particle size of 0.25 μm regardless of the rubber phase content. On the other hand, by the three points bending test, the blends containing 2 μm particles showed quite a large improvement in fracture toughness. They finally concluded that the fracture mechanics of rubber toughened PMMA is governed by the strain rate and the loading behavior. In the three points bending tests, the toughening mechanism of rubber modified PMMA is mainly by multiple crazing, whereas shear yielding induced by rubber particle cavitation is predominant in the impact tests. Jiang et al. (95) established a correlation between cavitational volume and brittle to ductile transition (BDT) for particle toughened thermoplastics. They have established that the volume of cavitation was a key parameter that determined the BDT of the particle toughened thermoplastics. Their results showed that the larger the volume of cavitation, the lower the BDT temperature (T_{bd}). They also found that the lower the cavitation ability of the particle, the smaller the interparticle distance.

The influence of the interfacial adhesion on the rubber toughening has been studied by Liu et al. (96) using rubber modified PVC. They have used three types of blends having a different level of interfacial interaction. The blend of PVC/NBR (acrylonitrile-butadiene rubber) 18 with 18 wt% of acrylonitrile (ACN) has a medium interfacial adhesion strength and exhibited a BDT at a critical matrix ligament thickness $T_c = 0.059$ μm. The blend of NBR 26 with 26 wt% ACN has a stronger interfacial adhesion and exhibited a BDT at $T_c = 0.041$ μm. The blend exhibiting a weaker interfacial interaction showed debonding at the interface upon impact, which induces shear yielding of the matrix. However, the PVC/NBR 26 blend had a stronger interface and showed no microvoid formation, and so the occurrence of shear yielding was delayed.

The incorporation of rubber into semicrystalline engineering polymers shifts the T_{bd} to lower temperatures. The T_{bd} shift is controlled by various parameters such as the rubber concentration, rubber particle size, particle dispersion, and morphology of the particles. In the blends, the onset of rubber cavitation occurs at low strains and before the yield point of the neat polymer is reached (97). Volume strain at low rubber concentration increases linearly with the concentration. Model calculations indicate that the cavitational process at high concentration is influenced by the

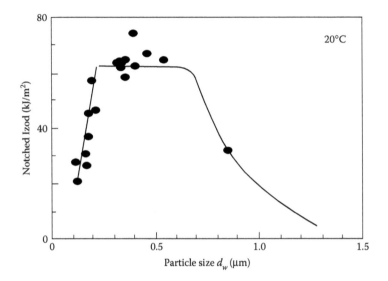

FIGURE 1.8 Notched Izod measurements as a function of the weight average particle size for PA 6/EPR (20 vol%). (Reproduced from Gaymans, R.J., in "Toughening of Semicrystalline Thermoplastics," in *Polymer Blends, Vol. 2. Performance*, Paul, D.R. and Bucknall, C.B., Eds., John Wiley & Sons, New York., 2000, 177–224. With permission.)

neighboring particles. A few percent of rubber is not sufficient to relieve all of the volume strain. The most important effect of an increase in rubber concentration is the shift in the T_{bd} to lower temperature. The decrease in T_{bd} is approximately linear with the increase in rubber concentration. The function of rubber particles is two-fold: to cavitate and to change the stress state around the particles. The cavitation of the particles is a function of the rubber properties, the stress state, and the particle size. The notched Izod impact strength curves are strongly influenced by the particle size, as shown in Figure 1.8. With decreasing particle size, the brittle to ductile transition is shifted to lower temperatures (97). As the rubber particle size is reduced from 2 to 0.3 μm, the shift is about 50°C for both polyamide/ethylene-propylene-diene monomer (PA/EPDM) and PP/EPDM blends. Blends with very small particles show poor impact behavior. This indicates that as the weight average diameter of the rubber particles falls below 200 nm, the impact strength at room temperature decreases and the T_{db} increases. As mentioned earlier, the reason for the lower limit in particle size seems to be that very small particles are more difficult to cavitate.

The chemical structure of the rubber and the properties of the matrix have been shown to influence considerably the toughness of the blends (98). The study indicated that rubber particles of similar size formed from styrene-ethylene-butylene-styrene graft maleic anhydride (SEBS-g-MA) readily cavitate when the matrix is high molecular weight nylon but do not cavitate when the matrix is low molecular weight nylon 6 even though both nylons have nearly the same toughness values. The toughening of semicrystalline polyamides using maleated rubbers has received con-siderable attention because of the ability of tailoring the rubber phase morphology by *in situ* chemical reaction between the maleic anhydride grafted to the rubber and the polyamide end groups during melt compounding. It has been reported that the nature of the rubber and the matrix, the morphology of the dispersed phase, the rubber content, as well as the processing conditions play an important role in determining the extent of toughening of polyamides. The literature in the area of semicrystalline thermoplastics describes the toughening effect in terms of triaxial stresses generated ahead of the propagating crack while the matrix is regarded as an isotropic continuum (99). These stresses preclude ductile yielding of the matrix but may, under appropriate circumstances, cause cavitation of the rubber particles, which in turn relieves the triaxial state of the stress and triggers ductile yielding of the matrix. On the other hand, Argon and coworkers (100–102) have suggested that the changes in the matrix crystal structure induced by the proximity of dispersed particles play an important role in toughening of semicrystalline polyamides. They demonstrated that PA 66 forms a layer approximately 150 nm thick at the PA/rubber interface with its crystallographic axes parallel to the surface. Similar layers of different thicknesses are found in PA6 and high-density polyethylene (HDPE). On the basis of these observations, they concluded that the critical interparticle spacing is a permeation threshold, marking the point at which the material of enhanced mobility — crystallized in thin layers around the rubber particles —forms a continuous pathway through the polymer matrix. Recently, the deformation mechanisms of rubber toughened PET (polyethylene terephathalate) have been studied by Loyens and Groeninckx (103) using EPR rubber and a compatibilising agent. It was found that the ductile fracture behavior above the T_{bd} consists of a high degree of rubber cavitation and extensive matrix shear yielding, both in the impact fracture plane and the stress whitened zone surrounding the crack. Very recently, the morphology and micromechanical behavior of a nanoblend system based on styrene/butadiene block copolymer having different molecular architectures were studied by Adhikari et al. (104). Unlike the classical rubber modified or particle filled thermoplastics, neither debonding at the particle and matrix interface nor particle cavitation were observed in these nanostructured blends. The micromechanical deformation of the blends revealed plastic drawing of PS lamellae or PS struts dispersed in the rubbery matrix; in addition, they found that the orientation of the whole deformation structure occurs parallel to the strain direction.

It is important to mention that the sequence of steps in the mechanisms of rubber toughening is still being debated (105). One prominent argument is that rubber particle cavitation occurs first to alleviate hydrostatic tension and to promote a stress state favorable for shear yielding (74,83,106). Another opinion is that rubber particle cavitation occurs after shear band formation (107,108). Experimental studies investigating these issues are complicated by many factors. Investigations based on post-mortem analysis of fracture surfaces reveal the presence of shear bands and cavitated particles, but cannot elucidate the sequence of mechanisms (109,110). Studies analyzing the sequence of mechanisms often investigate the stress field ahead of a notch (111–113). The assumptions associated with the stress field approach, the positional variation of the stress field, and the amount of constraint in the specimen limit the effectiveness of this approach. Toughening mechanisms and their sequence in rubber modified thermoplastics were determined experimentally by Crawford and Lesser (105) through the use of multiaxial tensile tests of a rubber modified material with favorable properties. They found that the onset of whitening in these systems occurs at a constant octahedral shear stress under a positive mean stress prior to yielding.

Rubber toughening of epoxies has received a lot of attention. It was first studied by McGarry and coworkers in the late 1960s and early 1970s (114,115). Epoxies can be toughened significantly by the incorporation of rubbery particles. It has been shown that the fracture toughness of rubber modified epoxies increases with increasing rubber concentration up to 10 to 15 parts per hundred (phr) and depends strongly on the size of the rubber particles. Generally, small particles of the size of microns are more effective in toughening than large particles of the size of tens of microns. The most important toughening mechanism involves the cavitation of the rubber particles and subsequent hole growth by matrix deformation. Other secondary toughening mechanisms such as rubber particle bridging and crack bifurcation and/or deflection may play a dominant role when the particle sizes are relatively large (116–118).

In epoxies with high cross-linking densities, rubber toughening is usually not very effective. In such cases, addition of a rigid ductile thermoplastic can significantly toughen the epoxy resin (116,119–122). Moreover, the incorporation of a rigid thermoplastic will not adversely affect the modulus of the system. Mostly, the thermoplastics used are those having a high modulus, high T_g, and high level of toughness. The thermoplastic modifier should be phased in separately during curing in the form of either a particulate or a cocontinuous morphology (Figure 1.9.). This is one of the prerequisites for substantial improvement in toughness. Another important requirement for efficient toughening is good interfacial adhesion owing to either chemical or physical interactions. The fracture toughness of thermoplastic modified epoxies increases continuously with increasing content of the modifier even though the phase morphology changes abruptly from particulate to cocontinuous. It has been shown that the cocontinuous structure provides a higher toughening effect than dispersed phase morphology. The thermoplastic particle crack bridging is the major

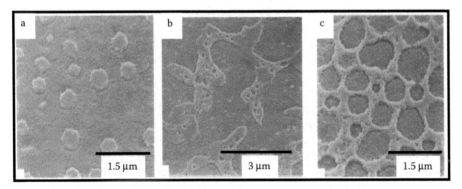

FIGURE 1.9 SEM micrographs of the etched surface of reactively terminated poly(ether sulfone) (PSF) modified epoxies. (a) Particulate microstructure, (b) cocontinuous microstructure, and (c) phase inverted microstructure. (Reproduced from Yee, A.F., Du, J., and Thouless, M.D, in Polymer Blends, Vol. 2, Performance, Paul, D.R. and Bucknall, C.B., Eds., John Wiley & Sons, New York, 2000, pp. 225–267. With permission.)

energy dissipating process while optimum toughening effects are usually obtained when several toughening mechanisms interact.

1.3.2 MORPHOLOGY AND OTHER ULTIMATE MECHANICAL PROPERTIES

The incorporation of rigid (amorphous or semicrystalline) polymers into thermoplastics will enhance stiffness particularly at temperatures above the T_g of the thermoplastic polymer. Of course, the inclusion of a rigid polymer phase is one of the well known ways of raising the heat distortion temperature (HDT) of many polymer systems. HDT is the temperature at which the deflection of a specified part exceeds a critical value under load and corresponds to a certain value of modulus. It has been shown that for automotive applications, it is important to increase the HDT to avoid excessive deformations. This can be achieved by the incorporation of rigid (amorphous or semicrystalline) polymers. However, unlike toughening, a good correlation has not been established between the microstructure and the tensile properties of polymer blends (123).

The inclusion of the hard phase also provides a route for balancing toughness versus stiffness. For example, rubber toughened nylon 6 can become both stiffer and tougher by adding a hard phase. It has been shown that ordinary brittle materials like SAN 25 may deform in a ductile manner provided the particles are small enough. The stress-strain behavior of PA6/SAN blends compatibilized by an imidized acrylic (IA) polymer has been studied as a function of the IA content and the dispersed SAN particle size (124). Nylon 6 showed ductile behavior with high elongation prior to breaking during slow speed tensile testing. However, SAN undergoes failure before yielding, i.e., at a low elongation at break. In the absence of IA, the strength of the blend is much lower than that of pure polyamide and SAN 25; this is associated with the poor adhesion between the phases. The addition of more than 0.5 wt% of IA leads to ductile blends whose yield strength reaches a plateau value that exceeds the fracture strength of pure SAN 25 and the yield strength of pure PA. A relationship was shown to exist between the elongation at break of the blends and the dispersed SAN25 particle size regardless of the composition or the IA content of the blend. It appears that the particle size should be below a critical value for the blend to exhibit plastic deformation. When the particle size exceeds the size range of 0.3 to 0.6 μm, the blends break in a brittle manner without yielding. The effect of the IA content on the deformation mode of the nylon6/SAN 25 blends has also been examined. Without IA compatibilization, the blends break before yielding, whereas blends that contain more than 1 wt% IA exhibit a stable plastic deformation beyond neck formation. Finally, the authors have attempted to make a correlation between the phase morphology and the deformation behavior of the blends. TEM was used to observe the phase morphology of deformed blends compatibilized by the progressive addition of IA before and after tensile testing. For blends that contained less than 0.5 wt% IA and that break before yielding, the SAN particle shape is essentially the same before and after the tensile test. On the other hand, blends compatibilized with 1 wt% or more of IA that are ductile show considerable plastic deformation of the dispersed SAN 25 particles upon tensile testing (124).

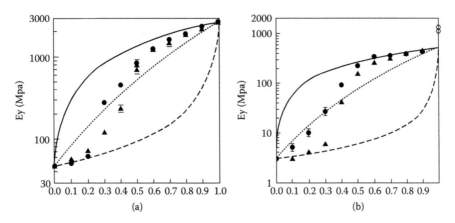

FIGURE 1.10 (a) Young's moduli for the PS/poly(ether-ester) systems Ia (▲) and Ib (●) as a function of volume fraction PS; (b) Young's moduli for the PP/SEBS systems IIa((▲) and IIb (●) as a function of volume fraction PP (when the error bars are not visible they are smaller than the marker). The experimental values are compared with the parallel (—), series (—–), and Davies (......) model. (Reproduced from Veenstra, H., Verkooijen, C.J.P., Lent, van J.J.B., Dam, van J., Posthuma de Boer, A., and Nijhof, H.J.P.A., *Polymer*, 41, 1817–1826, 2000. With permission.)

Similarly, the relationship between the phase morphology and the resulting mechanical properties of a binary blend of photocurable (2,2-bis(4-(acryloxy diethroxy)phynyl)propane (BP) and polysulphone (PSU) was investigated by Murata and Anazawa (125).

Veenstra et al. (126) and Joseph and Thomas (71) measured and compared the mechanical properties of polymer blends with cocontinuous phase morphologies to the properties of blends of the same polymers with a droplet-matrix morphology. In the studies of Veenstra et al., PS/poly(ether-ester) and polypropylene/styrene-ethylene-butylene-styrene (PP/SEBS) copolymer blends were prepared with both morphologies (dispersed-matrix and cocontinuous). The elastic moduli of the cocontinuous blends were significantly higher than the moduli of the dispersed blends. However, no significant difference in tensile or impact strength was found when the cocontinuous blends were compared to blends with droplet-matrix morphology. In Figure 1.10a, the Young's moduli for two blend systems of PS/poly(ether-ester), Ia (the blends processed at 230°C) and Ib (the blends processed at 200°C), are plotted as a function of volume fraction of PS. The morphology of the Ia blends showed cocontinuity over a small composition range (50 to 60 vol% PS), and the morphology of the Ib blends showed a broad range of cocontinuity (30 to 60 vol% PS). All other compositions of Ia and Ib blends showed droplet-matrix morphologies. It is likely that at a low volume fraction of PS, the Young's modulus is very close to that predicted by the series model. The moduli show a very sharp increase when PS becomes continuous throughout the sample. At high volume fraction of the PS, the moduli are found to be somewhat lower than those predicted by the parallel model. It is important to note that when the moduli of the cocontinuous blends are compared to those of the dispersed morphologies (with the same volume fractions), it becomes evident that at low volume fractions of PS, the cocontinuous blends show higher values for the Young's moduli than the dispersed blends. It is very clear that PS contributes more to the modulus of the blend when it is continuous than when it is dispersed in the poly(ether-ester) matrix. This is associated with the fact that in cocontinuous morphologies both phases take part in the load bearing process. At higher volume fractions of PS, the difference in modulus related to the morphology diminishes. Similarly, in Figure 1.10b, the Young's moduli for two blend systems of PP/SEBS blends, IIa (the blends processed at 250°C) and IIb (the blends processed at 190°C), are plotted as a function of the volume fraction of PP. The morphology of the IIa blends showed cocontinuity over a small composition range (50 to 60 vol% PP), and the morphology of the IIb blends showed a broad range of cocontinuity (40 to 80 vol% PP). From this figure, it is evident that cocontinuous blends show a much higher value for the Young's modulus than the dispersed blends. Joseph and Thomas (71) also noticed a similar behavior in the case of PS/PB blends. Since the Young's moduli of cocontinuous blends could not be predicted by any available models, Veenstra et al. (126) proposed a new model that depicts the basic element of cocontinuous structures as three orthogonal bars of one component embedded in a unit cube where the remaining volume was occupied by the other component. This model was quite successful in predicting the moduli of polymer blends with cocontinuous morphologies over the complete composition range.

The morphology and properties of thermoplastic-elastomer vulcanizates have been studied by Coran and Patel (127), Oderkerk et al. (128–130), and Thomas and coworkers (45,131–134). The stress-strain data of a 60:40 wt/wt blend of PP and EPDM as a function of particle size are given in Figure 1.11. It is very clear that

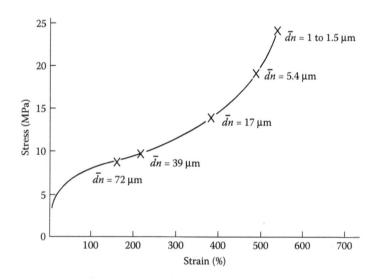

FIGURE 1.11 Stress-strain data of a 60:40 blend of PP and EPDM as a function of particle size. X denotes the fracture point. (Reproduced from Gaymans, R.J., in "Toughening of Semicrystalline Thermoplastics," in *Polymer Blends, Vol. 2, Performance,* Paul, D.R. and Bucknall, C.B., Eds., John Wiley & Sons, New York, 2000. With permission.)

the smaller particles provide the higher modulus, tensile strength, and failure strain. A significant change of the tensile properties has been observed at the region of phase cocontinuity.

1.4 RECENT ADVANCES IN NANOSTRUCTURED POLYMER BLEND SYSTEMS

In recent years, several strategies have been developed to form well defined and predictable multicomponent polymer structures with phase separation at the nanoscale (135). The most straightforward approach is to use linear block copolymers with two components A and B. However the most important drawback of this approach is that both components with their different chemical and electronic structures have to be connected by a covalent bond, which limits the probability that the components A and B will bond. Only a few examples of block copolymers containing two semiconducting polymers have been reported (136).

Nanostructured polymer morphologies have been prepared by reactive blending by Hu et al. (137) and Pernot et al. (138). Hu et al. have developed the concept of *in situ* polymerization and *in situ* compatibilization to obtain stabilized nanoblends and have shown the feasibility of their approach by using PP and PA6. Their method consists of polymerizing a monomer of PA6, ε-caprolactam (εCL), in a matrix of PP. A fraction of the PP bears 3-isopropenyl-α,α-dimethyl benzene isocyanate (TMI), which acts as a growing center to initiate PA6 chain growth. In fact, the polymerization of PA6 and the grafting reaction between PA6 and PP took place simultaneously in the matrix of PP, leading to the formation of compatibilized nanoblends. The size of the dispersed phase was between 10 and 100 nm, as shown in Figure 1.12. Pernot et al. (138) also adopted a similar methodology to produce a cocontinuous nanoscale product. In this approach, one component bears reactive groups along the backbone and the second component possesses complementary reactive moieties only at the end. Even though this novel strategy is expected to be versatile, it has yet to be applied to a wide range of polymers including semiconducting or fluorescent materials and to thin layer applications. Kietzke et al. (135) developed two new approaches to synthesize nanoscale polymer blend systems. In both cases, thin spin coated layers are used. In the first approach, two dispersions of single component nanospheres are mixed and processed into thin layers. In the second approach, the nanospheres are prepared from a mixture of two polymers in a suitable solvent. In this case, both polymers are contained in individual nanoparticles, where the particle size determines the upper limit to the dimension of the phase separation.

A fine dispersion in fluoroelastomer/semicrystalline perfluoropolymer nanoblends has been obtained with an innovative mixing technology based on microemulsion polymerization (139). The properties of the products were found to be optimized when the dispersed phase dimension is in the nanoscale, i.e., well below 0.1 μm. A further increase in the properties can be obtained by generating chemical links between the fluoroelastomer and semicrystalline perfluoropolymer. These nanoblends combine the performance properties of fluoroelastomers with those of semicrystalline perfluoropolymers. Very specifically, these nanoblends exhibit the sealing and mechanical properties of fluoroelastomers and the exceptional thermal

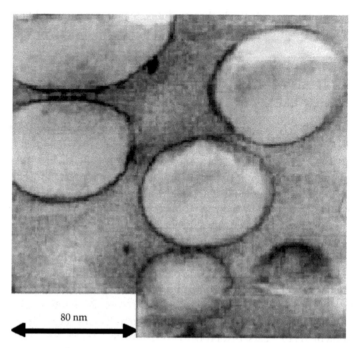

80 nm

FIGURE 1.12 Morphology of PP-g-TMI/εCL/NaCL microactivator system. (Reproduced from Hu, G.H., Cartier, H., and Plummer, C., *Macromolecules*, 32, 4713–4718, 1999. With permission.)

and chemical resistance, low permeability, and low coefficient of friction of semi-crystalline perfluoropolymers. Additionally, the finished products made out of these nanoblends were found to be optically transparent even at a concentration of 40 wt% semicrystalline perfluoropolymer because the size of the dispersed phase is below the visible light wavelength. The surface properties of nanostructured PS/PMMA blends have been reported by Prosycevas et al. (140). The blends were prepared by solvent casting using toluene as the mutual casting solvent. Chemically cleaned crystalline silicon and amorphous silica surfaces were vertically dipped into the dissolved polymer solution for approximately 10 to 20 sec, pulled out, and then dried in a horizontal position at room temperature in a chamber with a humidity of 60%. In fact, sorption of polymer molecules from the dilute solutions on the substrate allows for the formation of a very thin polymer layer whose thickness is comparable in size to that of the macromolecules. The AFM (atomic force microscopy) image of the dip coated PS/PMMA (75:25) film on crystalline silicon demonstrates that the morphology of the surface has a regular nanostructure. The nanostructured shaping of the film occurs in several stages. At first, macromolecules of PMMA and PS are adsorbed onto the surface of the substrates; the macromolecules are located on the active adsorption centers of the surface. During the evaporation of the solvent from the surface of the substrate, the macromolecules contribute to two dimensional displacement to form a film with a volume structure that is related to the regions occupied by the remaining solvent. When the remainder of the solvent is vaporized from the surface, a crater of certain size and depth in the nanometer range is formed. This structure development is associated with the formation and subsequent growing or "breathing" of the film. The formation of the blend morphology has been explained by two factors: differences in the solubility of the two polymers in the solvent and the rewetting of PMMA rich domains by the PS rich phase.

The nanoscale morphologies of polymer blends based on poly(2-methoxy-5(2-ethyl-hexyloxy)-1,4-phenylene vinylene) (MEH-PPV) have been studied using transmission electron microscopy by Yang et al. (141). It is shown that by controlling the preparation conditions, the phase separation in the blend results in a nanoscale network structure. These nanostructured polymer blends are a novel class of self-assembled blend systems. They offer enhanced electronic and optical properties. A series of blends have been prepared; these include blends of MEH-PPV with polyquinoline(PQ), polyaniline(PANI), polyethylene oxide (PEO), and Buckminster fullerene (C60). The blends were prepared by spin casting from the relevant solutions. All these blend systems show unusual electronic and optical properties. For example, bicontinuous PQ/MEH-PPV nanostructured network systems showed remarkable increase in electronic efficiency by a factor of more than 20 over pure MEH-PPV. Nanostructured PEO/MEH-PPV systems with additives resulted in efficient and fast response light-emitting electrochemical cells (142). The homogeneous nanostructured network found in the C60/MEH-PPV system (with a composition of approximately 1:1) produced a bulk heterojunction material that exhibited photovoltaic efficiencies more than two orders of magnitude larger than those that had been achieved with pure MEH-PPV alone (143).

The evolution in the morphology of nanostructured polymer blends consisting of a light emitting conjugated polymer MEH-PPV with PMMA has been reported by Iyengar et al. (144). Results of AFM, fluorescence microscopy, and photographs of the emission from light emitting devices illustrate the morphological evolution. The evolution of the morphology as a function of composition is given in Figure 1.13. The blends exhibit a phase segregated structure that is characterized by domains with length scales ranging form 200 to 900 nm. The nanostructure of the blends is strongly dependent on the composition, and as a consequence, the morphology can be tuned according to the particular requirement of a given electrooptical application.

The nanostructured domain formation at the surface of PS/PMMA ultrathin blend films upon annealing has been analyzed by Kailas et al. (145) using nano-SIMS (secondary ion mass spectrometry) imaging. In fact, PS/PMMA ultrathin films undergo phase segregation upon annealing, resulting in a decrease in the concentration of PS and an increase in the concentration of PMMA at the surface. Busby et al. (146) reported on the preparation of novel nanostructured polymer blends of ultrahigh molecular weight polyethylene and PMMA using supercritical carbon dioxide. Four alkyl functionalized methacrylate polymers have been studied, namely polymethymethacrylate (PMMA), polyethylmethacrylate (PEMA), polypropylmethacrylate (PPMA), and polybutylmethacrylate (PBMA). The blends obtained did not exhibit the usual gross phase separation that results when polymers are blended by conventional means. A combination of tapping mode atomic force microscopy, thermal characterization, and spectroscopy analysis revealed that the methacrylates reside in a nanometer scale phase as separated domains within the polyethylene.

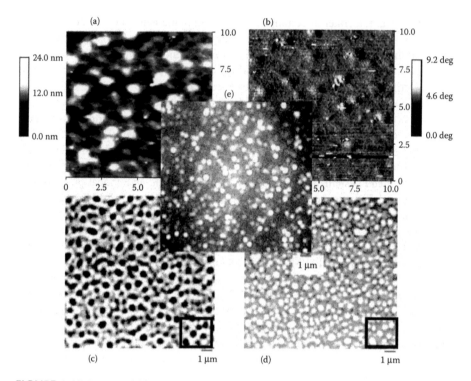

FIGURE 1.13 Images of 10 wt% MEH-PPV in PMMA: (a) tapping mode AFM topographic and (b) phase images; (c) bright field and (d) photoluminescence microscopic images; and (e) electroluminescence microscopic image. (Reproduced from Iyenger, N.A., Harrison, B., Duran, R.S., Schanz, K.S., Reynoldo, J.R., *Macromolecules*, 36, 8978–8995, 2003. With permission.)

Blends of high molecular weight PS and PVME showed nanoscale heterogeneity that was characterized by the coexistence of two thermally stimulated depolarization current (TSDC) peaks derived from the glass transition. These peaks were found to shift to higher temperatures as a function of increasing amounts of PS (147). Further, TEM analysis of the nanoscale phase separation products revealed that the domain sizes of the high molecular weight PS blend were three times larger than those of the low molecular weight PS blend.

Nanostructures can also be developed by the thermal treatment of initially miscible polymer blends. The concept of forming structures by combining phase separation and crystallization is becoming increasingly more important. The crystallization, phase structure, and semicrystalline morphology of phase separated nanostructured binary blends of PEO and poly(ether sulphone) (PES) have been reported by Dreezen et al. (148). The authors have shown that 75:25 and 50:50 PEO/PES blends show a clear cocontinuous structure with a characteristic dimension of approximately 400 and 200 nm, respectively. Nanostructured full and semi-interpenetrating polymer networks have been developed from natural rubber and polystyrene by Mathew et al. (149,150). The morphology of the nanostructured system was analyzed by transmission electron microcopy after staining the natural rubber phase by osmium tetroxide. The effects of varying the initiating system, blend ratio, and cross-linking density on the morphology and properties were analyzed in detail for these interpenetrating polymer network (IPN) systems. A cocontinuous nano-structure composed of phenolic resin rich phase and PMMA rich phase was prepared by Yamazaki et al. (151) through the reaction-induced phase separation process occurring during the curing process of miscible blends of a phenolic resin and PMMA. The sample was later thermally treated to generate a carbonaceous material with continuous nanopores. The sizes of the continuous pores were from tens to hundreds of nanometers, depending on the type of the phenolic resin used.

The properties of conducting polyaniline based nanoblends with extremely low percolation threshold was reported by Banerjee and Mondal (152). A series of polymers such as PS, PMMA, polyvinylacetate (PVAc) and polyvinylalcohol (PVA) have been blended with HCl-doped polyaniline (PANI) (PANI.HCL). These systems showed very low percolation threshold volume fractions. Transmission electron microscopy of the PANI.HCL/PVA blends films showed connectivity at compositions close to the percolation threshold value. Self-assembly of the nanoparticles was evident from the TEM pictures (Figure 1.14). Similarly, blends of PANI doped with camphor sulphonic acid (PANI-Csa) with PMMA showed a very low percolation threshold value, which was attributed to the self-assembly of PANI-Csa molecules into a nanofibrillar network morphology during liquid-liquid phase separation.

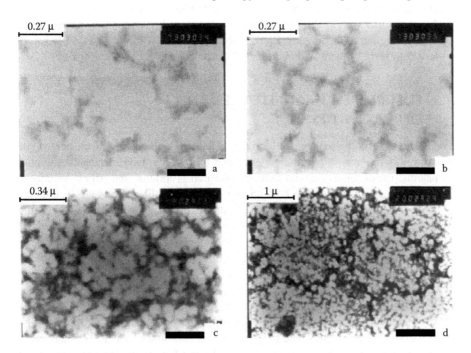

FIGURE 1.14 TEM micrographs of PANI.HCL/PVA blend films containing different amounts of PANI.HCl: (a) 0.035 wt%, (b) 0.045 wt%, (c) 0.05 wt%, and (d) 0.5 wt%. (Reproduced from Banerjee, P. and Mondal, M.J., *Macromolecules*, 28, 3940–3943, 1995. With permission.)

Nanoscale polymer blends have also been prepared by controlled evaporation techniques. Sumpter et al. (6) reported on the suppression of phase separation in mixed polymer systems by very rapid solvent evaporation from small (less than 10 um diameter) droplets of a dilute polymer solution. According to these authors, the primary condition for suppression of phase separation in these systems is that solvent evaporation must occur on a time scale that is fast compared to the self-organization times of the polymers. This implies time scales for particle drying on the order of a few milliseconds, which in most cases requires that the droplet sizes be less than 10 um (depending on the solvent, droplet environment, etc.). In addition to being a new route for forming nanoscale polymer systems, this microparticle format offers a new tool for studying multicomponent polymer blend systems confined to femtoliter and attoliter volumes where high-surface-area-to-volume ratios play a significant role in phase separation dynamics. Formation of homogeneous polymer blend components from bulk immiscible codissolved components using droplet techniques has two requirements. First, solvent evaporation must occur on a relatively short time scale compared to polymer translation diffusion. Second, the polymer mobility must be low enough so that once the solvent has evaporated, the polymers cannot overcome the surface energy barrier and phase separate. In order to explore the effects of polymer mobility in detail, Sumpter et al. (6) looked at blend systems of polyethylene glycol (PEG) oligomers of varying molecular weights (200, 400, 1000, and 3400 Da) with medium molecular weight (14K Da) atactic PVA. They have observed that the higher molecular weight PEG blend systems are more homogeneous as determined by a large number of techniques. Blend systems prepared with the 200 Da PEG were observed to form "sphere within a sphere" particles with a PVA central core.

1.5 FUTURE TRENDS, CHALLENGES, AND OPPORTUNITIES

For the past several decades, the domain of polymer blends has been one of the most important areas in polymer science and technology. Considering the number of patents, major university and industry research programs, and papers submitted in the last several years, it can be seen that the field is still growing. In fact, as the domain of polymer science grows by the introduction of new polymers and by the understanding of structure-property relationships of these new polymers, polymer blend technology will become more and more important. Of course, there are a large number of unsolved problems, and emerging opportunities exist in the polymer blend field, as indicated in Table 1.1.

The more recent developments in polymer blend technology include molecular composites, microfibrillar composites (MFC), electrically conducting polymer (ECP) blends, nanostructured polymer blends, biodegradable polymer blends, high temperature polymer blends, and polymer blends as biomaterials (153). Many commercial blend systems have appeared during the last few years. The most important high temperature polymer blends include those based on, for example, poly(aryl ketones), polyimides, and polyamide imides (154). One of the major areas for high temperature polymers is aerospace and military applications. Molecular composites

TABLE 1.1
Unsolved Problems and Potential Applications

Commercial Blend Opportunities

PVC/PO[a] (PVC/PP, PVC/HDPE)

PVC/PS

PP/PS, HDPE/PS

PPS[a]/other engineering polymers (e.g., PPS/PSF, PPS/PEI,[a] PPS/polyamides)

Environmental Concerns

Compatibilization of postconsumer polymer scrap

Reuse of diverse polymer blends (automotive)

Biodegradable blends

Low to no VOC (volatile organic compounds) based coatings

Emerging Blend Technology

Liquid crystalline polymers: (1) translation of potential properties into blends and (2) solve weld line problems

Molecular composites: go beyond the concept stage and develop commodity and engineering polymers

Electrically conductive polymer (ECP) blends: maximize performance while minimizing the levels of ECP incorporation

Electroresponsive polymers

Nonlinear optics polymers

Nanomolecular assemblies

Processing

Adapt metal forming (forging) concepts to polymer blends

Solve weld line problems for injection molded two-phase blends

Develop improved *in situ* polymerization techniques for diverse polyolefin blends

Develop methods for predicting morphology during processing

Develop rheology-morphology models

Theory and Science

Develop improved predictive methodology for phase behavior

Improve computer modeling and software for predicting phase behavior and morphology

Bridge molecular dynamics and mechanics; develop *ab initio* approaches to polymer blend problems

Establish relationships between morphology and performance

Develop methods for characterization of the interfacial region

Develop comprehensive methods for morphology characterization

[a] PO = polyolefins; PPS = poly(phenylene sulphide); and PEI = poly(etherimine).

Source: From Robeson, L.M., Perspectives in polymer blend technology, in Polymer Blends Handbook, Vol. 2, Utracki, L.A., Ed., Kluwer Academic Publishers, Dodrecht, 2002.

are formed from polymer blends by dispersing rigid rod macromolecules in a more flexible matrix; consequently, the resultant blend is reinforced at the molecular level. Examples include blends of rigid rod polybenzoxazoles with flexible chain polybenzoxazoles and LCPs with flexible matrix polymers (153). A similar approach has been used in the manufacture of MFCs too, as discussed earlier.

Electrically conducting polymer blends have achieved some commercial success (155,156). Blends of polythiophene or polypyrrole with PS and PC were reported to show excellent conductivity (155). The thiophene or pyrrole diffuses into the film and polymerizes *in situ* in the film. Threshold conductivity occurs at 18 wt% of both conducting polymers in PS. A large number of conducting polymer blends based on polyaniline has also been prepared; these blends showed excellent efficiency even at 1 wt% of PANI because of the special network structure morphology.

Another important emerging area in the polymer blend field concerns blends based on biodegradable polymers with commercial polymers. Important commercial biodegradable thermoplastics include PCL, PHB, PVAL, certain polyesters based on polyurethane (PU), and thermoplastic starch (153). These polymers can be blended with a range of commercial polymers. The biodegradable constituents in polymer blends offer the advantage of combining the desired property balance with environmental assimilation. Blends of cellulose with other synthetic polymers are good examples of this (154,157).

A novel concept has achieved a large amount of attention (153,158–162). This approach yielded alternating layers of anionic and cationic polyelctrolytes by alternatively dipping a substrate into a polycation solution followed by a polyanion solution to build monomolecular layers of polyelctrolyte complex. Due to charge repulsion, each coat consists of a monomolecular layer (151). Hillmyer et al. (159) have studied the self-assembly and polymerization of the epoxy resin-amphibilic block copolymer of poly(ethylene oxide) (PEO)/poly(ethyl ethylene) (PEE) (also known as OE nanocomposites). Figure 1.15 shows TEM micrographs of the core shell morphology of a cured OE-7/epoxy matrix (25 wt% OE-7). The authors have shown that adding a PEO compatible precure epoxy resin to an ordered polyalkaline/polyethylene oxide block copolymers leads to an ordered morphology swollen with a reactive diluent. Gelation of the epoxy resin forces an expulsion of the PEO from the matrix. Lipic et al. (160) have also introduced a novel method of templating an ordered structure, on a nanometer scale, in a thermoset resin by blending an amphibilic diblock copolymer, PEO/poly(ethylene-alt-propylene) (PEP), with a polymerisable epoxy resin. It was shown using TEM that in a combination of polyethylene oxide–epoxy polymer (PEO–EP-5) (OP5)/poly(bisphenol-A-co-epichlorohydrin (BPA 348)/methylene dianiline (MDA) blend (52 wt% OP5) cured, the epoxy resin selectively swells the PEO chains, creating a spontaneous interfacial curvature between the PEP and PEO/epoxy domains that controls the morphology. As the amount of epoxy resin added to the block copolymer is increased, the blend microstructure evolves from lamellar, to gyroid, to cylindrical, to body centered cubic packed spheres, and ultimately to disordered micelles. When the hardener is added and the epoxy is cured, the system retains the nanostructure and macrophase separation between the block copolymer and the epoxy is avoided.

Recently, Ritzenthaler et al. (161) achieved transparent nanostructured thermosets by blending and reacting an epoxy system with polystyrene block/polybutadiene block/poly(methyl methacrylate) (SBM) triblock synthesized anionically at an industrial scale. The morphology of diglycidyl ether of bisphenol A/4,4′-methylene bis(3-chloro 2,6-diethyl aniline) (DGEBA/MCDEA) 50 wt% $S_{22}^{27}B_9M_{69}$ (where the subscripts represent the mass percent of the blocks and the superscript represents the

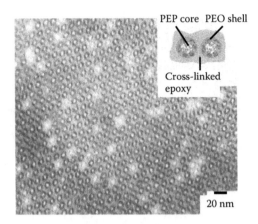

FIGURE 1.15 TEM micrograph of a cured OE-7/epoxy mixture (25 wt% OE-7). The sample was microtomed at room temperature and stained with RuO_4. The inset depicts the idealized nanostructure. (Reproduced from Lipic, P.M., Bates, F., and Hillmayer, M., *J. Am. Chem. Soc.*, 120, 8963–8970, 1998. With permission.)

PS block number average molecular mass in kg/mol) blend is presented in Figure 1.16 *top* (before reaction) and Figure 1.16 *bottom* (after reaction). The blend is transparent both before and after the reaction. It is worth noting that the ordered nanostructures before and after the reaction are quite similar. Very recently, Dean et al. (162), in their studies on nanostructure toughened epoxy resins, have reported some preliminary findings that reveal remarkable toughening associated with the formation of nanostructure in the poly(bisphenol A-co-epichlorohydrin) epoxies with and without brominated aromatic rings/phenol novolac (DER/PN) system, especially for the flame retardant brominated resin. These authors have introduced a new type of modifier morphology, self–assembled, wormlike micelles (see Chapter 11) and demonstrated a remarkable enhancement in fracture resistance with bisphenol A epoxy cured with PN, even for an epoxy formulation that contained brominated epoxies to improve the flame retardance.

Conductive polymer blends were successfully designed by the preferential location of carbon black at the interface of a PE/PS blend yielding a percolation threshold of 0.4 wt% (163).

The use of polymer blends in light-emitting diodes has been reported (164). A mixture of high- and low-band gap optically active polymers enhances the efficiency and brightness of the energy transfer between the blend polymers; poly(phenylene vinylene) and poly(vinyl carbazole).

Still, more work needs to be done on the prediction and modeling of miscibility, phase behavior, and microstructure of polymer blends. Group contribution methods for assessing the miscibility of polymers are gaining importance. An effort needs to be made in the field of computational modeling aiming at predicting the behavior of polymer blends. Computer software has been developed to predict the miscibility and phase behavior of binary polymer blends. The procedure involves the calculation of pairwise interactions via a Monte Carlo approach for the determination of the heat of mixing. Using this approach, the temperature dependence of the Flory-

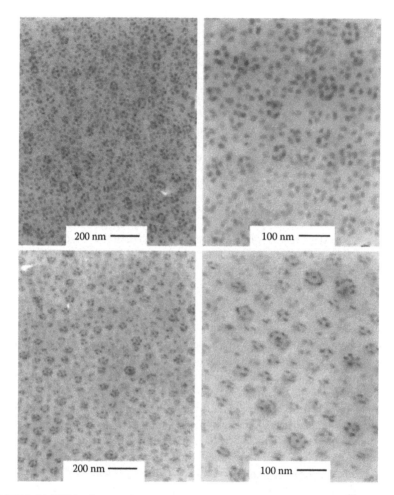

FIGURE 1.16 TEM micrographs obtained for DGEBA/MCDEA 50 wt% $S_{22}^{27}B_9M_{69}$ blend (*top*) before reaction and (*bottom*) after reaction. (Reproduced from Gubbels, F., Jérôme, R, Teyssié, Ph., Vanlathem, E., Deltour, R., Calderone, A., Parenete, V., and Brédas, J.L., *Macromolecules*, 27, 1972–1974, 1994. With permission.)

Huggins interaction parameter has been successfully calculated, which will allow the construction of phase diagrams for a given polymer pair. The different techniques, which are currently utilized for predicting polymer miscibility, including QSPR (Quantitative Structure Property Relationship) and PRISM (Statistical Mechanical Polymer Reference Interaction Site Model), have been recently reviewed (153). In the coming years, approaches employing molecular mechanics, molecular dynamics, and Monte Carlo methods will likely significantly improve and be able to predict polymer-polymer interactions.

The formation of novel morphologies as well as the ability to predict the morphology of polymer mixtures are areas of tremendous interest to both academia and industry. The phase separation process and the resultant morphology of block copolymers are well understood. The morphology seen in immiscible polymer blends under shear flow has been well studied. However, the morphology that forms as a result of phase separation of miscible polymer blends during a transition across a phase boundary is an area of considerable interest. One of the major questions that needs to be researched is the relative rate of spinodal decomposition versus nucleation and growth. In the metastable region, nucleation and growth are the phase separation processes that determine the resultant morphology. In the spinodal regime, both spinodal decomposition as well as nucleation and growth can take place. By operating close to the consolate point or by the use of rapid quenching, spinodal decomposition may become the preferred process. However, once spinodal decomposition occurs, the phase size will reach a point where nucleation occurs, and thus nucleation and growth can become competing phase separation processes. That complex phenomenon needs to be addressed both theoretically and experimentally.

Recycling of polymer blends and multilayered structures is still a major challenge. The recycling of automobile parts will probably be an active area in which blending know-how will be a key to success. The compatibilisation of postconsumer polymer scrap will continue to be a subject of interest for the coming decades too. Efficient multifunctional compatibilisers, which are able to compatibilise a polymer waste or scrap containing several types of polymers, will be needed. Finally, the real time monitoring of polymer blend morphology during processing is an important area that requires a lot of attention in the future. This information will supply enormous feedback for improving the uniformity of mixing and consequently the quality of the blend. Online morphology characterization, rheological measurements, and shear induced phase behavior, i.e., mixing and demixing behavior, are additional subjects that need further research. Of course, rheooptical and light scattering techniques under shear have been applied to transparent polymer blend systems such as low molecular weight PS/PIB and PS/PVME model systems (165,166). However, application of these techniques to real polymer blend systems is still a challenging task.

ACKNOWLEDGMENTS

The authors are indebted to the Research Council KULeuven for providing a senior research fellowship to C. Harrats and S. Thomas (GOA project 98/06), as well as to the Fund for Scientific Research, Flanders, Belgium, for the financial support given to the MSC laboratory.

REFERENCES

1. Utracki, L.A., *Introduction to Polymer Blends, in Polymer Blends Handbook*, Utracki, L.A., Ed., Kluwer Academic Publishers, Dordrecht, The Netherlands, 2002.
2. Macoscko, C.W., Morphology development and control in immiscible polymer blends, *Macromol. Symp.*, 149, 171–184, 2000.
3. Paul, D.R., in *Polymer Blends*, Paul, D.R. and Newman, S., Eds., Academic Press, New York, 1978.

4. Mark, H.F., Bikales, M., Overberger, C.G., Menges, G., and Kroschwitz, J.I., Eds., *Encyclopedia of Polymer Science and Engineering*, 2nd Ed., John Wiley & Sons, Inc., New York, 1987.

5. Paul, D.R. and Bucknall, C.B., Eds., *Polymer Blends (Vol. 1), Formulation (Vol. 2, Performance*, John Wiley & Sons, Inc, New York, 2000.

6. Sumpter, B.G., Noid, D.W., and Barnes, M.D., Recent developments in the formation and simulation of micron and nano-scale droplets of amorphous polymer blends and semicrystalline polymers, *Polymer*, 44, 43894403, 2003.

7. Macosko, C.W., Guegan, Ph., Khandpür, K., Nakayama, A., Maréchal, Ph., and Inoue, T., Compatibilizers for melt blending: premade block copolymers, *Macromolecules*, 29, 5590–5598, 1996.

8. Favis, B.D., Factors influencing the morphology of immiscible polymer blends in melt processing, in Polymer Blends, Vol. 1, Formulation, Paul, D.R. and Bucknall, C.B., Eds., John Wiley & Sons, New York, 2000, pp. 501–537.

9. Taylor, G.I., The deformation of emulsion in definable fields of flow, *Proc. Royal. Soc., London*, A138, 41–48, 1932.

10. Taylor, G.I., The formation of emulsions in definable fields of flow, *Proc. Royal Soc., London*, A146, 501–523, 1934.

11. Grace, H.P., Dispersion phenomena in high viscosity immiscible fluid systems and applications of static mixers as dispersion devices in such systems, *Chem. Eng. Commun.*, 14,225, 1982.

12. Rumscheidt, F.D. and Mason, S.G., Particle motions in sheared suspensions. XII. Deformation and burst of fluid drops in shear and hyperbolic flow, *J. Colloid. Sci.*, 16, 238–261, 1961.

13. Otiino, J.M., DeRoussel, P., Hansen, S., and Khakhar, D.V., Mixing and dispersion of viscous liquids and powdered solids, *Adv. Chem. Eng.*, 25, 105–204, 1999.

14. Van Oene, H.J., Modes of dispersion of viscoelastic fluids in flow, *J. Colloid Interface Sci.*, 40, 448–467, 1972.

15. Wu, S.H., Formation of dispersed phase in incompatible polymer blends — interfacial and rheological effects, *Polym. Eng. Sci.*, 27, 335–343, 1987.

16. Serpe, G., Jarrin, J., and Dawans, F., Morphology-processing relationships in poly-ethylene-polyamide blends, *Polym. Eng. Sci.*, 30, 553–565, 1990.

17. Karger-Kocsis, J., Kallo, A., and Kuleznev, V.N., Phase structure of impact-modified polypropylene blends, *Polymer*, 25, 279–286, 1984.

18. Nelson, C.J., Avgeropoulos, G.N., Weissert, F.C., et al., Relationship between rheology, morphology, and physical properties in heterogeneous blends, *Angew. Makromol. Chem.*, 60, 49, 1977.

19. Favis, B.D. and Chalifoux, J.P., The effect of viscosity ratio on the morphology of polypropylene/polycarbonate blends during processing, *Polym. Eng. Sci.*, 927, 1591–1600, 1987.

20. Tokita, N., Analysis of morphology formation in elastomer blends, *Rubber Chem. Technol.*, 50, 292–300, 1977.

21. Favis, B.D. and Willis, J.M., Phase size/composition dependence in immiscible blends: experimental and theoretical considerations, *J. Polym. Sci., Part B, Polym. Phys.*, 28, 2259–2269, 1990.

22. Elmendorp, J.J. and Van der Vegt, A.K., A study on polymer blending microrheology. 4. The influence of coalescence on blend morphology origination, *Polym. Eng. Sci.*, 926, 1332–1338, 1986.

23. Utracki, L.A. and Shi, Z.H., Development of polymer blend morphology during compounding in a twin-screw extruder. 1. Droplet dispersion and coalescence — a review, *Polym. Eng. Sci.*, 32, 1824–1833, 1992.

24. Levitt, L., Macosko, C.W., and Pearson, S.D., Influence of normal stress difference on polymer drop deformation, *Polym. Eng. Sci.*, 36, 1647–1655, 1996.

25. Migler, K.B., Droplet vorticity alignment in model polymer blends, *J. Rheol.*, 44, 277–290, 2000.

26. Hobbie, E.K. and Migler, K.B., Vorticity elongation in polymeric emulsions, *Phys. Rev. Lett.*, 82, 5393–5396, 1999.

27. Park, C.C., Baldessari, F., and Leal, L.G., Study of molecular weight effects on coalescence: interface slip layer, *J. Rheol.*, 47, 911–942, 2003.

28. Min, K., White, J.L., and Fellers, J.F., High density polyethylene/polystyrene blends: phase distribution, morphology, rheological measurements, extrusion, and melt spinning behavior, *J. Appl. Polym. Sci.*, 29, 2117–2142, 1984.

29. Favis, B.D., The effect of processing parameters on the morphology of an immiscible binary blend, *J. Appl. Polym. Sci.*, 39, 285–300, 1990.

30. Thomas, S. and Groeninckx, G., Nylon 6/ethylene propylene rubber (EPM) blends: phase morphology development during processing and comparison with literature data, *J. Appl. Polym. Sci.*, 71, 1405–1429, 1999.

31. Joseph, S. and Thomas, S., Morphology, morphology development and mechanical properties of polystyrene/polybutadiene blends, *Eur. Polym. J.*, 39, 115–125, 2003.

32. Sundararaj, U. and Macosko, C.W., Drop breakup and coalescence in polymer blends — the effects of concentration and compatibilization, *Macromolecules*, 28, 2647–2657, 1995.

33. Cigana, P., Favis, B.D., and Jérôme, R., Diblock copolymers as emulsifying agents in polymer blends: influence of molecular weight, architecture, and chemical composition, *J. Polym. Sci. Polym.Phys.*, Part B, 34, 1691–1700, 1996.

34. Thomas, S. and Prud'homme, R.E., Compatibilising effect of block copolymers in heterogeneous polystyrene/poly(methyl methacrylate) blends, *Polymer*, 33, 4260–4268, 1992.

35. Oommen, Z., Thomas, S., and Nair, M.R.G., Compatibilising effect of NR-g-PMMA in heterogeneous NR/PMMA blends, *Polym. Eng. Sci.*, 6, 151–160, 1996.

36. Ashaletha, R., Kumaran, M.G., and Thomas, S., The technological compatibilisation of natural rubber/polystyrene blends by the addition of natural rubber-graft-polystyrene, *Rubber Chem. Technol.*, 68, 671–687, 1995.

37. Geroge, J., Ramamurthy, K., Varughese, K.T., and Thomas, S., Melt rheology and morphology of thermoplastic elastomers from polyethylene /nitrile rubber blends: the effect of blend ratio, reactive compatiblisation and dynamic vulcanisation, *J. Polym. Sci. Polym. Phys.*, Part B, 38, 1104–1122, 2000.

38. Aravind, I., Albert, P., Ranganathaiah, C., Kurian, J.V., and Thomas, S., Compatiblising effect of EPM-g-MA in EPDM/poly(trimethylene terephthalate) incompatible blends, *Polymer*, 45, 4925–4937, 2004.

39. Lyu, S.P., Jones, T.D., Bates, F.S., and Macosko, C.W., Role of block copolymers on suppression of droplet coalescence, *Macromolecules*, 35, 7845–7855, 2002.

40. Van Puyvelde, P., Velankar, S., and Moldenaers, P., Effect of Marangoni stresses on the deformation and coalescence in compatibilised immiscible polymer blends, *Polym. Eng. Sci.*, 42, 1956–1964, 2002.

41. Willis, J.M., Favis, B.D., and Lunt, J., Reactive processing of polystyrene-co-maleic anhydride/elastomer blends: processing-morphology-property relationships, *Polym. Eng. Sci.*, 30, 1073–1184, 1990.

42. Velanakr, S., Van Puyvelde, P., Mewis, J., and Moldenaers, P., Steady-shear rheological properties of model compatibilised blends, *J. Rheol.*, 48, 725–744, 2004.

43. Van Puyvelde, P., Velankar, S., and Moldenaers, P., Rheology and morphology of compatibilised blends, *Curr. Opinion Coll. Interface Sci.*, 6, 457–463, 2001.

44. Moan, M., Huitric, J., Mederic, P., and Jarrin, J., Rheological properties and reactive compatibilisation of polymer blends, *J. Rheol.*, 44, 1227–1245, 2000.

45. Iza, M., Bousmina, M., and Jérôme, R., Rheology of compatibilised immiscible viscoelastic polymer blends, *Rheol. Acta*, 40, 10–22, 2001.

46. Velankar, S., Van Puyvelde, P., Mewis, J., and Moldenaers, P., Effect of compatibilisation on the break up of polymeric drops in shear flow, *J. Rheol.*, 45, 1007–1019, 2001.

47. Chapleau, N. and Favis, B.D., Droplet fiber transitions in immiscible polymer blends generated during melt processing, *J. Mater. Sci.*, 30,142–150, 1995.

48. Gonzalez–Nunez, R., DeKee, D., and Favis, B.D., The influence of coalescence on the morphology of the minor phase in melt-drawn polyamide-6/HDPE blends, *Polymer*, 37, 4689–4693, 1996.

49. Dreval, V.E., Vinogradov, G.V., Plotonikova, E.P., Kotova, E.V., and Pelzbauer, Z., Deformation of melts of mixtures of incompatible polymers in a uniform shear field and the process of their fibrillation, *Rheol. Acta*, 22, 102–107, 1983.

50. Delaby, I., Ernst, B., Froelich, D., and Muller, R., Droplet deformation in immiscible polymer blends during transient uniaxial elongational flow, *Polym. Eng. Sci.*, 36, 1627–1635, 1996.

51. Elmendorp, J.J.. and Maalcke, R.J., A study on polymer blending microrheology. *Polym. Eng. Sci.*, 25, 1041–1047, 1985.

52. LaMantia, F.P., Valenza, A., Paci, M., and Magagnini, P.L., Rheology-morphology relationships in nylon-6 liquid-crystalline polymer blends, *Polym. Eng. Sci.*, 30, 7–12, 1990.

53. Tsebrenko, M.V., Yudin, A.V., Ablazova, T.I, and Vinogradov G.V., Mechanism of fibrillation in flow of molten polymer mixtures, *Polymer*, 17, 831–834, 1976.

54. Tsebrenko, M.V., Danilova, G.P., and Malkin, A.Y., Fracture of ultrafine fibers in the flow of mixtures of non-newtonian polymer melts, *J. Non-Newt. Fl. Mech.*, 31, 1–26, 1989.

55. Sarkissova, M., Harrats, C., Groeninckx,G., and Thomas, S., Design and characterisation of microfibrillar reinforced composite materials, *Composites Part A*, 35, 489–499, 2004.

56. Campbell, J.R., Hobbs, S.Y., Shea, T.J., and Watkins, V.H., Poly(phenylene oxide)/polyamide blends via reactive extrusion, *Polym. Eng. Sci.*, 30, 1056–1062, 1990.

57. Favis, B.D. and Therrien, D., Factors influencing structure formation and phase size in an immiscible polymer blend of polycarbonate and polypropylene prepared by twin-screw extrusion, *Polymer*, 32, 1474–1481, 1991.

58. Hobbs, S.Y., Dekkers, M.E., and Watkins, V.H., Effect of interfacial forces on polymer blend morphologies, *Polymer*, 29, 1598–1602, 1988.

59. Guo, H.F., Packirisamy, S., Gvozdic, N.V., and Meier, D.J., Prediction and manipulation of the phase morphologies of multiphase polymer blends. 1. Ternary systems, *Polymer*, 38, 785–794, 1997.

60. Pagnoulle, C. and Jérôme, R., Particle-in-particle morphology for the dispersed phase formed in reactive compatibilization of SAN/EPDM blends, *Polymer*, 42, 1893–1906, 2001.

61. Martin, P., Maquet, C., Legras, R., Bailly, C., Leemans, L., van Gurp, M., and van Duin, M., Particle-in-particle morphology in reactively compatibilised PBT/epoxide containing rubber blend, *Polymer*, 45, 3277–3284, 2004.

62. Galloway, J.A., Koester, K.J., Paasch, B.J, and Macosko, C.W., Effects of sample size on the solvent extraction for detecting co-continuity in polymer blends, *Polymer*, 45, 423–428, 2004.

63. Lyngaae-Jorgensen, J. and Utracki, L.A., Structuring polymer blends with bicontinuous phase morphology. Part II. Tailoring blends with ultra critical volume fraction, *Polymer*, 44, 1661–1669, 2003.

64. Veenstra, H., Van Dam, J., and Posthuma de Boer, A., On the coarsening morphologies in polymer blends: effect of interfacial tension, viscosity and physical crosslinks, *Polymer*, 40, 1119–1130, 1999.

65. Willemse, R.C, Posthuma de Boer, A., van Dam, J., and Gotsis, A.D., Co-continuous morphologies in polymer blends: the influence of interfacial tension, *Polymer*, 40, 827–834, 1999.

66. Lee, J.K. and Han, C.D., Evolution of a dispersed morphology from a co-continuous morphology in immiscible polymer blends, *Polymer*, 40, 2521–2536, 1999.

67. Willemse, R.C., Posthuma de Boer, A., van Dam J., and Gotsis, A.D., Co-continuous morphologies in polymer blends: a new model, *Polymer*, 39, 5879–5887, 1998.

68. Tol, R.T., Groeninckx, G., Vinckier, I., Moldenaers, P., and Mewis, J., Phase morphology and stability of co-continuous (PPE/PS)/PA6 and PS/PA blends: effect of rheology and reactive compatibilisation, *Polymer*, 45, 2587–2601, 2004.

69. Marin, N. and Favis, B.D., Co-continuous morphology development in partially miscible PMMA/PC blends, *Polymer*, 43, 4723–4731, 2002.

70. Li, J. and Favis, B.D., Characterising co-continuous high density polyethylene/polystyrene blends, *Polymer*, 42, 5047–5053, 2001.

71. Joseph, S. and Thomas, S., Modelling of tensile moduli of polystyrene/polybutadiene blends, *J. Polym. Sci. Polym. Phys.*, Part B, 40, 755–764, 2002.

72. Li, J. and Favis, B.D., The role of the blend interface type on morphology in co-continuous polymer blends, *Macromolecules*, 35, 2005–2016, 2002.

73. Bucknall, C.B., Deformation mechanisms in rubber toughened polymers, in *Polymer Blends, Vol. 2, Performance,* Paul, D.R. and Bucknall, C.B., John Wiley & Sons, New York, 2000, pp. 83–117

74. Groeninckx, G. and Dompas, D., Plastic deformation mechanisms of polymers and rubber modified thermoplastic polymers: molecular and morphological aspects, in Structure and Properties of Multiphase Polymer Materials, Araki, T., Tran–Cong, Q., and Shibayama, M., Eds., Marcel Dekker Inc., New York, 1998, pp. 423–452.

75. Bubeck, R.A., Buckley, D.J., Kramer, E.J., and Brown, H.R., Modes of deformation in rubber-modified thermoplastics during tensile impact, *J. Mater. Sci.*, 26, 6249–6259, 1991.

76. Wu, S., Phase structure and adhesion in polymer blends: a criterion for rubber toughening, *Polymer*, 26, 1855–1863, 1985.

77. Wu, S., A generalized criterion for rubber toughening: the critical matrix ligament thickness, *J. Appl. Polym. Sci.*, 35, 549–561, 1988.

78. Okamoto, Y., Miyagi, H., Kakugo, M., and Takahashi, K., Impact improvement mechanism of HIPS with bimodal distribution of rubber particle size, *Macromolecules*, 24, 5639–5644, 1991.

79. Dompas, D. and Groeninckx, G., Toughening behaviour of rubber-modified thermoplastic polymers involving very small rubber particles. 1. A criterion for internal rubber cavitation, *Polymer*, 35, 4743–4749, 1994.

80. Dompas, D., Groeninckx, G, Isogawa, M., Hasegawa, T., and Kadokura, M., Toughening behaviour of rubber-modified thermoplastic polymers involving very small rubber particles. 2. Rubber cavitation behaviour in poly(vinyl chloride)/methyl methacrylate-butadiene-styrene graft copolymer blends, *Polymer*, 35, 4750–4759, 1994.

81. Dompas, D., Groeninckx, G, Isogawa, M., Hasegawa, T., and Kadokura, M., Toughening behaviour of rubber-modified thermoplastic polymers involving very small rubber particles. 3. Impact mechanical behaviour of poly(vinyl chloride)/methyl methacrylate-butadiene-styrene graft copolymer blends, *Polymer*, 35, 4760–4765, 1994.

82. Dompas, D., Groeninckx, G., Isogawa, M., Hasegawa, T., and Kadokura, M., Cavitation versus debonding during deformation of rubber-modified poly(vinyl chloride), *Polymer*, 36, 437–441, 1995.

83. Lazzeri, A. and Bucknall, C.B., Dilatational bands in rubber-toughened polymers, *J. Mater. Sci.*, 28, 6799–6808, 1993.

84. Ayre, D.S. and Bucknall, C.B., Particle cavitation in rubber-toughened PMMA: experimental testing of the energy-balance criterion, *Polymer*, 39, 4785–4791, 1998.

85. Bucknall, C.B, Rizzeri, R., and Loore, D.R., Detection of rubber particle cavitation in toughened plastics using thermal contraction tests, *Polymer*, 41, 4149–4156, 2000.

86. Steenbrink, A.C., Litvinov, V.M, and Gaymans, R.J, Toughening of SAN with acrylic coreshell rubber particles: particle size effect or corsslink density, *Polymer*, 39, 4817–4825, 1998.

87. Bucknall, C.B., Toughened Plastics, Applied Science Publishers, London, 1977.

88. Donald, A.M. and Kramer, E.J., Plastic deformation mechanisms in poly(acrylonitrile-butadiene styrene) (ABS), *J. Mater. Sci.*, 17, 1765–1772, 1982.

89. Bubeck, R.A., Buckley, D.J., Kramer, E.J., and Brown, H.R., Modes of deformation in rubber-modified thermoplastics during tensile impact, *J. Mater. Sci.*, 26, 6249–6259, 1991.

90. Magalhaes, A.M.L. and Borggreve, R.J.M., Contribution of the crazing process to the toughness of rubber-modified polystyrene, *Macromolecules*, 28, 5841–5851, 1995.

91. Rios-Guerrero, L., Keskkula, H., and Paul, D.R., Deformation process in high impact polystyrene as revealed by the analysis of arrested cracks, *Polymer*, 41, 5415–5421, 2000.

92. Lazzeri, A. and Bucknall, C.B., Applications of a dilatational yielding model to rubber-toughened polymers, *Polymer*, 15, 2895–2902, 1995.

93. Schirrer, R, Fond, C, and Lobrecht, A., Volume change and light scattering during mechanical damage in poly (methyl methacrylate) toughened with core shell rubber particles, *J. Mater. Sci.*, 31, 6409–6422, 1996.

94. Cho, K., Yang, J., and Park, C.E., The effect of rubber particle size on toughening behaviour of rubber-modified poly(methyl methacrylate) with different test methods, *Polymer*, 39, 3073–3081, 1998.

95. Jinag, W., Yuan, Q., An, L., and Jiang, B., Effect of cavitations on brittle-ductile transition of particle toughened thermoplastics, *Polymer*, 43, 1555–1558, 2002.

96. Liu, Z., Zhu, X., Wu, L., Li, Y., Qi, Z., Choy, C., Wang, F., Effects of interfacial adhesion on the rubber toughening of PVC. 1. Impact test, *Polymer*, 42, 737–746, 2001.

97. Gaymans, R.J., Toughening of semicrystalline thermoplastics, in *Polymer Blends, Vol. 2, Performance*, Paul, D.R. and Bucknall, C.B., Eds., John Wiley & Sons, New York., 2000, pp. 177–224.

98. Kayano, Y., Keskkula, H., and Paul, D.R., Fracure behaviour of some rubber modified nylon 6 blends, *Polymer*, 39, 2835–2845, 1998.

99. Huang, J.J, Keskkula, H., and Paul, D.R., Rubber toughening of amorphous polya-mide by functionalised SEBS copolymers: morphology and izod impact behaviour, *Polymer*, 45, 4203–4215, 2004.

100. Muratoglu, O.K, Argon, A.S., Cohen, R.E., and Weinberg, M., Toughening mecha-nism of rubber-modified polyamides, *Polymer*, 36, 921–930, 1995.

101. Argon, A.S., Bartczak, Z., Cohen, R.E, and Muratoglu, O.K., in Toughening of Plastics: Advances in Modelling and Experiments, ACS Symposium Series, Vol. 759, Pearson, R.A., Sue, H.J., and Yee, A.F., Eds., ACS, Washington D.C., 2000, pp. 98–124.

102. Argon, A.S. and Cohen R.E., Toughenability of polymers, *Polymer*, 44, 6013–6032, 2003.

103. Loyens, W. and Groeninckx, G., Deformation mechanisms in rubber toughened semi-crystalline polyethylene terephathalate, *Polymer*, 44, 4929–4941, 2003.

104. Adhikari, R., Michler., G.H., and Knoll., K., Morphology and micromechanical behaviour of binary blends comprising block copolymers having different architec-tures, *Polymer*, 45, 241–246, 2004.

105. Crawford, E. and Lesser, A.J., Mechanics of rubber particle cavitaion in toughened polyvinylchloride (PVC), *Polymer*, 41, 5865–5870, 2000.

106. Lu, F., Cantwell, W.J., and Kausch, H.H., The role of cavitation and debonding in the toughening of core–shell rubber modified epoxy systems, *J. Mater. Sci.*, 32, 3055–3059, 1997.

107. Evans, A.G., Ahamad, Z.B., Gilbert, D.G., and Beaumont, P.W., Mechanisms of toughening in rubber-toughened polymers, *Acta Metall.*, 34, 79–87, 1986.

108. Dekkers, M.E., Hobbs, S., and Watkins, V.H., Toughened blends of poly (butylene terephthalate) and BPA polycarbonate. 2. Toughening mechanisms, *J. Mater. Sci.*, 23, 1225–1230, 1998.

109. Yee, A.F. and Pearson, R.A., Toughening mechanisms in elastomer-modified epoxies. 1. Mechanical studies, *J. Mater. Sci.*, 21, 2462–2474, 1986.

110. Pearson, R.A. and Yee, A.F., Toughening mechanisms in elastomer-modified epoxies. 3. The effect of cross-link density, *J. Mater. Sci.*, 24, 2571–2580, 1989.

111. Parker, D.S., Sue, H.-J., Huang, J., and Yee, A.F., Toughening mechanisms in core shell rubber modified polycarbonate, *Polymer*, 31, 2267–2277, 1989.

112. Bensason, S., Hiltner, A., and Baer, E., Damage zone in PVC and PVC/MBS blends. 2. Analysis of the stress-whitened zone, *J. Appl. Polym. Sci.*, 63, 715–723, 1997.

113. Tse, A., Shin, E., Hiltenr, A., and Baer, E., Damage zone development in PVC under a multiaxial tensile-stress state, *J. Mater. Sci.*, 26, 5374–5382, 1991.

114. McGarry, F.J., Building design with fiber reinforced materials, *Proc. Royal. Soc. London, Part A*, 319, 59–68, 1970.

115. Sultan, J.N. and McGarry, F.J., Effect of rubber particle-size on deformation mech-anisms in glassy epoxy, *Polym. Eng. Sci.*, 13, 29–34, 1973.

116. Yee, A.F., Du, J., and Thouless, M.D, Toughening of epoxies, in Polymer Blends, Vol. II, Performance, Paul, D.R. and Bucknall, C.B., Eds.; John Wiley & Sons, New York, 2000, pp. 225–267.

117. Bagheri, R. and Pearson R.A., Role of particle cavitation in rubber-toughened epoxies. 1. Microvoid toughening, *Polymer*, 37, 4529–4538, 1996.

118. Pearson,R.A. and Yee, A.F., Influence of particle-size and particle-size distribution on toughening mechanisms in rubber-modified epoxies, *J. Mater. Sci,*. 26, 3828–3844, 1991.

119. Bagheri, R. and Pearson, R.A., Role of particle cavitation in rubber-toughened epoxies. II. Inter-particle distance, *Polymer*, 41, 269–276, 2000.

120. Kishi, H., Shi, Y.B., Huang, J., and Yee, A.F., Shear ductility and toughenability study of highly cross-linked epoxy/polyethersulphone, *J. Mater. Sci.*, 32, 761–771, 1997.
121. Kinloch, A.J., Yuen, M.L., and Jenkins, S.D., Thermoplastic-toughened epoxy polymers *J. Mater. Sci.*, 29, 3781–3790, 1994.
122. Francis, B., Vanden Poel, G., Posada, F., Groeninckx, G., Rao, V.L., Ramaswamy, R., and Thomas, S., Cure kinetics and morphology of blends of epoxy resin with poly(ether ether ketone) containing pendant tertiary butyl groups, *Polymer*, 44, 3687–3699, 2003.
123. Paul, D.R., High performance engineering thermoplastics via reactive compatiblisation, in Macromolecules 2003, Chemical Modification and Blending of Synthetic and Natural Macromolecules for Preparing Multiphase Structural and Functional Materials: Principles, Methods and Properties, Ciardelli, F. and Penczek, S., Eds., NATO Advanced Study Institute, Pisa, Italy, October, 6–16, 2003, pp. 27–28.
124. Kitayama, N., Keskkula, H., and Paul, D.R., Reactive compatibilisation of nylon 6/styrene — acrylonitrile copolymer blends. 3. Tensile stress-strain behaviour, *Polymer*, 42, 3751–3759, 2001.
125. Murata, K. and Anazawa, T., Morphology and mechanical properties of polymer blends with photochemical reaction for photocurable/linear polymers, *Polymer*, 43, 6575–6583, 2002.
126. Veenstra, H., Verkooijen, C.J.P., Lent, van J.J.B., Dam, van J., Posthuma de Boer, A., and Nijhof, H.J.P.A., On the mechanical properties of co-continuous polymer blends: experimental and modelling, *Polymer*, 41, 1817–1826, 2000.
127. Coran, A.Y. and Patel, R., Rubber-thermoplastic compositions. Part I. EPDM-polypropylene thermoplastic vulcanizates, *Rubber Chem. Technol.*, 53, 141–150, 1980.
128. Oderkerk, J. and Groeninckx, G., Morphology development by reactive compatibilisation and dynamic vulcanization of nylon 6/EPDM blends with high rubber fraction, *Polymer*, 43, 2219–2228, 2002.
129. Oderkerk, J., de Schaetzen, G., Goderis, B., Hellemans, L., and Groeninckx, G., Micromechanical deformation and recovery processes of nylon 6/rubber thermoplastic vulcanizates as studied by atomic force microscopy and transmission electron microscopy, *Macromolecules*, 35, 6623–6629, 2002.
130. Oderkerk, J., Groeninckx, G., and Soliman, M., Investigation of the deformation and recovery behavior of nylon-6/rubber thermoplastic vulcanizates on the molecular level by infrared-strain recovery measurements, *Macromolecules*, 35, 3946–3954, 2002.
131. Oommen, Z. and Thomas, S., Mechanical properties and failure mode of thermoplastic elastomers from natural rubber/poly(methyl methacrylate)/natural rubber-g-poly(methyl methacrylate) blends, *J. Appl. Polym. Sci.*, 65, 1245–1255, 1997.
132. George, S., Joseph, R., Varughese, K.T., and Thomas, S., Blends of nitrile rubber and polypropylene: morphology, mechanical properties and compatibilisation, *Polymer*, 36, 4405–4416, 1995.
133. George, J., Joseph, R., Thomas, S., and Varughese, K.T., High density polyethylene/acrylonitrile butadiene rubber blends: morphology, mechanical properties, and compatibilisation, *J. Appl. Polym. Sci.*, 57, 449–465, 1995.
134. Mathew, M. and Thomas, S., Compatibilisation of heterogeneous acrylonitrile butadiene rubber/polystyrene blends by the addition of styrene acrylonitrile copolymer: effect on morphology and mechanical properties, *Polymer*, 44, 1295–1306, 2003.
135. Kietzke, T., Neher, D., Landfester, K., et al., Novel approaches to polymer blends based on polymer nanoparticles, *Nature Mater.*, 2, 48–412, 2003.
136. Schmitt, C., Nothofer, H.-G., Falcou, A., et al., Conjugated polyfluorene/polyaniline block copolymers, *Macromol. Rapid Commun.*, 22, 624–628, 2001.

137. Hu, G.H., Cartier, H., and Plummer, C., Reactive extrusion: towards nanoblends, *Macromolecules*, 32, 4713–4718, 1999.
138. Pernot, H., Baumert, M., Court, F., and Leibler, L., Design and properties of co-continuous nanostructured polymers by reactive blending, *Nature Mater.*, 1, 54–58, 2002.
139. Apostolo, M. and Triulzi, F., Properties of fluoroelastomer/semicrystalline perfluoropolymer nano-blends, *J. Fluorine Chem.*, 125, 303–314, 2004.
140. Prosycevas I., Tamulevicius, S., and Guobiene, A., The surface properties of PS/PMMA blends: nanostructured polymeric layers, *Thin Solid Films*, 453, 304–311, 2004.
141. Yang, C.Y, Hide, F., Heeger, A.J., and Cao, Y., Nanostructured polymer blends: novel materials with enhanced optical and electronic properties, *Synthetic Metals*, 84, 895–896, 1997.
142. Coa, Y., Yu, G., Heeger, A.J., and Yang, C.Y., Efficient, fast response light-emitting electrochemical cells: electroluminescent and solid electrolyte polymers with interpenetrating network morphology, *Appl. Phys. Lett.*, 68, 3218–3220, 1996.
143. Yu, G., Gao, J., Hummelen, J.C., Wudl, F., and Heeger, A.J., Polymer photovoltaic cells: enhanced efficiencies via a network of internal donor-acceptor heterojunctions, *Science*, 270, 1789–1791, 1995.
144. Iyenger, N.A., Harrison, B., Duran, R.S., et al., Morphology evolution in nanoscale light-emitting domains in MEH-PPV/PMMA blends, *Macromolecules*, 36, 8978–8995, 2003.
145. Kailas, L., Audinot, J.-N., Migeon, H.N., and Betrand, P., TOF-SIMS molecular characterisation and nano-SIMS imaging of submicron domain formation at the surface of PS/PMMA blend and copolymer thin films, *Applied Surf. Sci.*, 231, 289–295, 2004.
146. Busby, A.J., Zhang, J.X., Naylor, A., Roberts, C.J., Davies, M.C., Tendler, S.J.B., and Howdle, S.M., The preparation of novel nano-structured polymer blends of ultra high molecular weight polyethylene with polymethacrylates using supercritical carbon dioxide, *J. Mater. Chem.*, 13, 2838–2834, 2003.
147. Shimizu H., Horiuchi S., and Kitano, T., An appearance of heterogeneous structure in a single phase state of the miscible PVME/PS blends, *Macromolecules*, 32, 537–540, 1999.
148. Dreezen, G., Ivanov, D.A., Nysten, B., and Groeninckx, G., Nano-structured polymer blends: phase structure, crystallisation behaviour and semi-crystalline morphology of phase separated binary blends of poly(ethylene oxide) and poly(either sulphone), *Polymer*, 41, 1395–1407, 2000.
149. Mathew, A.P., Packirisamy, S., Radusch, H.J., and Thomas, S., Effect of initiating system and blend ratio and crosslink density on the mechanical properties and failure topography of nano-structured full-interpenetrating polymer networks from natural rubber and polystyrene, *Eur. Polym. J.*, 37, 1921–1934, 2001.
150. Mathew, A.P., Groeninckx, G., Michler, G.J., Radusch, H.J., and Thomas, S., Viscoelastic properties of nanostructured natural rubber/polystyrene interpenetrating polymer networks, *J. Polym. Sci. Part B Polym. Phys.*, 39, 1451–1460, 2003.
151. Yamazaki, M., Kayama, K., Ikeda, K., Anil T., and Ichihara, S., Nanostructured carbonaceous material with continuous pores obtained from reaction induced phase separation of miscible polymer blends, *Carbon*, 42, 1641–1649, 2004.
152. Banerjee, P. and Mondal, M.J., Conducting polyaniline nanoparticle blends with extremely low percolation thresholds, *Macromolecules*, 28, 3940–3943, 1995.
153. Robeson, L.M., Perspectives in polymer blend technology, in Polymer Blends Handbook, Vol. 2, Utracki, L.A., Ed., Kluwer Academic Publishers, 2002.

154. Harris, J.E. and Robeson, L.M., in Mulitphase Macromolecular Systems, Culberston, B.M., Ed., Plenum Press, New York, 1989, pp.

155. Wang, H.-L, Toppare, L., and Fernandez, J.E., Conducting polymer blends: polythiophene and polypyrrole blends with polystyrene and poly(bisphenol A carbonate), *Macromolecules*, 23, 1053–1059, 1990.

156. Narkis, M., Ziebermann, M., and Siegman, A., On the "curiosity" of electrically conductive melt processed doped-polyaniline/polymer blends versus carbon-black/polymer compounds, *Polym. Adv. Technol.*, 8, 525–528, 1997.

157. Masson, J.-F. and St. John, M., Miscible blends of cellulose and poly(vinyl pyrrolidone), *Macromolecules*, 24, 6670–6679, 1991.

158. Rubner, M.F., Symposium on Advances in Conducting Polymers. A Tribute to MacDiarmid, A.G., U. Pennsylvania, Dec. 4, 1993.

159. Hillmyer, M., Lipic, P.M., Hajduk, D., Almdal, K., and Bates, F.C., Self assembly and polymerisation of epoxy resin–amphiphilic block copolymer nanocomposites, *J. Am. Chem. Soc.*, 119, 2749–2750, 1997.

160. Lipic, P.M., Bates, F., and Hillmayer, M., Nanostructured thermosets from self-assembled amphiphilic block copolymer nanocomposites, *J. Am. Chem. Soc.*, 120, 8963–8970, 1998.

161. Ritzenthaler, S., Court, F., David, L., Girard-Reydet, E., Leibler, L., and Pascault, J.P., ABC triblock copolymers/epoxy-diamine blends. 1. Keys to achieve nanostructured thermosets, *Macromolecules*, 35, 6245–6254, 2002.

162. Dean, J., Verghese, N., Pharm, H. and Bates, F., Nanostructured toughened epoxy resins, *Macromolecules*, 36, 9267–9270, 2003.

163. Gubbels, F., Jérôme, R, Teyssié, Ph., Vanlathem, E., Deltour, R., Calderone, A., Parenete, V., and Brédas, J.L., Selective localization of carbon-black in immiscible polymer blends — a useful tool to design electrically conductive composites, *Macromolecules*, 27, 1972–1974, 1994.

164. Karasz, F.E., Hu, B., and Wang, L., New copolymer blends for LEDs, *Polymer Preprints*, 38, 343–344, 1997.

165. Wu, R.-J., Shaw, M.T., and Weiss, R.A., Rheooptical studies of stress-induced phase-changes in blends exhibiting UCST — polystyrene and polyisobutylene, *J. Rheol.*, 36, 1605, 1992.

166. Chen, Z.J., Shaw, M.T., and Weiss, R.A., Two-dimensional light scattering studies of polystyrene/poly (vinyl methyl ether) blends under simple shear flow, *Polym. Prepr. Am. Chem. Soc. Polym. Div.*, 34, 838–839, 1993.

2 Theoretical Aspects of Phase Morphology Development

Ivan Fortelný

CONTENTS

2.1 INTRODUCTION

The present chapter deals with a theoretical description of the phase morphology development of immiscible polymer blends during their melt mixing and processing. Because the phase structure of polymer blends is a function of a large number of parameters, attempts to formulate general empirical rules for its prediction have not been successful. Therefore, theoretical analysis is a necessary tool in our attempts at controlling polymer blend morphology. The ultimate aim of the theory is a detailed prediction of the phase structure from a knowledge of the blend composition, properties of the components, and thermomechanical history during the blend preparation and processing. To achieve this aim is very difficult because polymer melts are viscoelastic substances displaying complicated rheological behavior; flow fields in mixing and processing equipment are complex; and thermomechanical history is complicated by substantial heat production. Instantaneous morphology of a polymer blend affects the flow fields during mixing and processing, and the amount of dissipated energy and, consequently, temperature profile in the sample. Thus, the morphology and thermomechanical history are interdependent, and it is not surprising that the state of the art is still far from the final desired product. Yet a number of results achieved in the theoretical description of phase structure evolution in mixtures of immiscible liquids can substantially improve the control of the phase structure in polymer blends.

Most theories of the phase structure development in a mixture of immiscible liquids describe a system on a microrheological level, i.e., as domains of individual components (e.g., droplets) with characteristic dimensions of the order of micrometers. These domains are characterized by the rheological properties of the components. An additional parameter that is used at the microrheological level is interfacial tension. Most theoretical analyses concentrate on binary blends of immiscible polymers. Less theoretical attention has been given to the phase structure development in blends containing a small amount of block or graft copolymers as a compatibilizer and still less attention has been given to the description of the phase structure development in multicomponent polymer blends.

The two main characteristics of the phase structure are the type and fineness. The *type* of the phase structure specifies whether the structure is cocontinuous or dispersed, whether the component becomes part of the matrix or the dispersed phase, and what shape is characteristic of the dispersed particles. The *fineness* of the phase structure describes the size of phase domains, e.g., the size of the dispersed particles. A theory to predict satisfactorily the type of the phase structure has not yet been formulated.

The droplet size in blends with dispersed structure can be better predicted. Consensus exists in the literature (1–6) that the droplet size is controlled by the competition between droplet breakup and coalescence in flow. Both the breakup and coalescence are complicated processes not easily described even with blends of Newtonian liquids in simple flow fields. The complexity of the description further increases if factors such as viscoelasticity of the droplets and matrix, and hydrodynamic interaction among droplets in concentrated systems are considered, or if realistic models of flow fields in mixing and processing devices are used. Quite sophisticated models have been suggested for individual aspects of the problem. Unfortunately, even approximate and partial descriptions lead mostly to equations that are solvable only numerically, complicating their utilization at further analysis. When an analytical solution of these equations exists, it is usually only for limited intervals of the parameters, a fact that is frequently ignored.

Substantial improvement in the control of the phase structure and properties of polymer blends does not require precise quantitative prediction of the morphology. Rather, qualitative knowledge of the dependencies of the phase structure on parameters of the system is much more important, e.g., will an increase in the viscosity of the dispersed phase or matrix or in the rate of mixing lead to an increase or decrease in the size of the dispersed droplets? Therefore, this chapter will look at the question of whether theoretical models provide such reliable predictions for real blends of immiscible polymers.

Section 2.2 deals with theories of the phase structure formed during melt mixing of polymers. Because the formation of cocontinuous structure is discussed in detail in Chapter 3, only the most important theoretical approaches to the prediction of dependence of the type of phase structure on the blend composition are discussed here. A short discussion of the theories predicting the phase structure of multicomponent polymer blends is included. In Section 2.3, the results of theoretical studies of droplet deformation and breakup in flow fields are summarized and a choice of an adequate model of droplet breakup during mixing of polymer blends is discussed. Section 2.4 presents theories of flow induced coalescence; the approximations that are used are compared, and limits in the applicability of the theories are discussed. Section 2.5 deals with the competition between flow induced breakup and coalescence. Assumptions and approximations used by various theories are shown, and the reliability of predictions of the dependence of the size of dispersed droplets on system parameters is discussed.

In Section 2.6, the results of the theories describing the phase structure evolution in molten quiescent polymer blends are summarized. Shape relaxation and breakup of deformed droplets, coarsening and breakup of a cocontinuous structure, and growth of the droplets in blends with dispersed structure are considered. Results of the theories of coalescence and of Ostwald ripening are compared, and their validity for molten polymer blends is discussed. Section 2.7 focuses on the effect of a compatibilizer (block or graft copolymer) on droplet breakup and coalescence in flow, on the competition between breakup and coalescence, and on the changes in the blend morphology at rest. The problems with predicting compatibilizer distribution between the interface and bulk phases are briefly discussed. Finally, Section 2.8 contains general conclusions drawn from the preceding parts. The main unsolved problems are stressed, and challenges for future investigation are pointed out.

2.2 PREDICTION OF THE TYPE OF PHASE STRUCTURE FORMED DURING MIXING OF POLYMER BLENDS

The type of the phase structure of blends of immiscible polymers 1 and 2 is dependent on the composition. It was found experimentally for most polymer blends that at low concentration of 2, particles of component 2 are dispersed in the matrix of component 1. With increasing concentration of 2, a partially continuous structure of 2 appears at first, and then, a fully cocontinuous structure is formed. After that, phase inversion occurs, and component 2 forms the matrix and component 1 the dispersed phase (7,8). The theory is expected to predict the composition at which partial or full continuity of individual components appears as a function of rheological properties of the components, interfacial tension, and mixing or processing conditions. The attempts to formulate such a rule based on the most important results of experimental studies are summarized in Chapter 3. So far, no theory or empirical rule predicts conditions for the formation of cocontinuous morphology quite satisfactorily.

Besides cocontinuous morphology, also droplet within droplet (i.e., droplets of dispersed phase containing small droplets of the matrix) are sometimes formed in blends with a higher content of a minor component (1). In some systems, ribbonlike or stratified morphology are detected rather than the classical cocontinuous one (9). No theory for formation of these types of morphology is available.

If steady morphology is established in flow, then a description of a blend by interfacial area per unit volume Q and interface tensor $\mathbf{q}$ (characterizing orientation of the interface) (10) could be combined with a proper condition for extremes of thermodynamic functions in steady state (11). Q and $\mathbf{q}$ are defined as

$$Q = \frac{1}{V} \int_A d A \tag{2.1}$$

and

$$\mathbf{q} = \frac{1}{V} \int_A \left(\mathbf{nn} - \frac{1}{3} \mathbf{I} \right) d A \tag{2.2}$$

where V is the volume of the system, A is the interfacial area, $\mathbf{n}$ is a unit normal vector of the interfacial area increment dA, and $\mathbf{I}$ is the unit tensor. For a system where linear relations between thermodynamic forces and relating fluxes are valid, the entropy production has its minimum value at a steady state that is compatible with prescribed conditions (11). In spite of the problems with linearity of the relations between thermodynamic forces and fluxes for viscoelastic systems, the expression of entropy production using the variables Q and $\mathbf{q}$ seems to be a useful tool for estimating the morphology that is formed in steady flow of polymer blends. The expression can elucidate the role of various factors and can show which of the different possible structures are stable in steady state. In an ideal case, the requirement of minimum entropy production could lead to equations that allow us to calculate parameters of the phase structure directly.

If the cocontinuous phase structure is transitory, continuity of individual blend components would be dependent explicitly on the mixing time. An adequate theory describing the dependence of continuity of individual components would have to reflect this fact and be of a kinetic nature.

Prediction of the type of morphology in polymer blends containing three or more immiscible components is a more complicated task than the prediction for binary blends. Most studies have dealt with systems where component 2 forms the matrix and minor components 1 and 3 are dispersed. Hobbs et al. (12) proposed that the encapsulation of components 1 and 3 is controlled by the spreading coefficient λ_{31} defined as

$$\lambda_{31} = \sigma_{12} - \sigma_{32} - \sigma_{13} \tag{2.3}$$

where σ_{12}, σ_{32}, and σ_{13} are the interfacial tensions for each component pair. If spreading coefficient λ_{31} or λ_{13} is positive, then component 3 encapsulates component 1 (particles with spherical core 1 and spherical shell 3 are formed) or component 1 encapsulates component 3, respectively. If both λ_{31} and λ_{13} are negative, separated droplets of 1 and 3 are formed. The concept of spreading coefficients was extended by Guo et al. (13). They assumed that morphology leading to the lowest Gibbs energy, G, of the blend is realized. G can be expressed as

$$G = G_0 + \sum_{i \neq j} A_i \sigma_{ij} \tag{2.4}$$

where G_0 is the part of G independent of the morphology, A_i is the interfacial area of the ith dispersed phase, and σ_{ij} is the interfacial tension. The authors assumed that dispersed particles are spherical and neglected their polydispersity. Using these assumptions, the following equations were derived for blends with separated droplets, Eq. (2.5), encapsulated droplets of 3 by 1, Eq. (2.6), and encapsulated droplets of 1 by 3, Eq. (2.7):

$$\left(\sum A_i \sigma_{ij} \right)_{1+3} = \left(4\pi \right)^{1/3} \left[n_1^{1/3} x^{2/3} \sigma_{12} + n_3^{1/3} \sigma_{23} \right] \left(3V_3 \right)^{2/3} \tag{2.5}$$

$$\left(\sum A_i \sigma_{ij} \right)_{1/3} = \left(4\pi \right)^{1/3} \left[n_1^{1/3} \left(1 + x \right)^{2/3} \sigma_{12} + n_3^{1/3} \sigma_{13} \right] \left(3V_3 \right)^{2/3} \tag{2.6}$$

$$\left(\sum A_i \sigma_{ij} \right)_{3/1} = \left(4\pi \right)^{1/3} \left[n_1^{1/3} x^{2/3} \sigma_{13} + n_3^{1/3} \left(1 + x \right)^{2/3} \sigma_{23} \right] \left(3V_3 \right)^{2/3} \tag{2.7}$$

where V_3 is the total volume of component 3, $x = V_1/V_3$, and n_1 and n_3 are the number of particles of components 1 and 3 in the system, respectively. For an evaluation of experimental data, the authors assumed that $n_1 = n_3$. Generally, the assumption is not correct. For an exact determination of n_i, the relation between the particle size and their volume fraction in a system must be known, which is difficult already for binary blends. Moreover, strong effect by a third component on the size of dispersed droplets was detected for some systems (14).

The predictive scheme based on Eq. (2.4) was applied also on quarternary systems (15). A substantial part of experimental results qualitatively agrees with predictions based on Eq. (2.3) and/or Eqs. (2.4) to (2.7) (1,12,13,16,17), but the effect of rheological properties of the components on the type of the phase structure established was detected in some papers (18–20). Reignier et al. (20) suggested using in Eqs. (2.5) to (2.7) the effective values of σ_{ij} calculated according to Van Oene's theory (9,21). Van Oene considered the Helmholtz energy of a liquid deformation in flow to be proportional to the first–normal stress difference. He assumed that the Helmholtz energy associated with the formation of a droplet with radius R is a product of the droplet volume and of the difference between the first–normal stress differences in the dispersed phase (droplet), $N_{1,d}$, and in the matrix, $N_{1,m}$. This energy is equal to the Helmholtz energy due to the interfacial tension on the droplet surface. The effective interfacial tension in flow, σ_{ef}, is given by the equation

$$\sigma_{ef} = \sigma_0 + \frac{1}{6} R \left(N_{1,d} - N_{1,m} \right) \tag{2.8}$$

where σ_0 is interfacial tension at rest. Substitution of σ_{ij} by the related σ_{ef} improved the agreement with experimental data (20). Unfortunately, determination of the normal stress difference at mixing condition is not easy, and, therefore, extensive comparison of experimental data with the above predictive schemes, where σ_{ij} are substituted by the related σ_{ef}, is not available.

The above considerations can be helpful for the prediction of encapsulation of two minor components in the matrix of the major component, but the type of the phase structure cannot be predicted with certainty. No attempt to formulate a theory predicting the type of the phase structure in ternary blends, where two or three phases can be fully or partially continuous, is found in the literature. It seems that progress in the theories of phase structure in binary blends is necessary prior to their extension to the case of multicomponent blends.

2.3 DROPLET DEFORMATION AND BREAKUP IN FLOW FIELDS

Droplet breakup is a decisive event in the formation of morphology. Its understanding and proper description are necessary conditions for the derivation of a theory of droplet size in immiscible polymer blends. Droplet breakup has been intensively studied both theoretically and experimentally, and the results have been summarized in a number of reviews and monographs (1–6,10,22–24). Most theoretical works have dealt with Newtonian droplets in Newtonian matrixes, where complicating factors such as elasticity and dependence of the viscosity on the deformation rate are absent. Two basic approaches can be found in the literature. The first one starts from spherical droplets and describes their deformation and subsequent breakup. The second one studies the stability of highly elongated droplets.

In flow, the shape of a droplet is given by the competition between the flow stress, which has the tendency to deform the droplet, and the interfacial stress, equal to the Laplace pressure ($2\sigma/R$ for spherical droplet), which has the tendency to minimize the interfacial area. Taylor (25) recognized that shear flow induced deformation of a Newtonian droplet in a Newtonian matrix can be expressed as a function of the ratio, p, of the viscosity of the dispersed phase, η_d, and the matrix, η_m, and the dimensionless capillary number Ca, defined as

$$Ca = \frac{\eta_m \gamma R}{\sigma} \tag{2.9}$$

where γ is the shear rate, R is the radius of undeformed droplet, and σ is the interfacial tension. For other flow fields, such as uniaxial extensional flow, plane hyperbolic flow, etc., Ca can be defined analogically to Eq. (2.9) using the relevant component of the rate of deformation tensor. The droplet deformation, D, is defined by

$$D = \frac{L - B}{L + B} \tag{2.10}$$

in which L and B are the length and breadth of deformed droplets, respectively. The theories discussed below neglect inertial effects (low Reynolds numbers) and buoyancy driven motion. This is fully justified for polymer blends of high viscosity components with inhomogeneities (droplets) with characteristic dimensions from 0.1 μm to tens of micrometers and without pronounced differences in density. When a velocity field is externally imposed at large distances from the droplet, with typical shear rate in the neighborhood of the droplet, the following equation describes the stress boundary condition on the droplet surface (22)

$$\mathbf{n} \cdot \mathbf{T}_m - p\mathbf{n} \cdot \mathbf{T}_d = \frac{1}{Ca}\left(\nabla_s \cdot \mathbf{n}\right)\mathbf{n} \tag{2.11}$$

where $\mathbf{T}_m$ and $\mathbf{T}_d$ are the stress tensors in the matrix and dispersed phase, respectively, $\mathbf{n}$ is the unit normal directed outward from the droplet, and $\nabla_s \cdot \mathbf{n}$ is the local interface curvature. To predict the droplet size in flow, the most important parameters characterizing the droplet breakup are the critical capillary number, Ca_c, i.e., the lowest Ca at which the breakup occurs, the breakup time t_B, i.e., the time necessary for droplet deformation from equilibrium spherical shape to breakup, and the number and size of the droplet fragments formed. The following discussion concentrates on predicting these parameters. The discussion of prediction and experimental determination of the droplet shape in flow can be found in Chapter 13 and ref. (10).

Taylor (25) showed that in steady uniform shear flow, the droplet deforms into a spheroid. He calculated the droplet deformation D using the perturbation method with spherical shape as a zero approximation. He obtained first-order solutions for the case when the interfacial tension effect dominates over the viscous effect (i.e., $p = O(1)$, $Ca \ll 1$)

$$D = Ca\,\frac{19p + 16}{16p + 16} \tag{2.12}$$

and for the case when interfacial tension effect is negligible compared to the viscosity effect (i.e., $Ca = O(1)$, $p \gg 1$)

$$D = \frac{5}{4p} \tag{2.13}$$

Taylor (25) proposed that breakup appears when D achieves the value 0.5. It should be mentioned that the equation

$$Ca_c = \frac{1}{2}\frac{16p + 16}{19p + 16} \tag{2.14}$$

which follows from the combination of the above condition and Eq. (2.12) and which is frequently used in discussing experimental results, is in no case applicable to systems with $p > 1$. Taylor's theory was extended by including second-order solutions (26) and generalization of the first-order (27) and second-order (28) theories to general time dependent flow. Cox (27) derived the equation for D in steady shear flow

$$D = \frac{5(19p+16)}{4(p+1)\left[(19p)^2 + \left(\frac{20}{Ca}\right)^2\right]^{1/2}} \tag{2.15}$$

For steady plane hyperbolic flow ($v_x = \varepsilon x$, $v_y = \varepsilon y$, $v_z = 0$) and for steady axisymmetric extensional flow ($v_x = -(\varepsilon/2)x$, $v_y = -(\varepsilon/2)y$, $v_z = \varepsilon z$), Cox's theory gives

$$D = ACa\frac{19p+16}{16p+16} \tag{2.16}$$

where $A = 2$ for the plane hyperbolic flow and $A = 1.5$ for the axisymmetric extensional flow. It is evident from Eq. (2.15) that the value $D = 0.5$ cannot be achieved and, therefore, breakup does not occur for a high p even for $Ca \to \infty$. On the other hand, D for hyperbolic and extensional flow is proportional to Ca, and, hence, a droplet bursts at high enough Ca for any p. The theories using somewhat different approximations (25,26,28) lead to the qualitatively same conclusions; however, they somewhat differ at maximum p at which droplet breakup in shear flow can appear (about 2.5 follows from Eq. (2.15); Taylor (25) proposed the value 4). The reason for the different droplet behavior in shear and extensional flows is the rotation of viscous droplets in shear flow, limiting the maximum loading.

The results of rather extensive experimental studies of the dependence Ca_c vs. p, summarized in a number of articles (5,10,22,24,29), qualitatively agree with the above theories. For shear flow, Ca_c achieves a minimum for $0.1 < p < 1$, steeply increases with increasing p for $p > 1$, and go to infinity at about $p = 4$. For $p < 0.1$, Ca_c somewhat increases with decreasing p. For small p, tip streaming (i.e., the breaking off of small droplets from the ends of an elongated drop) was observed at smaller Ca rather than the breakup into two daughter droplets. De Bruijn suggested an empirical equation based on Grace's data (29) for the dependence of Ca_c on p (22,30)

$$\log Ca_c = -0.506 - 0.0995 \log p + 0.124(\log p)^2 - \frac{0.115}{\log p - \log 4.08} \tag{2.17}$$

In extensional flow, a minimum for Ca_c was observed at about $p = 1$. A very slight increase in Ca_c with increasing or decreasing p was observed for $p > 1$ or $p < 1$, respectively. Generally, flow fields in mixing and processing devices contain both shear and extensional components. Therefore, it is important to know the dependence of Ca_c on a flow type parameter α_f (4,22). The velocity gradient tensor $\partial v_i/\partial x_j$ for two-dimensional (2-D) linear flow can be expressed as

$$\frac{\partial v_i}{\partial x_j} = \frac{\gamma}{2} \begin{pmatrix} 1+\alpha_f & 1-\alpha_f & 0 \\ -1+\alpha_f & -1-\alpha_f & 0 \\ 0 & 0 & 0 \end{pmatrix} \tag{2.18}$$

where $\alpha_f = 0$ for simple shear flow and $\alpha_f = 1$ for a two-dimensional irrotational flow. It should be mentioned that the common expression for shear flow is obtained from Eq. (2.18) in a basis rotated over 45°. The dependence of Ca_c on α_f gradually changes from that typical for extensional flow to that typical for shear flow with a decrease of α_f from 1 to 0 (22). Therefore, it can be assumed that also the Ca_c vs. p dependence for real mixing and processing devices lies somewhere between those for shear and irrotational flows.

The effect of elasticity of droplets and of a matrix on Ca_c has been studied quite intensively, especially during recent years. The results are less conclusive than those for the dependence of Ca_c on p because: (i) the description of droplet breakup in systems with viscoelastic droplets and/or matrix is more complicated than in systems with Newtonian components only, (ii) it is difficult to separate the effect of elasticity and that of shear thinning, and (iii) of possible interrelations between the effect of p and elasticity. Nevertheless, the results show (24,31–33) that an increase in elasticity of droplets leads to an increase in Ca_c. On the other hand, matrix elasticity seems to promote droplet deformation and, therefore, decreases Ca_c.

The second approach to the problem of droplet breakup is based on the assumption that a droplet with $Ca \gg Ca_c$ is almost affinely deformed into a long slender filament. When the radius of the filament decreases to the value for which the flow field stress and the interfacial stress attain the same order of magnitude, small disturbances at the interface of the filament grow and finally result in the disintegration of the filament into a line of droplets. Descriptions of the growth of disturbances are based on Tomotika's theory (34) that describes the disintegration of a Newtonian thread in a quiescent Newtonian continuous phase. The results of the theory are summarized in several reviews and monographs (1,3–5,10,24). Tomotika studied the development of a sinusoidal Rayleigh disturbance with an initial amplitude α_0 on a liquid cylinder with initial radius R_0. He showed that distortions with a wavelength λ greater than the initial circumference of the cylinder will grow exponentially with time:

$$\alpha(t) = \alpha_0 e^{qt} \tag{2.19}$$

where $\alpha(t)$ is distorsion amplitude at time t, and q is the growth rate defined as

$$q = \frac{\sigma \Omega(\lambda, p)}{2\eta_m R_0} \qquad (2.20)$$

where $\Omega(\lambda, p)$, a decreasing function of p, can be determined from equations in refs. (34,35) or from graphs presented in (3,4). At the beginning, small amplitudes of all wavelengths are present. For a given value of p, however, one disturbance with the dominant wavelength, λ_m, grows most rapidly and finally results in disintegration of the cylinder. At a given R_0, λ_m has minimum for p between 0.1 and 1 (3,4). Breakup time, t_B, follows from Eq. (2.19)

$$t_B = \frac{1}{q} \ln \frac{\alpha_B}{\alpha_0} \qquad (2.21)$$

where α_B is the amplitude at breakup. The breakup occurs when α_B is equal to the average radius of the thread. This condition leads to the equation (3,4)

$$\alpha_B = (2/3)^{1/2} R_0 \approx 0.82 R_0 \qquad (2.22)$$

For calculation of t_B from Eq. (2.21) the initial amplitude of the disturbance must be known. Kuhn (36) proposed an estimation of α_0 based on temperature fluctuation due to Brownian motion

$$\alpha_0 = \left(\frac{21 k T}{8 \pi^{3/2} \sigma} \right)^{1/2} \qquad (2.23)$$

where k is Boltzmann's constant and T is the absolute temperature. The use of Eq. (2.23) is questionable since experimental data match better to higher values of α_0 (3,4). After substitution from Eqs. (2.20) and (2.22), Eq. (2.21) can be rewritten as

$$t_B = \frac{2\eta_m R_0}{\sigma \Omega} \ln \frac{0.82 R_0}{\alpha_0} \qquad (2.24)$$

where α_0 can be taken as an adjustable parameter. It is seen that t_B increases with η_m, p, and R_0 and decreases with σ.

Experimental studies showed that Tomotika's theory describes Newtonian systems satisfactorily; numerical simulations showed differences only for large α, where nonlinear effects become important (3). The effect of the elastic properties of a thread and matrix is discussed in (24). Palierne and Lequeux (35) provided very general analysis of the development of a Rayleigh disturbance for a viscoelastic thread in quiescent viscoelastic continuous phase. They also considered the presence of a surfactant, as is discussed later in Section 2.7. It was shown that an increase in the relaxation times of the matrix and thread enhance the rate of growth of instability

(24). For some systems with viscoelastic components, good agreement was found with the predictions of Tomotika's theory (3). For others, especially polymer solutions, dumbbell shapes rather than sinusoidal disturbances appeared on a stretched drop after some time. This effect was explained as a consequence of the buildup of orientational stresses (3). Irregular disturbances were found also during numerical modeling of viscoelastic systems (3).

The analysis of the breakup of an extending liquid cylinder immersed in another flowing immiscible liquid was carried out in several papers (24,37,38). The results are summarized in ref. (3). In contrast to the quiescent state, the cylinder is continuously stretched. It was shown (38) that any linear flow can be decomposed into three-dimensional (3-D) elongation and shear components. Under some circumstances, the shear part may be neglected, and the only relevant parameter is the (orientation dependent) stretching rate $\varepsilon(= \mathbf{D}: \mathbf{mm}$, where $\mathbf{D}$ is the rate of deformation tensor and $\mathbf{m}$ is the orientation vector). The stretching efficiency, e_f, is defined as $e_f = (2)^{1/2}\varepsilon/\gamma$, where $\gamma = (2\mathbf{D}:\mathbf{D})^{1/2}$. It should be mentioned that e_f is constant in steady 2-D or 3-D elongational flow ($e_f = (1/2)^{1/2}$ or $(2/3)^{1/2}$, respectively) and decays to zero in steady simple shear flow (due to rotation of the extending thread toward the streamline direction). Because of the stretching of the thread, the wavelength of a disturbance and the thread radius changes according to

$$\lambda = \lambda_0 e^{\varepsilon t} \tag{2.25}$$

$$R = R_0 e^{-\varepsilon t/2} \tag{2.26}$$

where λ_0 and R_0 are the values of λ and R at $t = 0$, respectively. In order to find the evolution of α, the balance equation of mass and motion of elongated thread with Newtonian behavior and prescribed boundary conditions have to be solved. The resulting growth rate of small sinusoidal disturbance amplitude α can be expressed as (3)

$$\frac{d(\ln \alpha)}{dt^*} = 2^{-3/2} e_f \left[\frac{1-x^2}{\eta_m \varepsilon R / \sigma} \Phi(x, p) - 3(p-1)\bar{\Phi}(x, p) - 1 \right] \tag{2.27}$$

where $t^* = \gamma t$, $x = 2\pi R/\lambda$ is the dimensionless wave number, and $\Phi(x,p)$ and $\bar{\Phi}(x, p)$ are functions defined in refs. (3,37). Integration of Eq. (2.27) leads to

$$\ln \frac{\alpha}{\alpha_0} = \int_{x_0}^{x} \left[\frac{-x_0^{1/3}(1+x^2)}{3x^{4/3}R_0\eta_m\varepsilon/\sigma} \Phi(x, p) + \frac{p-1}{x}\bar{\Phi}(x, p) + \frac{1}{3x} \right] dx \tag{2.28}$$

where x_0 is the value of x for R_0 and λ_0. According to Eq. (2.28), the amplitude is damped at first, then grows for a while, and finally continues to be damped. The first damping stage can be omitted because the amplitude cannot damp below the

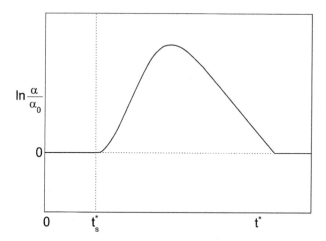

FIGURE 2.1 The dependence of the disturbance amplitude α on the reduced time t^*. t_s^* is the reduced time at which α starts to grow from its "noise" level, α_0. The curve is calculated by integration of Eq. (2.28) from x_s (related to t_s) to x for $\alpha > \alpha_0$; otherwise $\alpha = \alpha_0$.

initial "noise" level α_0, i.e., α remain equal to α_0. The second stage starts at the time t_s, relating to a local minimum of Eq. (2.28). The value of α at $t > t_s$ can be obtained by integration from x corresponding to t_s with the condition that $\alpha(t_s) = \alpha_0$. In the third stage, the amplitude damps again, but in reality, it cannot decrease bellow α_0 (see Figure 2.1). The breakup occurs for $\alpha(t) = R(t)$. Graphic solution of this equation is presented in (3). t_s and the maximum amplitude of a disturbance increases with decreasing λ. The breakup time t_B is a sum of t_s and a time t_g is needed for the growth of α to a critical value. The total breakup time, including the time necessary for the stretching of the droplet into an elongated thread, should be known for a description of the time development of the droplet size in flow. The problem was studied by Van Puyvelde et al. (39) who proposed

$$t_B \propto \gamma^{-1} Ca^{2/3} \tag{2.29}$$

Evidently, the breakup time of the thread increases and the size of fragments decreases in stretching flow with respect to the quiescent state. Analysis of the results shows (3) that for flow with constant ε (various types of elongational flow), the ratios of the thread radius at breakup to α_0 and, therefore, also of the radius of formed droplets, R_f, to α_0 decrease with increasing p and $\eta_m \varepsilon \alpha_0 / \sigma$. For values of $\eta_m \varepsilon \alpha_0 / \sigma$ not too large, the relation $R_f \propto (\eta_m \eta_d)^{-0.45}$ holds. It should be mentioned that R_f can be substantially smaller than the limit radius of droplets formed by the first breakup mechanism $(\sigma Ca_c / [2^{1/3} \eta_m \varepsilon])$. In steady simple shear flow, the results are more complicated (3,38). Generally, the decrease in R_f with increasing γ is less steep, and its dependence on p is less pronounced than in elongational flow. The trends predicted theoretically match the experiments quite well. However, the actual drop sizes are generally larger than those predicted by theory.

The third possible breakup mechanism is end pinching — two droplets (larger than that formed by the preceding mechanism) are formed at the ends of a finite droplet and tear off (22). After that, the "neck" either continues to breakup or relaxes in shape. The mechanism is typical for the step changes in flow. First, it was observed experimentally (25,29) and later explained as a consequence of an interfacial tension–driven flow associated with curvature variation along the surface of a finite droplet (22). The droplet attempts to return to a spherical shape, and fluid motion, produced by internal pressure gradients, leads to breakup. The effect was studied experimentally and by numerical simulation. The larger the p of the system, the longer a droplet must elongate for breakup by this mechanism. At a step change to subcritical conditions with $Ca \neq 0$, droplets can break up without large-scale stretching. This type of breakup adds just a few rather large droplets of uniform size to the droplet size distribution. It is possible to fragment very viscous droplets (with $p > 4$) in flows with significant vorticity (e.g., a simple shear flow). Formation of small satellite droplets and still smaller subsatellite droplets was found experimentally (10,22,29). The effect cannot be predicted by linear stability theory, but it can be predicted by numerical modeling (22).

Recently, Cristini et al. (40) studied the deformation and breakup of droplets in an instantaneously started shear flow using boundary integral simulations and videomicroscopy experiments. They found quite good agreement of the dependence of Ca_c vs. p with theories based on the perturbation methods. For breakup time t_B of the droplets with Ca not much larger than Ca_c,

$$t_B \propto \frac{1}{\gamma}\left(\frac{R}{R_c} - 1\right)^{-1/2}$$
(2.30)

where

$$R_c = \frac{\sigma}{\eta_m \gamma} Ca_c(p)$$
(2.31)

For $R \gg R_c$, they proposed

$$t_B \propto \frac{1}{\gamma}\left(\frac{R}{R_c}\right)^{2}$$
(2.32)

However in this case, the droplet disintegrates by the Tomotika mechanism in shorter time. They found that the radius, R_f, of a daughter droplets increase with the initial droplet radius, R_0, up to a value $R_f = 0.90R_c$ and keeps this value for $R_0 > 1.4R_c$. Also the existence of satellite and subsatellite droplets is assumed in this analysis, and the distribution of fragments is proposed.

Attempts to study droplet breakup in more complex 3-D flows (such as converging and diverging flows (24), 2-D low Reynolds number chaotic flows, etc.) have been made. It is difficult to describe three-dimensional flows typical of industrial applications so that the fundamental studies of droplet breakup will lead to a useful picture of real mixing processes.

All the above considerations are about dilute systems where the effect of other droplets on the breakup of a droplet can be neglected. In real polymer blends, where the volume fraction of the dispersed phase is not negligible, the effect of other droplets on the droplet breakup should be considered. Choi and Schowalter (41) extended Cox's theory (27) to describe the deformation of droplets in a moderately concentrated emulsion of Newtonian liquids. They found that deformations are larger due to an increase in viscosity of the emulsion above the viscosity of the matrix. Jansen et al. (42) proposed to substitute η_m by the viscosity of the emulsion in the definition of p and Ca for concentrated emulsions. This mean field concept satisfactorily explained experimental results but slightly overpredicted the stress at which breakup occurred. The deviation was explained by the existence of the continuous phase layer around the droplet, which impairs the transmission of the stresses from the bulk to the droplet interface. Unfortunately, the dependence of viscosity of polymer blends on their composition is complicated and rarely can be *a priori* predicted (24,43). Therefore, also the possibilities of using the mean field approach to predict droplet breakup in concentrated polymer blends are limited.

Dependencies of the breakup time and a number and size distribution of the droplet fragments on system parameters differ qualitatively for various breakup mechanisms. Therefore, the recognition of a decisive breakup mechanism is essential for the correct prediction of blend morphology. It is generally accepted that breakup through capillary wave instability is operative for $Ca \gg Ca_c$, and breakup into two parts is operative for Ca not much higher than Ca_c (10). However, the choice of a proper mechanism in modeling the morphology development in mixing and processing equipment is not an easy task.

2.4 FLOW INDUCED COALESCENCE

Flow induced coalescence is the second fundamental event that should be understood and properly described prior to the derivation of a theory of droplet size evolution in immiscible polymer blends. Coalescence is a consequence of droplet collisions brought on by the difference in velocity (see Figure 2.2). While droplets approach each other, their hydrodynamic interaction increases. When the droplets are close enough, droplet deformation due to the axial force (which is important if the driving force of the coalescence is strong enough) and the removal of the trapped continuous phase start. If the critical distance of the droplets is achieved, rupture of the remainder of the continuous phase, usually by the formation of a "hole" on the thinnest spot, occurs. An initial contact of droplets is followed by evolution of a "neck" and formation of a coalesced droplet. Usually, it is assumed (23,44) that the coalescence is controlled by droplet approach and drainage of the continuous phase trapped between droplets. The rupture of the remainder of the continuous phase is assumed to be much quicker (23), and the neck evolution is mostly not considered. A discussion

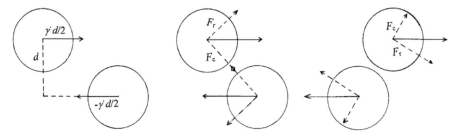

FIGURE 2.2 The approach and collision of the droplets with the same R and the same coordinate in neutral direction in the shear flow. The coordinate system with the origin moving with the center of inertia of the droplet pair is considered. The picture relates to droplet fusion not being realized.

of its effect on the phase structure development is given later. In most of the theories, hydrodynamic interaction between droplets is neglected until they are close enough ("ballistic" approximation) (45,46). Droplets flattening occurs at a short distance. For a radius of the flattened part of a droplet, r_f, the following equation holds (44)

$$r_f = \left(\frac{RF_c}{2\pi\sigma} \right)^{1/2}$$ (2.33)

where F_c is the driving force of the coalescence. For Stokes flow of droplets, the average driving force of the coalescence in shear flow can be estimated as (3,4,47)

$$F_c = 6\pi\eta_m \gamma R^2$$ (2.34)

Substitution from Eq. (2.34) into Eq. (2.33) leads to

$$r_f = \left(\frac{3}{\pi} Ca \right)^{1/2} R$$ (2.35)

From Eq. (2.35), it follows that $r_f < 0.1\, R$ for $Ca < 0.01$, which means that the droplet flattening can be neglected only at the conditions of rheological measurements at very low shear rates but not in other cases (mixing, processing, rheometry at high deformation rates).

As shown in Figure 2.2, the forces on colliding droplets in shear flow ($v_0 = (\gamma y, 0, 0)$) in the coordinate system moving with the center of inertia of the droplet pair can be split into radial, F_r, and axial, F_c, components that cause the pair to rotate around their center of inertia and to approach each other, respectively. A similar situation exists also for collisions induced by uniaxial extensional flow $[v_0 = \varepsilon(2x, -y, -z)]$ (48), where the axial component of the force depends on the orientation of the vector connecting droplet centers with respect to the flow. After achieving a certain orientation (e.g., the same x-coordinate of both droplets at shear flow), F_c

drives the droplets from each other. During the approach of the droplets, the matrix phase trapped between them must be drained away. If the droplets touch (i.e., get to the critical distance of matrix rupture) before achieving the critical orientation, then fusion occurs.

The rate of coalescence, J, can be calculated from the flux of a pair through the collision surface, which is equal to the flux at $r \rightarrow \infty$ through a cross section A_c (upstream interception area) (49–51)

$$J = -n_i n_j \int_{A_c} \mathbf{v_0} \cdot \mathbf{n} \, d \, S \tag{2.36}$$

where n_i and n_j are the number of droplets with radii R_i and R_j, respectively, and $\mathbf{n}$ is the outward unit normal to the spherical contact surface. When any interaction between droplets before touching is neglected (Smoluchowski approximation), the following equations can be derived using geometric considerations from Eq. (2.36)

$$J_0 = \frac{4}{3} \left(R_i + R_j \right)^3 \gamma n_i n_j \tag{2.37}$$

for shear flow and

$$J_0 = \frac{8\pi}{3\sqrt{3}} \left(R_i + R_j \right)^3 \varepsilon n_i n_j \tag{2.38}$$

for uniaxial extensional flow.

Coalescence is frequently characterized by probability, P_c, that collision of droplets in geometric sense (Smoluchowski approximation) is followed by their fusion. P_c is defined as

$$P_c = J/J_0 \tag{2.39}$$

Wang et al. (50) derived a theory for the coalescence in shear flow and uniaxial extensional and compressional flow for a system of Newtonian droplets in Newtonian matrix. They assumed that the droplets keep spherical shape through the whole approach. Hydrodynamic interaction between droplets is considered through the whole process, and van der Waals forces between droplets are included directly in equations describing droplet motion. They calculated numerically droplet trajectories, and using their analysis, they determined A_c required in Eq. (2.36). They showed that P_c is a decreasing function of the viscosity ratio, p, and the ratio of the radius of a larger droplet to the radius of a smaller one. On the other hand, P_c is independent of viscosity of the matrix, η_m, the average radius of the droplets, and the deformation rate (γ or ε).

Theories considering flattening of the droplets mostly use the assumption that interaction between droplets can be neglected until a certain distance h_0 (3,23), usually estimated as $R/2$. Description of drainage of the continuous phase between flattened droplets is a difficult task because the boundary conditions for the equation describing this effect depend on fluid circulation inside the droplets. For a constant driving force of the coalescence and with r_f substantially larger than the distance between droplets, h, the equations for the rate of droplet approach are available in the literature for systems differing in their interfacial mobility (3,44,47,52). The following equations were derived for pairs of droplets with the same R for systems with mobile, partially mobile, and immobile interface, respectively, (3,44,47)

$$-\frac{d h}{d t} = \frac{2\sigma h}{3\eta_m R} \tag{2.40}$$

$$-\frac{d h}{d t} = \frac{4(2\pi)^{1/2} \sigma^{3/2} h^2}{\eta_d R^{3/2} F_c^{1/2}} \tag{2.41}$$

$$-\frac{d h}{d t} = \frac{8\pi\sigma^2 h^3}{3\eta_m R^2 F_c} \tag{2.42}$$

Unfortunately, the above equations predict different rate dependences on system parameters without any transition between the equations upon change in the mobility of the interface. The most plausible equation, at least for the dependence of dh/dt on viscosity of the components, seems to be the equation derived by Jeelani and Hartland (52)

$$-\frac{d h}{d t} = \frac{8\pi\sigma^2 h^3}{3\eta_m R^2 F_c}\left(1+3C\,\frac{\eta_m}{\eta_d}\right) \tag{2.43}$$

where C is the ratio of circulation length and droplet distance and is the order of 1. For $\eta_d \gg \eta_m$, Eq. (2.43) becomes Eq. (2.42) for immobile interface, and for $\eta_d \ll \eta_m$, Eq. (2.43) predicts the same dependence on η_d as Eq. (2.41) for partially mobile interface. In the latter case, Eqs. (2.43) and (2.41) differ in the dependence of the rate of approach on F_c and h. The above results were derived for the approach of droplets with the same R. It was proposed that the above results can be used also for unequal droplets if an equivalent droplet radius, R_{eq}, is used (3), where

$$R_{eq}^{-1} = \frac{R_1^{-1} + R_2^{-1}}{2} \tag{2.44}$$

For drainage of the continuous phase between the droplets keeping their spherical shape, the following equation was derived (53)

$$-\frac{dh}{dt} = \frac{2F_c h}{3\pi\eta_m R^2 g(m)}$$
(2.45)

where

$$g(m) = \frac{1+0.402m}{1+1.711m+0.461m^2}$$
(2.46)

and

$$m = \frac{\eta_m}{\eta_d}\left(\frac{R}{2h}\right)^{1/2}$$
(2.47)

It should be noted that the dependence of dh/dt on F_c in Eq. (2.45) and in Eq. (2.43) are opposite. Therefore, it is clear that theories for flattened droplets [Eqs. (2.41) to (2.43)] overestimate the rate of film drainage for small driving forces (the equations diverge for $F_c \rightarrow 0$). On the other hand, Eq. (2.45) overestimates the droplet approach for higher F_c, where flattening is remarkable. Choosing an improper equation for the droplet approach can have dire consequences.

For the force F_c in shear flow, the following equation is valid if Stokes flow of the droplets is assumed

$$F_c = 12\pi\eta_m \gamma R(R' + h/2)\sin^2\Theta\sin\varphi\cos\varphi$$
(2.48)

where R' is the distance between the center of a droplet from the center of its flattened part, and φ and Θ are Euler angles determining orientation of the connection of droplet centers with respect to flow. During the coalescence, R' and h changes due to droplet motion and flattening. However, the approximation

$$R' + h/2 = R$$
(2.49)

seems to be reasonable during the whole coalescence event. Elmendorp (23) used for F_c the hydrodynamic force acting on the doublet of touching spheres (54), and he obtained in Eq. (2.48) a constant of 12.24 instead of 12. For uniaxial extensional flow, F_c can be expressed as

$$F_c = 12\pi\eta_m \varepsilon R^2(1 - \cos^2\Theta)$$
(2.50)

Equations (2.48) and (2.50) show that the force driving droplets together and, therefore, also the range of droplet flattening, change strongly during coalescence. It illustrates that attention must be paid to the application of an adequate equation for matrix drainage.

Elmendorp and coworkers (23,55,56) solved simultaneously equations describing rotation and approach of the droplets. They assumed that the resistance against the approach of the droplets is the sum of the resistance against drainage of the matrix film between flattened parts of droplets and the resistance against approach of hard spheres, i.e.,

$$\left(\frac{dh}{dt}\right)^{-1} = \left(\frac{dh}{dt}\right)_f^{-1} + \left(\frac{dh}{dt}\right)_{Sp}^{-1} \tag{2.51}$$

where for the first term, Eq. (2.40) is used for a mobile interface and Eq. (2.42) is used for an immobile interface. For the second term, Eq. (2.45) with $g(m) = 1$ (limit for rigid spheres) was used. By combining the equations for dh/dt with equations for $d\varphi/dt$ and $d\Theta/dt$, the equation describing the dependence of h on φ and Θ was derived. From its solution, the limit values of initial angles φ_0 and Θ_0 (at h_0) for which h at $\varphi = \pi/2$ is smaller than the critical distance for the film rupture, h_c, were calculated. These angles were used for calculation of J from Eq. (2.36). Elmendorp found that the results using the expression for a mobile interface, Eq. (2.40), agree better with the experimental results than using the expression for an immobile interface, Eq. (2.42). The value of h_c is about 50 nm.

The approach by Janssen and coworkers (3,4,47) is based on preaveraging of both the driving force of the coalescence and the time of droplet rotation. It is assumed that the driving force of the coalescence is given by the equation

$$F_c = 6\pi\eta_m \gamma R^2 \tag{2.52}$$

The time of coalescence, t_c, can be determined by integration Eqs. (2.40) to (2.43) from h_0 to h_c. Integration of Eqs. (2.41) and (2.43) leads to the equations for t_c for systems with partially mobile interface and for the Jeelani-Hartland model

$$t_c = \frac{\sqrt{3}\eta_d \eta_m^{1/2} \gamma^{1/2} R^{5/2}}{4\sigma^{3/2}} \left(\frac{1}{h_c} - \frac{1}{h_0}\right) \tag{2.53}$$

$$t_c = \frac{9\eta_m^2 \gamma R^4}{8\sigma^2 (1+3C\eta_m / \eta_d)} \left(\frac{1}{h_c^2} - \frac{1}{h_0^2}\right) \tag{2.54}$$

h_c can be approximated by (3,44)

$$h_c \approx \left(\frac{AR}{8\pi\sigma} \right)^{1/3}$$ (2.55)

where A is the effective Hamaker constant, given by the properties of the blend components. Typical value of h_c for polymer blends with R about 1 μm is about 5 nm. Therefore, $1/h_0$ can be neglected with respect to $1/h_c$. It is assumed that P_c can be expressed as

$$P_c = \exp\{-t_c/t_i\}$$ (2.56)

where t_i the interaction time defined as

$$t_i = \gamma^{-1}$$ (2.57)

From the above, it follows for P_c of a system described by models of partially mobile interface and the Jeelani-Hartland model

$$P_c = \exp\left\{ -\frac{\sqrt{3}}{4} \frac{R}{h_c} pCa^{3/2} \right\}$$ (2.58)

$$P_c = \exp\left\{ -\frac{9Ca^2 R^2}{8h_c^2 \left(1+3C\eta_m / \eta_d\right)} \right\}$$ (2.59)

Results obtained by Elmendorp's and Janssen's procedures using Eq. (2.43) for film drainage between flattened droplets were compared for shear flow–induced coalescence for a system with the same coordinate of the droplets in neutral direction (46) and for extensional flow–induced coalescence (48). For both types of flow, P_c was calculated also by another procedure that considers changes in flattening of the droplets and that is based on the modification by Elmendorp (55, 56). In the procedure, droplets are considered as spherical if the ratio r_f/R, with r_f from Eq. (2.33), is smaller than a given small number. In the opposite case, the droplets are considered to be flattened. From this condition, the angles are determined that divide the trajectory of the particles to regions where the matrix drainage is described by Eq. (2.43) or (2.45). The condition r_f/R = constant is not identical to the condition that the velocity of the approach of spherical and flattened droplets, calculated from Eqs. (2.43) and (2.45), be equal. The latter condition cannot be used for dividing the trajectory because of the interdependence of the parameters. Therefore, the underestimation of the matrix resistance against drainage for small or large h leads to unrealistic dependence of P_c on R (48). The trouble was avoided by the assumption that the approach of deformed droplets cannot be faster (therefore P_c cannot be higher) than that of the related spherical ones.

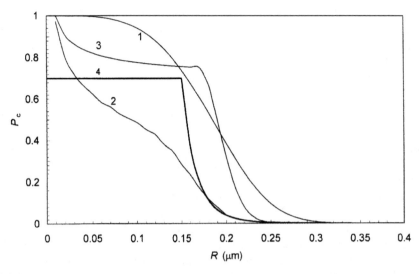

FIGURE 2.3 The dependence of P_c on R in extensional flow calculated by (1) Janssen's method, (2) Elmendorp's method, (3) the method that considers changes in droplet flattening, and (4), Rother and Davis' method. Curves 1 to 3 are calculated using Eq. (2.43) for the approach of flattened droplets. The same parameters as for the related curves in Figure 3 of ref. (48) were used in the calculations. Curve 4 is constructed using results presented in Figure 8 in ref. (51). The parameters of the curve 4 differ from those of curves 1 to 3 since different parameters are used in the theories derived in (48) and (51).

Rother and Davis (51) generalized the theory of Wang et al. (50) for deformable droplets, considering droplet deformation as a small but singular perturbation. They found that P_c for small Ca is identical to that predicted (50) for spherical droplets with the same parameters. At a certain Ca, P_c steeply decreases to a very low value. It should be noted that the shapes of the curves P_c vs. Ca and P_c vs. R are the same for systems with constant deformation rate and interfacial tension. The dependencies of P_c on R obtained by Elmendorp's and Janssen's procedures, by the procedure considering changes in flattening of the droplets, and by Rother's and Davis' theory for extensional flow are compared in Figure 2.3. The shapes of P_c vs. R curves for shear flow are similar to those for elongational flow (the curve calculated by the third method is not available) (46,48). All the curves show a steep decrease to a very low value and after that an asymptotic approach to zero. Elmendorp's procedure apparently underestimates the values of P_c. It is a consequence of underestimating $(dh/dt)_{Sp}$ since the expression for hard spheres — instead of Eq. (2.45) with the relevant p — was used. Janssen's procedure clearly overestimates P_c in the region of small Ca (R), where it predicts larger P_c than the theory for spherical droplets (50). The third and fourth procedure predict $P_c < 1$ for small R (an increase in P_c for the third procedure with decreasing R in a region of very small R is a consequence of the approach of R to h_c). The very sharp break of the curve at a certain R seems to be caused by combining the equations for highly flattened and spherical droplets in the third procedure and by treatment of the deformation as a singular perturbation in the fourth procedure. Values of P_c calculated by the third and fourth procedure differ due to the differences in the model and approximations used. It follows from the above discussion that for small Ca (R), P_c can be described by the theory for spherical droplets. Janssen's approach provides explicit expression for P_c and predicts a decrease in P_c in qualitative agreement with other procedures. Therefore, for qualitative or semiquantitative considerations, it was suggested (57)

$$P_c = P_s \exp\{-\gamma t_c\} \tag{2.60}$$

where P_s is a decreasing function of p and of the ratio of the radii of larger to smaller droplet, which could be obtained from ref. (50), and t_c can be substituted from Eq. (2.54) or (2.53).

It should be mentioned that in all the theories discussed above, the end of the coalescence is assumed at the start of continuous phase rupture. It is apparent that the breakup of the thin neck between the droplets is much easier than the breakup of relaxed droplets. Therefore, using P_c to describe the competition between droplet breakup and coalescence somewhat overestimates coalescence contribution.

All the above theories were derived using the assumptions that droplets and matrix are Newtonian liquids, the droplets are spherical till collision, and only binary collisions, not affected by the presence of other droplets, are important. None of these assumptions is fulfilled in typical polymer blends. However, because a description of flow-induced coalescence is very complicated even for systems complying with the above assumptions, it is not surprising that generalization of the above theories is rare. No theory describing the effect of viscoelastic properties of droplets and/or matrix is available. The effect of droplets deformation in flow was studied by Patlazhan and Lindt (58). They modified Janssen's approach (3,4,47) and found that t_c is smaller for ellipsoidal than for spherical droplets due to a smaller radius of curvature of the ellipsoid. They estimated that t_i is larger for ellipsoids than for the related spheres. Therefore, P_c increases because of droplet deformation. Patlazhan and Lindt studied systems with a low p, and they did not consider rotation of the ellipsoids around their common center of inertia. Hence, the plausibility of their results for systems with a higher p needs further investigation. Consideration of simultaneous collisions of three, four, etc., droplets leads to the appearance of terms proportional to the third, fourth, etc., powers of the droplet volume fraction in the expression for J [see Eqs. (2.37) and (2.38)] (6). Analysis of the coalescence of a larger number of droplets is extremely difficult because, aside from drainage of their matrix, other effects such as shape relaxation of coalescing droplets on droplets in their neighborhood have to be considered (59). Moreover, the whole trajectory of a droplet is affected by the presence of other droplets in blends containing a high volume fraction of the dispersed phase. Adequate treatment of these problems remains a challenge for the future.

2.5 COMPETITION BETWEEN DROPLET BREAKUP AND COALESCENCE

The competition between droplet breakup and coalescence can be described either for a discrete model where the volume of a dispersed droplet is a product of an integer and the elementary volume, V_1, or for a continuous model where such restriction does not apply. In the former case, the equation for a temporal change in n_k, i.e., in the number of droplets with volume kV_1, can be expressed as (6)

$$\frac{d n_k}{d t} = \frac{1}{2} \sum_{i+j=k} \left[K(i,j)n_i n_j - F(i,j)n_{i+j} \right] -$$

$$\sum_{j=1} \left[K(k,j)n_k n_j - F(k,j)n_{k+j} \right] + ... \tag{2.61}$$

where the first sum is over all i and j for which $i+j=k$, $K(i,j)$ is coagulation kernel, giving the probability of coalescence of the droplets with volumes iV_1 and jV_1 in a time unit, and $F(i,j)$ is the fragmentation kernel describing the probability of the fragmentation of a droplet with volume $(i+j)V_1$ into fragments with volumes iV_1 and jV_1. Omitted further terms would describe simultaneous coalescence of more than two droplets and disintegration into more than two fragments. Equation (2.61) represents a set of differential equations. For the continuous model, Eq. (2.61) becomes the integrodifferential equation

$$\frac{d n(V)}{d t} = \frac{1}{2} \int \left[K(x,V-x)n(x)n(V-x) - F(x,V-x)n(V) \right] d x -$$

$$\int \left[K(V,x)n(V)n(x) - F(V,x)n(V+x) \right] d x \tag{2.62}$$

Equations (2.61) and (2.62) describe systems in which droplets break up into two fragments (the first breakup mechanism in Section 2.3) and only the coalescence of two droplets is important, as may be expected for small volume fractions of the droplets. For breakup mechanisms resulting in a larger number of fragments, the terms describing a droplet fragmentation into a given number of fragments must substitute for $F(i,j)$ in Eqs. (2.61) and (2.62). For a system where a droplet with volume iV_1 forms $n_f(i)$ fragments and only binary droplet collisions occur, Eq. (2.61) becomes (58)

$$\frac{d n_k}{d t} = \frac{1}{2} \sum_{i+j=k} K(i,j)n_i n_j - G(k)n_k - \sum_{j=1} K(k,j)n_k n_j +$$

$$\sum_{j=k+1} \beta(k,j)n_f(j)G(j)n_j \tag{2.63}$$

where $G(i)$ is the overall breakup frequency (i.e., into any number of fragments of any size) and $\beta(i,j)$ is the probability that a fragment formed by the breakup of a droplet with volume jV_1 will have volume iV_1. Due to incompressibility of polymer melts, solutions of Eqs. (2.61) to (2.63) must satisfy the condition

$$V_1 \sum_i in_i = \phi \qquad (2.64)$$

where ϕ is the volume fraction of the droplets and numbers n_i are related to the volume unit.

At steady rheological measurements or after long time mixing in batch mixers (if degradation of the blend components is avoided), steady distribution of the droplet sizes is achieved. In this case, dynamic equilibrium between droplet formation and disappearance is established, and, consequently, the left sides of Eqs. (2.61) to (2.63) are zero. Obviously, this assumption is not plausible for transient rheological measurements or for the description of the phase structure development during rapid blend throughput through an extruder with various mixing zones. Though solving the above equations is not easy, a more serious problem from the point of view of the phase structure in polymer blends is the proper modeling of the coagulation and fragmentation kernels.

Many studies, using various approximations in formulating and solving the equation for the evolution of the droplet population and/or in determining coagulation and fragmentation kernels, deal with systems where dynamic equilibrium between breakup and coalescence is established (45,55,58,60–64). In other papers, the path of a droplet through an extruder was modeled and breakup and/or coalescence in individual zones were considered (46,64–66) using expressions for coalescence in shear flow and applying varied approximations.

The calculations in refs. (60,61) focused on the average radius of a droplet in steady shear flow using the assumption that the system of droplets remains monodispersed, i.e., the breakup leads only to an increase and coalescence to a decrease in the droplet number. Then the equation describing dynamic equilibrium between the droplet breakup and coalescence simplifies to (6,60,61,67)

$$Fn = Kn^2 \qquad (2.65)$$

where for shear flow

$$K = \frac{16}{3} \gamma R^3 P_c \qquad (2.66)$$

Equation (2.66) follows from Eqs. (2.37) and (2.39) if duplicate inclusion of the contribution of a droplet pair is avoided. Tokita (60) used Eq. (2.66) with constant P_c (independent of R) for K, and he proposed

$$F = \frac{\eta \gamma^2}{E_{DK} + \dfrac{3\sigma}{R}} \qquad (2.67)$$

where η is the apparent viscosity of the blend and E_{DK} is the volume energy (60). Equation (2.67) is based on a notion that the total breakup energy consists of the volume and interfacial energy of the droplets. Substitution from Eqs. (2.67) and (2.66) into Eq. (2.65) leads to

$$R = \frac{12\sigma P_c \phi}{\pi\eta\gamma - 4P_c\phi E_{DK}} \tag{2.68}$$

if the relation

$$\frac{4}{3}\pi R^3 n = \phi \tag{2.69}$$

i.e., the condition described in Eq. (2.64) for monodispersed spheres is used. Equation (2.68) gives a reasonable shape of R vs. ϕ dependence but predicts an unrealistic relation $R = 0$ for $\phi \to 0$. Equations (2.65) and (2.66) with constant P_c were used also in ref. (61). The terms for the collision of more droplets were formally included in the expression for K. F was considered to be a positive function of Ca, which can be expanded in a Taylor series on the right of Ca_c. Using Eq. (2.66) and the first term of the Taylor series, the following equation was derived for R

$$R = R_c + \frac{4\sigma P_c}{\pi\eta_m f_F}\phi \tag{2.70}$$

where f_F is a function of the rheological properties of the blend components, independent of R and ϕ.

Elmendorp and coworkers (55,56) tried to estimate the range of possible droplet sizes in steady shear flow. The smallest size of droplets, which can disintegrate, was determined using Taylor's theory [cf. Eq. (2.14)]. The upper limit of the size of coalescing droplets was determined from the condition that P_c decreases to 0 in this model of coalescence (23,55,56). These conditions correspond to straight lines in the plot $\ln R$ vs. $\ln (\eta_m\gamma/\sigma)$, dividing it into four regions. Steady droplet radii must lie either in the region where the dynamic equilibrium between the droplet breakup and coalescence is established or in the region where both breakup and coalescence are absent. From the regions where only coalescence or only breakup are operating, the droplet radii must pass to any of the former regions (23,55,56). The Lyngaae-Jørgensen and Valenza model (62) tried to describe a situation where strongly elongated droplets burst into a large number of small droplets. The model assumes that dynamic equilibrium between small spherical and large ellipsoidal droplets is established in steady state. For a system with a low p, Lyngaae-Jørgensen and Valenza derived a set of three equations for the volume fraction of large particles and length of their long and short semiaxis. Huneault et al. (64) derived the following equation for R in a system where dynamic equilibrium between the droplet breakup and coalescence is established:

$$R = R_0 + (1.5 C_{\mathrm{H}} Ca_c t_{\mathrm{B}}^* \phi^{8/3})^{1/2} \tag{2.71}$$

where R_0 is the droplet radius at $\phi = 0$, C_{H} is the coalescence constant, which can be calculated from the slope of the dependence R vs. ϕ, and t_{B}^* is the dimensionless breakup time defined as

$$t_{\mathrm{B}}^* = t_{\mathrm{B}} \frac{\gamma}{Ca} \tag{2.72}$$

It is assumed that t_{B}^* is a function of p and independent of Ca. The authors used a theory of coalescence differing from the one derived for shear flow–induced coalescence [see Eq. (2.66)].

In ref. (63), Eq. (2.61) was solved using the assumption that a droplet breaks up into two halves or with the same probability into two pieces having any volume. It was assumed that F is proportional to $(Ca - Ca_c)^n$, where n is a positive number. A P_c that decreases with a power of the average droplet radius was chosen. Equation (2.61) was solved using the scaling rules for the droplet size distribution based on the assumption that coagulation and fragmentation kernels are homogeneous. Algebraic equations, for different forms of F, were derived for the average droplet radius. Patlazhan and Lindt (58) solved Eq. (2.63) using expressions for droplet breakup derived by a combination of Tomotika's theory and experimental results for systems with a low p. They used Janssen's theory modified for ellipsoidal droplets (cf. Section 2.4) for a description of coalescence. Size distribution functions for various breakup mechanisms and initial droplet size distributions were obtained by numerical solution of Eq. (2.63). Milner and Xi (45) assumed that a mixer contains two regions: high shear and low shear. In the high shear region, droplets break in two fragments with t_{B} independent of the droplet size. In the low shear region, only shear flow–induced coalescence is assumed. P_c was calculated using the Wang et al. (50) theory for coalescence of spherical droplets (cf. Section 2.4). Also, this theory provides the droplet size distribution by a numerical solution of the related equation. The theory is focused especially on the effect of a compatibilizer, which is discussed in Section 2.7. Breakup frequency independent of R for $R > R_c$ was assumed also by Lyu et al. (68) in their study of the effect of the used coalescence theory [Smoluchowski's using Eq. (2.37), trajectory theory (50), and Janssen's theory (3,4,47) for systems with a partially mobile interface] on the evaluation of the dependence of an experimentally determined average droplet size on strain in shear flow of polymer blends.

Huneault et al. (64) developed a computational model for droplet size evolution during mixing in a twin-screw extruder. They considered that the process is isothermal and the blend components show power law relations between shear stress and shear rate. They assumed that the drop deformation takes place only within the pressurized screw zones. No droplet size evolution was assumed for $Ca < Ca_c$. Droplet split up and coalescence was assumed for $Ca_c < Ca < 4Ca_c$ and its fibrillation and disintegration for $Ca > 4Ca_c$. It is assumed that the coalescence mechanism is the same as in a batch mixer and C_H can be determined from experimental data using Eq. (2.71). A computational model was used for the determination of the dependence of the dispersed droplet diameter on its position in an extruder. Janssen and Meijer (47) described the development of droplet size in an extruder with a two zone model. They assumed that in the "strong" zone, affine stretching and thread breakup in flow occurs and that thread breakup at rest and coalescence are typical in the "weak" zone. The strong zone is modeled by elongational flow (it also may be represented as a sequence of stretching and folding in a simple shear dominated region). They assumed that the development of the disturbance amplitude in the stretched thread is described by Eq. (2.27). Simple shear flow with a low γ was assumed in the weak zone. The probability that noninteracting droplets collide there was calculated using Eq. (2.37) and Eq. (2.58) for P_c. The residence time in the weak zone was distributed as a cascade of ideal mixers. In numerical calculations, η_m was substituted by effective viscosity for emulsion of Newtonian liquids, except in the calculation of the film drainage during coalescence. The results are discussed in (3,47). Delamare and Vergnes (65) calculated evolution of the droplet size distribution in the twin screw extrusion process. Their model of the breakup mechanism was very similar to that used in ref. (64), only $2Ca_c$ instead of $4Ca_c$ was chosen as the boundary between regions of Ca in which breakup into two pieces or droplet fibrillation proceeded. $t_B{}^*$ independent of Ca (i.e. R) for $Ca > Ca_c$ was assumed in the region where droplets broke up into two pieces. The coalescence probability was calculated in the same manner as in ref. (47). Mean droplet diameters and local distributions of the droplet sizes were calculated as functions of parameters of the extrusion process. Potente and Bastian (66) described droplet size evolution during extrusion using the boundary element and finite element methods for flow simulation, particularly for an analysis of the stress that acts on the particles during their trajectories. They calculated Ca_c by combining the results for shear and elongational flows, using the α_f parameter, see Eq. (2.18), calculated for a given zone in the extruder. They describe the droplet breakup using σ_{ef} given by Eq. (2.8) for calculation of Ca. Repeated breakup into two fragments was assumed with the breakup time slowly increasing with Ca for $Ca > Ca_c$. Coalescence was treated by Janssen's method (47,65).

The above theories differ in a number of various assumptions and approximations. Therefore, an analysis of the effect of various assumptions is not easy. Especially, different types of dependence of F (t_B) on Ca (i.e. R) were used in the above theories. The effect of the dependence of F on R on the shape of the average R vs. ϕ curve for a monodispersed system, described by Eq. (2.65), in the steady state was discussed in (67). It was shown that the form of a function $F(R)$ has a fundamental effect on the R vs. ϕ dependence. Graphic solution of Eq. (2.65) gives good insight into the effects of the form of the functions $F(R)$ and $P_c(R)$ on the dependence of the average R on the volume fraction of the dispersed phase. For shear flow, Eq. (2.65) can be rewritten as

$$F(R) = \frac{4}{\pi} \gamma \phi P_c(R) \qquad (2.73)$$

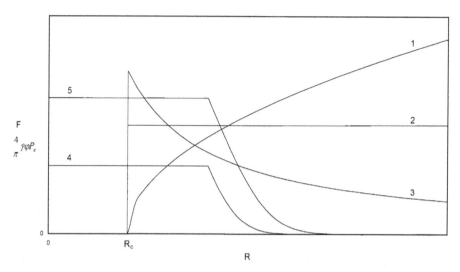

FIGURE 2.4 The scheme for the graphic solution of Eq. (2.73) for various forms of the $F(R)$ function: (1) F increasing with $R - R_c$, (2) F constant for $R \geq R_c$, and (3) $F(R_c) > 0$ and $F(R)$ decreasing with R for $R > R_c$. Curves 4 and 5 show $(4/\pi)\gamma \phi P_c(R)$ with the same γ but with a larger ϕ for curve 5. The solution of Eq. (2.73) is R at the intersection of curve 1, 2, or 3 with curve 4 or 5.

For the breakup into two fragments

$$F = t_B^{-1} \tag{2.74}$$

For the breakup into more than two fragments, F calculated according to Eq. (2.74) should be multiplied by half of the number of fragments. It follows from Section 2.4 that $P_c(R)$ is constant or a very slowly decreasing function of R up to a certain R, at which point, a steep decrease of P_c starts. Within a narrow range of R, P_c falls to a very low value and asymptotically approaches 0 with increasing R. A graphic method for solving Eq. (2.73) with this type of $P_c(R)$ and with different $F(R)$ is shown in Figure 2.4. It is clear that for $F(R_c) \neq 0$, a region of ϕ exists for which $R = R_c$. A step increase in R appears when ϕ achieves a critical value. Equation (2.73) has two solutions for a certain set of parameters if $F(R)$ is a decreasing function or shows a maximum for $R > R_c$. If R_c is larger than R at which P_c falls to a small value, only slow coalescence occurs in a blend and, therefore, the droplet radius very slowly approaches R_c.

It follows from modeling and experimental studies of droplet breakup that the breakup into two fragments (a stepwise breakup mechanism) is dominant for Ca not much larger than Ca_c. This condition should be fulfilled for steady flow of blends with a low ϕ. In this case, a steady droplet size distribution can be obtained by solution of Eq. (2.61). For construction of fragmentation kernel F, Eq. (2.30) can be used. Proper expression for a coagulation kernel can be chosen using results from Section 2.4, e.g., Eq. (2.60). Using the approximation that the system is monodispersed, an average droplet radius can be found by solving Eq. (2.73). If we assume that the average R is smaller than the radius at which P_c starts to decrease rapidly, P_c can be approximated by P_s, which is independent of R. In this case, an explicit expression for the average droplet radius, R, is obtained

$$R = R_c + \left(\frac{4}{\pi}\right)^2 f_c R_c P_s^2 \phi^2 \qquad (2.75)$$

where R_c can be calculated from Eq. (2.31) and f_c is a function of p for blends of Newtonian liquids. For polymer blends, f_c generally is a function of rheological properties of the blend components as well. For steady mixing of polymer blends, Eq. (2.75) describes R as a function of ϕ only qualitatively because of the differences between simple shear flow and flow in the mixer. Equation (2.75) seems to describe quite well the shape of the R vs. ϕ dependence found experimentally (1,6). It should be mentioned that steady blend morphology in the whole batch mixer is established if the residence time in places with different flow rates is substantially shorter than the time needed to attain dynamic equilibrium between droplet breakup and coalescence. Otherwise, the local dynamic equilibrium is established and position dependent morphology is formed, i.e., blends with nonuniform morphology are obtained.

2.6 THE PHASE STRUCTURE EVOLUTION IN QUIESCENT POLYMER BLENDS

In the quiescent state, the general tendency of immiscible polymer blends is to form the morphology with minimum interfacial area. The behavior of elongated droplets in molten quiescent polymer blends was discussed already in Section 2.3. Highly elongated drops breakup into a chain of droplets by the Tomotika mechanism [cf. Eqs. (2.19) to (2.24)]. As discussed in Section 2.3, the flow stabilizes the drops against breakup by this mechanism, and highly stretched threads are frequently obtained in flow. Hence, thread breakup by the Tomotika mechanism has frequently been observed immediately after the flow has stopped. It was found that this mechanism is decisive if the ratio of long and short axes of the ellipsoid, a/b, which is formed by drop deformation, is higher than about 60 (10). At intermediate aspect ratios, breakup is dominated by end pinching. The drop forms bulbous ends, which pinch off into separate droplets. If the remaining drop is long enough, bulbs form again on new ends, and the process repeats itself until the remaining portion of the drop is short enough to retract to a sphere (10). The retraction to a sphere appears for a/b smaller than about 10 (10). The droplet shape relaxes exponentially with a relaxation time that is dependent on the rheological model of the ellipsoid. Small deformation theory for a single Newtonian droplet gives for the relaxation time, τ_R, as (10,69)

$$\tau_R = \frac{\eta_m R}{\sigma} \frac{(2p+3)(19p+16)}{40(p+1)} \qquad (2.76)$$

Palierne's theory (70) for finite ϕ provides

$$\tau_R = \frac{\eta_m R_v}{4\sigma} \frac{(19p+16)\left[2p+3-2\phi(p-1)\right]}{10(p+1)-2\phi(5p+2)} \tag{2.77}$$

where R_v is the volume average of droplet radius. Further theories of τ_R and experimental results on the droplet shape and excess stress relaxation are discussed in Chapter 13 and in ref. (10).

Breakup of the minor phase into droplets or coarsening was observed for blends with the cocontinuous structure. If the length of the threads between crossing points of physical networks of the minor phase is much larger than their thickness, the threads burst by the Tomotika mechanism and the cocontinuous structure changes in the dispersed one. As discussed in Section 2.3, the rate of breakup increases with increasing interfacial tension and decreases with viscosity of both the phases and thickness of the thread. In practice, a rapid change of cocontinuous to dispersed structure appears for blends with a low amount of the continuous minor phase that obviously forms a "loose" net. For blends with composition near 50/50, cocontinuous structure formed in flow is unfavorable for thread breakup, and coarsening of the cocontinuous structure occurs in quiescent molten blends. Using the idea that the rate of the coarsening process is proportional to the same variables as the growth rates for retraction and breakup, Veenstra et al. (71) proposed a semiempirical equation for the time evolution of the average thickness of the network strand, R_t

$$\frac{d R_t}{d t} = c_c \frac{\sigma}{\eta_e} \tag{2.78}$$

where η_e is the effective viscosity, calculated as the weight average of viscosities of the blend components, and c_c is a dimensionless factor that should be determined from experiments. Generally, both thread breakup and coarsening of the cocontinuous structure can contribute to the time-dependent evolution in morphology and to the local morphology of individual parts of the physical network.

In contrast to thread breakup and contraction, there is not full consensus on mechanism of the growth of dispersed droplets of minor phase in quiescent molten polymer blends. Generally, there are two main mechanisms for droplet growth in a blend of immiscible liquids: Ostwald ripening and coalescence. In the course of Ostwald ripening, small droplets dissolve and large droplets grow due to the concentration gradients of the minor component dissolved in the matrix, which are dependent on radius of droplet (72). Dissolution of small droplets and growth of large droplets is controlled by the diffusion of the molecules of the minor phase through the matrix. Time development of the average droplet radius, R, due to Ostwald ripening is described by the equation (72,73)

$$R^3 = R_0^3 + \frac{8D_d v_m \sigma C_e^m}{9kT}\left(1+0.74\sqrt{\phi}\right)^3 t \tag{2.79}$$

where R_0 is the average droplet radius at $t = 0$, D_d is the diffusion coefficient of molecules of the dispersed component in the matrix, C_e^m is the equilibrium molecular concentration of the dispersed component dissolved in the matrix, and v_m is the molecular volume of the diffusing species. Analysis of Eq. (2.79) showed that the rate of growth of R should decrease with increasing interfacial tension (Flory-Huggins χ parameter) between the blend components (73). In agreement with physical intuition, Ostwald ripening is negligible in systems where conjugated phases are almost identical with blend components, i.e., contain negligible amounts of the other component. On the other hand, a pronounced increase in the rate of growth of R with increasing σ was found experimentally (73). Therefore, the Ostwald ripening approach is not a decisive mechanism, at least for systems with high and moderate interfacial tension.

Several potential forces drive coalescence in quiescent polymer blends. It was recognized many years ago that the Brownian motion–induced coalescence is the main mechanism of an increase of droplet size in dispersions of low molecular weight liquids (74). However, application of Smoluchowski's theory of Brownian motion–induced coalescence (which reasonably describes coalescence in usual colloid systems) to polymer blends leads to the conclusion that an increase in R is negligible for hours due to the high viscosity of the polymer melts (75). It was shown that the approximation for diffusion flux of particles commonly used for usual colloid systems is not applicable in polymer blends. An approximation that was better suited to polymer blends was proposed and the theory of the Brownian motion coalescence was derived, but this theory also predicts much slower coalescence than was found experimentally (75). The reason for the inability of Smoluchowski's theory and its modification to explain diffusion in polymer blends can be attributed to either the existence of other driving forces behind coalescence in polymer blends or to the fact that Smoluchowski's theory was derived for dilute systems where only binary collisions were considered whereas typical polymer blends, for which substantial growth of the droplet size is observed, are concentrated systems (76).

van Gisbergen and Meijer (77) assumed that the driving force of coalescence is gravity as exemplified by the difference in the density of the dispersed phase and matrix. They assumed that the rate of coalescence is controlled by the coalescence time, t_c, for flattened droplets with a mobile, partially mobile, or immobile interface [see Eqs. (2.40) to (2.42), (2.53), and (2.54)] and that the volume fraction of the dispersed phase has no effect on the rate of coalescence. van Gisbergen and Meijer's approach was generalized by including the assumption that a droplet can coalesce only if its distance from at least one of its neighbors is shorter than $R/2$ (78). The probability of a droplet coalescencing is a function of ϕ. It was assumed that the driving force of the coalescence can be molecular forces, Brownian motion (treated in an approximate manner), and gravity. The time dependence of average R was calculated for systems with a mobile, partially mobile, and immobile interface and for all three mentioned driving forces. The results depended on the choice of both the driving force and mobility of the interface. As mentioned in Section 2.4, the theories of the matrix film drainage between flattened droplets incorrectly predict the approach of the droplets for the weak driving forces of coalescence.

where

$$H_B = \frac{1}{R} \int_0^{Rf} \frac{2h + 0.402\sqrt{2}\,p^{-1}R^{1/2}h^{1/2}}{2h + 1.711\sqrt{2}\,p^{-1}R^{1/2}h^{1/2} + 0.461p^{-2}R}\,dh \qquad (2.86)$$

Besides a system of Newtonian droplets in Newtonian matrix, coalescence of Newtonian droplets in viscoelastic matrix, described by the Maxwell model, was also studied in ref. (79). It was shown that coalescence is quicker in a viscoelastic matrix than in a Newtonian one with the same zero shear viscosity. The rate of coalescence increases with increasing relaxation time of the Maxwell model; however, the increase is not pronounced. For typical polymer blend parameters, a steeper increase in R is predicted for van der Waals' force–induced coalescence than for Brownian motion–induced coalescence. The theory of van der Waals' force–induced coalescence predicts an increase in the rate of coalescence with increasing interfacial tension (proportional to A) and decreasing η_d and η_m, in qualitative agreement with experimental results (73,79).

Yu et al. (80,81) assumed that the time of coalescence is a sum of the time of drainage of the matrix trapped between the droplets and the time of the shape relaxation of coalescing droplets to a sphere. They considered film drainage between flattened droplets. As discussed above, the rate of drainage for the weak driving forces of coalescence is overestimated by this model, which is probably why they found in their interpretation of the experimental data that the coalescence is controlled by the shape relaxation of joined droplets (the drainage time is substantially shorter than the shape relaxation time). For polymer blends with a high content of the dispersed phase, where the distance between a droplet and its neighbors is small, "collision-induced collision *via* flow" and/or "geometrical collision-induced collision" can contribute to the coalescence (59,82). The "collision-induced collision *via* flow" mechanism is a consequence of the matrix flow induced by the shape retraction of coalescing droplets (see Figure 2.5) that attracts droplets in the direction of the line through the centers of colliding droplets. On the other hand, the flow causes repulsion of the droplets in the perpendicular direction. The flow velocity is proportional to σ/η_{ef} [cf. Eqs. (2.76) to (2.78)]. In "geometrical collision-induced collision," relaxing droplets formed by coalescence touch a neighbor droplet located in the direction perpendicular to the line connecting the centers of original coalescing droplets (see Figure 2.5). It should be mentioned that the above mechanisms are negatively correlated.

Generally, it can be concluded that all the forces contributing to droplet collisions in quiescent polymer blends are weak, and, therefore, drainage of the matrix film between *spherical* droplets should be considered as the description of coalescence. Coalescence in quiescent blends seems to be a typical many-body problem and, in contrast to the above described theories, an adequate theory of the effect should consider simultaneous interaction of more than two droplets. A plausible model of the coalescence is as follows: Molecular forces (van der Waals, etc.) with a certain contribution from Brownian motion are the driving forces of the coalescence. During

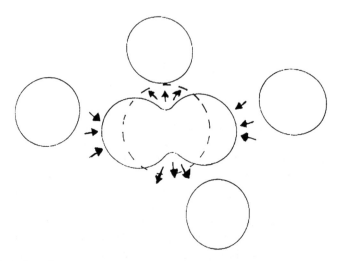

FIGURE 2.5 The effect of a shape relaxation of coalescing droplets in collision-induced collision *via* flow and geometrical collision-induced collision. Arrows show the direction of the induced flow, and the broken circle shows the relaxed shape of the coalesced droplets.

the process, collision-induced collision *via* flow and/or geometrical collision-induced collision contributes to coarsening of the phase structure. This model leads to a qualitative prediction that the rate of coalescence increases with interfacial tension and decreases with increasing viscosity of the droplets and matrix. A pronounced increase of the coalescence rate with increasing volume fraction of the dispersed phase is predicted by the model. These predictions agree with the results of experimental studies (73). For derivation of a reliable theory of coalescence in quiescent blends, contributions of individual mechanisms should be estimated. The theory should consider all the relevant mechanisms and should adequately describe multi-droplet interaction (mean field theory, interaction of a droplet with its neighbor in the first coordination sphere with proper boundary conditions, etc.).

2.7   EFFECT OF A COMPATIBILIZER ON THE PHASE STRUCTURE DEVELOPMENT

The presence of compatibilizers (block or graft copolymer, premade or formed in flow) at the interface has fundamental effect on phase structure development in polymer blends. A compatibilizer affects the type of the phase structure by two competing mechanisms. A decrease in the interfacial tension caused by a compatibilizer favors the formation and stability of cocontinuous structures. On the other hand, a compatibilizer suppresses coalescence, a supposed reason of cocontinuity formation (7). So far, no quantitative theory describing the effect of a compatibilizer on the type of the phase structure has been proposed.

The effect of a compatibilizer on the size of dispersed droplets in a matrix can be predicted if the compatibilizer effect on droplet breakup and coalescence and its distribution between the interface and bulk phases are known. The effect of copolymers on droplet breakup and coalescence share features with the related effect of surfactants in water-oil mixtures, which is broadly studied in the literature. The approach to the description of both types of systems is similar, but the plausibility of various approximations and the importance of various contributions to the effects can differ strongly because of the different molecular nature of the components and many orders difference in the values of rheological functions. Therefore, the results for low molecular weight emulsions containing surfactants should be applied to compatibilized polymer blends only with great care. A quite recent summary and discussion of the results of the studies of the effect of compatibilizers on the morphology of polymer blends can be found in (83).

Block or graft copolymers at the interface reduce the interfacial tension between the dispersed and matrix phases (84). For the blends of homopolymers A and B compatibilized with poly(A-b-B) block copolymer, the following equations were derived (84) for the interfacial tension in the dry brush limit (in which N_A/P_A and N_B/P_B are small, where N_i and P_i are numbers of segments in a block and a homopolymer i)

$$\sigma = \sigma_0 - 3q^3 N \frac{kT}{a^2} = \sigma_0 - f_{DB}q^3 \tag{2.87}$$

and in the wet brush limit (N_A/P_A and N_B/P_B are large)

$$\sigma = \sigma_0 - \left[2 + \frac{3^{1/3}}{2^{2/3}}\left(\frac{N_A}{P_A^{2/3}} + \frac{N_B}{P_B^{2/3}}\right)q^{2/3}\right]q\frac{kT}{a^2} \approx \sigma_0 - f_{WB}q^{5/3} \tag{2.88}$$

where σ_0 is the interfacial tension of the blend without a compatibilizer, q is the density of a copolymer at the interface, and a is the segment length that is assumed to be the same for A and B species. Equation (2.87) with a somewhat different f_{DB} was derived also by Lyatskaya et al. (85). It should be mentioned that both Eqs. (2.87) and (2.88) differ from the equation commonly used for the interfacial tension in systems containing surfactants, where the second term in the equation for σ is assumed to be proportional to q (22). At high enough q, zero interfacial tension can be achieved. However, in real polymer blends, this q is not achieved, and compatibilized blends show positive σ (84).

In flow, besides the effect of a copolymer on the equilibrium value of interfacial tension, gradients in q along the droplet surface are induced, which affects deformation and breakup characteristics of the droplets (22,83). In this case, Eq. (2.11) describing the stress boundary condition on the droplet surface must be generalized to (22,86)

$$\mathbf{n} \cdot \mathbf{T}_m - p\mathbf{n} \cdot \mathbf{T}_d = \frac{\sigma(q)}{\eta_m \gamma R_0}\mathbf{n}(\nabla_s \cdot \mathbf{n}) - \frac{1}{\eta_m \gamma R_0}\nabla_s \sigma(q) \tag{2.89}$$

where $\sigma(q)$ is the actual interfacial tension [its equilibrium value is given by Eq. (2.87) or (2.88)], R_0 is the radius of an undeformed spherical droplet, and ∇_s is the surface gradient operator $\nabla_s = (\mathbf{I} - \mathbf{nn})\cdot\nabla$, where $\mathbf{I}$ is the unit tensor. In contrast to Eq. (2.11), Eq. (2.89) reflects the Marangoni stress (the consequence of nonuniform distribution of a compatibilizer along the interface, which produces a tangential stress jump) and a change in interfacial tension due to compatibilizer dilution upon droplet deformation. If transport of a copolymer between the interface and bulk phases can be neglected, the changes in q at the interface can be described by the equation (22,86–88)

$$\frac{\partial q}{\partial t} + \nabla_s\left(q\mathbf{u}_s - \frac{1}{Ca_{eq}\mu}\nabla_s q\right) + q(\nabla_s\cdot\mathbf{n})(\mathbf{u}\cdot\mathbf{n}) = 0 \qquad (2.90)$$

where

$$\mu = \frac{\sigma_{eq}R_0}{\eta_m D_s} \qquad (2.91)$$

where Ca_{eq} is the capillary number for a droplet with equilibrium interfacial tension σ_{eq} determined from Eq. (2.87) or (2.88), $\mathbf{u}$ is the flow velocity, $\mathbf{u}_s$ is the velocity component tangential to the interface [$\mathbf{u}_s = (\mathbf{I} - \mathbf{nn})\cdot\mathbf{u}$], and D_s is the surface diffusivity of the copolymer. Solution of Eqs. (2.89) and (2.90) is much more complicated than the solution of Eq. (2.11) because the copolymer distribution is coupled to the droplet shape, which, of course, depends upon the detailed copolymer distribution. Equations (2.89) and (2.90) were solved using various approximations, especially those for small deformations (22,87,88). Using the assumptions that $Ca_{eq} \ll 1$, $p \ll 1$, σ_{eq} is given by Eq. (2.87), and the difference between σ and σ_{eq} is small, the following equation was derived for the deformation parameter, D, of a droplet in the extensional flow (86)

$$D = \frac{3Ca_{eq}b_r}{4 + Ca_{eq}b_r} \qquad (2.92)$$

where

$$b_r = \frac{5}{4}\frac{(19p+16)+\dfrac{12\psi\mu}{1-\psi}}{10(p+1)+\dfrac{6\psi\mu}{1-\psi}} \qquad (2.93)$$

and

$$\psi = \frac{f_{DB} q_{eq}^3}{\sigma_0} \qquad (2.94)$$

For $Ca_{eq} \ll 1$ and either $\psi \to 0$ or $\mu \to 0$, Eq. (2.92) reduces to Eq. (2.16) for a system without a compatibilizer. The first case corresponds to no effect of the compatibilizer on interfacial tension. In the second case, the gradient of the compatibilizer concentration is not established because diffusion eliminates all flow-induced changes in the compatibilizer concentration.

The deformation of droplets in extensional flows can be qualitatively understood by accounting for (i) convection of copolymers toward stagnation points at the end of droplets — this tends to lower the interfacial tension and hence increases the observed deformation; and (ii) dilution of a copolymer due to an increase in the droplet surface upon deformation — this tends to increase the interfacial tension and hence decreases the observed deformation (22). Generally, additional adsorption or desorption of the copolymer to and from the interface as a result of fluxes from the bulk phases can appear during droplet deformation. However, it is usually assumed that the last effect can be neglected for polymer blends (83). Marangoni stresses reduce the interfacial velocity and cause the droplet to behave as if it has higher viscosity. Therefore, compatibilizer effects are more pronounced for blends of a low p (22). The convection of the copolymer towards the end of droplets induces tip streaming or asymmetric breakup (see Figure 1 in ref. 83). At tip streaming, small droplets are breaking off the ends of the "mother drop." Interfacial tension of daughter droplets is smaller, sometimes substantially, than that of the mother drops. Tip streaming is typical for systems with a low p and can occur also at $Ca < Ca_c$ (22). For theoretical evaluation of the effect of a compatibilizer on droplet breakup in polymer blends, the value of the diffusivity, D_s, of the copolymer along the droplet surface must be known. These values are not available so far. Generally, theoretical considerations, numerical simulations, and experimental results show that the critical capillary number for blends containing a compatibilizer is higher than that for the related uncompatibilized blends with the same p (83). Therefore, the critical droplet radius R_c under certain flow conditions is not scaled by the ratio of interfacial tension of blends with and without the compatibilizer. It seems to be a reasonable assumption that the addition of a compatibilizer reduces R_c but less than it reduces interfacial tension.

The effect of a compatibilizer on the breakup of an extended thread is described in ref. (35). The first-order effect is a decrease in the rate of the perturbation wave and, therefore, an increase in t_B caused by the reduction of interfacial tension [cf. Eq. (2.24)]. An analysis (86) of Palierne and Leblanc's theory (35) showed that an increase in t_B for compatibilized blends is somewhat greater than the decrease in σ. Undoubtedly, a compatibilizer efficiently stabilizes highly elongated thread against breakup, a fact that was also confirmed by numerical analysis and experiments (83). It should be mentioned that the compatibilizer density at the droplet surface can change substantially and the effect of tip streaming can appear during the droplet deformation into an elongated thread.

Two other basic mechanisms of suppression of flow induced coalescence by a compatibilizer were proposed besides a decrease in interfacial tension: a decrease in mobility of the interface due to the Marangoni effect (45,52,83,89) and the effect of the steric repulsive force due to compression of block copolymer layers attached to the surfaces of two approaching droplets (90–92) (see Figure 2.6). The Marangoni effect appears during drainage of the matrix trapped between coalescing droplets. It is caused by the drag of compatibilizer molecules to the periphery of two approaching droplets (45,52,83,86,89). Nonuniform distribution of compatibilizer molecules results from the competition between their drag and diffusion; it leads to a decrease in mobility of the interface, i.e., suppression of a interdroplet circulation. Milner and Xi (45) considered the Marangoni effect for collision of spheres covered with a block copolymer. They assumed that the copolymer is compressed to a half of the droplet surface and did not consider diffusion of the copolymer along the interface. The resulting repulsive force from the copolymer compression was incorporated into the Wang et al. theory of coalescence (50). The theory predicts that the minimum coverage with the copolymer fully suppressing coalescence increases with the shear rate. It seems that this theory somewhat overestimates the Marangoni effect because the area from which the copolymer is displaced (or diluted if diffusion exists) is for $h \ll R$ (which is decisive for the drainage time) less than half of the droplet surface, and diffusion of the copolymer along the interface cannot be neglected *a priori*. A number of studies on various aspects of the Marangoni effect at the approach of spherical droplets, mostly using assumptions plausible for low molecular weight emulsions, exist in the literature (93–95). The results can be summarized as follows (57,86): The magnitude of the Marangoni effect depends on D_s. Generally, an increase in the copolymer diffusion along the interface reduces an increase of the contribution of the Marangoni effect with the shear rate. Due to the Marangoni effect, the probability, P_s, for spherical droplets in compatibilized blends is a function of γ, in contrast to systems without a compatibilizer. At constant γ, the Marangoni effect manifests itself in P_s as an increase in η_d.

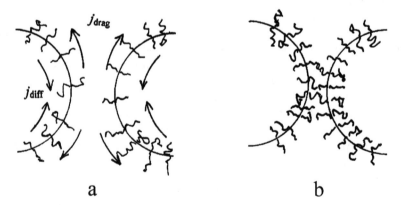

a   b

FIGURE 2.6 The two mechanisms proposed for block copolymer suppression of coalescence: (a) Marangoni effect (j_{drag} and j_{diff} show copolymer fluxes because of its drag and diffusion) and (b) steric repulsion.

The possibility of droplet flattening at coalescence should also be considered for compatibilized polymer blends. For substantially flattened droplets, the equation derived by Jeelani and Hartland (52) can be used for the time of coalescence, t_c,

$$t_c = \frac{9\eta_m^2 \gamma R^4}{8\sigma^2 h_c^2 \left[1 + \frac{3C}{p} \left(1 + \frac{\sqrt{3}Ca^{1/2}R^2}{4h_0\sigma} \frac{\partial\sigma}{\partial r} \right) \right]} \qquad (2.95)$$

where $\partial\sigma/\partial r$ is the gradient of interfacial tension at the periphery of the film. $\partial\sigma/\partial r$ is always negative and can be calculated by combination of Eq. (2.90) with Eq. (2.87) or (2.88). If we assume that Eq. (2.60) holds also for compatibilized polymer blends, the probability P_c can be expressed as

$$P_c = P_s \exp\left\{ -\frac{9Ca^2R^2}{8h_c^2 \left[1 + \frac{3C}{p} \left(1 + \sqrt{3}\frac{Ca^{1/2}R^2}{4h_0\sigma} \frac{\partial\sigma}{\partial r} \right) \right]} \right\} \qquad (2.96)$$

According to Eq. (2.96) for a certain blend, R (for which P_c falls to a negligible value) decreases with increasing γ in spite of an increase in P_s with γ, which should be considered while evaluating experimental results (57). Dependences of P_c on R for blends with and without a compatibilizer are compared in Figure 2.7. For compatibilized blends, P_s is lower as a result of the suppressed mobility of the interface. Because of the combined decrease in the interfacial tension and in the mobility of the interface, a steep decrease in P_c starts at lower R.

Steric repulsion of droplets manifests itself only at an interdroplet distance that is comparable to the radius of gyration of a copolymer. Compressed layers of the copolymer on the surface oppose the approach of the droplets to the distance required for matrix film rupture. It was found that the minimum block copolymer coverage for complete suppression of coalescence is independent of γ and reciprocally proportional to the molecular weight of the outside blocks (91,92). It should be mentioned that a negative relation exists between the steric repulsion and the Marangoni effect because a reduction in the block copolymer coverage on the "poles" of approaching droplets attenuates the steric repulsion (the Marangoni effect results from the squeezing flow of the matrix from the gap between the droplets). Lyu (92) estimated that the reduction of the block copolymer coverage is negligible up to γ = 10 s^{-1} for block copolymers with block molecular weights of 20,000 Da. Nevertheless, the problem needs further study. Another point, which should be elucidated, is the effect of the steric repulsion on coalescence in systems where the block copolymer coverage is lower than that necessary for elimination of the droplet coalescence. It is practically impossible to distinguish between coalescence sup-

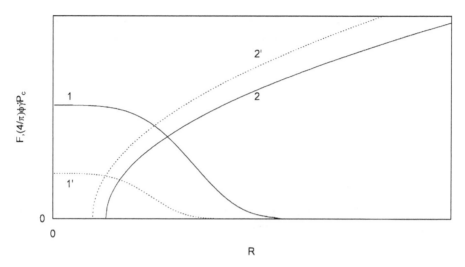

FIGURE 2.7 Comparison of the graphic solution of Eq. (2.73) for blends without (F — curve 1, $(4/\pi)\gamma\,\phi P_c$ — curve 2) and with (F — curve 1', $(4/\pi)\gamma\,\phi P_c$ — curve 2') a compatibilizer. For compatibilized blend, a decrease in the interfacial tension and the Marangoni effect are assumed.

pressed by the Marangoni effect and by the steric repulsion that arises from the experimental dependence of droplet size on shear rate (57).

P_c in compatibilized polymer blends can be further lowered, especially at high γ, by slowing down the development of the neck between coalescing droplets. A thin neck between coalescing droplets can break down immediately after the axial force on the coalescing droplets changes its sign.

The competition between the breakup and coalescence of droplets in uncompatibilized and compatibilized blends is compared in Figure 2.7 where the graphic solution of Eq. (2.73) for monodispersed systems is presented. The coalescence in a compatibilized blend with a constant q (density of a copolymer at the interface) is assumed to be adversely affected by the Marangoni effect as described by Eq. (2.96). In compatibilized blends, P_s is decreased by a decrease in mobility of the interface. Due to a decrease in the interfacial tension and in the mobility of the interface, the steep decrease in P_c appears at substantially smaller R for compatibilized blends than for blends without a compatibilizer. Because of the higher Ca_c in compatibilized blends (see the discussion above), a decrease in R_c can be assumed to be substantially smaller than a decrease in the interfacial tension. Therefore, in compatibilized blends (rather than in those without a compatibilizer), R_c is more likely larger than the value of R at which P_c falls to a very small value. Of course, coalescence can be eliminated for compatibilized polymer blends if q is independent of R. To keep the value of q constant, the amount of a copolymer at the interface must increase with decreasing R.

The assumption that the amount of a copolymer at the droplet surface is constant during individual breakup and coalescence events seems to be plausible for polymer blends. On the other hand, a steady droplet distribution between the interface and bulk phases, dependent on the droplet size, should be established in the steady state (i.e., after long enough steady flow) (96). The effect of competition between droplet breakup and coalescence was studied for a monodisperse system described by Eq. (2.73) with F proportional to $Ca - Ca_c$ [see the derivation of Eq. (2.70)] using Janssen's theory of coalescence with P_c given by Eq. (2.96). Equation (2.87) was used for σ, and Ca_c was considered to be a known parameter, independent of R. The dependence of the droplet size on the volume fraction of the dispersed phase was determined for the following cases: (i) a constant density of the compatibilizer at the interface, q; (ii) the compatibilizer was localized at the interface; and (iii) the maximum density of the compatibilizer at the interface is known and is characteristic of the studied system. The shape of the dependence of R on ϕ for case (i) matches the one for blends without a compatibilizer. The shapes of the dependence of R on ϕ for cases (ii) and (iii) differ from those for blends without a compatibilizer and depend on the choice of the parameter kept independent of ϕ, i.e., the amount of the copolymer or the ratio of the amount of the copolymer and the volume fraction of the dispersed phase (96).

In the quiescent state, the shape relaxation of the droplets in compatibilized polymer blends is slower due to a decrease in the interfacial tension. The same holds for the breakup of highly extended threads and breakup and coarsening of cocontinuous structures. On the other hand, nonuniform distribution of a copolymer along the interface can accelerate the shape relaxation due to the drag induced by diffusion of the copolymer; nonuniform distribution of a copolymer can also induce asymmetric modes of droplet breakup: end pinching and tip streaming. The coalescence in quiescent compatibilized blends is suppressed (i) by a decrease in the driving forces of the main modes of coalescence, i.e., a decrease in the effective Hamaker constant and in the rate of shape relaxation of coalescing droplets, (ii) by a decrease in mobility of the interface, and (iii) eventually by steric repulsion of the droplets.

All the above considerations were based on the assumption that the amount of a copolymer at the interface, or at least the distribution of a copolymer between the interface and bulk phases, is known. Various aspects of the ability of block copolymers to compatibilized polymer blends were considered from the point of view of equilibrium thermodynamics, as summarized, for example, in ref. (84). Nevertheless, a rule is still not available for predicting the equilibrium amount of a copolymer located at the interface from the concentrations of the components, their molecular characteristics, and interfacial area. Moreover, it is not clear whether the copolymer distribution between the interface and bulk phases at thermodynamic equilibrium is a reasonable approximation for the distribution in steady flow. Pronounced migration of a copolymer between the interface and bulk phases detected during annealing of compatibilized blends (97) shows that the copolymer distributions in steady flow and in thermodynamic equilibrium (at rest) can differ significantly. If such is the case, the distribution of a copolymer in the interface and bulk phases should be calculated simultaneously with the development of droplets, strongly complicating the description of phase structure development in compatibilized polymer blends.

The lack of knowledge about the distribution of a compatibilizer between the interface and bulk phases, and about compatibilizer diffusivity at the interface is the main obstacle in the application of the above theory for quantitative prediction of the phase structure development in compatibilized polymer blends.

2.8 CONCLUSIONS

As shown in Section 2.2, a theory is not available for predicting the type of the phase structure formed in polymer blends from knowledge of the blend composition, properties of the components, and flow conditions even for binary polymer blends. The key remaining question in the development of a theory is whether the cocontinuous structure (and other structures such as the droplet within droplet structure) is present under steady state conditions (i.e., still exists after long steady flow, e.g., mixing) or only in transition (i.e., gradually changes to the dispersed structure under long enough steady flow). If the first case is correct, a theory can be formulated by combining a description of a blend structure by the interfacial area and orientation of the interface together with a proper rule of nonequilibrium thermodynamics. In the second case, kinetic theory that describes the time development of the phase structure should be derived. For multicomponent polymer blends, the present theories consider only interfacial energy. It is to be expected that a general theory for multicomponent blends will follow only after a reliable theory for binary polymer blends is put forward.

The development of the dispersed structure of blends is much better understood. The basic idea that the droplet size is given by competition between droplet breakup and coalescence is generally accepted. On the other hand, droplet breakup and coalescence are complicated events, especially in complex flow fields of mixing and processing devices, and their description is not easy. A number of approximate descriptions of these events is available, but the problem is choosing an approximate theory that reflects all the important features of the event and provides the results for use in further analysis. Most studies of droplet breakup deal with Newtonian droplets in a Newtonian matrix. There should be more study of (i) droplet deformation and breakup in viscoelastic systems where the interfacial tension reflects the normal stress differences seen between droplets and their surrounding matrix [see Eq. (2.8)], (ii) the conditions under which individual droplets break up, and (iii) the dependence of breakup frequency on system parameters. The recent theories of coalescence, which combine expressions for the rate of matrix drainage between spherical and flattened droplets, seem to give a reasonable description of the phenomena; the further generalization of these theories by including parameters that have been omitted so far, e.g., elastic properties of the components, is desirable. Most theories describing droplet breakup and coalescence were derived for dilute systems where simultaneous interactions of more than two droplets were not considered. So the effect of hydrodynamic interaction of adjacent droplets on the deformation and breakup of a droplet begs for further investigation. The study of the simultaneous collisions of three or more droplets is especially important for predicting the dependence of the coalescence frequency on the volume fraction of the dispersed phase. The reliability of the solution of the equations describing the competition between droplet breakup and coalescence is limited by the uncertainty in the expressions for the breakup and coalescence frequency and by the fact that terms for simultaneous coalescence of three or more droplets are not included in the equations.

Relaxation of the droplet shape and breakup of a highly elongated thread in quiescent polymer blends are well understood and satisfactorily described. Breakup and/or coarsening of the cocontinuous structure can be explained by mechanisms of the thread breakup and droplet shape relaxation. The quantitative theory needs a proper model of a cocontinuous structure. The coarsening of the phase structure in blends with droplets in matrix morphology can be qualitatively explained as a consequence of coalescence induced by molecular forces between droplets. The effect needs a new quantitative theory that will consider (i) interaction of a droplet with all of its neighbors (at least in the first coordination sphere) and (ii) the contribution of droplet shape relaxation and Brownian motion. A number of theories describing the effects of a compatibilizer on droplet breakup and coalescence can be found in the literature. It should be mentioned that the expressions derived for low molecular weight emulsions and polymer blends used different relations to describe the effect of compatibilizer concentration on interfacial tension. Improvement in our knowledge of copolymer diffusivity at an interface and of the dependence of the copolymer distribution between an interface and both bulk phases on parameters of the system seems to be the most important step that we need to take to adequately describe the effect of a compatibilizer on phase structure development in polymer blends.

ACKNOWLEDGMENTS

The author is very grateful to Dr. A. Zivný and Mr. J. Pelcman for their help with the preparation of figures, to Dr. Z. Horák for valuable comments, and to the Academy of Sciences of the Czech Republic (project AVOZ4050913) and the Grant Agency of the Czech Republic (grant No. 106/02/1248) for financial support.

REFERENCES

1. Favis, B.D., Factors influencing the morphology of immiscible polymer blends in melt processing, in *Polymer Blends, Vol. 1: Formulations*, Paul, D.R. and Bucknall, C.B., Eds., J. Wiley & Sons, New York, 2000, pp. 501–537.
2. Lyngaae-Jørgensen, J., Rheology of polymer blends, in *Polymer Blends and Alloys*, Folkes, M.J. and Hope, P.S., Eds., Blackie Academic & Professional, London, 1993, pp. 75–102.
3. Janssen, J.M.H., Emulsions: the dynamic of liquid-liquid mixing, in *Materials Science and Technology, Vol. 18, Processing of Polymers*, Meijer, H.E.H., Ed., Wiley-VCH, Weinheim, Germany, 1997, pp. 115–188.
4. Meijer, H.E.H. and Janssen, J.M.H., Mixing of immiscible liquids, in *Mixing and Compounding of Polymers*, Manas-Zloczower, I. and Tadmor, Z., Eds., Hanser Publishers, Munich, 1994, pp. 85–147.
5. Utracki, L.A. and Shi, Z.H., Development of polymer blend morphology during compounding in a twin-screw extruder. Part I. Droplet dispersion and coalescence — a review. *Polym. Eng. Sci.*, 32, 1824–1833, 1992.

6. Fortelný, I., Kovář, J., and Stephan, M., Analysis of the phase structure development during the melt mixing in polymer blends, *J. Elastomers Plastics*, 28, 106–139, 1996.
7. Pötschke, P. and Paul, D.R., Formation of cocontinuous structures in melt-mixed immiscible polymer blends, *J. Macromol. Sci.*, C43, 87–141, 2003.
8. Lyngaae-Jørgensen, J. and Utracki, L.A., Dual phase continuity in polymer blends, *Makromol. Chem., Macromol. Symp.*, 48/49, 189–209, 1991.
9. Van Oene, H., Rheology of polymer blends and suspensions, in *Polymer Blends, Vol. 1,* Paul, D.R. and Newman, S., Eds., Academic Press, New York, 1978, pp. 295–352.
10. Tucker, C.L. and Moldenaers, P., Microstructural evolution in polymer blends, *Annu. Rev. Fluid Mech.*, 34, 177–210, 2002.
11. Glansdorff, P. and Prigogine, I., *Thermodynamic Theory of Structure, Stability and Fluctuations*, Wiley-Interscience, London, 1971.
12. Hobbs, S.Y., Dekkers, M.E.J., and Watkins, W.H., Effect of interfacial forces on polymer blends, *Polymer*, 29, 1598–1602, 1988.
13. Guo, H.F., Packirisamy, S., Gvozdic, N.V., et al., Prediction and manipulation of the phase morphologies of multiphase polymer blends. I. Ternary systems. *Polymer*, 38, 785–794, 1997.
14. Miroshnikov, Y.P., Kozlova, G.S., and Voloshina, Y.N., Osobennosti processov dispergirovania v mnogokomponentnych polimernych sistemach, *Vysokomol. Soed*, 31, 767–771, 1989.
15. Guo, H.-F., Gvozdic, N.V., and Meier, D.J., Prediction and manipulation of the phase morphologies of multiphase polymer blends. II. Quarternary systems, *Polymer*, 38, 4915–4923, 1997.
16. Luzinov, I., Pagnoulle, C., and Jérôme, R., Ternary polymer blend with core-shell dispersed phases: effect of the core-forming polymer on phase morphology and mechanical properties, *Polymer*, 41, 7099–7109, 2000.
17. Henmati, N., Nazokdast, H., and Panahi, H.S., Study on morphology of ternary polymer blends. I. Effects of melt viscosity and interfacial interaction, *J. Appl. Polym. Sci.*, 82, 1129–1137, 2001.
18. Nemirovski, N., Siegmann, A., and Narkis, M. Morphology of ternary immiscible polymer blends, *J. Macromol. Sci. Phys.*, B34, 459–475, 1995.
19. Gupta, A.K. and Shrinivasan, K.R., Melt rheology and morphology of PP/SEBS/PC ternary blends, *J. Appl. Polym. Sci.*, 47, 167–184, 1993.
20. Reignier, J., Favis, B.D., and Heuzey, M.-C., Factors influencing encapsulation behavior in composite droplet-type polymer blends, *Polymer*, 44, 49–59, 2003.
21. Van Oene, H.J., Modes of dispersion of viscoelastic fluids in flow, *J. Colloid Interface Sci.*, 40, 448–467, 1972.
22. Stone, H.A., Dynamics of drop deformation and breakup in viscous fluids, *Annu. Rev. Fluid Mech.*, 26, 65–102, 1994.
23. Elmendorp, J.J., Dispersive mixing in liquid systems, in *Mixing in Polymer Processing*, Rauwendaal, C., Ed., Marcel Dekker, New York, 1991, pp. 17–100.
24. Han, C.D., *Multiphase Flow in Polymer Processing*, Academic Press, New York, 1981.
25. Taylor, G.I., The formation of emulsions in definable fields of flow, *Proc. R. Soc. London, Ser. A*, 146, 501–523, 1934.
26. Chaffey, C.E. and Brenner, H., A second-order theory for shear deformation of drops, *J. Colloid Interface Sci.*, 24, 258–269, 1967.
27. Cox, R.G., The deformation of a drop in a general time-dependent fluid flow, *J. Fluid Mech.*, 37, 601–623, 1969.
28. Barthès-Biesel, D. and Acrivos, A., Deformation and burst of a liquid droplet freely suspended in a linear shear field, *J. Fluid Mech.*, 61, 1–21, 1973.

29. Grace, H.P., Dispersion phenomena in high viscosity immiscible fluid systems and application of static mixers as dispersion devices in such systems, *Chem. Eng. Commun.*, 14, 225–277, 1982.

30. deBruijn, R.A., *Deformation and Break-up of Drops in Simple Shear Flows,* Ph.D. thesis, Eindhoven Univ. Technol., The Netherlands, 1989.

31. Levitt, L., Macosko, C.W., and Pearson, S.D., Influence of normal stress difference on polymer drop deformation, *Polym. Eng. Sci.*, 36, 1647–1655, 1996.

32. Mighri, F., Carreau, P.J., and Ajji, A., Influence of elastic properties on drop deformation and breakup in shear flow, *J. Rheol.*, 42, 1477–1490, 1998.

33. Lerdwijitjarud, W., Larson, R.G., Sirivat, A., et al., Influence of weak elasticity of dispersed phase on droplet behavior in sheared polybutadiene/poly(dimethyl siloxane) blends, *J. Rheol.*, 47, 37–58, 2003.

34. Tomotika, S., On the instability of a cylindrical thread of a viscous liquid surrounded by another viscous fluid, *R. Soc. London, Ser. A*, 150, 322–337, 1935.

35. Palierne, J.F. and Lequeux, F., Sausage instability of a thread in a matrix; linear theory for viscoelastic fluids and interface, *J. Non-Newtonian Fluid. Mech.*, 40, 289–306, 1991.

36. Kuhn, W., Spontane Aufteilung von Flüssigkeitszylindern in kleine Kugeln, *Kolloid Z.*, 132, 84–99, 1953.

37. Mikami, T., Cox, R.G., and Mason, R.G., Breakup of extending liquid threads, *Int. J. Multiphase Flow*, 2, 113–138, 1975.

38. Khakhar, D. and Ottino, J.M., Breakup of liquid threads in linear flows, *Int. J. Multiphase Flow*, 13, 71–86, 1987.

39. Van Puyvelde, P., Yang, H., Mewis, J., et al., Breakup of filaments in blends during simple shear flow, *J. Rheol.*, 44, 1401–1415, 2000.

40. Cristini, V., Guido, S., Alfani, A., et al., Drop breakup and fragment distribution in shear flow, *J. Rheol.*, 47, 1283–1298, 2003.

41. Choi, S.J. and Schowalter, W.R., Rheological properties of nondilute suspensions of deformable particles, *Phys. Fluids*, 18, 420–427, 1975.

42. Jansen, K.M.B., Agterof, W.G.M., and Mellema, J., Droplet breakup in concentrated emulsions, *J. Rheol.*, 45, 227–236, 2001.

43. Utracki, L.A., On the viscosity-concentration dependence of immiscible polymer blends, *J. Rheol.*, 35, 1615–1637, 1991.

44. Chesters, A.K., The modeling of coalescence processes in fluid-liquid dispersions: a review of current understanding, *Trans. Inst. Chem. Eng. (A)*, 69, 259–270, 1991.

45. Milner, T. and Xi, H., How copolymers promote mixing of immiscible homopolymers, *J. Rheol.*, 40, 663–687, 1996.

46. Fortelný, I., Coalescence in polymer blends: solved and open problems, *Macromol. Symp.* 158, 137–147, 2000.

47. Janssen, M.H. and Meijer, H.E.H., Dynamic of liquid-liquid mixing: a 2-zone model, *Polym. Eng. Sci.*, 35, 1766–1780, 1995.

48. Fortelný, I. and Živný, A., Extensional flow induced coalescence in polymer blends, *Rheol. Acta*, 42, 454–461, 2003.

49. Zeichner, G.R. and Schowalter, W.R., Use of trajectory analysis to study stability of colloidal dispersions in flow fields, *AIChE J.*, 23, 243–254, 1977.

50. Wang, H., Zinchenko A.K., and Davis, R.H., The collision rate of small drops in linear flow fields, *J. Fluid Mech.*, 265, 161–188, 1994.

51. Rother, M.A. and Davis, R.H., The effect of slight deformation on droplet coalescence in linear flow, *Phys. Fluids*, 13, 1178–1190, 2001.

52. Jeelani, S.A.K. and Hartland, S., Effect of interfacial mobility on thin film drainage, *J. Colloid Interface Sci.*, 164, 296–308, 1994.

53. Zhang, X. and Davis, R.H., The collision of small drops due to Brownian and gravitational motion, *J. Fluid Mech.*, 230, 479–504, 1991.
54. Nir, A. and Acrivos A., On the creeping motion of two arbitrary-sized touching spheres in linear shear fields, *J. Fluid Mech.*, 59, 209–223, 1973.
55. Elmendorp, J.J. and Van der Vegt, A.K., A study of polymer blending microrheology. Part IV. The influence of coalescence on blend morphology origination, *Polym. Eng. Sci.*, 26, 1332–1338, 1986.
56. Elmendorp, J.J., *A Study on Polymer Blending Microrheology*, Ph.D. thesis, Tech. University Delft, The Netherlands, 1986.
57. Fortelný, I., An analysis of the origin of coalescence suppression in compatibilized polymer blends, *Europ. Polym. J.*, 40, 2161–2166, 2004.
58. Patlazhan, S.A. and Lindt, J.T., Kinetics of structure development in liquid-liquid dispersion under simple shear flow, *Theory. J. Rheol.*, 40, 1095–1113, 1996.
59. Tanaka, H., Coarsening mechanisms of droplet spinodal decomposition in binary fluid mixtures, *J. Chem. Phys.*, 105, 10099–10114, 1996.
60. Tokita, N., Analysis of morphology formation in elastomer blends, *Rubber Chem. Technol.*, 50, 292–300, 1977.
61. Fortelný, I. and Ková, J., Droplet size of the minor component in the mixing of melts of immiscible polymers, *Eur. Polym. J.*, 25, 317–319, 1989.
62. Lyngaae-Jørgensen, J. and Valenza, A., Structuring of polymer blends in simple shear flow, *Makromol. Chem., Macromol. Symp.*, 38, 43–60, 1990.
63. Fortelný, I. and Živný, A., Theory of competition between breakup and coalescence in flowing polymer blends, *Polym. Eng. Sci.*, 35, 1872–1877, 1995.
64. Huneault, M.A., Shi, Z.H., and Utracki, L.A., Development of polymer blend morphology during compounding in a twin-screw extruder. Part IV. A new computational model with coalescence, *Polym. Eng. Sci.*, 35, 115–127, 1995.
65. Delamare, L. and Vergnes, B., Computation of morphological changes of a polymer blend along a twin-screw extruder, *Polym. Eng. Sci.*, 36, 1685–1693, 1996.
66. Potente, H. and Bastian, M., Calculating morphology development of polymer blends in extruders on the basis of results of boundary and finite element simulations using the sigma simulation software, *Polym. Eng. Sci.*, 40, 727–737, 2000.
67. Fortelný, I., Analysis of the effect of breakup frequency on the steady droplet size in flowing polymer blends, *Rheol. Acta*, 40, 485–489, 2001.
68. Lyu, S.P., Bates, F.S., and Macosko, C.W., Modeling of coalescence in polymer blends, *AIChE J.*, 48, 7–14, 2002.
69. Maffettone, P.L. and Minale, M., Equation of change for ellipsoidal drops in viscous flow, *J. Non-Newton. Fluid Mech.*, 78, 227–241, 1998.
70. Palierne, J.F., Linear rheology of viscoelastic emulsions with interfacial tension, *Rheol. Acta*, 29, 204–214, 1990; erratum, *Rheol. Acta*, 30, 497, 1991.
71. Veenstra, H., Van Damm, J., and Posthuma de Boer, A., On the coarsening of cocontinuous morphologies in polymer blends: effect of interfacial tension, viscosity and physical cross-links, *Polymer*, 41, 3037–3045, 2000.
72. Crist, B. and Nesarikar, A.R., Coarsening in polyethylene-copolymer blends, *Macromolecules*, 28, 890–896, 1995.
73. Fortelný, I., Živný, A., and Jůza, J., Coarsening of the phase structure in immiscible polymer blends. Coalescence or Ostwald ripening? *J. Polym. Sci., Part B, Polym. Phys.*, 37, 181–187, 1999.
74. Overbeek, J.T.G., Kinetics of flocculation, in *Colloid Science*, Vol. I, Kruyt, H.K., Ed., Elsevier, Amsterdam, 1952, pp. 278–301.

75. Fortelný, I. and Ková, J., Theory of coalescence in immiscible polymer blends, *Polym. Compos.*, 9, 119–124, 1988.
76. Willemse, R.C., Ramaker, E.J.J., Van Dam, J. et al., Coarsening in molten quiescent polymer blends: the role of initial morphology, *Polym. Eng. Sci.*, 39, 1717–1725, 1999.
77. van Gisbergen, J.G.M. and Meijer, H.E.H., Influence of electron beam irradiation on the microrheology of incompatible polymer blends: thread break-up and coalescence, *J. Rheol.*, 35, 63–87, 1991.
78. Fortelný, I. and Živný, A., Coalescence in molten quiescent polymer blends, *Polymer*, 36, 4113–4118, 1995.
79. Fortelný, I. and Živný, A., Film drainage between droplets during their coalescence in quiescent polymer blends, *Polymer*, 39, 2669–2675, 1998.
80. Yu, W., Zhou, C., and Inoue, T., A coalescence mechanism for the coarsening behavior of polymer blends during a quiescent annealing process. I. Monodispersed particle system, *J. Polym. Sci., Part B, Polym. Phys.*, 38, 2378–2389, 2000.
81. Yu, W., Zhou, C., and Inoue, T. A coalescence mechanism for the coarsening behavior of polymer blends during a quiescent annealing process. II. Polydispersed particle system, *J. Polym. Sci., Part B, Polym. Phys.*, 38, 2390–2399, 2000.
82. Wallheinke, K., Pötschke, P., Macosko, C. W., et al., Coalescence in blends of thermoplastic polyurethane with polyolefins, *Polym. Eng. Sci.*, 39, 1022–1034, 1999.
83. Van Puyvelde, P., Velankar, S., and Moldenaers, P., Rheology and morphology of compatibilized polymer blends, *Curr. Opin. Colloid Interface Sci.*, 6, 457–463, 2001.
84. Hudson, S.D. and Jamieson, A.M., Morphology and properties of blends containing block copolymers, in *Polymer Blends, Vol. 1: Formulations*, Paul, D.R. and Bucknall, C.B., Eds., J. Wiley& Sons, New York, 2000, pp. 461–499.
85. Lyatskaya, Yu., Gersappe, D., Gross, N.A., et al., Designing compatibilizers to reduce interfacial tension in polymer blends, *J. Phys. Chem.*, 100, 1449–1458, 1996.
86. Fortelný, I., Breakup and coalescence of dispersed droplets in compatibilized polymer blends, *J. Macromol. Sci. Phys.*, B39, 67–78, 2000.
87. Stone, H.A. and Leal, L.G., The effects of surfactants on drop deformation and breakup, *J. Fluid Mech.*, 220, 161–186, 1990.
88. Flumerfelt, R.W., Effects of dynamic interfacial properties on drop deformation and orientation in shear and extensional flow fields, *J. Colloid Interface Sci.*, 76, 330–349, 1980.
89. Van Puyvelde, P., Velankar, S., Mewis, J., et al., Effect of Marangoni stresses on the deformation and coalescence in compatibilized immiscible blends, *Polym. Eng. Sci.*, 42, 1956–1964, 2002.
90. Sundararaj, U. and Macosko, C.W., Drop break-up and coalescence in polymer blends: the effects of concentration and compatibilization, *Macromolecules*, 28, 2647–2657, 1995.
91. Lyu, S.-P., Jones, T.D., Bates, F.S., and Macosko, C.W., Role of block copolymers on suppression of droplet coalescence, *Macromolecules*, 35, 7845–7855, 2002.
92. Lyu, S.-P., Block copolymers suppressing coalescence through stopping film rupture, *Macromolecules*, 36, 10052–10055, 2003.
93. Cristini, V., Blawzdziewicz, J., and Loewenberg, M., Near-contact motion of surfactant-covered spherical drops, *J. Fluid Mech.*, 366, 259–287, 1998.
94. Blawzdziewicz, J., Cristini, V., and Loewenberg, M., Near-contact motion of surfactant-covered spherical drops: Ionic surfactants, *J. Colloid Interface Sci.*, 211, 355–366, 1999.

95. Chesters, A.K. and Bazhlekov, I.B., Effect of insoluble surfactants on drainage and rupture of a film between drops interacting under a constant force, *J. Colloid Interface Sci.*, 230, 229–243, 2000.
96. Fortelný, I. and Živný, A., Theoretical description of steady droplet size in polymer blends containing a compatibilizer, *Polymer*, 41, 6865–6873, 2000.
97. Hlavatá, D., Hromádková, J., Fortelný, I., et al., Compatibilization efficiency of styrene-butadiene triblock copolymers in polystyrene-polypropylene blends with varying compositions, *J. Appl. Polym. Sci.*, 92, 2431–2441, 2004.

3 Cocontinuous Phase Morphologies: Predictions, Generation, and Practical Applications

Charef Harrats and Nafaa Mekhilef

CONTENTS

3.1   INTRODUCTION

The properties of immiscible polymer blends are mainly controlled by their phase morphology. Two major classes of phase morphologies can be distinguished in a binary two-phase blend: isolated particles dispersed in a matrix and a two-phase cocontinuous morphology. In the former, nonconnected droplets, platelets, rods, or fibers of one phase are distributed in the matrix of the other phase. The physical properties of the blends having phase morphology are mainly controlled by the properties of the matrix except in toughened plastics and in oriented structures where the characteristics of the dispersion (size, volume fraction, and spatial distribution) play a key role in controlling the toughness of the blend. The cocontinuous two-phase morphology consists of two coexisting, continuous, and interconnected phases throughout the whole blend volume. In contrast, in isotropic cocontinuous two-phase morphology, the two blend components contribute simultaneously to the properties of the blend in all directions. Cocontinuous microphase morphologies can be frozen-in from partially miscible blends during the spinodal demixing process as a result of the thermodynamic instability of the induced molecular miscibility; whereas, cocontinuous micro- or macrophase morphologies can be generated via melt-blending of immiscible polymers under a particular set of conditions, including the blend composition, component characteristics, and process parameters selected for their melt-processing.

3.2   COCONTINUOUS PHASE MORPHOLOGY IN PARTIALLY MISCIBLE BLENDS

A large number of homopolymer pairs exhibit partial miscibility when mixed together either in a melt-blending process or via precipitation from a single solution. Usually the partial miscibility is revealed only upon thermal treatment (heating and cooling operations) to the molecularly miscible mixture. A characteristic temperature-composition phase diagram is then identified for each pair of partially miscible homopolymers. As shown in Figure 3.1, typical phase diagrams of a partially miscible pair of polymers that contain three distinct zones of miscibility: a stable homogeneous region, an unstable homogeneous region, and, in between, a two-phase metastable zone. When the curve delimiting the homogeneous and the heterogeneous

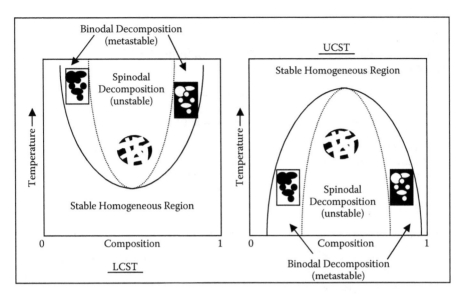

FIGURE 3.1 Scheme showing typical temperature-composition, LCST and UCST, phase diagrams and the binodal and spinodal decomposition ranges.

zones is convex as in Figure 3.1, the mixture is identified as exhibiting a lower critical solution temperature (LCST). If a miscible A/B blend having a composition slightly different from the diluted extreme cases is heated above any temperature of the LCST curve, the blend demixes into two distinct phases, A and B. Some polymer mixtures are miscible at high temperatures but demix at lower ones as is shown Figure 3.1b. In this case the phase diagram of the miscibility of the blend obeys an upper critical solution temperature (UCST), i.e., a concave curve. Indeed, a stable A/B miscible blend generated at a temperature above the UCST can demix into two A and B phases if cooled down below any temperature on the UCST curve.

Two different main mechanisms of decomposition are known — binodal and spinodal mechanisms — by which a partially miscible blend evolves to an immiscible state. In the metastable (binodal decomposition) region, phase separation occurs via nucleation and growth as in the crystallization process in semicrystalline polymers. A nucleus is formed, on which both phases A and B in a droplet-in-matrix phase morphology–type grow independently. Depending on the composition of the blend, A dispersed in B matrix or B dispersed in A matrix phase morphologies can be obtained. In the unstable region, a spinodal decomposition mechanism prevails. This is an unstable process that is initiated by thermal fluctuations already present in the miscible blend and lead to demixed morphologies via spinodal decomposition. The final phase morphology resulting from this process of phase separation is a cocontinuous two-phase morphology where both A and B phases coexist in an interconnected network throughout the whole blend volume. Cahn and Hilliard were the first to develop, using the mean field approach, a theory about the dynamics of phase separation for metallurgical applications (1). For the spinodal decomposition, three stages exhibiting three mechanisms of domain growth have been identified, including

TABLE 3.1
**Examples of Partially Immiscible Blends Exhibiting UCST and
LCST Phase Diagrams**

Blend	Miscibility Conditions and Critical Solution Temperature (°C)	Reference
PMMA/α-methyl-styrene-acrylonitrile	LCST, 185	Goh et al. (1982)
PMMA/chlorinated polyethylene	LCST, 100	Walsh et al. (1982)
PMMA/polyvinyl chloride	LCST, 190	Jagger et al. (1983)
PES/PEO[a]	LCST, 80	Walsh and Rostami (1985)
PS/PMPS	UCST, 103	Takahashi et al. (1986)
SBR/BR	UCST, 140	Ougizawa et al. (1985)
NBR/SAN	UCST, 140	Ougizawa and Inoue (1986)

[a] PES = polyethersulphone, PEO = polyethylene oxide, PS = polystyrene, PMPS = poly(methylphenyl-siloxane), SBR = styrene-butadiene rubber, BR = butadiene rubber, NBR = acrylonitrile-butadiene rubber, and SAN = styrene-acrylonitrile.

molecular diffusion, liquid flow, and phase coalescence (2). The first stage follows the Oswald ripening law and holds for small scale windows comprised within 2 to 9 nm up to a scale window 5 times larger. Flow regime dominates when the phase size is about 1 μm. Above that scale, coalescence dominates, leading to irregular structure. As most of the investigations carried out on spinodal decomposition have focused on the effects of temperature and blend composition on phase equilibriums, the effect of shear rate and pressure, which are important in industrial processes, has been neglected. For example, the LCST of a polycarbonate/polybutylene tereph-thalate (PC/PBT) miscible blend was increased by 60°C upon extrusion, causing miscibility. The extruded blend exhibited phase separation via a spinodal decompo-sition, leading to a cocontinuous phase morphology, which imparted to the blend excellent performance (3).

Polymethylmethacrylate/styrene-acrylonitrile (PMMA/SAN) copolymer having an acrylonitrile content within a determined range of 5.7 to 38.7 wt% is a typical example of a partially miscible blend exhibiting a LSCT miscibility phase diagram that has been largely studied and reported over the years (4–8). However, many other partially miscible blends exist that can generate cocontinuous phase morphology by controlling the cooling rate within the phase diagram and that obey phase separation via a spinodal decomposition mechanism (Table 3.1).

Recently, there has been intensive study of thermoplastics and thermosets blend systems where a thermoplastic polymer, initially miscible with the precursor (a curable monomer) of the thermosetting polymer, phase separates into a two-phase blend upon curing (9–11). In this case, the phase separation is induced by chemical reaction between the species of the initially miscible blend. Depending on the curing conditions of the base monomer of the thermoset resin (concentration of the curing agent, the curing temperature, the kinetics of curing versus phase-separation), it is possible to generate cocontinuous phase morphology via a spinodal decomposition process. Kim et al. (12) investigated the phase morphology development in a thermoset and thermoplastic blend system containing triglycidyl p-aminophenol/polyethersulfone (PES) and using 4,4′-diaminophenylsulfone as the curing agent. When a nonreactive polyethersulfone was used, a spherical domain structure was generated. In contrast, a cocontinuous two-phase morphology was obtained when reactive polyethersulfone bearing amine end groups were employed. That clearly suggests that the spinodal decomposition rate was suppressed by the curing reaction, and a delayed phase coarsening resulted from the *in situ* generation of polyethersulfone-epoxy block copolymer (the phase separated morphology was fixed at an early stage of the spinodal decomposition). A nanoscale cocontinuous phase morphology was generated with a characteristic periodic length of the cocontinuity of about 20 nm. In the nonreactive polymer, this dimension was increased from 20 nm to 1 μm. In a mixed system, i.e., a 20:10 reactive to nonreactive polymer, the dimension was reduced from 0.7 μm to 0.1 μm.

A study of the phase separation process in miscible poly(ε-caprolactone)/epoxy blends containing 15 wt% poly(ε-caprolactone) (PCL) revealed that upon curing using 4,4′-diaminophenylsulfone revealed that a fine cocontinuous phase morphology was generated via a spinodal decomposition mechanism after 70 min of epoxy curing at a fixed temperature of 150°C (13). Longer curing time (from 70 to 90 min) caused a progressive evolution of the cocontinuous phase morphology into a macrodispersed epoxy phase in the PCL matrix. In Figure 3.2, a phase diagram of the epoxy conversion as a function of the PCL content is shown. The cocontinuous phase

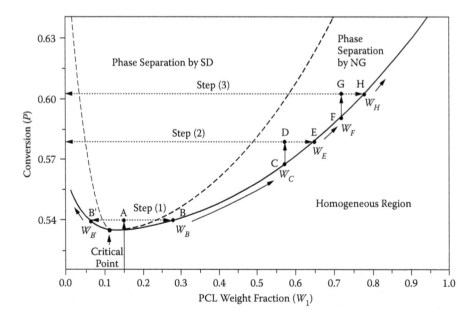

FIGURE 3.2 Phase diagram of the conversion P vs. weight fraction of PCL (w1) in PCL/epoxy blends. (Reprinted with permission from Chen and Chang, *Macromolecules*, 32, 5348, 1999.)

morphology develops via a spinodal decomposition mechanism above the delimiting dashed curve. Below the solid curve, the mixture is miscible and no visible phase separation occurs, whereas, between the two curves the mixture phase separate via a nucleation and growth mechanism resulting in a dispersed phase morphology.

Oyanguren et al. (14) reported on the development of cocontinuous phase morphologies in a polysulfone-epoxy/methyl tetrahydrophthalic anhydride blend containing 10 wt% polysulfone. Figure 3.3 shows optical micrographs obtained at various time intervals of the curing at 80°C. The overall picture is characteristic of a spinodal decomposition mechanism upon epoxy curing, resulting in a cocontinuous phase morphology. In Figure 3.3a, a dispersed morphology is observed after 28 min of curing, which, according to the author, is generated via a percolation to cluster transition from a cocontinuous morphology developed during the early stages of the spinodal decomposition. As the curing proceeds, the volume fraction of the epoxy-rich phase increases due to the continuous segregation of epoxy-anhydride species from the polysulfone-rich phase. This leads to coalescence of the epoxy-rich domains and the formation of larger scale cocontinuous phase morphology. The morphology is frozen in because of the gelation of the epoxy phase as shown in Figures 3.3c and 3.3d.

Other situations exist where one polymer of the pairs exhibiting partial miscibility is crystallizable. The cocontinuous phase morphology can be frozen in by controlling the crystallization so that it occurs below the melting depression curves of the crystalline polymer or in the temperature range situated below the UCST as is the case for the polycaprolactone/polystyrene (PCL/PS) blend (15–17). Polyvinylidene fluoride/polymethylmethacrylate (PVDF/PMMA) blends have also been reported to exhibit a UCST phase diagram below the melting point of the PVDF (18,19).

3.3 COCONTINUOUS PHASE MORPHOLOGY IN MELT-MIXED IMMISCIBLE POLYMER BLEND

In their review on cocontinuous structures in immiscible melt-mixed blends, Pötschke and Paul (20) gave an excellent overview of the open literature dealing with cocontinuous phase morphology as an intermediate stage in polymer blending. In their attempt at understanding the initial stages of phase morphology development during compounding, Shih et al. and Sundararaj et al. (28–33) visualized an initial phase morphology composed of softened sheets of the major phase that exhibit a higher melting point than that of the minor phase and that form an interconnected structure in the continuous minor phase. As these sheets melt, they progressively break apart and coalesce to form the final matrix, thereby causing an intensive breaking up of the minor phase. The ultimate and final phase morphology is composed of a major phase with intersdispersed particles of the minor phase. An illustrative practical example of the formation of a cocontinuous structure as an intermediate step was reported in a blend of a low melting point polyethylene and a high viscosity polystyrene grade (34). A polyethylene content as low as 0.5–7.8 wt% was able to form the continuous phase before the polystyrene major phase completely softened. After that, the polyethylene minor phase was broken apart by the matrix, causing an inverted phase morphology.

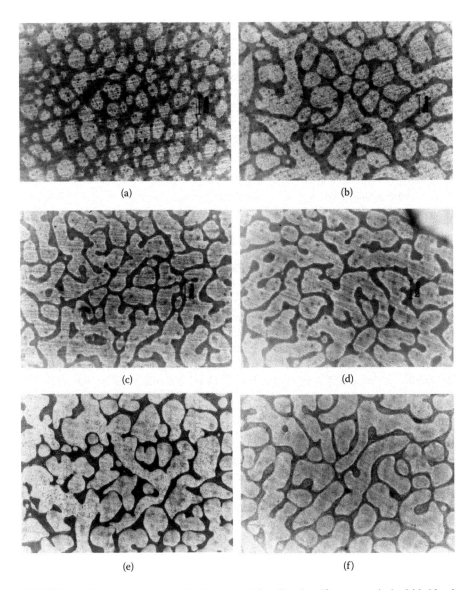

(a)   (b)

(c)   (d)

(e)   (f)

FIGURE 3.3 Development of morphologies in a 10 wt% polysulfone-epoxy/anhydride blend cured at 80°C, followed by optical microscopy. (a) 30 min, (b) 40 min, (c) 60 min, (d) 90 min, (e) after a postcure at 120°C, and (f) after a postcure at 200°C. (Reprinted with permission from Oyanguren et al., *Polymer*, 40, 5249, 1999.)

Processing conditions or reactions occurring in one or both phases of the blend can strongly affect phase inversion. Of course, these two parameters have a direct effect on the viscosity ratio of the components. The same blend of polyamide 6 (PA6)/styrene-acrylonitrile copolymer developed a phase morphology where PA6 formed the matrix when processed using a single-screw extruder while the inverse happened when the blend was mixed several times in a laboratory mixer.

A cocontinuous phase morphology develops in melt-mixed blends under a particular set of processing conditions that depend on components characteristics such as:

The viscosity ratio of the components, which depends on the melting temperature at which the blend is formed and on the shear rate, which itself depends on the mixing r/min of the mixer used
The interfacial tension
The blend composition (content of the A and B components)

3.3.1 PREDICTION OF ONSET OF PHASE INVERSION IN IMMISCIBLE BLENDS

In order to better understand the formation of cocontinuous structures and their relationship with composition and viscoelastic properties and in the absence of a general rule or theory to predict these morphologies, it is only possible to examine the relationship between rheology-interfacial tension and morphology based on experimental observations. The effect of the viscosity ratio on the morphology of polymer blends below the percolation threshold has been long studied and is relatively well understood both for Newtonian and non-Newtonian liquids (35,36). For cocontinuous systems, the prediction of the onset of phase inversion is solely based on experimental observations and on the definition established by Lyngaae-Jorgensen, Utracki, and coworkers (37–40). In the following section, a description of the models is developed in chronological order (also summarized in Table 3.2).

3.3.1.1 Avgeropoulos's Model

Avgeropoulos et al. (41) were the first to propose an empirical relationship to predict the point of phase inversion in an immiscible polymer blend. The expression proposed describes the relationship between the ratio of the volume fraction of component ϕ_i forming the blend and the ratio of their individual torque T_i measured in the molten state when a steady state is reached.

$$\frac{T_1}{T_2} = \frac{\phi_1}{\phi_2} \qquad (3.1)$$

The torque can be conveniently measured in a Brabender-type mixer for each polymer at a specific temperature. The use of the torque is a better reflection of the mixing conditions since it includes all the viscous and elastic deformations in shear and extension. However, the accuracy of its measurement is only as good as the torque transducer and temperature controls used in the equipment. Many authors used this expression to predict the point of phase inversion in immiscible polymer blends (42). Good agreement between the predictions of Eq. (3.1) and the experimental observations of the phase morphology was obtained with blends having a torque ratio close to unity.

TABLE 3.2
Summary of Empirical and Semiempirical Models for the Prediction of Phase Inversion and Dual Phase Continuity in Immiscible Polymer Blends

Reference	Equation	Comments	Conditions
Avgeropoulos et al. (1976)	$\dfrac{T_1}{T_2} = \dfrac{\phi_1}{\phi_2}$	Torque as measured by Branbender-type machinery provides a ratio of the steady state viscosity. Accuracy of the measurement depends on torque transducer.	Valid only when the torque ratio is close to unity.
Paul and Barlow (1980) Jordhamo and Manson (1986) Gergen et al. (1987) Miles and Zurek (1988)	$\dfrac{\eta(\gamma)_1}{\eta(\gamma)_2} = \dfrac{\phi_1}{\phi_2}$	Initially proposed by Paul. Viscosity ratio can be determined by capillary rheometry, for example. Measurement similar to the torque ratio but more accurate.	Viscosity ratio close to unity and measured at low shear rate. Typically zero shear viscosity.
Metelkin and Blekht (1984)	$\phi_2 = \dfrac{1}{1 + \lambda F(\lambda)}$ $F(\lambda) = 1.25 \log \lambda + 1.81 \, (\log \lambda)^2$	Model based on the filament instability concept. λ is the viscosity ratio at the blending shear rate.	Viscosity ratio measured at blending shear rate.
Utracki (1991)	$\lambda = \left[\dfrac{(\phi_m - \phi_{2i})}{(\phi_m - \phi_{1i})} \right]^{[\eta]\phi_m}$	Model based on the Krieger-Dougherty model for monosisperse concentrated suspensions.	Viscosity ratio away from unity and measured at low shear rate.
Miles and Zurek (1988) Everaert et al. (1999)	$\alpha \left[\dfrac{\eta(\gamma)_1}{\eta(\gamma)_2} \right]^{\beta} = \dfrac{\phi_1}{\phi_2}$	Model based on viscosity ratio [Jordhamo et al. (1986)] and modified to better fit various systems. α and β are adjustable parameters for different blend systems.	Model takes into account the viscosity ratio being close or far from unity.
Willemse et al. (1988 and 1999)	$\dfrac{1}{\Phi_2} = 1.38 + 0.0213 \left(\dfrac{\eta_1 \gamma_p R_0}{\sigma} \right)^{4.2}$	Model based on the continuity of elongated rods and maximum packing density.	First model to predict the range of cocontinuity rather than the point of phase inversion.
Lyngaae-Jorgensen and Utracki (1991)	$\Phi_1 = k \left(\phi - \phi_{cr} \right)^{0.45}$	Model based on percolation theory.	Model predicts the range of cocontinuity.

3.3.1.2 Miles and Zurek's Model

Paul and Barlow (43) proposed a similar empirical prediction of the onset of phase inversion as Eq. (3.1). Jordhamo and Manson (44) and Gergen et al. (45) expressed this relation in a semiempirical equation. Instead of the torque ratio, the viscosity ratio measured at low shear rate was used:

$$\frac{\eta(\dot{\gamma})_1}{\eta(\dot{\gamma})_2} = \frac{\phi_1}{\phi_2} \tag{3.2}$$

Based on experimental observation made on many case studies, this equation was then generalized by Miles and Zurek (46). Many authors then validated this relationship for blend systems with a viscosity ratio close to unity (44,46–48). However, the use of the viscosity ratio only reflects one component, which is the shear viscosity of the blend components. Instead, by using the torque ratio, additional components such as elasticity and extensional viscosity, which may play a major role in the formation of cocontinuous structures, are also accounted for. Indeed, this expression was shown to be inadequate for systems with a viscosity ratio away from unity (35,49–51).

3.3.1.3 Metelkin and Blekht's Model

Metelkin and Blekht (52) were the first to introduce an empirical relationship to predict the onset of phase inversion using a different concept. Instead of using viscoelastic parameters, they based their model on an interfacial parameter that used Rayleigh instabilities, also known as capillary instabilities. These perturbations are caused by the difference in the surface tension of a liquid dispersed in a matrix and then subjected to a simple shear deformation. In the case of a Newtonian system, both the matrix and the dispersed phase are shear independent. In such a system, a particle subjected to a deformation is elongated and shows sinusoidal oscillations growing exponentially with time and occurring at an amplitude of:

$$\alpha = \alpha_0 \exp qt \tag{3.3}$$

Where q is the rate at which the sinusoidal shape grows and is a function of the viscosity of the matrix, the interfacial tension, and the Tomotika function Ω (l,λ):

$$q = \left[\frac{\sigma}{2\eta_m R_0}\right]\Omega(l,\lambda) \tag{3.4}$$

While these instability equations apply only to Newtonian systems, they were also found to be valid for viscoelastic systems (52). Combining Eqs. (3.3) and (3.4), the breakup time can be calculated as:

$$t_b = \frac{1}{q} \ln \left(\frac{k}{\alpha_0} \right) \tag{3.5}$$

The main assumption made by Metelkin and Blekht is that phase inversion starts to occur when the breakup time t_b is reached (52). They developed a relationship between a function of the viscosity ratio λ and the volume fraction at which phase inversion occurs:

$$\phi_2 = \frac{1}{1 + \lambda F(\lambda)} \tag{3.6}$$

where

$$F(\lambda) = 1.25 \, Log \, \lambda + 1.81 (Log \, \lambda)^2 \tag{3.7}$$

However Eq. (3.7) was obtained based on experimental data for low density poly-ethylene/ethylene propylene rubber (LDPE/EPR) blends for which, as we have seen in the previous paragraph, one component of the phases, in this case, EPR, does not exhibit a terminal viscoelastic region. Therefore, this relationship may not be valid for a thermoplastic polymer pair with a well-defined Newtonian plateau (42,53).

3.3.1.4 Willemse's Model

So far we have reviewed the empirical models that describe the volume fraction at which phase inversion occurs in immiscible polymer blends. As reported by many, cocontinuous morphologies are hardly stable and are formed over a wide range of composition. This range seems to be dependent on the viscosity ratio and the interfacial tension between the two phases and the process conditions. In all the cases, cocontinuity was observed over at least a 10% range. Therefore, the models described above are not adequate to predict the cocontinuity range but simply the point at which cocontinuity begins. Recently, Willemse et al. (54–56) proposed a model based on the mechanism proposed by Scott and Macosko (57) on the devel-opment of polymer morphology as it relates to processing conditions. The model is based on the assumption that elongated particles are stretched, owing to shear forces, and then break up to form rods, which in turn coalesce to form a continuum within a matrix polymer. A schematic description of this mechanism is shown in Figure 3.4. Continuity is achieved when the packing density, whose value is dependent on the aspect ratio of the rod-like structure, is achieved. If this structure is stable enough,

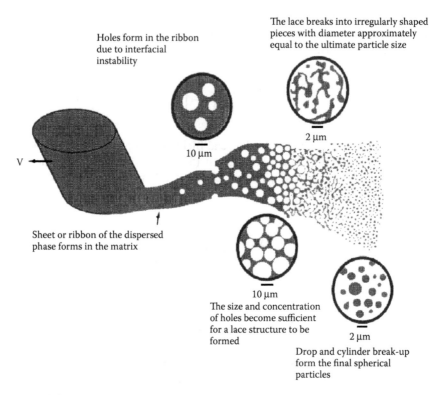

FIGURE 3.4 Morphology development mechanism in an immiscible polymer blend as proposed by Scott and Macosko (1995). (Reprinted with permission from Scott and Macosko, *Polymer*, 36, 461, 1995)

the maximum packing volume corresponds to the smallest volume fraction of the minor phase before phase inversion occurs:

$$\frac{1}{\Phi_2} = 1.38 + 0.0213 \left(\frac{\eta_1 \gamma_p R_0}{\sigma} \right)^{4.2} \tag{3.8}$$

where ϕ_2 is the volume fraction at the onset of phase inversion, η_1 is the viscosity of the matrix, γ_p is the shear rate corresponding at the mixing conditions, R_0 is the particle diameter, and σ is the interfacial tension. This model predicts well the occurrence of phase continuity for thermoplastic blends of PE/PP (polypropylene), PE/PS and PE/PA6 prepared by melt extrusion. However, the model cannot be used as a predictive tool as was the case for the models described above because the particle radius R_0 of the dispersed phase can only be measured once the blend has been prepared. In addition, estimating the shear rate in a twin-screw extruder is not trivial. An approximation of an average shear rate is always possible though. However, the model can be useful in at least describing the importance of the various parameters in a defined blend system.

3.3.1.5 Lyngaae-Jorgensen and Utracki's Model

Based on the percolation theory, Lyngaae-Jorgensen, Utracki, and coworkers (37,38) derived an equation to predict the range where cocontinuity occurs in immiscible polymer blends. This is the only model that describes a range of composition at which cocontinuity occurs and reflects better the experimental observations made on multiple blend systems.

$$\Phi_1 = k\left(\phi - \phi_{cr}\right)^{0.45} \tag{3.9}$$

While the parameter k is empirical, the exponent 0.45 is determined from numerical simulations and is valid depending on the desired final properties. This type of relationship has been widely used in a search of the end-use properties of many systems, in particular, electrical and mechanical properties (58–62). Figure 3.5 shows a representation of this model in terms of the composition range. Below 10%, most if not all immiscible blend systems exhibit matrix-droplet morphology. In the

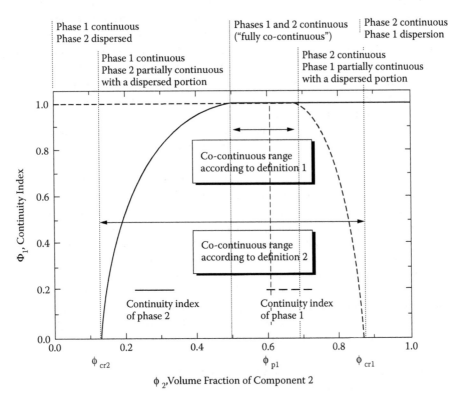

FIGURE 3.5 Dual phase continuity prediction as proposed by the Lyngaae-Jorgensen and Utracki's model. (Reprinted with permission from Lyngaae-Jorgensen and Utracki, *Makromol. Chem. Macromol. Symp.*, 48–49, 189, 1991, and Lyngaae-Jorgensen et al., *Poly. Eng. Sci.*, 39(6), 1060, 1999.)

10 to 50% range, it is possible to obtain matrix-droplet morphology if the viscosity ratio is close to one, whereas, away from unity, the dispersed phase can be elongated to the extent where it can form a partial continuum. This morphology development is almost symmetrical in terms of composition and depends enormously on the viscosity of the individual components. Fully cocontinuous morphology, while possible within a wide range of composition, is often found in the 50 to 70% composition range.

3.3.1.6 Summary

We have seen a number of empirical models that have been developed over the past three decades. These models are summarized in Table 3.2 with their conditions of use and applicability. The earlier models rely essentially on the viscosity or a parameter related to the rheological properties of the polymers used, which includes shear, extensional viscosities, as well as elasticity. These models were found to be valid when the viscosities of the polymers were nearly identical. However, it is still difficult to predict the exact composition at which phase inversion occurs since dual phase continuity occurs over a wide range of composition, typically 10 to 15% (42). Recent models are based on interfacial stability owing to a better understanding of the rheological properties of immiscible blends as well as the role of the interface in the stability of polymer blends (53,63). In the majority of the cases, the models predict relatively well the occurrence of phase inversion and its dependence on factors such as viscosity, shear rate, and interfacial tension. When the viscosity ratio is near unity, all the models seem to converge to the same composition of 50:50. When the viscosity ratio deviates from unity, the basic models do not accurately predict the point of phase inversion while the more recent models are more effective. A summary of the predicted volume fraction for dual phase continuity is shown in Figure 3.6.

Despite all the progress made in this area, there remains significant challenges in understanding how these structures are formed. Questions related to the role of the crystallization thermal expansion coefficient during cooling of a blend of a crystalline polymer with an amorphous polymer, or of the yield stress typically encountered with thermoplastic elastomers and rubbers on the formation of cocontinuous morphologies, have not yet been addressed. None of the models described here take into account these parameters; it appears that the area of new applications provides the driving force to better understand the formation of these structures. Furthermore, because cocontinuity is typically observed with a fairly wide range of composition, it is easily achievable in practice and is currently extensively used in the plastics industry. This is discussed more in the applications section.

3.3.2 METHODS OF COCONTINUOUS MORPHOLOGIES CHARACTERIZATION

It is well recognized that cocontinuous morphologies do not occur at a single threshold point but rather within a range of composition that could extend over 10 to 20%. This is reasonable considering that beyond the composition at which the

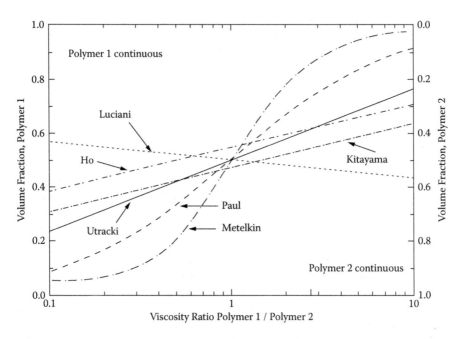

FIGURE 3.6 Diagram showing the occurrence of the point of phase inversion as predicted by the various models. Note that all models predict a volume fraction of 50% when the viscosity ratio is equal to 1. (Reprinted with permission from Pötschke and Paul, *J. Macrool. Sci., Part C, Polym. Rev.*, 43, 87, 2003.)

morphology is no longer of the droplet-matrix-type, the two phases can form a blend in partial continuum. One can only wonder whether each phase forms a continuum throughout the structure. In order to answer this question, scientists have examined cocontinuous morphologies using several methodologies. The most systematic ones, such as solvent extraction, are a good indicator of the spread of continuity over a wide composition range. More sophisticated techniques, such as confocal microscopy, have also been used for quantitative analysis of these types of morphologies. No matter what technique is used, it is still a challenge to predict or to quantify these morphologies, especially in the presence of compatibilizers and other modifiers. Whether it is important to achieve total continuity rather than, say, 95% cocontinuity, with the remaining 5% being droplets of elongated particles, remains a mystery. Indeed, there is no indication as to whether the final properties of a polymer blend exhibiting cocontinuity are dependent on a fully cocontinuous structure. Once again, this depends on the blend's application and use in practical terms.

3.3.2.1 Direct Methods

Cocontinuous morphologies are not trivial to characterize quantitatively. Whereas optical microscopy, SEM (scanning electron microscopy), TEM (transmission electron microscopy), and AFM (atomic force microscopy) are among the most popular techniques for qualitative analysis, they only provide part of the answer because all microscopy techniques provide mostly two-dimensional measurements and may not reveal the entire composition gradient throughout a three-dimensional cocontinuous structure. It is common to use these techniques to look for the presence of cocontinuity even while assuming that the material is anisotropic; samples are often checked at various locations (64). More tedious techniques have been developed over the years using the slicing technique, which consists of observing the morphology within a certain depth. This ensures that within a realistic thickness the suggested morphology is cocontinuous. A more sophisticated technique is confocal laser scanning microscopy that is typically used in biological sciences. This technique is very similar to the slicing technique except that it is performed digitally owing to the powerful confocal optics and image processing (47). It was only recently that scientists have started to look at methods to quantify these structures. The primary tool to achieve this is image analysis, which allows the conversion of micrographs into data. While it is not the scope of this chapter to discuss image analysis in details, it is important to briefly mention the most common techniques. First, it is worthwhile to mention that image analysis is a powerful technique to extract information from pictures, provided adequate programming and analysis is available. Today, image analysis software is a common off-the-shelf tool. However more powerful software such as Visilog® is available, but it requires good programming skills. The software is also commonly used on a PC platform, and owing to rapid microprocessors, analysis can be done within a few minutes.

However, sample preparation can be tedious and often requires many hours of work. The contrast between phases usually requires some kind of chemical treatment such as etching with heavy metal oxides or extraction for long periods of time. The basic methodology relies on the determination of a shape factor and the statistical analysis of the data. The shape factor for a specific morphology is a function of the area to perimeter ratio which can be captured on a micrograph when the structure changes from droplet-matrix to cocontinuous or even when cocontinuous systems of immiscible blends are subjected to annealing. In the first case, a significant drop in the shape factor is obtained as the point of phase inversion is approached, while in the second case, the shape factor increases as the continuous phase coarsens (53,65). A similar technique is based on that of Chalky (66), Colman (67), and Quintens et al. (68) where a stochastic analysis of the volume to surface ratio is used; the technique is also used for tracking annealing and stability processes. Fractal analysis is more powerful and expresses the results in a form of a distribution function that show the intensity of the fractal distribution, which is indicative of the structure continuity. This analysis is more complex from a mathematical aspect but provides significant information, especially in terms of phase stability (69,70).

3.3.2.2 Indirect Methods

Many other techniques have been used to characterize cocontinuous morphologies. These are mainly indirect and qualitative methods that allow a correlation with morphology in some way. The most relevant one is probably solvent extraction where a sample is immersed in a solvent that is appropriate for one of the phases of the blend. After adequate extraction time, the sample is weighed again and a value

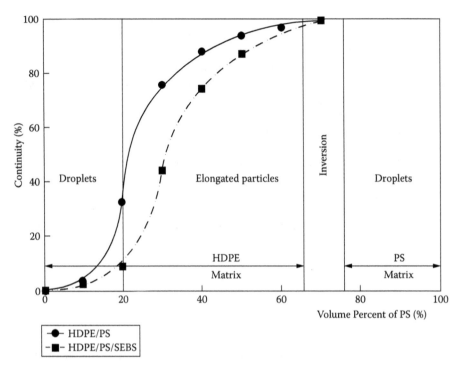

FIGURE 3.7 Dual phase continuity as determined by the solvent extraction method on PS/HDPE blends in THF for 5 days. (Reprinted with permission from Boury and Favis, *J. Polym. Sci., Phys. Ed.*, 36, 1889, 1998.)

representing the degree of extraction is calculated. If, for example, the whole amount of the dissolved sample is extracted, then 100% cocontinuity is achieved (38). If one of the components is not totally extracted, this indicates that a partial amount of the sample is encapsulated in the other phase of the blend, suggesting discrete phases. This technique has been successfully used for systems for which the two phases are not soluble in the same solvent (54,72,73). Figure 3.7 shows the results of an extraction technique used in PS/HDPE (high density polyethylene) blends. The blends have been prepared in increments of 10% and dissolved in THF (tetrahydro-furan) to extract the polystyrene phase. The continuous line represents the continuity profile of a nonmodified blend whereas the discontinuous line shows the profile in the presence of SEBS as a compatibilizer. When the volume fraction is less than 20% by volume, the morphology is of a droplet-matrix-type, whereas in between, nonspherical domains are also present. In both cases, 100% cocontinuity is achieved at about 70%; the compatibilizer only helped narrow the cocontinuity domain. This is in agreement with the fact that compatibilizers help stabilize the morphology of immiscible blends and extend the range of droplet-matrix morphology (72,74).

Favis et al. (75,76) have recently introduced a new method to characterize cocontinuous phase morphologies. It consists of dissolving one of the two phases of the blend and dealing with the rest as a porous material. The Brunawer-Emmett-Teller (BET) gas adsorption technique was used to characterize the porosity, pore size, and size distribution. This method remains quantitatively useful for fully cocontinuous blends and in blend systems where it is practically impossible to selectively extract one phase without affecting the state of the other.

Measuring the mechanical properties has also been used as a means to characterize cocontinuous morphologies. Changes in the mechanical properties are more a consequence of this type of morphology, especially in immiscible polymer blends, rather than a targeted goal. The typical tensile strength, elastic modulus, or toughness as a function of blend composition show a dramatic dip near percolation due to the much smaller interfacial areas compared to droplet-matrix morphologies (77–80). The tensile modulus of polymer blends has been determined for a variety of systems with and without considering the morphology. For systems exhibiting matrix-droplet morphology, the tensile modulus of the blends can be described by several expressions, which include the volume fraction, the individual modulus of the blend components, and the aspect ratio of the dispersed phase. Davies (81) gave the tensile modulus of a cocontinuous blend for a homogenous and isotropic blend as:

$$E_b = \left[\frac{1}{\phi_1 \left[E_1^{1/5} - E_2^{1/5} \right] + E_2^{1/5}} \right]^{1/5} \tag{3.10}$$

Equation (3.10) is the simplest relationship as it only depends on the modulus of the individual components and volume fraction. More complex relationships were developed by others to include the critical volume fraction for phase inversion or relationships based on the percolation theory (82–84). In every case, the tensile modulus of polymer blends exhibiting cocontinuous morphologies is higher than that of matrix-droplet type morphologies. For example, in a blend of PA6 with poly(acrylonitrile butadiene styrene) (ABS), the mechanical properties (tensile strength and modulus) of ABS can be improved by the addition of 30% PA6 while the elongation at break increases dramatically as shown in Figure 3.8. This type of synergistic effect opens new doors of opportunities for the development of new materials with enhanced mechanical properties.

While tensile properties show an improvement in the cocontinuous region of PA6/ABS blends, this trend may not be generalized for all immiscible blend systems. This has been shown in the case of PEE /PS and PP/SEBS blends (85).

Other mechanical properties such as impact strength show similar effects for a variety of blend systems. The improved impact strength in the region of cocontinuity is attributed to the high elongation at break induced by deformed continuous structures. This effect is even more pronounced when compatibilizers are present. An example was shown by the work of Niebergall et al. (86) who studied the PP/HDPE blend (Figure 3.9).

Other mechanical testing methods such as dynamic mechanical analysis (DMA) are also among the most relevant techniques to characterize cocontinuity (77,85). DMA is known to be sensitive to the type of dispersion and can discriminate between droplet-matrix versus cocontinuous. Because of the linear deformations applied in this test, the structural information is most relevant for the cocontinuous region. An

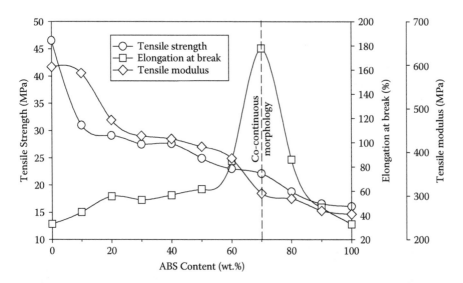

FIGURE 3.8 Tensile properties of PA6/ABS blends at 100 mm/min. Note that cocontinuous morphology occurs at 70 wt% ABS content. (Replotted from Marin and Favis, *Polymer*, 43, 4723, 2002.)

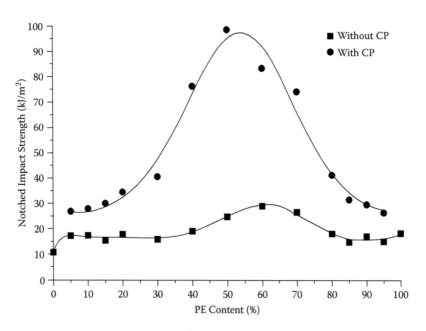

FIGURE 3.9 Impact strength of PP/HDPE blend with and without compatibilizer (CP). (Reproduced from Kolarik, *Poly. Comp.*, 18, 433, 1997.)

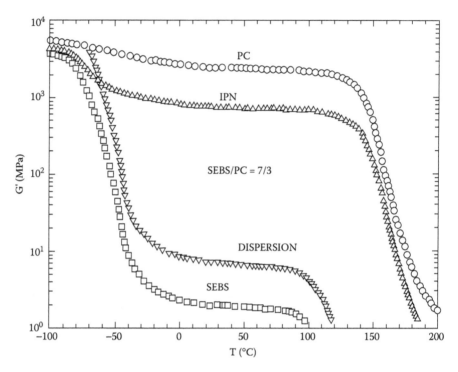

FIGURE 3.10 Storage shear modulus of PC/SEBS 70:30 blends showing mechanical properties of a droplet morphology vs. cocontinuous morphology. (Reprinted with permission from Mamat et al., *J. Poly. Sci., Poly. Phys. Ed.*, 35, 2583, 1997.)

example is shown in Figure 3.10 for a blend system of PC containing SEBS, where the modulus of the matrix is not dramatically changed in the cocontinuous region, whereas in the droplet-matrix region, the modulus drops significantly (80).

The methods described above allow for the correlation between the structure of polymer blends and their end-use properties. These techniques and measurements are relevant as they are performed at room temperature where the materials are in the solid state. In the melt state, there is generally correlation between the structure and the end-use application. However, knowing the properties and the melt behavior of polymer blends is essential to understanding morphology formation. This is why so much effort is placed on the rheological characterization of polymer blends. Rotational rheometry is a practical method to investigate cocontinuous morphologies in polymer blends. The basis for comparing the viscoelastic properties of polymer blends to their morphology originated after Palierne (87,88) developed an adequate emulsion model for dispersed droplets in a non-Newtonian media. The model relies essentially on the complex modulus of the dispersed phase and the matrix but also takes into account the interfacial tension between the two polymers and the size of the droplets. As observed in many immiscible polymer blend systems, the storage modulus exhibits an increase in the low frequency domain due to the balance between the strain applied to the system versus the interfacial tension acting against it (87,88).

$$G_b^*(\omega) = G_m^*(\omega)\frac{1+3\sum_i \varphi_i H_i^*(\omega)}{1-2\sum_i \varphi_i H_i^*(\omega)} \qquad (3.11)$$

where

$$H_i^*(\omega) =$$

$$\frac{8(\alpha/d_i)[2G_m^*(\omega)+5G_d^m(\omega)]+[G_d^*(\omega)-G_m^*(\omega)][16G_m^*(\omega)+19G_d^*(\omega)]}{80(\alpha/d_i)[G_m^*(\omega)+G_d^*(\omega)]+[2G_d^*(\omega)-3G_m^*(\omega)][16G_m^*(\omega)+19G_d^*(\omega)]} \qquad (3.12)$$

where G_m^* is the complex modulus of the matrix, G_d^* is the complex modulus of the dispersion, ω is the angular frequency, α is the interfacial tension between the matrix and the dispersion, and the dispersion, and d_i is the particle diameter. This model has been successfully used to predict the interfacial tension between immiscible polymer blends as well as to correlate the morphology to the viscoelastic properties (89–92). However, it is less accurate when a compatibilizer is present at the interface because of the rubberlike character of compatibilizers, which exhibit a yield stress contributing to an imbalance between shear deformation and interfacial forces (93,94). Another limitation is the fact that the model only applies to emulsion-type systems, i.e., the dispersed phase needs to remain spherical in shape and more or less monodisperse in size.

When the composition of the blend exceeds the threshold of percolation and the dispersed phase is no longer spherical, elongated droplets will form, some of which will percolate. Under quiescent conditions, the elongated particles will only exhibit percolation, whereas if the blend is subjected to deformation, then both percolation or coalescence and breakup occur simultaneously. These changes in the morphology of the blends can be qualitatively examined using microscopy techniques. Figure 3.11 shows a PS/PE blend with a composition of 70:30. After annealing for over an hour, the size of the phases increased by several fold as a result of a quiescent coalescence. In the presence of SEBS as a compatibilizer, the blend morphology is stabilized for long periods, ensuring stability during the processing and ultimately in the end product (see Figure 3.12).

The stability of the cocontinuous morphology can also be monitored using small amplitude oscillatory measurements. Since the storage modulus is very sensitive to changes in morphology, Mekhilef et al. (94) were able to show that for uncompatibilized PS/PE blend the storage modulus decreases over time due to a decrease in the surface area at the interface. However, for the same blend containing SEBS, the blend was shown to be stable and the modulus remained constant over time. Other reports show that the storage modulus G' shows a minima in the cocontinuous region because of an increase in the interfacial area (94,95). However, there are some contradictory results when elasticity is used as a parameter to detect cocontinuous morphologies. This is attributed to the fact that the response to deformation in immiscible blends not only depends on their morphology but also on the type of

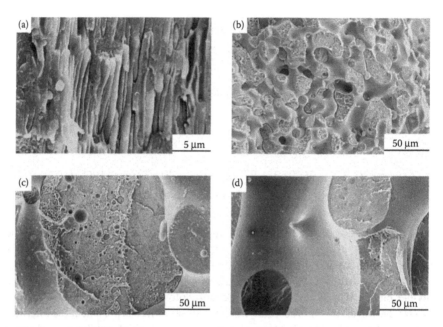

FIGURE 3.11 SEM micrographs for the annealed PS/PE 30:70 blend. (a) 0 min, (b) 1 min, (c) 16 min, and (d) 75 min. (Reproduced with permission from Bousmina and Muller, *J. Rheol.*, 37, 663, 1993.)

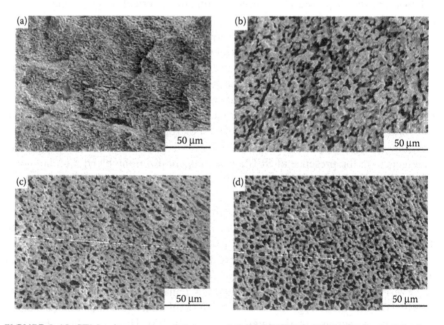

FIGURE 3.12 SEM micrographs of the annealed PS/PE/SEBS 30:70:20 blend. (PS phase has been extracted.) (a) 0 min, (b) 1 min, (c) 16 min, and (d) 75 min. (Reproduced with permission from Bousmina and Muller, *J. Rheol.*, 37, 663, 1993.)

interactions and interfacial tension between their phases. Therefore, the response of the blend to deformation is only partially reflected in the modulus.

3.4 ROLE OF COMPATIBILIZATION IN THE DEVELOPMENT AND STABILITY OF COCONTINUOUS PHASE MORPHOLOGIES

The development and structural stability of cocontinuous phase morphologies in compatibilized immiscible polymer blends has been the subject of numerous studies. In addition to the determination of the cocontinuity diagram, which is the composition window at which the blend exhibits phase cocontinuity, the structural and dimensional stability of the generated morphology upon blend annealing at high temperatures under quiescent conditions and also under shear has been an important research topic during the past decade.

Various observations were reported on the role the compatibilizer plays in the formation of cocontinuous phase morphologies in compatibilized immiscible blends. Fast kinetics of cocontinuity formation has been reported when a styrene-ethylene-butadiene (SEB) copolymer has been added to an immiscible low density polyethylene/polystyrene (LDPE/PS) blend (97). Cocontinuous phase morphologies were generated faster and were more stable in compatibilized LDPE/PS blends than in uncompatibilized ones. Similar effects of fast development of cocontinuity in polyamides/PMMA blends compatibilized using reactive styrene-co-maleic anhydride (SMA) copolymer were also reported (98).

A general tendency toward delaying cocontinuity formation (cocontinuity is formed at a higher concentration of the minor phase compared to uncompatibilized blends) and a narrowing of the composition range at which cocontinuity occurs can be clearly seen in published reports dealing with cocontinuous blends. Illustrative examples include physically compatibilized PS/PDMS (polydimethysilocane) and PS/PMMA using PS-b-PDMS and polystyrene-block-polymethylmethacrylate (PS-b-PMMA), respectively (99), physically compatibilized HDPE/PS using SEBS (72,100), reactively compatibilized PA6/PMMA using SMA (101), and reactively compatibilized PA66/PP using polypropylene-graft-maleic anhydride (PP-g-MA) (102). Figure 3.13 clearly shows how the composition range of cocontinuity is narrowed when immiscible PS/PA6 or (PPE/PS)/PA6 (polyamide 6) blends are compatibilized using an SMA reactive precursor containing 2 wt% of maleic anhydride groups. Upon compatibilization, the composition range of the phase cocontinuity was narrowed almost by a factor of three (103), which can be explained by a reduction of the interfacial tension between the phases acting against the formation of a cocontinuous structure.

Although these reports agree on their description of the effect, they remain nevertheless far from providing a fundamental understanding of how the compatibilizer acts during the first stages of cocontinuity generation. It is not clear whether the contribution of the copolymer originates from a reduction of the interfacial tension between the phases, as in the droplet-in-matrix morphology, or from a modification of the viscoelastic ratio of the phases. This situation is particularly crucial in reactively compatibilized blends where the reaction kinetics are very important for the phase morphology development. The formation of cocontinuous

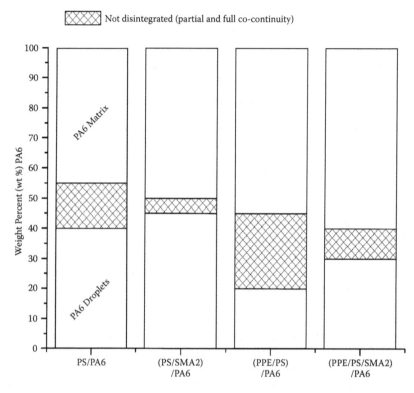

FIGURE 3.13 Cocontinuity intervals determined with dissolution tests on uncompatibilized and compatibilized blends. (Reproduced with permission from Dedecker and Groeninckx, *Polymer*, 39, 4985, 1998.)

phase morphologies does not necessarily require a reduction of the interfacial tension between the phases because this is known to favor a breakup of the dispersed phase, depending on the capillary number. The copolymer limits the coalescence of particles, which tends to oppose the breakup process. To favor cocontinuity of the phases, the breakup process should be dominated by extensional "fibrillization" of the minor phase and coalescence of the generated particles, thereby ensuring network buildup. The first steps of the formation of a cocontinuous structure consist of the shear-driven transformation of the flat sheets of the minor phase into spaghettilike rods interconnected throughout the whole volume of the blend. It can be suggested that any set of conditions (operating conditions or material characteristics) that favors extensional fibrillization of the particles under deformation and is not favorable for their breakup is expected to lead to the formation of a cocontinuous structure.

Compatibilized blends produce much finer cocontinuous structure than those generated in the absence of compatibilizer (104–107). As seen in Figure 3.14 for a 20:80 PS/LDPE blends, in the absence of compatibilizer, the cocontinuous phase morphology is not stable and evolves to a droplet-in-matrix phase morphology just after molding (within 3 min). (Compare the SEM pictures on the top-left to the top-right). The middle pictures are obtained from the same blend but compatibilized using 10 wt% SEBS

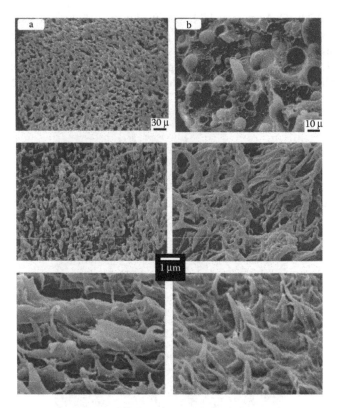

FIGURE 3.14 SEM pictures showing the effect of thermal annealing on the stability of the cocontinuous phase morphology (before annealing, *left column*, after 150 min of annealing at 180°C, *right column*, uncompatibilized (top) and physically compatibilized LDPE/PS 20:80 blends using a triblock polystyrene-hydrogenated polybutadiene-polystyrene copolymer (PS-hPB-PS) (middle) and a tapered PS-hPB (bottom) copolymer.

triblock copolymer. The phase morphology is finer than for uncompatibilized blend, but upon annealing, the phase morphology coarsens without a major alteration of cocontinuity (compare the pictures in the middle-left to middle-right). When the blend is modified using 10 wt% of a tapered hydrogenated polybutadiene-block polystyrene (HPB-b-PS) diblock copolymer, the cocontinuous phase morphology generated is much more stable when the blend is annealed at the same temperature of 180°C for 150 min. This figure illustrates the significant differences between compatibilizers having different molecular architecture in stabilizing the cocontinuous phase morphology against annealing. This behavior has been ascribed to the efficiency of the tapered diblock copolymer, owing to its intermediate mixed HPB/PS sequence, to form a stable interphase between the PS and the LDPE phases.

Compatibilization and stabilization (against annealing) of blends having cocontinuous phase morphology and containing one biodegradable component using pure diblock copolymer has been recently reported by Sarazin et al. (107). For example, in the absence of compatibilizer, a significant extent of coarsening is observed for a 50:50 PS/PCL blend upon annealing during 60 min at three various temperatures;

190, 200, and 220°C. As expected, because the kinetics of phase coarsening depend also on the viscosity of the phases, the higher the temperature, the faster the coarsening process (Figure 3.15). Figure 3.16 shows clearly how the addition of 15 wt% of PS-co-PLA copolymer (based on PS phase) to the 50:50 PS/PLA blend retards the coarsening process upon annealing during 90 minutes at 200 and 220°C. Note

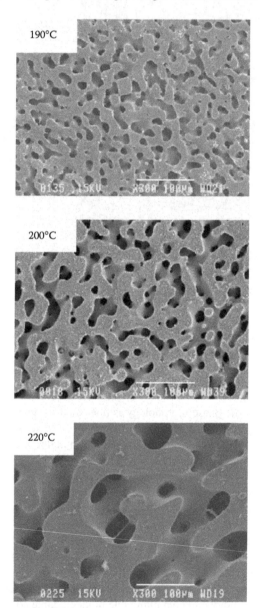

FIGURE 3.15 SEM micrographs for 50:50 PS/PLLA poly(L-lactide) annealed for 60 min at 190, 200, and 220°C, respectively (the white bars denotes 100 µm). (Reproduced with permission from Jafari et al., *J. Appl. Polym. Sci.*, 84, 2753, 2002.)

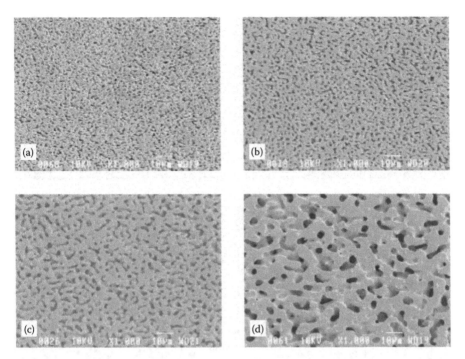

FIGURE 3.16 SEM micrographs for 50:50 PS/PLLA compatibilized with 15% polystyrene-co-polylactide (PS-co-PLA) based on the PS phase. (a) Nonannealed sample; (b and c) annealed at 200°C for 60 and 90 min, respectively; and (d) annealed at 220°C for 90 min. The white bar indicates 10 μm for all micrographs. (Reproduced with permission from Jafari et al., *J. Appl. Polym. Sci.*, 84, 2753, 2002.)

that the cocontinuous phase morphology is not completely stabilized, but compared to uncompatibilized blends, a significant resistance to coarsening is observed.

In Figure 3.17, SEM micrographs of a 50:50 PP/PS blend compatibilized using two reactive precursors, maleic anhydride grafted polypropylene (PP-MA) and amine functional polystyrene (PS-NH2), are shown. When the *in situ* formed compatibilizer is not efficient, the stabilization effect upon annealing is poor (note that PP-MA8 containing 8 wt% maleic anhydride groups is much more efficient than the one containing only 1 wt%). The phase morphology gets coarser after 10 minutes of annealing whereas, in the case of PP-MA8, no significant phase coarsening is observed.

Another aspect, which has been recently observed during the extrusion of blends having cocontinuous phase morphology, is the extensive orientation of the cocontinuous structure without any loss of phase cocontinuity (108). Microscopy observation has been carried out on cocontinuous polypropylene/polycyclohexylmethacrylate (PP/PCHMA) blends in two sampling directions: one cut perpendicular to the extrusion direction and the other one parallel to it. As seen in Figure 3.18a, in the perpendicular direction, the phase morphology appears as a classical droplet-in-matrix phase morphology. The extraction experiment, where about 100% of the PCHMA phase is extracted using chloroform, reveals that the

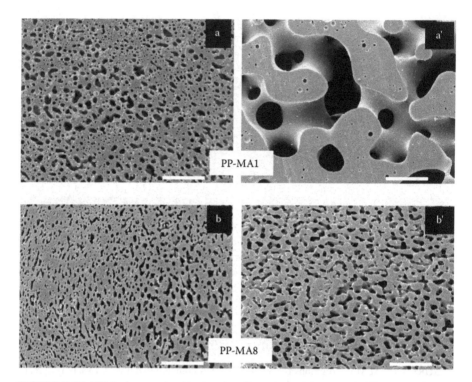

FIGURE 3.17 SEM micrographs showing the effect of reactive compatibilization, using two reactive compatibilizers, PP-MA and PS-NH2, on stabilization against thermal annealing (20 min) of PP/PS 50:50 blends that have cocontinuous phase morphologies.

minor PCHMA phase (black holes in the picture) is continuous. As presented in Figure 3.18b, the SEM micrograph of a parallel cut confirms the extraction data and reveals that the rodlike structures are interconnected throughout the volume of the blend. A close inspection of the picture clearly reveals the existence of inter-bridging connections between the many, variously oriented strands. Upon addition of reactive precursors PS-NH2 and PP-g-MA8 containing 8 wt% maleic anhydride groups, which are generated *in situ* during the melt-blending of a PP-g-PS com-patibilizer, the cocontinuity is significantly altered and the oriented strands are almost isolated in the PP matrix.

3.5 PRACTICAL ASPECTS AND APPLICATIONS

A survey of the literature indicates extensive work in the area of cocontinuous morphologies in polymer blends. However, this approach to the formation of novel structures is not limited to polymers. In fact, cocontinuous morphologies were initially used in the area of metallurgy by combining two or more metals to achieve combined properties of the individual materials. Only until the past half century have scientists become interested in polymeric systems. While extensive work has been done on thermosets, thermoplastics have only been studied for the last 30 years.

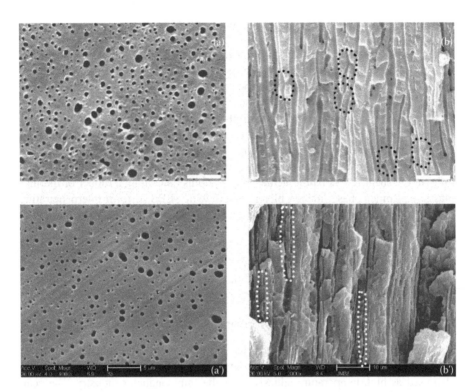

FIGURE 3.18 SEM micrographs of PP/PCHMA 80:20 that have cocontinuous phase morphology compatibilized with PP-MA1 (top pictures) or with PP-MA8 (bottom pictures). The samples are observed parallel to the direction of extrusion (left column) and perpendicular to the direction of extrusion (right column).

3.5.1 MATERIALS

Throughout this chapter, and possibly this book, the focus is mainly on thermoplastics-based blends, however, there is really no limit to the type of material that can be used. In the following few paragraphs, we will describe in some detail the different types of materials that have and are currently being used commercially to develop complex structures for various applications.

3.5.1.1 Immiscible Polymer Blends

This category consists exclusively of polymer blends based on thermoplastic resins. Examples in the open literature and patents include polymeric systems such as poly(phenylene ether)/polyamide 6 (PPE/PA6) and polycarbonate/poly(ethylene terephtalate) (PC/PET) as conductive paints (109, 110) and polyamide 6/poly(acrylonitrile butadiene styrene) blends (PA6/ABS), a commercial blend under the trade name Triax©, which are noted for their improved impact properties and paintability (111). Many of these blends have enhanced properties such as tensile and impact strength, barrier resistance, and electrical properties, and many are currently being used in the industry for LCP/PPO and EVOH/PE systems (112,113).

3.5.1.2 Immiscible Thermoplastics with Rubbers or Elastomers

In this class of blends, the miscibility criteria are still valid, i.e., the rubbers and thermoplastic elastomers that are used as modifiers in thermoplastic resins are typically immiscible with the matrix they are being blended with. The difference lies in the rheological properties of the elastomeric phase. Typically, an infinite viscosity at low shear rate associted with the presence of a yield stress characterize these materials. These result in a different mechanism for breakup during processing, but more importantly, they produce the most stable morphologies. Furthermore, the models used to predict the onset of phase inversion do not apply since the viscosity ratio at low shear rate is hardly achievable. Systems containing a rubber or thermoplastic elastomer phase include blends of PC with SEBS to form interpenetrating polymer networks (IPNs) or polypropylene/ethylene-propylene-diene monomer (PP/EPDM) and PP/SEBS (88,114,115). In every case, the products have improved mechanical properties while their impact strength is improved in the presence of compatibilizers. The blends of rubber modified SMA with PBT that are produced commercially have a cocontinuous morphology. The resulting material exhibits a higher ductility and allows for a better blow molding of the blend resin (116).

3.5.1.3 Thermosets with Resin or Fillers

Polymeric systems are typically used with an inorganic filler to enhance properties such as dielectric loss, electrical conductivity, and dimensional stability. They are also used as high temperature adhesives for the semiconductor industry. They can also be used as IPNs, for example, polyether/epoxy blends that are typically used in sheet molding compound (SMC) products (117, 118).

3.5.2 Processes

3.5.2.1 Melt Processing

The most popular route to producing cocontinuous structures remains melt-processing. Extrusion and injection molding are the most common processes since they provide high production capacity and low cost. This can be achieved either by melt-blending two polymers in a twin-screw extruder or by grafting one polymer onto the other using free radical formation or simply by "reactive extrusion" where the morphology of the blend can be controlled by the balance between the reactivity ratio and residence time in the extruder. Another alternative is the synthesis of IPNs. This methodology allows for better control of the domain size while maintaining the properties of both phases.

3.5.2.2 Curing and Reactive Extrusion

In these processes, one or both phases forming the cocontinuous structure is inherently unstable due to an increase in viscosity over time or due to a chemical reaction between the phases. In the first case, the minor phase is dispersed in a matrix of one polymer in the melt state. During compounding, the viscosity of the minor phase builds up to the point where the viscosity ratio is reversed, at which point the blend undergoes phase inversion and the minor phase becomes the continuous one. In this case, the two phases may initially be immiscible, but in the presence of a functional group or peroxide capable of creating free radicals, the phases now become partially miscible, a high degree of mixing is achieved, and cocontinuity is obtained at a much smaller scale. When this type of process is properly controlled, one of the phases can be crosslinked, thus providing for a more stable morphology. From a practical standpoint, this type of process can be advantageous in, for example, injection molding.

3.5.2.3 Solution Blending and Casting

Solution blending is also practiced in the industry to produce new materials with cocontinuous morphologies and exhibiting improved properties. This process has a long history and is more frequently used than one might think. It is mostly used in the fabrication of porous membranes for filtration and ophthalmic applications. For these systems, cocontinuity can be defined as the coexistence of the two phases or the presence of only one phase while the second phase has been extracted. This is the case for porous membranes where the polymer is dissolved in a good solvent and then blended with a nonsolvent. After phase inversion occurs, the solution is cast and the solvent is extracted, leaving a microscopic and continuous structure. This procedure has been long practiced in the membrane industry using polyolefins, polyamides, and fluoropolymers but is now controversial because of the environmental implications of using large amounts of solvents. Another application gaining ground in the last decade is in the energy sector where the fabrication of membranes as electrolyte carriers has driven the battery and fuel cell markets to produce lighter and more efficient energy storage capacitors. Even though the use of solvent is detrimental from an environmental standpoint, this technology will continue to grow, especially in the energy sector (119).

3.5.2.4 *In Situ* Polymerization

In the next chapters of this book, a much broader description of IPN formation will be presented. This paragraph is limited to some industrial examples where this methodology is used to produce commercial materials such as conductive resins and structural adhesives. There are a few applications where cocontinuous structures are commercially produced. An example is the polymerization of dicyclopentadiene in a poor solvent such as liquid paraffin in the presence of a catalyst such as 2,6-di-tert-butyl-4-methylphenol (BHT). The solution is cast on a polyester film followed by soaking in n-hexane under ultrasonication. The result is a porous film with a pore size of approximately 1 μm and that possesses good heat and solvent resistance and good mechanical properties (120). Another example is the polymerization of methylmethacrylate (MMA) in the presence of ethylene vinyl acetate (EVA) copolymer in a composition range where cocontinuity is obtained. The product possesses mechanical properties such as tensile strength, hardness, and impact resistance that are higher than either component (117). This route is also useful in the preparation of adhesives for semiconductors by blending an acrylated phenolic resin with polyethersulphone in the region where an IPN can be formed during curing of the thermoset in the presence of a curing agent (118).

3.5.3 APPLICATIONS

Many applications have been developed that use complex structures such as cocontinuous morphologies. While fundamental work to understand the mechanism of their formation and their characterization and quantification have been the topic of the last decade in the academic arena, industrialists have taken advantage of this type of morphology to develop new and, on many occasions, sophisticated applications. The patent literature, for example, reveals over 150 inventions related to this area. The applications cover many areas and are summarized here.

3.5.3.1 Membranes, Porous Polymers, Separation, and Filtration

In this area, the most predominant application is by far the use of porous polymeric materials as filtration media in a variety of applications such as the chemical and food industries. These filtration media are made out of solvent resistant polymers such as polyvinylidene fluoride (PVDF), polyamides, and polyethylenes, which can also be radiation crosslinked for best performance. One familiar application in the polymer field is chromatography and electrophoresis where dilute polymer solutions are filtered prior to injection in the columns (121,122). In the food industry, beverage producers use many types of filtration media for the production of flavored water and fruit juices (123). In all these cases, the performance of the filters is dependent on the morphology of the filters being used and the process by which they are made. Other applications in this area include the use of nylon/polyester blends exhibiting cocontinuous structure with domain sizes ranging between 0.01 to 10 μm. They are used in hydrophilic separating membranes, filtering membranes, photographic films, and microfilms (124).

3.5.3.2 Medical and Biomedical Applications

In the medical and biomedical fields, cocontinuous structures have been mostly used in the controlled release applications of drugs and in ophthalmic applications. Cocontinuous structures have been developed using the interpenetrating networks approach by which perfluoropolyethyer monomer is polymerized in an emulsion in the presence of progens. This technique allows for the formation of a porous structure, which can contain a large volume of water and can therefore be used for ophthalmic lenses (125). Other methods are used to produce high performance lenses and are produced by the copolymerization of an oxyperm macromer and an ionoperm monomer containing siloxane (126). Another area is in tissue engineering where two biodegradable polymers are combined into a cocontinuous structure in the presence of an active substance, usually a drug that is dispensed into the human body at controlled rates. The device mimics the extracellular matrices of the body by providing scaffolds with controlled biodegradation kinetics (127). These structures, e.g., the one produced by hydroxyapaptite blended with poly(lactic acid) and PMMA in a melt state using conventional extrusion techniques, can be biocompatible and can be used as animal cell tissue and in prosthetic applications (128). Prosthetics is also another area where there is a growing need for human body part replacements. Blends of PP with SBR and/or SEBS Kraton® are commonly used to prepare a superstructure by using a complex procedure and achieving the needed high compression set and tensile strength to sustain body weight and movement (129). Another high volume usage is in the area of absorbent materials mainly produced from starch or starch-modified derivatives produced by mixing them with water and trisodium trimetaphosphate and oil to produce a material capable of absorbing ten times its own weight (130).

The applications described above are only examples of what is practiced in the biomedical field. Patent applications are on the rise and will probably continue to grow as more complex and sophisticated applications are developed in the area of prosthetics, drug release, and organ replacement.

3.5.3.3 Conductive and Semiconductive Materials

In the absence of inexpensive conductive polymers, an alternative is to combine these materials with commodity polymers and produce complex structures containing only a fraction of the conductive material dispersed in an organized fashion within a thermoplastic. This practice has become more popular in recent years using two thermoplastic resins. In contrast, the combination of a thermoplastic and thermoset has been practiced for a longer time. For example, blends of a thermoplastic resin such as polyether sulfone with a thermoset resin such as phenolic novolak for which the functional group has been substituted with a photosensitive group such as an acrylate are blended in a curing agent, resulting in a sealing material used for circuit boards and semiconductors (118,131). Liquid crystal displays (LCDs) are also another type of material where cocontinuous morphologies are needed and are produced by spinodal decomposition of the liquid phase, usually in a solvent containing polymers with different refractive indices. When the solvent is evaporated, a cocontinuous structure is formed on a reflecting plate, resulting in high intensity light scattering (132). Other display devices, such as liquid crystal panel and electroluminescence element touch panels, solar batteries, and polymer batteries, are produced from a mixture of an electroconductive resin and a crosslinkable polymer. Phase separation of the liquid mixture is induced by temperature or e-beam radiation. The structure is achieved by hardening the crosslinkable resin (133). Conductive polymeric materials are also produced by blending EPDM with polypropylene, linear low-density polyethylene (LLDPE), polyesters, and polyamides (134). The rubber phase can be urethane or EPR and carbon black, which is typically dispersed in either the thermoplastic or at the interface of the two phases.

3.5.3.4 Automotive Applications and Enhanced
           Physical Properties

The automotive industry also takes advantage of these types of materials. Indeed this industry has the largest number of patents covering a wide range of applications, including the automotive body, engine, and interior. Automotive body parts are made using a cocontinuous system of polycarbonate with polyethylene terephathalate and conductive filler, which makes the part electrostatically chargeable and paintable. The two phases are cocontinuous in the presence of an impact modifier and are quenched following a transesterification reaction (135,136). Rubber modified SMA and PBT are blended together to produce blow moldable containers with high ductility that are required in low temperature environments (137). Other automotive parts are made by blending LCPs with polyphenylene sulfide (PPS) in proportions such that a cocontinuous morphology is obtained. The resulting material appears to have a synergistic effect on the mechanical properties; this is not obtained with matrix-droplet type morphologies (112). Gas barrier properties are also of interest in auto-motive applications. Thermoplastic polyester is used in combination with a hydro-genated block copolymer containing blocks of vinyl compounds and conjugated diene. The blends have dual phase continuity and exhibit excellent barrier properties to gases while maintaining good chemical, water, and heat resistance (138). Other synergistic effects in mechanical properties have been claimed for blends of thermo-plastic elastomers with polyester resins when exposed to radiation or crosslinked, or for blends of PPS with PC and polyether ether ketone (PEEK). The improved properties are high temperature dimensional stability, solvent resistance, high bending strength, and improved impact strength (139,140). Heat distortion and impact resistance are also improved in PVC (polyvinyl chloride)/polyester and thermoplastic polyurethane/acrylnitrile butadiene styrene (TPU/ABS) blends (141,142).

3.5.3.5 Insulation and Adhesive Applications

In these applications, typically the superstructures are formed permanently. In other words, the morphology is frozen in owing to a chemical reaction between two phases and/or the curing of one phase for which the viscosity becomes infinite, thereby preventing further coalescence. The majority of the applications found in these areas are in thermosetting adhesives used in the manufacturing of circuit boards. This is achieved by dispersing a cured heat resistant resin powder soluble in an acid into an uncured heat resistant resin matrix hardly soluble in the acid. After curing, one of the phases is cured and the other is not, thus forming the adhesive layer (143,144). Marine adhesives are, for example, fabricated using polyaromatics containing reactive end groups and a thermoset resin with the help of a reactive catalyst containing a Lewis acid with amine functionality. After curing, the amine functionality reacts with the polymer to form a high temperature curable adhesive with excellent mechanical properties and environmental resistance (145). In insulating circuit boards, cocontinuous structures are also used in combination with curable thermosets. These structures impart high heat resistance to circuit boards and high toughness, little thermal deformation, and good adhesion to copper wiring; they make the formation of fine patterns possible (146,147).

3.6 CONCLUSIONS

The cocontinuous phase morphology remains the most complex and challenging type of morphology encountered in immiscible polymer blends. The conditions under which phase cocontinuity occurs continue to be a topic of scientific debate. Many empirical and semiempirical models have been proposed, but they need to be considered on a case by case basis. To date, model generalization is far from being achieved. Each model suffers from a lack of applicability. An effort is necessary to experimentally fit each of the series of models proposed and bring about a resolution. The main problems are the set of experimental conditions used and the multiplicity of materials employed. That makes judicious comparison of models practically impossible.

The experimental design and development of blends having cocontinuous phase morphology continue to be a subject of research for many research groups. The number of practical applications reviewed in this chapter clearly reveals that industries utilize cocontinuous phase morphologies in very demanding and very targeted applications. Particular application requirements can only be fulfilled through cocontinuity of the phases in a blend.

Characterization techniques of the cocontinuous phase morphology are being improved further to include selective extraction of one phase in combination with direct observation via various microscopes. Recently, gas adsorption (BET) techniques are being used since the cocontinuous blend is a porous material (of course, after the selective extraction of one of the two phases).

REFERENCES

1. Cahn, J.W. and Hilliard, J.E., Free energy of uniform system. I. Interfacial energy, *J. Chem. Phys.*, 31, 688, 1958.
2. Siggia, E.D., Late stages of spinodal decomposition in binary mixtures, *Phys. Rev. A*, 20, 595, 1979.
3. Utracki, L.A., Ed., *Polymer Blends Handbook, Vol. 2,* Kluwer Academic Publishers, Dordrecht, 2002.
4. Kessler, J., Kammer, H.W., and Klostermann, K., Phase-behavior of poly(methylmethacrylate) and poly(styrene-co-acrylonitrile) blends, *Polym. Bull.*, 15, 113, 1986.
5. Macbrierty, V.J., Douglass, D.C, and Kwei, T.K., Compatibility in blends of poly(methyl methacrylate) and poly(styrene-co-acrylonitrile). 2. NMR study, *Macromolecules*, 11, 1265, 1978.
6. Cho, K., Kressler, J., and Inoue,T., Adhesion and welding in the system san pmma, *Polymer*, 35, 1332, 1994.
7. Higashida, N., Kressler, J., and Inoue, T., Lower critical solution temperature and upper critical solution temperature phase-behavior in random copolymer blends — poly(styrene-co-acrylonitrile)/poly(methyl methacrylate) and poly(styrene-co-acrylonitrile)/poly(epsilon-caprolactone), *Polymer*, 36, 2761, 1995.
8. McMaster, L.P. and Olabisi, O., *ACS Org. Coat. Plast. Chem. Prepr.*, 35, 322, 1975.
9. Yamanaka, K. and Inoue, T., Structure development in epoxy-resin modified with poly(ether sulfone), *Polymer*, 30, 662, 1989.
10. Yamanaka, K., Takagi, Y., and Inoue, T., Reaction-induced phase-separation in rubber-modified epoxy-resins, *Polymer*, 30, 1839, 1989.
11. Yamanaka, K. and Inoue, T., Phase-separation mechanism of rubber-modified epoxy, *J. Mater. Sci.*, 25, 241, 1990.
12. Kim, B.S., Chiba, T., and Inoue, T., Morphology development via reaction-induced phase separation in epoxy/poly(ether sulfone) blends: morphology control using poly(ether sulfone) with functional end-groups, *Polymer*, 36, 43, 1995.

13. Chen, J.-L. and Chang, F.-C., Phase separation in poly(ε-caprolactone)-epoxy blends, *Macromolecules*, 32, 5348, 1999.
14. Oyanguren, P.A., Galante, M.J., Andromaque, K., et al., Development of bicontinuous morphologies in polysulfone-epoxy blends, *Polymer*, 40, 5249, 1999.
15. Tanaka, H. and Nishi, T., New types of phase-separation behavior during the crystallization process in polymer blends with phase-diagram, *Phys. Rev. Lett.*, 55, 1102, 1985.
16. Tanaka, H. and Nishi, T., Local phase-separation at the growth front of a polymer spherulite during crystallization and nonlinear spherulitic growth in a polymer mixture with a phase-diagram, *Phys. Rev. A.*, 39, 783, 1989.
17. Li, Y. and Jungnickel, B.-J., The competition between crystallization and phase-separation in polymer blends. 2. Small-angle x-ray-scattering studies on the crystalline morphology of poly-(epsilon-caprolactone) in its blends with polystyrene, *Polymer*, 34, 9, 1993.
18. Saito, H., Fujita,Y., and Inoue, T., Upper critical solution temperature behavior in poly(vinylidene fluoride) poly(methyl methacrylate) blends, *Polym. J.*, 19, 405, 1987.
19. Saito, H., Fujita, Y., and Inoue, T., Light-scattering analysis of upper critical solution temperature behavior in a poly(vinylidene fluoride) poly(methyl methacrylate) blend, *Macromolecules*, 25, 1611, 1992.
20. Pötschke, P. and Paul, D.R., Formation of co-continuous structures in melt-mixed immiscible polymer blends, *J. Macrool. Sci., Part C, Polym. Rev.*, 43, 87, 2003.
21. Goh, S.H., Paul, D.R., and Barlow, J.W., Miscibility of poly(alpha-methyl styrene-co-acrylonitrile) with polyacrylates and polymethacrylates, *Polym. Eng. Sci.*, 22, 34, 1982.
22. Walsh, D.J., Higgins, J.S., and Zhikuan, Z.K., A pair of high molecular-weight compatible polymers showing upper critical solution temperature behavior, *Polymer*, 22, 1005, 1982.
23. Jagger, H., Vorenkamp, E.J., and Challa, G., LCST behaviour in blends of PMMA with PVC, *Polym. Commun.*, 24, 290, 1983.
24. Walsh, D.J. and Rostami, S., Effect of pressure on polymer phase-separation behavior, *Macromolecules*, 18, 216, 1985.
25. Takahashi, M., Hirouchi, H., Kinoshita, S., and Nose, T., A test of the scaling law for structure-function in phase separation process of polystyrene poly (methylphenylsiloxane) liquid mixtures, *J. Phys. Soc. Jap.*, 55, 2687, 1986.
26. Ougizawa, T., Inoue, T., and Kammou, H.W., Upper and lower critical solution temperature behavior in polymer blends, *Macromolecules*, 18, 2089, 1985.
27. Ougizawa, T. and Inoue, T., UCST and LCST behavior in polymer blends and its thermodynamic interpretation, *Polym. J.*, 18, 521, 1986.
28. Shih, C.K., Mixing and morphological transformations in the compounding process for polymer blends — the phase inversion mechanism, *Polym. Eng. Sci.*, 35, 1688, 1995.
29. Sundararaj, U., Macosko, C.W., and Shih, C.K., Evidence for inversion of phase continuity during morphology development in polymer blending, *Polym. Eng. Sci.*, 36, 1769, 1996.
30. Shih, C.K., *Fundamentals of Polymer Compounding. Part II. Simulation of Polymer Compounding Process*, 49th ANTEC, Montreal, SPE,1991, p. 99.
31. Shih, et al., *Rheological Properties of Multicomponent Polymer System Undergoing Melting or Softening during Compounding*, 48th ANTEC, Dallas, SPE, 1990, 951.
32. Shih, C.K., *Advances Polym. Techn.*, 11, 223, 1992.
33. Shih, C.K., Tynan, D.G., and Denelsbeck, D.A., Rheological properties of multicomponent polymer systems undergoing melting or softening during compounding, *Polym. Eng. Sci.*, 31, 1670, 1991.

34. Scott, C.E. and Joung, S.K., Viscosity ratio effects in the compounding of low viscosity, immiscible fluids into polymeric matrices, *Polym. Eng. Sci.*, 36, 1666, 1996.
35. Favis, B.D. and Chalifoux, J.P., Influence of composition on the morphology of polypropylene polycarbonate blends, *Polymer*, 29, 1761, 1988.
36. Favis, B.D. and Therrien, D., Factors influencing structure formation and phase size in an immiscible polymer blend of polycarbonate and polypropylene prepared by twin-screw extrusion, *Polymer*, 32, 1474, 1991.
37. Lyngaae-Jorgensen, J. and Utracki, L.A., *Makromol. Chem. Macromol. Symp.*, 48–49, 189, 1991.
38. Lyngaae-Jorgensen, J., Rasmusen, K.L., Chtcherbakova, E.A., and Utracki, L.A., Flow induced deformation of dual-phase continuity in polymer blends and alloys. Part I, *Poly. Eng. Sci.*, 39(6), 1060, 1999.
39. Utracki, L.A., On the viscosity-concentration dependence of immiscible polymer blends, *J. Rheol.*, 35, 1615, 1991.
40. Lyngaae-Jorgensen, J. and Utracki, L.A., Structuring polymer blends with bicontinuous phase morphology. Part II. Tailoring blends with ultralow critical volume fraction, *Polymer*, 44, 1661, 2003.
41. Nelson, C.J., Avgeropoulos, G.N., Weissert, F.C., and Bohm, G.G.A., Relationship between rheology, morphology, and physical properties in heterogeneous blends, *Angew. Makrom. Chemie*, 60–61(4), 49, 1977.
42. Mekhilef, N. and Verhoogt, H., Phase inversion and dual-phase continuity in polymer blends: theoretical predictions and experimental results, *Polymer*, 37(18), 4069, 1996.
43. Paul, D.R. and Barlow, J., *Macromol. Sci. Rev. Macomol. Chem. C*, 18, 109, 1980.
44. Jordhamo, G.M., Manson, J.A., and Sperling, L.H., Phase continuity and inversion in polymer blends and simultaneous interprenetrating networks, *Poly. Eng. Sci.*, 26, 517, 1986.
45. Gergen, W.P., Lutz, R.G., and Davidson, S., Thermoplastic Elastomers: A Comprehensive Review, Legge, N.R., Holden, G., and Davidson, S., Eds., Hanser, Munich, 1987.
46. Miles, I.S. and Zurek, A., Preparation, structure, and properties of 2-phase co-continuous polymer blends, *Poly. Eng. Sci.*, 28, 796, 1988.
47. Verhoogt, H., Morphology, properties and stability of thermoplastic polymer blends, Ph.D. thesis, Delft University of Technology, The Netherlands, 1992.
48. Levij, M. and Maurer, F.H.J., Morphology and rheological properties of polypropylene-linear low density polyethylene blends, *Polym. Eng. Sci.*, 28, 670, 1988.
49. Utracki, L.A., On the viscosity-concentration dependence of immiscible PP/CPSI/PPE blends' influence of the melt-viscosity ratio and blend composition, *J. Rheolo.*, 35, 1615, 1991.
50. Everaert, V., Aerst, L., and Groeninckx, G., *Polymer*, 40, 6627, 1999.
51. Elmendorp, J.J., A study on polymer blending microrheology, *Polym. Eng. Sci.*, 26(6), 418, 1986.
52. Metelkin, V.I. and Blekht, V.P., Formation of a continuous phase in heterogeneous polymer mixtures, Col. J. USSR, 46, 425, 1984.
53. Steinmann, S., Gronski, W., and Friedrich, C., Quantitative rheological evaluation of phase inversion in two-phase polymer blends with cocontinuous morphology, *Polymer*, 42(15), 6619, 2001.
54. Willemse, R.C., Posthuma DeBoer, A., Van Dam, J., and Gotsis, A.D., Co-continuous morphologies in polymer blends: a new model, *Polymer*, 39, 5879, 1998.
55. Willemse, R.C., Speijer, A., Langeraar, A.E., et al., Tensile moduli of co-continuous polymer blends, *Polymer*, 40, 6645, 1999.

56. Willemse, R.C., Co-continuous morphologies in polymer blends: stability, *Polymer*, 40, 2175, 1999.
57. Scott, C.E. and Macosko, C.W., Morphology development during the initial-stages of polymer-polymer blending, *Polymer*, 36, 461, 1995.
58. De Gennes, P.G., Title of chapter appears here, in Percolation, Localization and Super-conductivity, Goldman, A.M. and Wolf, S.A., Eds., Plenum: New York, 1984, p. 83.
59. Hsu, W.Y. and Wu, S., Percolation behavior in morphology and modulus of polymer blends, *Polym. Eng. Sci.*, 33, 293, 1993.
60. Margolina, A. and Wu, S., Percolation model for brittle-tough transition in nylon rubber blends, *Polymer*, 29, 2170, 1988.
61. Margolina, A. and Wu, S., Percolation model for brittle tough transition in nylon rubber blends — reply, *Polymer*, 31, 971, 1990.
62. Feng, S. and Sen, P.N., Percolation on elastic networks — new exponent and threshold, *Phys. Rev. Letter*, 52, 216, 1984.
63. Veenstra, H., Norder, B., Van Dam, B., and Posthuma de Boer, A., Stability of co-continuous polystyrene/poly(ether-ester) blends in shear flow, *Polymer*, 40, 5223, 1999.
64. Arns, C.H., Knackstedt, M.A., Roberts, A.P., and Pinczewski, V.W., Morphology, cocontinuity, and conductive properties of anisotropic polymer blends, *Macromolecules*, 31, 5964, 1999.
65. Weis, C., Leukel, J., Borkenstein, K., Maier, D., Gronski, W., Friedrich, C., and Honerkamp, J., Morphological and rheological detection of the phase inversion of PMMA/PS polymer blends, *J. Polym. Bull.*, 40, 235, 1998.
66. Chalkey, H. W. and Park, J., *Science*, 110, 295, 1949.
67. Colman, R., in *Stochastic Geometry, Geometric Statistics and Stochastic Statistics Stereology*, Ambartzumion, R. and Weil, W., Eds., Teubner, Leipzig, 1984, p..
68. Quintens, D., Groeninckx, G., Guest, M., and Aerts, L., Phase morphology coarsening and quantitative morphological characterization of a 60/40 blend of polycarbonate of bisphenol-a (PC) and poly(styrene-co-acrylonitrile) (SAN), *Poly. Eng. Sci.*, 30, 1484, 1990.
69. Blacher, S., Brouers, F., Fayt, R., et al., Multifractal analysis — a new method for the characterization of the morphology of multicomponent polymer systems, *J. Poly. Sci., Poly. Phys. Ed.*, B31, 655, 1993.
70. Harrats, C., Blacher, S., Fayt, R., Jerome, R., et al., Molecular design of multicomponent polymer systems. 19. Stability of cocontinuous phase morphologies in low-density polyethylene polystyrene blends emulsified by block-copolymers, *J. Poly. Sci., Poly. Phys. Ed.*, B33, 801, 1995.
71. Gubbels, F., Blacher, S., Vanlethem, E., et al., Design of electrical conductive composites — key role of the morphology on the electrical-properties of carbon-black filled polymer blends, *Macromolecules*, 28, 1559, 1995.
72. Boury, D. and Favis, B., Cocontinuity and phase inversion in HDPE/PS blends: influence of interfacial modification and elasticity, *J. Polym. Sci., Phys. Ed.*, 36, 1889, 1998.
73. Willemse, R.C., Posthuma DeBoer, A., Van Dam, J., et al., Morphology development in immiscible polymer blends: initial blend morphology and phase dimensions, *Polymer*, 40, 6651, 1999.
74. Veenstra, H., Van Lent, B.J.J., and Van Dam, J., Co-continuous morphologies in polymer blends with SEBS block copolymers, *Polymer*, 40, 6661, 1999.
75. Li, J. and Favis, B.D., Characterizing cocontinuous high density polyethylene/polystyrene blends, *Polymer*, 42, 5047, 2001.

76. Marin, N. and Favis, B.D., Cocontinuous morphology development in partially miscible PMMA/PC blends, *Polymer*, 43, 4723, 2002.

77. Paul, D.R. and Newman, S., *Polymer Blends*, Academic Press, New York, 1978.

78. Mamat, A., Vu-Khan, T., and Cigana, P., Impact fracture behavior of nylon-6/ABS blends, *J. Poly. Sci., Poly. Phys. Ed.*, 35, 2583, 1997.

79. Noel, O.F. and Carley, F., Properties of polypropylene-polyethylene blends, *Poly. Eng. Sci.*, 15, 117, 1975.

80. Gergen, W.P., Lutz, R.G., and Davidson, S., Hydrogenated block copolymers in thermoplastic elastomer interpenetrating polymer network, in *Thermoplastic Elastomers*, 2nd Ed., Holden, G., Legge, N.R., Quirk, R., and Schroeder, H.E., Eds., Hanser Publisher, Munich, 1996, p..

81. Davies, W., *J. Phys. D: Appl. Phys.*, 4, 1325, 1971.

82. Coran, A.Y. and Patel, R.J., Predicting elastic-moduli of heterogeneous polymer compositions, *J. Appl. Poly. Sci.*, 20, 3005, 1976.

83. Chuai, C.Z., Almdal, K., and Lyngaae-Jorgensen, J., Phase continuity and inversion in polystyrene/polymethylmethacrylate blends, *Polymer*, 44, 481, 2003.

84. Kolarik, J., Three-dimensional models for predicting the modulus and yield strength of polymer blends, foams, and particulate composites, *Poly. Comp.*, 18, 433, 1997.

85. Veenstra, H., Verkooijen, P.C.J., Van Lent, B.J.J., et al., On the mechanical properties of co-continuous polymer blends: experimental and modeling, *Polymer*, 41 1817, 2000.

86. Niebergall, U., Bohse, J., Seidler, S., et al., Relationship of fracture behavior and morphology in polyolefin blends, *Poly. Eng. Sci.*, 39, 1109, 1999.

87. Palierne, J.F., Linear rheology of viscoelastic emulsions with interfacial-tension, *Rheol. Acta*, 29, 204, 1990.

88. Palierne, J.F., *Rheol. Acta*, 30, 497, 1991.

89. Graebling, D., Benkira, A., Gallot, Y., et al., Dynamic viscoelastic behavior of polymer blends in the melt — experimental results for PDMS POE-DO, PS PMMA and PS PEMA blends, *Eur. Polym. J.*, 30, 301, 1994.

90. Bousmina, B. and Muller, R., Linear viscoelasticity in the melt of impact PMMA — influence of concentration and aggregation of dispersed rubber particles, *J. Rheol.*, 37, 663, 1993.

91. Carreau, P.J., Bousmina, M., and Ajji, A., in *Progress in Pacific Polymer Science*, Springer-Verlag, Berlin, 1994.

92. Bousmina, M., Bataille, P., Sapieha, S., et al., Comparing the effect of corona treatment and block-copolymer addition on rheological properties of polystyrene polyethylene blends, *J. Rheol.*, 39(3), 499, 1995.

93. Brahimi, B., Ait Kadi, A., Ajji, A., et al., Rheological properties of copolymer modified polyethylene polystyrene blends, *J. Rheol.*, 35(6), 1069, 1991.

94. Mekhilef, N., Carreau, P.J., Favis, B.D., et al., Viscoelastic properties and interfacial tension of polystyrene-polyethylene blends, *J. Poly. Sci., Phys. Ed.*, 38, 1359, 2000; Norphological stability, interfacial tension, and dual-phase continuity in polystyrene-2-polyethylene blends, *J. Poly. Sci., Phys. Ed.*, 35, 293, 1997.

95. Steinmann, S., Gronski, W., and Friedrich, C., Quantitative rheological evaluation of phase inversion in two-phase polymer blends with cocontinuous morphology, *Rheol. Acta*, 41, 77, 2002.

96. Galloway, J.A., Momtminy, M.D., and Macosko, C.W., Image analysis for interfacial area and cocontinuity detection in polymer blends, *Polymer*, 43, 4715, 2002.

97. David, B., Kozlowski, M., and Tadmor Z., The effect of mixing history on the morphology of immiscible polymer blends, *J. Polym. Eng.*, 33, 227, 1993.

98. Dedecker, K. and Groeninckx, G., Reactive compatibilization of A/(B/C) polymer blends. Part 1. Investigation of the phase morphology development and stabilization, *Polymer*, 39, 4985, 1998.
99. Lyngaae-Jorgensen, J., Structuring of interface-modified polymer blends, *Intern. Poly. Process.*, XIV(3), 213, 1999.
100. Li, J., Ma, P., and Favis, B.D., The role of the blend interface type on morphology in cocontinuous polymer blends, *Macromolecules*, 35, 2005, 2002.
101. Dedecker, K. and Groeninckx, G., Reactive compatibilization of A/(B/C) polymer blends. Part 2. Analysis of the phase inversion region and the co-continuous phase morphology, *Polymer*, 39, 4993, 1998.
102. Hietaoja, P.T., Holsti-Miettinen, R.M., Seppälä, J.V., and Ikkala, O.T., The effect of viscosity ratio on the phase inversion of polyamide-66 polypropylene blends, *J. Appl. Polym. Sci.*, 54, 1613, 1994.
103. Tol, R.T., Groeninckx, G., Vinckier, I, et al., Phase morphology and stability of co-continuous (PPE/PS)/PA6 and PS/PA6 blends: effect of rheology and reactive compatibilization, *Polymer*, 45, 2587, 2004.
104. Harrats, C,S. Blacher, Fayt, R., et al., Phase co-continuity stabilization by tapered diblock and triblock copolymers in PS rich (LDPE/PS) blends, *J. Poly. Sci. Poly. Phys. Ed.*, 41, 202, 2003.
105. Jafari, S.H., Potschke, P., Stephan, M., et al., Thermal behavior and morphology of polyamide 6 based multicomponent blends, *J. Appl. Polym. Sci.*, 84, 2753, 2002.
106. Zhenhua, Y. and Favis, B.D., Macroporous poly(L-lactide) of controlled pore size derived from annealing of co-continuous polystyrene/poly(L-lactide) blends, *Biomaterials*, 25, 2161, 2004.
107. Sarazin, P., Roy X., and Favis B.D., Controlled preparation and properties of porous poly(L-lactide) obtained from a co-continuous blend of two biodegradable polymers, *Biomaterials*, 2004 (in press).
108. Harrats, C, Omonov T., Groeninckx, G., et al., Phase morphology development and stabilization in polycyclohexylmethacrylate/polypropylene blends: uncompatibilized and reactively compatibilized blends using two reactive precursors, *Polymer*, 2004 (in press).
109. Scobboe, J.J., Conductive polyphenylene ether/polyamide blends for electrostatic painting applications, in *Conductive Polymers and Plastics*, Rupprecht, L., Ed., Chemtech, Toronto, 1999, p. 181.
110. Stanbroek, C.E. and Patel, B.R., US patent 2002099128.
111. Anwendungstechnische Information zu Triax©, ATI 0501 d,e, Bayer AG, Geschafts-bereich Kunststof, 1998.
112. WO patent 09902607, *A Liquid Crystalline Polymer Blend Containing Poly-(Phenylene Oxide)*.
113. EP patent 518517, *Blends of Immiscible Polymers EVOH/PE Having Novel Phase Morphology*.
114. Gergen, W.P., Lutz, R.G., and Davidson, S, Chap. 14 in *Thermoplastic Elastomers: A Comprehensive Review*, Legge, N.R., Holden, G., and Schroeder, H.E., Eds. Hanser-Verlag, Munich, 1987.
115. Zumbrunnen, D.A. and Inamdar, S., Novel sub-micron highly multi-layered polymer films formed by continuous flow chaotic mixing, *Chem. Eng. Sci.*, 56, 3893, 2001.
116. US patent 4931502, 4912144, 4891405, *Moldable SMA/PBT Composition with High Ductility for Blow Molding*.
117. DE patent 2748751, *Impact-Resistant Polymer Mass*.
118. JP patent 07102175, *Resin Composition Useful for Sealing Semiconductors and Containing Thermosetting Resin and Thermoplastic Resin*.

119. M. Mulder, *Basic Principle of Membrane Technology*, Kluwer Academic Publishers, Dordrecht, 1996, p..

120. JP patent 2001253936, *Porous Resins Having Uniform and Fine Co-Continuous Phase Structures.*

121. US patent 5135627, *Mosaic Micro-Columns, Slabs and Separation Media for Electrophoresis and Chromatography.*

122. WO patent 2003099933, *Dispersion for Moldings Such As Porous Bodies and Particles Comprises Resin Component and Water-Soluble Auxiliary Component Comprises Oligosaccharide and Optionally Water-Soluble Plasticizer.*

123. JP Patent 2000001612, *Nylon-Polyester Composition Has Co-Continuous Phase-Separating Structure and a Polyester Phase.*

124. DE patent 4236935, *Anisotropic Porous Network with Accurately Controlled Pore Size Containing Acrylol Oxy-Benzoyloxy-Phenyl Acrylate and Liquid Crystalline Compounds Useful in Liquid Chromatography and Membrane Production.*

125. WO patent 9735906, *High Water Content Porous Polymer Made from Macro-Monomers Having Per-Fluoro-Polyether Units, Using Progens where Polymerization Starts from a Co-Continuous Micro-Emulsion and Display Discrete Phase Interpenetrating Polymer Network; Soft Contact Lens.*

126. US patent 5760100, *Extended Wear Ophthalmic Lens; Oxygen and Ion or Water Permeability.*

127. US patent 2003072790, *Device for Tissue Engineering Comprises Porous Polymeric Scaffold Having First Biodegradable Polymer with Continuous Large Network, Interconnected Pores and Second Biodegradable Polymer with Partially Interconnected Pores.*

128. WO patent 2003103925, *Co-Continuous Phase Composite Polymer Blends for Biomedical Applications.*

129. US patent 5093423, *Manufacture of Thermoplastic Elastomer Compositions from SBR.*

130. WO patent 9912976, *Manufacture of Absorbing Material Based on Starch Having Improved Absorbent Properties.*

131. JP patent 2003073649, *Adhesive Agent for Forming Adhesive Layer on Printed Circuit Board with Heat Resistant Matrix Formed by Homogenous Mixture of Specific Reins and Attains a Specific Structure by Hardening.*

132. JP patent 2002148414, *Light Scattering Sheet for Reflection Type Liquid Crystal Display Devices Contains Co-Continuous Phase Structure Formed by Spinodal Decomposition of Liquid Phase Containing Polymers.*

133. JP patent 2001316595, *Conductive Resin Composite Molding Has Co-Continuous Phase Structure Consisting of Cross-Linked Polymer and Electro-Conductive Chain Polymer and Has Network Structure Having Preset Average Network Diameter.*

134. WO patent 9941304, *Conductive Polymer Blends with Finely Divided Conductive Material Selectively Localized in Continuous Polymer Phase or Continuous Interface and Preparation Thereof.*

135. US patent 6673864, *Conductive Polyester/Polycarbonate Blends, Methods for Preparation Thereof and Articles Derived Therefrom.*

136. US patent 2002099128, *Conductive Polyester/Polycarbonate Blends, Methods for Preparation Thereof, and Articles Derived Therefrom, for Molding Rigid Electrostatically Painted Automobile Parts; Strength and Stiffness.*

137. US patents 4931502, US4912144, US4891405, *Ductile, Blow Moldable Composition Containing a Styrene-MMA Copolymers Having Pendant Carboxy Ester Groups, Blend with Styrene-MA Copolymer and Shell-Core Graft Polymer Containing PBT.*

138. JP patent 10212392, *Thermoplastic Copolymer Composition Comprises Thermoplastic Polyester-Type Polymers and Crosslinked Product of Hydrogenated Block Copolymer Consisting Mainly of Aromatic Vinyl Compounds and Isobutylene Containing Polymer Block.*
139. US patent 5300573, *Thermoplastic Elastomer Composition of High Temperature Dimensional Stability Containing Polyester Resin and Crosslinked Acrylic-Ester Copolymer Rubber.*
140. JP patent 2000248179, *Manufacture of PPS Compositions Having Co-Continuous Structure and Molded Products, Mats, and Fibrous Products Therefrom.*
141. EP patent 212449, *Blend of PVC and Segmented Poly-Ester of Terephtalic Acid, Diol and Aliphatic Polyester.*
142. WO patent 484367, *Thermally Processable Blends of High Modulus Polyurethane and Mass Polymerized ABS.*
143. US patent 5519177, *Adhesive Layers for Electroless Plating and Printed Circuit Boards; Dispersion of a Cured Heat Resistant Resin Powder Soluble in an Oxidizing Agent into an Uncured Heat Resistant Resin Matrix Hardly Soluble in Oxidizing Agent after Curing.*
144. JP patent 2003073649, *Adhesive Agent for Forming Adhesive Layer on Printed Circuit Boards.*
145. WO patent 2001032779, *Curable Composition Useful as an Adhesive in the Marine Industry.*
146. WO patent 2002008338, *Insulating Resin Composition for Multiplayer Printed Circuit Board Containing Thermoset/Thermoplastic Resins.*
147. JP patent 2001220515, *Resin Composite Used as Protective Film.*

4 Phase Morphology Development in Polymer Blends: Processing and Experimental Aspects

Uttandaraman Sundararaj

CONTENTS

4.1 INTRODUCTION

Polymer blending is an attractive method of creating new materials with more enhanced properties and greater performance than existing homopolymers, especially since conceiving and synthesizing new polymers is time consuming and involves significant costs. It is well known that the properties can be affected significantly by the structuring (or morphology) of the blend (1–8). Despite an increase in sales of blends by industry, a fundamental understanding of blend morphology and properties is still required. To do this, several experimental methods have been used on numerous processing techniques to understand the different aspects of morphology development in polymer blends.

In this chapter, we review the research on morphology development in polymer blends and describe the different experimental methodologies used to determine blend morphology in three different types of polymer processing equipment: twin screw extruder, single-screw extruder, and twin rotor batch mixer. We also describe some miniature mixing devices that have been used for specialty materials available in milligram quantities. To understand more about polymer blends and polymer blending, the reader is referred to Manson and Sperling (9), Paul and Bucknall (10), and Utracki (11,12), and for polymer compabilization, the reader is referred to Datta and Lohse (13) and Baker et al. (14).

4.2   EXPERIMENTAL METHODS

Two of the most difficult aspects of studying polymer blend morphology are that the processing experiments are typically done at very high temperatures (> 200°C) and it is difficult to devise equipment to view the structure directly. Further complicating the study of morphology is the requirement of describing useful blend properties at the submicron size scale. This small size precludes the use of optical methods except for the initial stages of mixing. Thus, online morphology analysis is difficult. Blend samples are typically quenched and then analyzed using scanning electron microscopy (SEM), transmission electron microscopy (TEM), and other such techniques that allow for high magnification imaging. However, it is difficult to know if the morphology of the quenched samples corresponds to the morphology of the sample in the melt state. This becomes especially important when deciphering how the morphology develops, particularly since the structures formed during morphology *development* are inherently unstable.

Nevertheless, researchers are continuing to come up with new and innovative procedures to study the morphology either during processing or to quickly freeze sample morphology that is representative of the in-process structure. This has involved finding better ways to sample while the process is in operation (15–19) or by using inline (20,21) or online analysis methods (21–23) to directly measure morphology during the process. Investigators have had some limited success using these methods but there is much room for improvement.

For postmortem morphology analysis, sample preparation may require a few minutes or a few hours for each sample, depending on the imaging technique. Many groups have used SEM for morphology analysis because it requires relatively easy sample preparation. SEM also gives a reasonable representation of the micron- and nanosize morphology. Two main types of samples are used: fracture surfaces or selectively extracted samples. For fracture surfaces, the polymer blend sample is quenched in liquid nitrogen to bring all components well below their respective glass transition temperatures and then the sample is broken to give a fracture surface. If the sample is fractured at room temperature, the polymer can deform and smear, resulting in a poor quality surface or even an unrepresentative morphology. This is

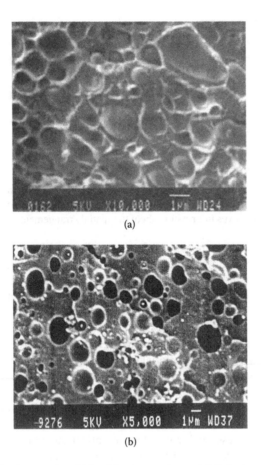

FIGURE 4.1 SEM micrographs of blend fracture surfaces. (a) Polystyrene/ethylene propylene rubber 80:20 blend. (b) Polyamide/polystyrene 80:20 blend. In (b), the polystyrene
dispersed phase has been etched away with methylene chloride.

also the case for microtomed surfaces when the soft polymer forms striations across
the surface, marring the true sample morphology. A *cryoultramicrotome* reduces the
streaking problem by bringing the polymer to well below its glass transition temperature. Cryogenic conditions are recommended for ultramicrotoming of most polymers. The dispersed component in the blend may also be selectively extracted for
additional contrast, but this is not necessary. Figure 4.1 shows SEM micrographs of
a fracture surface with and without extraction of the dispersed phase. Another method
to enhance the contrast is to stain with an agent like osmium tetraoxide or ruthenium
tetraoxide, assuming the reagent segregrates selectively into one of the phases. The
metal stain allows one to distinguish the phases in the SEM. When we look at surfaces,
only a two-dimensional (2-D) representation of the morphology is obtained, and a
perpendicular cross-section may be required to fully understand the morphology.

Selective extraction is typically used to remove the major phase. The major
component of the blend is dissolved using a selective solvent so that the complete

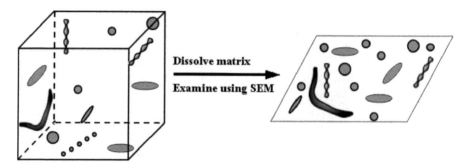

FIGURE 4.2 Effect of selective extraction. Schematic showing how dissolution of matrix phase can give us more detail about the 3-D morphology of the dispersed phase. Performing selective extraction removes information about the spatial arrangement of the dispersed phase.

structure of the other component is seen (see Figure 4.2). A 3-D picture of the dispersed phase morphology can be seen; however, we lose the spatial arrangement of these domains, and so it is a useful technique to understand deformation mechanisms and the type of structure in the blend. Still, it should not be used to assess distributive mixing. Samples where the matrix has been extracted to reveal dispersed phase domains are shown in Figure 4.3.

For most SEM analysis, the sample must be coated with carbon, gold, palladium, or other electrically conductive material before imaging the sample. A coating is not required if an extremely low SEM accelerating voltage is used to image the sample; however, this often compromises the image quality. Newer SEM instruments can be used to image uncoated organic samples, but a coating is often used even for these newer instruments. It is assumed that the coating is sufficiently thin (on the order of a nanometer thickness) and that it does not affect the morphology. However, as we get to polymer molecular dimensions such as in nanoblends (24,25), we must be careful to interpret SEM images that use coatings on the polymer sample.

For TEM samples, a cryoultramicrotome is typically used to cut thin sections of samples. Usually, the thin slices are selectively stained and one phase is tinted; if there are more than two phases, there may be three shades of tinting. For example, in the case of compatibilized blends, phase A may be dark, phase B may be light, and copolymer A-B will be grey. The staining agents include osmium tetraoxide and ruthenium tetraoxide that diffuse into various phases based on the level of bond saturation. Imaging can be done without staining; however, the sample will not be stable under the electron beam, and image contrast will be poorer. TEM analysis is complementary to that of SEM and gives much better contrast and resolution. Typical TEM micrographs of blends are shown in Figure 4.4. In Figure 4.4b, nanoscale micelles of block copolymer can be seen in both phases. Though TEM analysis is potentially powerful, sample preparation is cumbersome and is attempted only after SEM.

A relatively newer method in blend analysis is atomic force microscopy (AFM) that gives results similar to SEM but has the advantages of quantitative depth resolution and minimal sample preparation (does not require coating) (26,27). However, since the scanning is done more slowly, it takes much longer to image a sample.

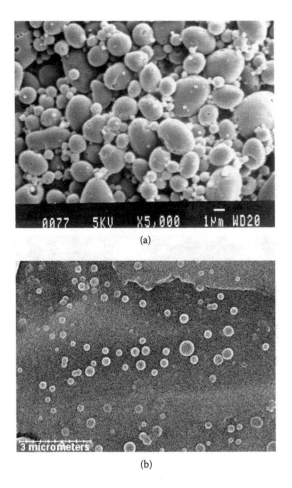

(a)

(b)

FIGURE 4.3 Micrographs of blend samples where the matrix has been extracted to reveal dispersed phase drops. (a) Polystyrene/polypropylene 80:20 blend mixed in twin-rotor batch mixer. (b) Same blend mixed in APAM (Alberta Polymer Asymmetric Minimixer). In both cases, the polystyrene (matrix) phase has been extracted.

In addition, it is more difficult to image rough samples. An AFM contrast between blend components is based on a difference in friction coefficient between the phases. Figure 4.5 shows an AFM image of a polymer blend. It is anticipated that characterization of blend morphology using AFM will be refined and will be used more often for blends characterization (28).

Light scattering has been used for morphology characterization for postmortem samples (29–33) and for analysis inline (21). The technique typically provides an average diameter for the blend, and so this method does not give as much insight into deformation mechanisms. However, it is a very fast measurement and can be performed immediately at the site of the operation. Light scattering can give information about the size and orientation of dispersed phase domains. Online light scattering analysis has been done for simulated blends (20), i.e., the blend consisted

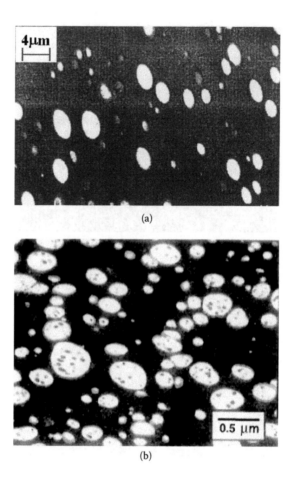

FIGURE 4.4 TEM micrographs of (a) polycarbonate/poly(ether imide) 80:20 blend (Lin et al., *Polym. Eng. Sci.*, 43, 891, 2003) and (b) polystyrene/poly(methyl methacrylate)/PS-PMMA diblock copolymer 67:28:5 (Macosko et al., *Macromolecules*, 29, 5590, 1996). In (a), the polycarbonate phase is stained by ruthenium tetraoxide and appears black; in (b), PS is stained by the same agent. It is interesting to note the nanoscale micelles in both phases in (b).

of glass fibers and glass beads dispersed in a polymer and then extruded. The contrast between polymer materials is low, and with current methods, the contrast is not sufficient to obtain reliable results.

In the last decade, there have been a few attempts to directly visualize the blending process on processing equipment. For extruders, Sakai (34) and White and coworkers (35,36) made the first visualization attempts. Sakai used a glass strip mounted on the side of a JSW twin screw extruder. Several authors (22,23) have performed visualization recently by using a contoured quartz insert on top of the barrel. Using high-zoom and high-speed video, some of the morphology development can be observed online. For batch mixers, Shih et al. (37) were one of the first to present visualization data, and several authors afterward have used this

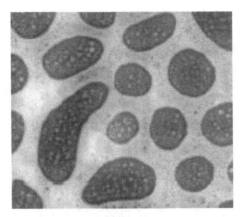

FIGURE 4.5 AFM image of poly(vinyl chloride)/polycarbonate 50:50 blend. The horizontal image width is 25 μm. (Gilmore et al., *Surf. Interface Anal.*, 35, 888, 2003.)

technique (including refs. 38–40). Migler and coworkers (41–43) have affixed a transparent slit die at the end of an extrusion process to visualize the morphology online. They have seen unique morphologies with this device and other parallel plate flow devices (44–46).

Other researchers have worked on methods to take samples out of extruders more quickly (15,16,21,47). Some of the early problems in studying morphology development in extruders stemmed from the fact that it took so long to obtain a sample from within the extruder. The most common technique used was modeled after Tadmor and Gogos' carcass analysis method (48) and is commonly referred to as "screw-pulling" experiments. Tadmor and Gogos applied this method to study melting of one component in a single-screw extruder, but the same technique can be used for polymer blends in a twin-screw extruder. In this method, the twin-screw extruder is stopped suddenly during operation and then cooled down. The die is removed so that the screws can be pulled out of the extruder and samples taken along the length for analysis (49,50). This process can take over 5 min and during that time, the polymer blend morphology can change significantly, particularly where the initial deformation occurs as a result of inherently unstable phase domains. Another method of sampling along the length of the extruder is the use of a split barrel or "suitcase" type of design (51,52) as shown in Figure 4.6. Here, the extruder is shut down, and the two piece barrel is opened up like a suitcase so that samples can be taken along the screw length. This technique is faster but still requires 30 sec or more to obtain samples. In these screw pulling or barrel opening experiments, a different color pellet can be used for each phase so that it is easy to distinguish the melting and dispersion mechanisms (52,53). If SEM or solvent extraction is to be used, the color is not essential.

Nishio and coworkers (54) and Sakai (34) developed a sampling technique that could take samples from pressurized zones of an extruder. However, the significant morphology development may occur in partially filled zones with relatively low pressures. Cartier and Hu (47) simply used a spoon to extract the sample through an opening in the extruder about 5 sec after stopping the motor. Stephan and

FIGURE 4.6 Baker-Perkins 50 mm twin-screw extruder with opening clamshell barrel. After reaching steady state for blend extrusion, the extruder is stopped and opened, and blend samples are taken along its length.

coworkers (15,16) used a novel sampling device (see Figure 4.7) that would take a sample from the extruder while the extruder was still running. Because of its piston design, this device did not rely on extruder pressure to move the melt, and therefore, it worked even in low pressure zones.

4.3 MORPHOLOGY DEVELOPMENT

There has been a flurry of activity in the area of polymer blend morphology development over the last few years with 90% of the articles being published in the past decade. Here, we will review some of the early work and summarize the findings of the more recent work. Figure 4.8 shows the important mechanisms during blending of immiscible polymers. Polymer-polymer melt blends usually start in the solid state as millimeter sized pellets, powder, or flakes, and are then heated and sheared inside an intensive mixing device such as an extruder or batch mixer. These millimeter size domains melt and break up into micrometer or nanometer sized domains during the initial softening stages.

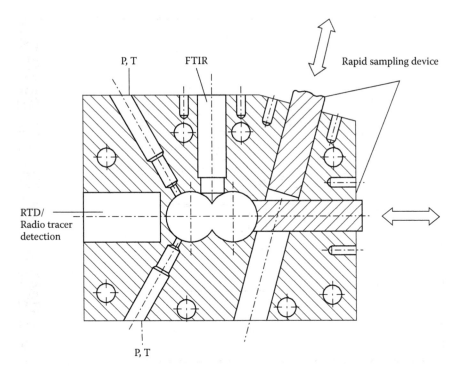

FIGURE 4.7 Sampling device developed to take samples at any location along the extruder. A horizontal piston pulls sample out of the extruder and a vertical piston ejects the sample. (Franzheim et al., *Adv. Polym. Technol.*, 16, 1, 1997; and Potente et al., *J. Appl. Polym. Sci.*, 82, 1986, 2001.)

Morphology development for polymer blends in the initial stages of softening for single-screw extruders and twin-screw extruders was not studied before 1990 because of the difficulty in obtaining representative samples. Initial work on morphology development used batch mixers and varied the mixing time (55,56–59). Relatively little change in particle size was seen since the studies focused on the later stages of mixing. Plochocki et al. (58) postulated an abrasion mechanism during the initial softening stages. An erosion mechanism similar to this was seen in shear visualization experiments (60). However, it was hard to comprehend how the phase size changed so quickly during initial softening. Scott and Macosko (55) showed that sheeting occurred during initial melting stages in twin-rotor batch mixers, Sundararaj et al. (61) reported sheet formation in miniature cup and rotor mixers, and Sundararaj et al. (62,63) showed pellets stretching into sheets in simple shear flow. It was thought that sheets formed when pellet melted on the hot barrel (55), but later studies showed that sheets formed in the melt state (62–70).

After this initial stage, depending on the composition of the two components, a processing phase inversion may occur, i.e., the minor phase, the component in lower concentration, softens first and becomes the matrix phase initially. After the major component softens, once it has sufficient melt concentration, the phases invert and

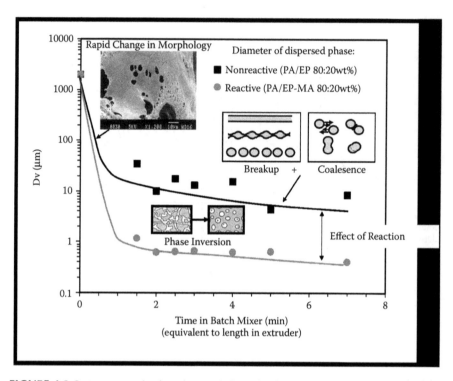

FIGURE 4.8 Important mechanisms in morphology development during polymer blending of nylon/ethylene-propylene rubber blends (data from C.E. Scott, *Characterization of Reactive Polymer Blending Process*, Ph.D. thesis, University of Minnesota, 1990). Initially a sheet morphology is formed, and the dispersed phase domain size decreases sharply (note logarithmic scale for y axis). A processing phase inversion may occur before breakup and coalescence mechanisms take over, leading to a stable drop size. The drop size can change by an order of magnitude because of compatibilization via reactive blending or by adding premade copolymer.

the minor phase becomes the dispersed phase. If the major phase softens first and initially forms the matrix, no phase inversion will be seen.

In the final stage of polymer blending, a balance is established between drop breakup and coalescence so that a stable morphology is formed. Wu (71) and Favis and Chalifoux (72) showed that the final morphology depended on polymer rheological properties and processing parameters. A typical Newtonian capillary number versus viscosity ratio type of relationship was found, but as will be explained later in this chapter, the viscoelastic parameters of polymers need to be considered to better model polymer blend morphology. At compositions near 50%, the morphology may be cocontinuous but unstable so that the cocontinuous morphology present during processing may break down upon annealing to give a droplet-in-matrix morphology (73). Even for lower concentrations, where droplet-in-matrix morphology is usually seen, the phase size is seen to increase upon postprocess annealing under quiescent conditions. Stabilization of the morphology during and after processing can be achieved using compatiblizers. Two main types of compatibilizers used in polymer blends are premade block copolymers and compolymers formed through *in situ* reaction between functionalized polymers. Of the two, the latter is practiced more in industry.

4.3.1 INITIAL MORPHOLOGY

It was found that the dispersed phase breaks up by stretching into sheets (51,55,74) and these sheets subsequently broke up into smaller domains. Figure 4.9 is a summary of work by Scott, Macosko, Sundararaj, and coworkers showing how a pellet deforms into a sheet and subsequently breaks up into droplets (38,51,55,64,75,76). The figure also shows the effect of stabilization on final drop size. Most of the significant morphology change occurred within one length/diameter (L/D) of extruder of the first point of melting in a twin-screw extruder (51) and within the first minute of mixing in a batch mixer (55). For single-screw extruders, the sheet breakup may occur over a longer section of the extruder before droplets are seen (77). Once domains reach the micron size scale, they may be stretched into filaments and are then broken into droplets. At this size scale, interfacial forces become important and pull in the edges of the stretched domains, giving cylindrical or filament type morphology. The stretching and breaking up of filaments is due to the combined effect of high shear zones and low shear zones in the mixer (78). The domains are stretched in the high shear zones and then relax and break up in the low shear zones via Rayleigh type instabilities. It may be possible to form drops directly from sheets but

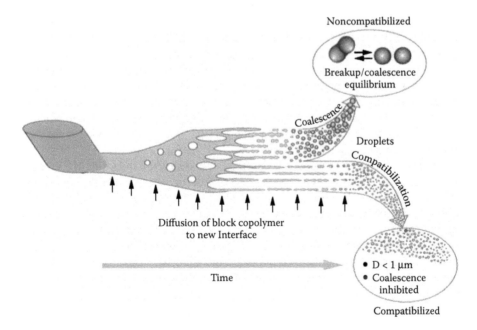

FIGURE 4.9 Schematic showing morphology development mechanisms during polymer blending process. Sheets are broken into droplets, and then the final morphology is dependent on whether compatibilization was used. (Reproduced with permission from Macosko et al., *Macromolecules,* 29, 5590, 1996.)

only at very low shear rates. Much polymer blending work reports filaments even at the end of the mixing process (79,51), and this may be a result of a high shear zone at the end of the process (78).

Breakup phenomena in polymer blends are very non-Newtonian in nature (37,51,55,74,59,80–82). Gogos et al. (49) used energy balances to deal with heat dissipative sources such as plastic energy dissipation and viscous energy dissipation in the initial melting region. They found that these dissipative heat sources led to different polymer deformation mechanisms (83).

In practice, reactive blending or addition of compatibilizers to modify the interface further complicates mixing (37,84–87). For example, Li and coworkers (88) have shown that the presence of a compatibilizer increases the rate of melting by reducing the interfacial thermal resistance and thus increasing the blend's overall thermal conductivity. The presence of a compatibilizer at the interface may also decrease the amount of interfacial slip, and thus, at equivalent shearing conditions, leads to greater deformation of the dispersed phase.

Sundararaj et al. (51) found sheeting in twin-screw extruders, and Lindt and Ghosh (74) saw lamellar morphology in single-screw extruders. It was found that deformation of millimeter size domains into thin micron or submicron size sheets permitted quick reduction in phase size. Bourry and Favis (89) indicated that the morphology development did not depend on melting but on the flow fields seen by the blend. It has been seen that the sheeting mechanism occurs in the melt state well above the polymer melting temperature (62–68), indicating that the phenomenon is more dependent on polymer rheological properties than the actual melting region. Modeling of this sheet formation and breakup has been done (62,66,67,90,91) and shows that the rapid decrease in morphology can be explained by deformation into sheets but not by deformation into filaments only (91). The polymer's viscoelastic character was also found to be important in determining the occurrence and size of sheets. Correlation of a stress ratio based on the polymer blend component properties and a Deborah number incorporating the processing shear rate gave a map of sheet breakup (62). Correlations of this type have been shown to be promising for prediction of morphology (81,68,92,93).

Recent studies of morphology development (60,94,95) show that the erosion mechanism is general for polymer systems and occurs even in simple shear flows. Figure 4.10 shows the erosion that takes place at the surface of the pellets of the polymer that melts at higher temperatures. Erosion occurs because the shear stress at the interface is several times greater than that in the bulk (96). During melting in extruders and batch mixers, small pieces of the dispersed phase pellet are shed at its surface in the form of sheets, ribbons, and cylinders (> 10 μm) into the matrix. These are subsequently broken into micron size droplets (see Figure 4.9).

4.3.2 PROCESSING PHASE INVERSION

When polymer blends are prepared from a mixture of solid pellets or powder, the melting order of the components is important in determining the morphology development (37–39,65,66,69,97–101). If the component in lower concentration, the minor component, softens first, then it will initially form the matrix surrounding the

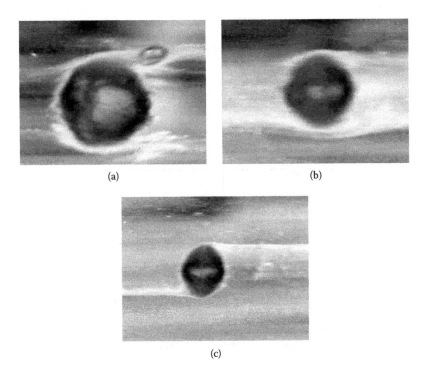

(a)　　　　　　　　(b)

(c)

FIGURE 4.10 Erosion phenomenon during polymer blending. A polycarbonate drop is sheared inside a polyethylene matrix using a transparent Couette cell at 230°C (Favis, *J. Appl. Polym. Sci.*, 39, 285, 1990). The viscosity ratio of this system is close to 9, showing that polymer drops can break up at viscosity ratios greater than 4.

pellets of the major component. As higher temperatures are reached during processing, the major component reaches sufficient concentration in the melt to envelop the minor component and the major component becomes the matrix. It has been reported that the blend component with the higher melting point in a differential scanning colorimetry (DSC) trace may melt sooner in the mixing process (99), and so it is important not to depend solely on equilibrium melting data to determine softening order in a mixer. The mechanism of the processing phase inversion (sometimes referred to as inversion of phase continuity during processing) is shown schematically in Figure 4.11. Lazo and Scott (66) showed a similar mechanism in 3-D; this mechanism incorporates the breakup of the minor phase domains.

Shih and coworkers (37,38,97,100) used visualization and mixer torque rheology and showed that a blend goes through four distinct rheological regimes during processing:

1. Elastic solid pellets
2. Deformable pellets
3. Transition material (powdery or doughy)
4. Viscoelastic fluid

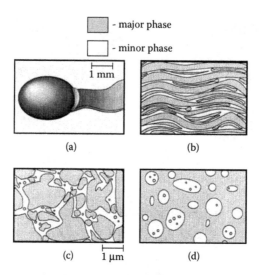

FIGURE 4.11 Phase inversion mechanism. (a) Sheets are pulled off the softening pellets near the softening transition of the major phase. (b) The sheets form lamellar domains inside the minor phase and (c) break up into irregular pieces (the minor phase is also breaking up at this point) until (d) phase inversion occurs during processing through coalescence of the major phase domains. (Sundararaj et al., *Polym. Eng. Sci.*, 36, 1769, 1996.)

Regime 4 is the preferred state for processing operations, but the material must pass through regimes 1 to 3 to enter regime 4. In regime 3, the blend can undergo the processing phase inversion, during which complex morphological and rheological phenomena occur. In some cases, during this inversion, an order of magnitude increase in torque is seen (37,38,97,98), and in other cases, an initial sharp decrease in torque followed by a sharp increase in torque is seen (99–101). Phase inversion is usually associated with a thermodynamic change where the dispersed phase inverts to become the major phase to minimize the free energy. Here, we are looking at dynamic phase inversion that occurs due to a change in melt concentration during processing. Since the change in phase continuity requires a certain amount of time, it may not be complete at the end of mixing. This mechanism may be responsible for the high polymer blend dispersed phase concentrations quoted in the literature — up to 95% dispersed phase concentration (102).

This type of phase inversion can also occur in Newtonian mixtures (103) by slowly adding one fluid into another. In Newtonian systems, it has been shown that the phase inversion is catastrophic and that there is hysteresis in the concentration for phase inversion, depending on the order the materials are added. In the same way, we expect different morphologies and rheologies based on which polymer melts first. We can change the melting order by melting the higher temperature material, first in the extruder and then side feeding the lower temperature material further downstream. The same effect can be achieved in a batch mixer by adding the second material after the first one has melted. There have been several studies in the literature studying the effect of mixing protocol or order of addition of components

(86,89,104), and differences in morphology are partly due to the fact that processing phase inversion occurs in some cases and not in others.

Shih (97) postulated that initially the minor component forms the matrix with closely packed pellets of the major component. As the major component begins to be dispersed in the minor component matrix, a packing arrangement or percolation threshold was reached, and the blend inverts in phase continuity. Sundararaj and coworkers (38) investigated the phenomenon in more detail by mixing reactive and nonreactive blends in a batch mixer and studying dispersed phase morphology by selectively extracting the matrixes and viewing the products under SEM (see Figure 4.2). This would correspond to extracting the minor phase during the initial stages of blending to view the dispersed major phase.

It was found that the major phase pellets are initially deformed into sheets, and when a critical composition of major phase domains is reached, the mixture undergoes phase inversion. The phenomenon is present in both reactive and nonreactive blends. It was reported that reactive blends required more time and mixing energy input because the occluded domains are stabilized against coalescence by copolymers formed at the interface. Scott and Joung (105) studied the effect of the viscosity ratio ($\eta_r = \eta_{minor}/\eta_{major}$) on phase inversion, particularly at early mixing times. They reported a qualitative difference in torque traces, depending on the viscosity ratio. For $\eta_r > 0.1$, the time to phase inversion was relatively constant; however, for $\eta_r < 0.1$, the time to phase inversion increased as the viscosity ratio decreased. Lazo and Scott (65,66) performed model isothermal experiments in a simple shear cell and showed that phase inversion can occur even in simple shear flows and that melting is not a requirement for the phase inversion phenomenon. They also indicated that, besides rheology and component concentration, the thickness of the minor component domains should also be considered since the minor component sheets need to break up before phase inversion.

4.3.3 FINAL MORPHOLOGY

By far the majority of the work in polymer blends has focused on the final morphology. This is partly a result of the fact that the final properties are dependent on the final morphology. For different applications, different types of morphology are sought. For example, for barrier properties, a lamellar structure (1,2,8) is required, whereas for impact properties, a droplet-in-matrix morphology is preferred (3,4). Filament morphology may be desired for transport applications or for applications where nonisotropic properties are required. Here, we will focus on droplet-in-matrix final morphologies.

4.3.3.1 Breakup and Coalescence Balance

The size and shape of the dispersed phase depends on several material and processing parameters, including rheology, interfacial properties, and blend composition. A review of early work in drop breakup and coalescence in polymer blends is given by Utracki and Shi (106). Tokita (107) suggested that an equilibrium drop diameter occurs from continuous breakup and coalescence of the dispersed phase. Drop breakup has been

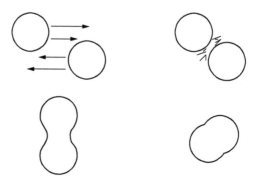

FIGURE 4.12 Idealized depiction of coalescence in polymer blends. Drops are brought close to each other by the shear field, and then the matrix film between the drops thins.

studied extensively (56–59,71,72,108–112). Coalescence during polymer blending has been studied more recently (84,30–32,113–125). Coalescence of drops is shown schematically in Figure 4.12. Fortelný and coworkers (126–128) found that the amount of coalescence in blends depends on matrix phase viscosity and dispersed phase volume fraction. However, this was for a quiescent blend without mechanical stress.

4.3.3.2 Effect of Copolymer on Coalescence

It has been shown that the particle size decreases when premade copolymers (84,129–134) or reactive polymers (84,86,87,121) are used. Similar to Newtonian emulsions, an "emulsification curve" concept (Figure 4.13) has been advanced for polymer blends using premade copolymers (132–134) or reactive polymers (86). The particle size is seen to decrease when low concentrations of copolymer compatibilizer are added to a blend, and a critical compatibilizer concentration exists above which there is no change in particle size. This suggests that there is a point where the copolymer saturates the interface and the addition of any more compatibilizer will have no effect. In the case of block copolymers, this extra amount can exist in the form of micelles.

When the dispersed phase concentration is increased in uncompatibilized blends, the particle size increases, but for blends containing premade or *in situ* copolymers, the particle size remains fairly constant. This observation suggests that coalescence is suppressed by the addition of copolymers. Some researchers have reported that a lowered interfacial tension was responsible for morphology stabilization (132). Though there is evidence that there exists an optimum block copolymer molecular weight for minimizing the interfacial tension, most research indicates that the stable morphology is due to a suppression of coalescence. There were two major theories proposed to describe the suppression of coalescence. Sundararaj and Macosko (84) described a steric interaction between droplets due to the copolymer at the interface, shown in Figure 4.14a. Milner and Xi (135) suggested that a local concentration gradient was created when two drops came together leading to a Marangoni stress that slowed the rate of film drainage between the drops, thereby suppressing coalescence (see Figure 4.14b).

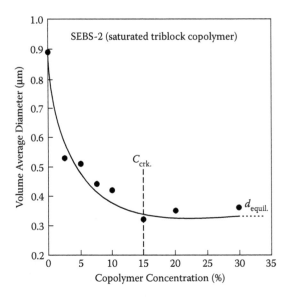

FIGURE 4.13 Emulsification curve for polystyrene/ethylene-propylene rubber 90:10 blend compatibilized with SEBS triblock copolymer. The particle size initially decreases with concentration of copolymer and then reaches a stable size after a critical copolymer concentration.

There have been many studies attempting to resolve which mechanism is prevalent in polymer blends. Recently, two reports using designed experiments have tried to find the answer. Lyu et al. (118) showed that the minimum coverage required for coalescence suppression in PS/HDPE 87:13 blends with polystyrene/polyethylene (PS-PE) block copolymer did not depend on shear rate but did depend on copolymer molecular weight. Both of these observations support the steric interaction mechanism but are in disagreement with the Marangoni force mechanism. However, Ha et al. (124) performed model experiments for coalescence of two polybutadiene drops in a poly(dimethylsiloxane) (PDMS) matrix, and they concluded that the most probable mechanism was Marangoni forces. However, the results were not definitive since they found the coalescence was insensitive to copolymer concentration for high viscosity ratio systems, which is not consistent with the Marangoni force mechanism but is consistent with the steric interaction mechanism. The debate on the mechanism of coalescence suppression continues, but the stabilizing effect that copolymers have on polymer blend morphology is not in question. It is accepted that some type of stabilization is required to create immiscible polymer blends with tailored morphology and properties.

4.3.4 Effect of Processing and Material Parameters on Morphology

Many researchers have tried to change the blend morphology using processing parameters such as shear rate (related to r/min), temperature, mixing time, mixer type, and composition. Among these, the most important parameter is blend com-

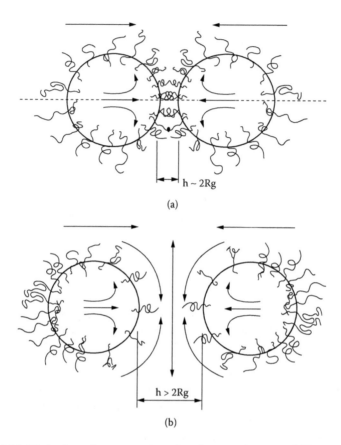

FIGURE 4.14 Mechanisms for suppression of coalescence in compatibilized blends. The copolymer may reduce coalescence effects by (a) steric repulsion of interacting drops or (b) creating a surface tension gradient (Marangoni forces). (Lyu et al., *Macromolecules*, 35, 7855, 2002.)

position. The other processing parameters, for the most part, have been found to have much less effect on morphology. Part of the reason is that the dispersed phase size scale changes from millimeters to microns within a very short period. The final domain size is about 1 μm for the majority of uncompatibilized polymer blends, regardless of processing conditions. However, subtle changes in particle size can have important consequences for mechanical properties, and therefore, there have been many studies about the effect of processing parameters on morphology.

The viscosity ratio and other material rheological parameters have been shown to be important in determining breakup. Wu (71) and Favis and Chalifoux (72) performed the first investigations in this area. Both concluded that domain size correlated with the viscosity ratio ($\eta_r = \eta_d/\eta_m$, where η_d is the dispersed phase viscosity), but the relationship was not exactly the same as reported in the Newtonian literature. Interfacial tension is another key parameter. Most polymer blends have an interfacial tension in the range 0.002 to 0.005 N/m, much lower than that of conventional fluid systems like oil/water. During drop deformation, viscous forces balance the interfacial forces, and the capillary number is a ratio of these:

$$Ca = \frac{\eta_m \gamma R}{\Gamma} \tag{4.1}$$

where η_m is the matrix viscosity, γ is the shear rate, R is the drop radius and Γ is the interfacial tension. When the viscous forces are much greater than the interfacial forces (i.e. at a high capillary number), drop breakup is expected.

Wu used these dimensionless parameters, capillary number and viscosity ratio, to study breakup and tried to predict final morphology in a twin-screw extruded blend. It should be pointed out that Wu mistakenly defined twice the capillary number (2Ca) as the Weber number (We). The Weber number is used widely in emulsions since inertial forces are important. The capillary number is more appropriate for polymer blends since, due to the high polymer viscosity, the flows are typically laminar flows, and viscous forces dominate. Wu defined a critical Capillary number (what he called the Weber number) for each viscosity ratio above which drop breakup occurred:

$$2Ca = 4\eta_r^{\pm 0.84} \tag{4.2a}$$

Favis and Chalifoux (72) showed similar trends in their work. Two major discrepancies in Wu's work were that: (1) the dispersed phase concentration in his blends was 15% so the correlation did not really apply to single drop breakup and (2) he used a twin-screw extruder, which produced a combination of shear and extensional flows, so the flow field was not well defined. However, it was an excellent first attempt to try to understand the influence of processing parameters on final morphology

4.3.4.1 Effect of Composition

Much of the early research on polymer blends concentrated on the effect of the composition since by changing the composition, one could easily affect final blend properties. It was found that increasing the composition typically increased particle size up to the cocontinuous morphology region (32,72,84,121,136). There have been a few exceptions to this general observation at higher concentrations (137) because of the influence of the dispersed phase viscosity on dispersed phase domain deformation. Sundararaj and Macosko (84) showed that the particle size increased with concentration of dispersed phase due to coalescence in uncompatibilized blends. At low concentrations, the particles reached a limiting size. This limiting size should correspond to the dilute solution limit; i.e., the stable size of a single drop in a matrix. Therefore, this dilute solution drop size should be able to provide correlations similar to those seen in Newtonian emulsions. It was found that the drop size was greater than that predicted by Taylor's limit (138,139) but lower than that predicted by Wu's correlation (71). It is logical that the drop size observed for the dilute solution or low concentration is lower than that obtained by Wu because coalescence was present in Wu's data. The discrepancy with the Taylor limit was explained to polymer elasticity of the drop.

For 1 wt% blends, Everaert et al. (140) found a correlation similar to Wu but with different fitting coefficients:

$$2Ca = 1.2\,\eta_r^{\pm0.45} \tag{4.2b}$$

In addition, they found that at low viscosity ratios, Taylor's limit is valid, but at higher viscosity ratios, the drop size is greater than Taylor's limit and lower than Wu's correlation [Eq. (4.1)]. Again, it makes sense that at 1% concentration, the drop size should be less than that predicted by Wu's correlation because of coalescence effects in Wu's work. However, it is not so clear why the Taylor limit is valid for some regions of viscosity ratio but not others. In particular, Taylor predicts no breakup for viscosity ratios above 4, and yet breakup is seen in many polymer blends with much higher viscosity ratios (71,72,84,140). Recent work using model mixers has shown that polymer drops can break up in simple shear at viscosity ratios above 4 (43,60,62,65,67,68,70). It is clear now that polymer drops behave very differently from Newtonian drops and that Ca or η_r alone may not be sufficient to describe the breakup.

At higher concentrations, cocontinuous blend morphology is found and has been studied widely (e.g., 55,141–146). The concentration range of the cocontinuous region depends on several factors, including viscosity ratio and interfacial tension. The scope of this chapter does not allow us to investigate this area in any detail. The reader is referred to a comprehensive review article on the area by Potschke and Paul (147).

4.3.4.2 Effect of Shear Rate

The customary expectation is that the dispersed phase domain size will decrease with increased shear rate. Even when considering the shear thinning nature of polymer viscosity, the shear stress continues to increase with increasing shear rate so drop size should decrease at higher shear rates. However, most works studying the effect of shear rate on morphology in polymer blends do not show this trend. As the shear rate was increased, Favis (59) found a modest decrease initially and then a leveling off in the particle size. White and coworkers found that changing the shear stress did affect the particle size (148) while Walczak found that changing the shear stress by a factor of three had little effect on structure (149). During mixing, Plochocki et al. (58) observed that the drop size decreased initially as the shear rate was increased, and then the drop size actually increased at higher shear rates. The authors explained that an optimum energy input or SEC (specific energy consumption) gave a minimum drop size. Ghodgaokar and Sundararaj (81) also observed a minimum drop size at an intermediate shear rate and used a force balance to predict this behavior.

4.3.4.3 Effect of Mixing Time

As previously mentioned, dispersed phase domain size decreases very rapidly in the initial mixing stages. After 1 min in the batch mixer or one L/D length of the first mixing section of an extruder, there is very little change in the morphology (16,51,55,150–153). Mixing for up to 10 min in the batch mixer showed very little change from the morphology seen at 1 min (55–57,). Mixing for longer times actually led to coarsening, and particle size increased between 10 and 20 min of mixing (121). The starting particle size for polymer mixing is usually dictated by feeding limitations — the particle must be small enough to be fed into the mixer but not too small or powdery that it is difficult to handle. Therefore, a millimeter size feed is normally used. If we use the viscous force to interfacial force balance as a guide to determine the particle size, an appropriate drop size is in the micron- or even nanosize range. However, to directly feed materials of this size scale would be tricky. Because the polymer blend inherently has high viscosities and low interfacial tension, the particle size changes rapidly from millimeter size to a micron- or nanosize. However, once it reaches this stable size, the morphology does not change significantly upon further mixing.

4.3.4.4 Effect of Temperature

Temperature can have varying effects on the polymer morphology. Thomas and Groeninckx (121) report an increase in particle size with temperature for a nylon/EP (ethylene-propylene) rubber blend but indicate that this may be due to a decrease in the matrix phase nylon viscosity at higher temperatures. In addition, at higher concentrations, the lower matrix viscosity may enhance coalescence. Temperature has a marked effect on component polymer properties so it is difficult to make any sweeping generalizations about the effect of temperature on morphology. However, temperature definitely influences the morphology, and it must be considered when developing new blends.

4.3.4.5 Effect of Mixer Type

We have chiefly discussed morphology development in two types of mixers in this chapter since the majority of the work in polymer blending has been performed on batch mixers or twin-screw extruders. Single-screw extruders do not have as much capability for mixing and do not have the folding-reorientation mechanism that has been found to be important for blending (78,154,155). For batch mixers and twin-screw extruders, morphology development mechanisms are similar and the final morphology at matched conditions is also similar (51,121). It is also important to note that researchers using batch mixers (e.g., 72) and those using twin-screw extruders (e.g., 71) achieved similar morphologies, mechanisms, and conclusions about the polymer blending process.

4.3.4.6 Miniature Mixers

However, many laboratories, especially those at universities, have used miniature mixers to simulate larger scale mixers. One of the most popular miniature mixers is the MiniMax mixer (156) that is named after its inventor, Bryce *Max*well. Its simple design consists of a cylindrical rotor that rotates in a stator cup filled with polymer. This mixer has been used widely (29–33,61,75,131). However, this mixer did not do an adequate job in simulating larger scale mixers for uncompatibilized blends (61). It did perform adequately for reactive blends (61,157) and copolymer compatibilized blends (75).

Some improvements have been made to miniature mixing in the past decade. Periodic lifting of the rotor in the MiniMax improves the mixing (61). Maric and Macosko (131) showed that adding steel balls to the mixer cup greatly ameliorated the blend morphology. Recently, Sundararaj and coworkers (157) developed an asymmetric miniature mixer [Alberta Polymer Asymmetric Minimixer (APAM)] that uses a novel asymmetric rotor that creates axial flow and folding flow in a single-rotor geometry. The morphology created in the APAM matched those found in batch mixer and twin-screw extruder, and was superior to that created in standard operation of a MiniMax mixer. Both DSM and Daca Instruments have sold a microtwin-screw extruder.

4.3.5 MODEL EXPERIMENTS

Because of the difficulty of following a particle in the complex flow paths of a twin-screw extruder or twin-rotor batch mixer, several researchers have used model flow fields to study polymer blending (46,64,62,63,67,76,92,93,123–125,158–160). Figure 4.15 shows that polymer pellet breakup via sheeting is seen even in simple shear flows, and it was confirmed that this type of deformation occurs in the melt state (62,64,67,161).

Using these model experiments, researchers have been able to investigate the critical parameters responsible for breakup and have developed rules to determine when particular breakup mechanisms will dominate (60,62,64,68,92,93). In particular, these studies have shown that the elasticity of the blend components, characterized by the first or second normal stress difference, can control the breakup behavior. A summary of four different breakup mechanisms seen in simple shear flow is shown in Figure 4.16. It will be important in future research to extend these studies on model geometries to practical mixers.

4.4 CONCLUSIONS

Morphology development in polymer blends is very complex. The major mechanisms have been described in this chapter, but there is still much scope for research in this area to achieve a better understanding of what controls morphology development in polymer blends. The area of morphology development is important because it relates to the final properties of a polymer blend. If we can understand how the structure is formed, it may be possible to control final morphology and properties.

A new discovery in polymer blend morphology is the ability to make controlled nanostructured materials by reactive blending (24,25) and these materials may afford new properties. Use of other media to enhance dispersion such as the use of CO_2 in blends extrusion (162–164) is another area that may lead to new types of structures. Several researchers have used chaotic mixing to create structured blends (165–169). Solid state pulverization is also being studied vigorously and has shown some advantages but may lead to polymer degradation (170,171).

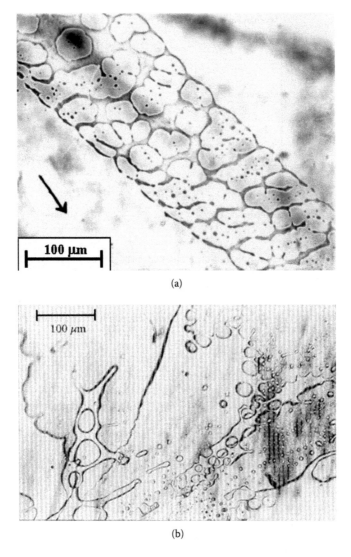

(a)

(b)

FIGURE 4.15 Sheeting mechanism occurs even in the melt state. (a) A sheet is formed when a polypropylene pellet in a polystyrene matrix is subjected to a shear flow in parallel plates. (b) Sheets form after mixing the same blend for 2 sec in a twin-rotor batch mixer. The blend was premelted before shearing in the batch mixer. (Sundararaj et al., *Polymer*, 36, 1957, 1995.)

The majority of morphology research has been performed on twin-rotor batch mixers or twin-screw extruders. New morphologies created by novel processes may give unanticipated benefits leading to new applications. Research by polymer blend scientists has led to the use of new or existing processes to create innovative materials simply and cost effectively.

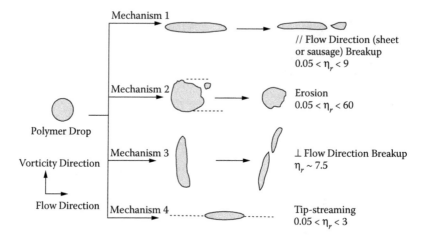

FIGURE 4.16 Four different polymer drop breakup mechanisms seen in simple shear flow. (a) The drop may form a sheet in the flow direction and break up, (b) the drop may erode at the surface slowly, (c) the drop may stretch in the vorticity direction and break up, or (d) the drop may spit out small droplets via a tip streaming mechanism. A combination of these breakup mechanisms can occur for the same blend depending on the temperature, shearing conditions, and drop size.

REFERENCES

1. Subramanian, P.M., Permeability barriers by controlled morphology of polymer blends, *Polym. Eng. Sci.*, 25, 483, 1985.
2. Subramanian, P.M. and Mehra, V., Laminar morphology in polymer blends — structure and properties, *Polym. Eng. Sci.*, 27, 663, 1987.
3. Wu, S., Phase structure and adhesion in polymer blends — a criterion for rubber toughening, *Polymer*, 26, 1855, 1985.
4. Borggreve, R.J.M., Gaymans, R.J., Schuijer J., and Housz, J.F.I., Brittle tough transition in nylon rubber blends — effect of rubber concentration and particle size, *Polymer*, 28, 1489, 1987.
5. Oshinski, A.J., Keskkula, H., and Paul, D.R., Rubber toughening of polyamides with functionalized block copolymers 2 — nylon 6,6, *Polymer*, 33, 284, 1992.
6. Lu, M., Keskkula H.G., and Paul, D.R., Toughening of nylon 6 with core-shell impact modifiers: effect of matrix molecular weight, *J. Appl. Polym. Sci.*, 59, 1467, 1996.
7. Macosko, C.W., Morphology development and control in immiscible polymer blends, *Macromol. Symp.*, 149, 171, 2000.
8. Kamal, M.R., Garmabi, H., Hozhabr, S., and Arghyris, L., The development of laminar morphology during extrusion of polymer blends, *Polym. Eng. Sci.*, 35, 41, 1995.
9. Manson J.A. and Sperling L.H., *Polymer Blends and Composites*, Plenum Press, New York, 1976.
10. Paul, D.R. and Bucknall, C.B., *Polymer Blends: Formulation and Performance*, John Wiley & Sons, New York, 2000.
11. Utracki, L.A., *Polymer Alloys and Blends*, Hanser Publishers, Munich, 1989.

12. Utracki, L.A., *Polymer Blends Handbook*, Kluwer Academic, Dordrecht, Netherlands, 2003.
13. Datta, S. and Lohse, D.J., *Polymer Compatibilizers*, Hanser Gardner Publications, Munich, 1996.
14. Baker, W.E., Scott, C.E., Hu, G.H., *Reactive Polymer Blending*, Hanser Gardner Publications, Munich, 2000.
15. Franzheim, O., Stephan, M., Rische, T., Heidemeyer, P., Burkhardt, U., and Kiani, A., Analysis of morphology development of immiscible blends in a twin screw extruder, *Adv. Polym. Technol.*, 16, 1, 1997.
16. Potente, H., Krawinkel, S., Bastian, M., Stephan, M., and Potschke, P., Investigation of the melting behavior and morphology development of polymer blends in the melting zone of twin-screw extruders, *J. Appl. Polym. Sci.*, 82, 1986, 2001.
17. Machado, A.V., Covas, J.A., and van Duin, M., Evolution of morphology and of chemical conversion along the screw in a corotating twin-screw extruder, *J. Appl. Polym. Sci.*, 71, 135, 1999.
18. Machado, A.V., Covas, J.A., Walet, M., and van Duin, M., Effect of composition and processing conditions on the chemical and morphological evolution of PA-6/EPM/EPM-g-MA blends in a corotating twin-screw extruder, *J. Appl. Polym. Sci.*, 80, 1535, 2001.
19. Machado, A.V., van Duin, M., and Covas, J.A., Monitoring the evolution of the properties of PA-6/EPM-g-MA blends in a twin-screw extruder, *Polym. Eng. Sci.*, 42, 2032, 2002.
20. Belanger, C., Cielo, P., Patterson, W.I., Favis, B.D., and Utracki, L.A., A polarized light scattering approach for dispersed phase morphology characterization in simulated polymer blends, *Polym. Eng. Sci.*, 34, 1589, 1994.
21. Schlatter, G., Serra, C., Bouquey, M., Muller, R., and Terrisse, J., Online light scattering measurements: a method to assess morphology development of polymer blends in a twin-screw extruder, *Polym. Eng. Sci.*, 42, 1965, 2002.
22. Chen, H., Sundararaj, U., Nandakumar, K., and Wetzel, M.D., Investigation of the melting mechanism in a twin-screw extruder using a pulse method and online measurement, *Ind. Eng. Chem. Res.*, 43, 6822, 2004.
23. Zhu, L.J. and Geng, X.Z., Experimental investigation of polymer pellets melting mechanisms in corotating twin-screw extrusion, *Adv. Polym. Technol.*, 21, 188, 2002.
24. Pernot, H., Baumert, M., Court, F., and Leibler, L., Design and properties of co-continuous nanostructured polymers by reactive blending, *Nature Materials*, 1, 54, 2002.
25. Ryan, A.J., Polymer science — designer polymer blends, *Nature Materials*, 1, 8, 2002.
26. Gilmore, I.S., Seah, M.P., and Johnstone, J.E., Quantification issues in ToF-SSIMS and AFM co-analysis in two-phase systems, exampled by a polymer blend, *Surf. Interface Anal.*, 35, 888, 2003.
27. Loyens, W. and Groeninckx, G., Phase morphology development in reactively compatibilised polyethylene terephthalate/elastomer blends, *Macromol. Chem. Phys.*, 203, 1702, 2002.
28. Mirabella, F.M., Atomic force microscopy observation of crystallization and melting of narrow composition distribution polyethylene/alpha-olefin copolymers, Society of Plastics Engineers Tech. Papers, Fundamentals Forum, 2004.
29. Okamoto, M. and Inoue, T., Reactive processing of polymer blends — analysis of the change in morphological and interfacial parameters with processing, *Polym. Eng. Sci.*, 33, 175, 1993.

30. Ibuki, J., Charoensirisomboon, P., Chiba, T., Ougizawa, T., Inoue, T., Weber, M., and Koch, E., Reactive blending of polysulfone with polyamide: a potential for solvent-free preparation of the block copolymer, *Polymer*, 40, 647, 1999.

31. Charoensirisomboon, P., Chiba, T., Solomko, S.I., Inoue, T., and Weber, M., Reactive blending of polysulfone with polyamide: a difference in interfacial behavior between in situ formed block and graft copolymers, *Polymer*, 40, 6803, 1999.

32. Charoensirisomboon, P., Inoue, T., Solomko, S.I., Sigalov, G.M., and Weber, M., Morphology of compatibilized polymer blends in terms of particle size-asphericity map, *Polymer*, 41, 7033, 2000.

33. Pan, L., Chiba, T., and Inoue, T., Reactive blending of polyamide with polyethylene: pull-out of in situ-formed graft copolymer, *Polymer*, 42, 8825, 2001.

34. Sakai, T., The development of online techniques and novel processing systems for the monitoring and handling of the evolution of microstructure in nonreactive and reactive polymer systems, *Adv. Polym. Tech.*, 14, 277, 1995.

35. Min, K., Kim, M.H., and White, J.L., Flow visualization and performance of a non-intermeshing counter-rotating twin screw extruder, *Intern. Polymer Process.*, 3, 165, 1988.

36. Kim, P.J. and White, J.L., Flow visualization and residence time distributions in a modular co-rotating twin-screw extruder, *Intern. Polymer Process.*, 9, 108, 1994.

37. Shih, C.K., Tynan, D.G., and Denelsbeck, D.A., Rheological properties of multicomponent polymer systems undergoing melting or softening during compounding, *Polym. Eng. Sci.*, 31, 1670, 1991.

38. Sundararaj, U., Macosko, C.W., and Shih, C.K., Evidence for inversion of phase continuity during morphology development in polymer blending, *Polym. Eng. Sci.*, 36, 1769, 1996.

39. Ratnagiri, R. and Scott, C.E., Visualization and microscopic modeling of phase inversion during compounding, *Polym. Eng. Sci.*, 41, 1310, 2001.

40. Lin, B. and Sundararaj, U., Visualization of poly(ether imide) and polycarbonate blending in an internal mixer, *J. Appl. Polym. Sci.*, 92, 1165, 2004.

41. Li, S., Migler, K.B., Hobbie, E.K., Kramer, H., Han, C.C., and Amis, E.J., Light-scattering photometer with optical microscope for the in-line study of polymer extrusion, *J. Polym. Sci.: Polym. Phys.*, 35, 2935, 1997.

42. Hobbie, E.K., Migler, K.B., Han, C.C., and Amis, E.J., Light scattering and optical microscopy as in-line probes of polymer blend extrusion, *Adv. Polym. Technol.*, 17, 307, 1998.

43. Migler, K.B., Hobbie, E.K., and Qiao, F., In line study of droplet deformation in polymer blends in channel flow, *Polym. Eng. Sci.*, 39, 2282, 1999.

44. Migler, K.B., Droplet vorticity alignment in model polymer blends, *J. Rheol.*, 44, 277, 2000.

45. Migler, K.B., String formation in sheared polymer blends: coalescence, breakup, and finite size effects, *Phys. Rev. Lett.*, 86, 1023, 2001.

46. Pathak, J.A., Davis, M.C., Hudson, S.D., and Migler, K.B., Layered droplet microstructures in sheared emulsions: finite-size effects, *J. Colloid Interface Sci.*, 255, 391, 2002.

47. Cartier, H. and Hu, G.H., Morphology development of in situ compatibilized semi-crystalline polymer blends in a co-rotating twin-screw extruder, *Polym. Eng. Sci.*, 39, 996, 1999.

48. Tadmor, Z. and Gogos, C.G., *Priniciples of Polymer Processing*, John Wiley & Sons, New York, 1979.

49. Tadmor, Z., Gogos, C.G., and Kim, M.H., Melting phenomena and mechanisms in polymer processing equipment, *Adv. Polym. Technol.*, 17, 285, 1998.
50. DeLoor, A., Cassagnau, P., Michel, A., and Vergnes, B., Morphological-changes of a polymer blend into a twin-screw extruder, *Intern. Polym. Proc.*, 9, 211, 1994.
51. Sundararaj, U., Macosko, C.W., Rolando, R.J., and Chan, H.T., Morphology development in polymer blends, *Polym. Eng. Sci.*, 32, 1814, 1992.
52. Curry, J., Melting mechanisms in ZSK extruders, *SPE Tech. Papers*, 53, 92, 1995.
53. Potente, H. and Melisch, U., Theoretical and experimental investigations of the melting of pellets in co-rotating twin-screw extruders, *Intern. Polym. Process.*, XI, 101, 1996.
54. Nishio, T., Suzuki, Y., Kojima, K, and Kakugo, M., Morphology of maleic-anhydride grafted polypropylene and polyamide alloy produced by reactive processing, *J. Polym. Eng.*, 10, 123, 1991.
55. Scott, C.E. and Macosko C.W., Model experiments concerning morphology development during the initial-stages of polymer blending, *Polym. Bull.*, 26, 341, 1991; *Polymer, 36,* 461, 1995.
56. Karger-Kocsis, J., Kallo, A., Szafner, A., Bodor, G., and Senyei, Z., Morphological-study on the effect of elastomeric impact modifiers in polypropylene systems, *Polymer*, 19, 448, 1978.
57. Schrieber, H.P. and Olguin, A., Aspects of dispersion and flow in thermoplastic-elastomer blends, *Polym. Eng. Sci.*, 23, 129, 1983.
58. Plochocki, A.P., Dagli, S.S., and Andrews, R.D., The interface in binary-mixtures of polymers containing a corresponding block copolymer — effects of industrial mixing processes and of coalescence, *Polym. Eng. Sci.*, 30, 741, 1990.
59. Favis, B.D., The effect of processing parameters on the morphology of an immisible binary blend, *J. Appl. Polym. Sci.*, 39, 285, 1990.
60. Lin, B., Sundararaj, U., Mighri, F., and Huneault, M.A., Erosion and breakup of polymer drops under simple shear in high viscosity ratio systems, *Polym. Eng. Sci.*, 43, 891, 2003.
61. Sundararaj, U., Macosko, C.W., Nakayama, A., and Inoue, T., Milligrams to kilograms — an evaluation of mixers for reactive polymer blending, *Polym. Eng. Sci.*, 35, 100, 1995.
62. Sundararaj, U., Dori, Y., and Macosko, C.W., Sheet formation in immiscible polymer blends, *Polymer*, 36, 1957, 1995.
63. Sundararaj, U., Dori, Y., and Macosko, C.W., Visualization of polymer blending, *SPE Tech. Papers*, 52, 2448, 1994.
64. Levitt, L., Macosko, C.W., and Pearson, S.D., Influence of normal stress difference on polymer drop deformation, *Polym. Eng. Sci.*, 36, 1647, 1996.
65. Lazo, N.D.B. and Scott, C.E., Morphology development during phase inversion of a PS PE blend in isothermal, steady shear flow, *Polymer*, 40, 5469, 1999.
66. Lazo, N.D.B. and Scott, C.E., Morphology development during phase inversion in isothermal, model experiments: steady simple-shear and quiescent flow fields, *Polymer*, 42, 4219, 2001.
67. Cristini, V., Hooper, R.W., Macosko, C.W., Simeone, M., and Guido, S., A numerical and experimental investigation of lamellar blend morphologies, *Ind. Eng. Chem. Res.*, 41, 6305, 2002.
68. Lin, B., Sundararaj, U., Mighri, F., and Huneault, M.A., Parallel breakup of polymer drops under simple shear, *Macromol. Rapid. Comm.*, 24, 783, 2003.
69. Yang, Y., Lee, L.J., and Koelling, K.W., Structure evolution in polymer blending using microfabricated samples, *Polymer*, 45, 1959, 2004.

70. Lin, B. and Sundararaj, U., Sheet formation during drop deformation in polyethylene/polycarbonate systems sheared between parallel plates, *Polymer*, 45, 7605, 2004.

71. Wu, S., Formation of dispersed phase in incompatible polymer blends — interfacial and rheological effects, *Polym. Eng. Sci.*, 27, 335, 1987.

72. Favis, B.D. and Chalifoux, J.P., The effect of viscosity ratio on the morphology of polypropylene polycarbonate blends during processing, *Polym. Eng. Sci.*, 27, 1591, 1987.

73. Galloway, J.A., Montminy, M.D., and Macosko, C.W., Image analysis for interfacial area and cocontinuity detection in polymer blends, *Polymer*, 43, 4715, 2002.

74. Lindt, J.T. and Ghosh, A.K., Fluid-mechanics of the formation of polymer blends. 1. Formation of lamellar structures, *Polym. Eng. Sci.*, 32, 1802, 1992.

75. Macosko, C.W., Guegan, P., Khandpur, A.K., Nakayama, A., Marechal, P., and Inoue, T., Compatibilizers for melt blending: premade block copolymers, *Macromolecules*, 29, 5590, 1996.

76. Levitt, L. and Macosko, C.W., Shearing of polymer drops with interface modification, *Macromolecules*, 32, 6270, 1999.

77. Ghosh, A.K. and Tyagi, S., Morphology development during blending of immiscible polymers in screw extruders, *Polym. Eng. Sci.*, 42, 1309, 2002.

78. Janssen, J.H.M. and Meijer, H.E.H., Dynamics of liquid-liquid mixing — a 2-zone model, *Polym. Eng. Sci.*, 35, 1766, 1995.

79. Elemans, P.H.M., Bos, H.L., Janssen, J.H.M., and Meijer, H.E.H., Transient phenomena in dispersive mixing, *Chem. Eng. Sci.*, 48, 267, 1993.

80. Bordereau, V., Shi, Z.H., Utracki, L.A., Sammut, P., and Carrega, M., Development of polymer blend morphology during compounding in a twin-screw extruder. 3. Experimental procedure and preliminary-results, *Polym. Eng. Sci.*, 32, 1846, 1992.

81. Ghodgaonkar, P. and Sundararaj, U., Prediction of drop size in polymer blends, *Polym. Eng. Sci.*, 36, 1656, 1996.

82. van Oene, H.J., Modes of dispersion of viscoelastic fluids in flow, *J. Colloid Interface Sci.*, 40, 448, 1972.

83. Qian, B., Todd, D.B., and Gogos, C.G., Plastic energy dissipation and its role on heating/melting of single-component polymers and multi-component polymer blends, *Adv. Polym. Technol.*, 22, 85, 2003.

84. Sundararaj, U. and Macosko, C.W., Drop breakup and coalescence in polymer blends, *Macromolecules*, 28, 2647, 1995.

85. Dedecker, K. and Groeninckx, G., Reactive compatibilisation of A/(B/C) polymer blends. Part 1. Investigation of the phase morphology development and stabilization, *Polymer*, 39, 4985, 1998.

86. Thomas, S. and Groeninckx, G., Reactive compatibilisation of heterogeneous ethylene propylene rubber (EPM)/nylon 6 blends by the addition of compatibiliser precursor EPM-g-MA, *Polymer*, 49, 5799, 1999.

87. Tol, R.T., Groeninckx, G., Vinckier, I., Moldenaers, P., and Mewis, J., Phase morphology and stability of co-continuous (PPE/PS)/PA6 and PS/PA6 blends: effect of rheology and reactive compatibilization, *Polymer*, 45, 2587, 2004.

88. Li, H., Hu, G.H., and Sousa, J.A., Morphology development of immiscible polymer blends during melt blending: effects of interfacial agents on the liquid-solid interfacial heat transfer, *J. Polym. Sci. Polym. Phys.*, 37, 3368, 1999.

89. Bourry, D. and Favis, B.D., Morphology development in a polyethylene/polystyrene binary blend during twin-screw extrusion, *Polymer*, 39, 1851, 1998.

90. Sundararaj, U., *Morphology Development during Polymer Blending*, Ph.D. thesis, University of Minnesota, Minneapolis, MN, 1994.

91. Willemse, R.C., Ramaker, E.J.J., van Dam, J., and de Boer, A.P., Morphology development in immiscible polymer blends: initial blend morphology and phase dimensions, *Polymer*, 40, 6651, 1999.

92. Lerdwijitjarud, W., Sirivat, A., and Larson, R.G., Influence of elasticity on dispersed-phase droplet size in immiscible polymer blends in simple shearing flow, *Polym. Eng. Sci.*, 42, 798, 2002.

93. Lerdwijitjarud, W., Larson, R.G., Sirivat, A., and Solomon, M.J., Influence of weak elasticity of dispersed phase on droplet behavior in sheared polybutadiene/poly(dimethyl siloxane) blends, *J. Rheol.*, 47, 37, 2003.

94. Potente, H., Bastian, M., Bergemann, K., Senge, M., Scheel, G., and Windelmann, T., Morphology of polymer blends in the melting section of co-rotating twin screw extruders, *Polym. Eng. Sci.*, 41, 222, 2001.

95. Potente, H. and Bastian, M., Polymer blends in co-rotating twin screw extruders, *Intern. Polym. Proc.*, 16, 14, 2001.

96. Chen, H., Sundararaj, U., and Nandakumar, K., Modeling of polymer melting, drop deformation and breakup under shear flow, *Polym. Eng. Sci.,* 44, 1258, 2004.

97. Shih, C.K., Mixing and morphological transformations in the compounding process for polymer blends — the phase inversion mechanism, *Polym. Eng. Sci.*, 35, 1688, 1995.

98. Sundararaj, U., Effect of processing parameters on phase inversion during polymer blending, *Macromol. Symp.*, 112, 85, 1996.

99. Ratnagiri, R. and Scott, C.E., Phase inversion during compounding with a low melting major component: polycaprolactone/polyethylene blends, *Polym. Eng. Sci.*, 38, 1751, 1998.

100. Ratnagiri, R., Scott, C.E., and Shih, C.K., The effect of scaleup on the processing behavior of a blend exhibiting phase inversion during compounding, *Polym. Eng. Sci.*, 41, 1019, 2001.

101. Burch, H.E. and Scott, C.E., Effect of viscosity ratio on structure evolution in miscible polymer blends, *Polymer*, 42, 7313, 2001.

102. Valenza A., Demma, G.B., and Acierno, D., Phase inversion and viscosity-composition dependence in PC/LLDPE blends, *Polym. Networks Blends*, 3, 15, 1993.

103. Salager, J.L., in *Encyclopedia of Emulsion Technology, Vol. 3*, Becher, P., Ed., Marcel Dekker Inc., New York, 1988..

104. Burch, H.E. and Scott, C.E., Effect of addition protocol on mixing in miscible and immiscible polymer blends, *Polym. Eng. Sci.*, 42, 1197, 2002.

105. Scott, C.E. and Joung, S.K., Viscosity ratio effects in the compounding of low viscosity, immiscible fluids into polymeric matrices, *Polym. Eng. Sci.*, 36, 1666, 1996.

106. Utracki, L.A. and Shi, Z.H., Development of polymer blend morphology during compounding in a twin-screw extruder. 1. Droplet dispersion and coalescence — a review, *Polym. Eng. Sci.*, 32, 1824, 1992.

107. Tokita, N., Analysis of morphology formation in elastomer blends, *Rubber Chem. Technol.*, 50, 292, 1977.

108. Roland, C.M. and Bohm, G.G.A., Shear-induced coalescence in 2-phase polymeric systems. 1. Determination from small-angle neutron-scattering measurements, *J. Polym. Sci. Polym. Phys.*, 22, 79, 1984.

109. Elmendorp, J.J. and van der Vegt, A.K., A study on polymer blending microrheology. 4. The influence of coalescence on blend morphology origination, *Polym. Eng. Sci.*, 26 1332, 1986.

110. Lyngaae-Jorgensen, J. and Valenza, A., Structuring of polymer blends in simple shear-flow, *Makromol. Chem., Macromol. Symp.*, 38, 43, 1990.
111. Chin, H.B. and Han, C.D., Studies on droplet deformation and breakup. 2. Breakup of a droplet in nonuniform shear-flow, *J. Rheol.*, 24, 1, 1980.
112. Min, K., White, J.L., and Fellers, J.F., High-density polyethylene polystyrene blends — phase distribution morphology, rheological measurements, extrusion, and melt spinning behaviour, *J. Appl. Polym. Sci.*, 29, 2117, 1984.
113. Schoolenberg, G.E., During, F., and Ingenbleek, G., Coalescence and interfacial tension measurements for polymer melts: experiments on a PS-PE model system, *Polymer*, 39, 765, 1998.
114. Grizzuti, N. and Bifulco, O., Effects of coalescence and breakup on the steady state morphology of an immiscible polymer blend in shear flow, *Rheol. Acta*, 36, 406, 1997.
115. Wallheinke, K., Potschke, P., Macosko, C.W., and Stutz, H., Coalescence in blends of thermoplastic polyurethane with polyolefins, *Polym. Eng. Sci.*, 39, 1022, 1999.
116. Lyu, S.P., Bates, F.S., and Macosko, C.W., Coalescence in polymer blends during shearing. *AIChE J.*, 46, 229, 2000.
117. Lyu, S.P., Bates, F.S., and Macosko, C.W., Modeling of coalescence in polymer blends, *AIChE J.*, 48, 7, 2002.
118. Lyu, S.P., Jones, T.D., Bates, F.S., and Macosko, C.W., Role of block copolymers on suppression of droplet coalescence, *Macromolecules*, 35, 7855, 2002.
119. van Puyvelde, P., Velankar, S., Mewis, J., and Moldenaers, P., Effect of Marangoni stresses on the deformation and coalescence in compatibilized immiscible polymer blends, *Polym. Eng. Sci.*, 42, 1956, 2002.
120. Lepers, J.C., Favis, B.D., and Lacroix, C., The influence of partial emulsification on coalescence suppression and interfacial tension reduction in PP/PET blends, *J. Polym. Sci. Polym. Phys.*, 37, 939, 1999.
121. Thomas, S. and Groeninckx, G., Nylon 6 ethylene propylene rubber (EPM) blends: phase morphology development during processing and comparison with literature data, *J. Appl. Polym. Sci.*, 71, 1405,1999.
122. Wildes, G., Keskkula, H., and Paul, D.R., Coalescence in PC/SAN blends: effect of reactive compatibilization and matrix phase viscosity, *Polymer*, 40, 5609, 1999.
123. Yang, H., Park, C.C., Hu, Y.T., and Leal, L.G., The coalescence of two equal-sized drops in a two-dimensional linear flow, *Phys. Fluids*, 13, 1087, 2001.
124. Ha, J.W., Yoon, Y., and Leal, L.G., The effect of compatibilizer on the coalescence of two drops in flow, *Phys. Fluids*, 15, 849, 2003.
125. Leal, L.G., Flow induced coalescence of drops in a viscous fluid, *Phys. Fluids*, 16, 1833, 2004.
126. Fortelny, I. and Kovar, J., Theory of coalescence in immiscible polymer blends, *Polym. Compos.*, 9, 119, 1988.
127. Fortelny, I. and Zivny, A., Coalescence in molten quiescent polymer blends, *Polymer*, 36, 4113, 1995.
128. Fortelny, I. and Zivny, A., Film drainage between droplets during their coalescence in quiescent polymer blends, *Polymer*, 39, 2669, 1998.
129. Fayt, R., Jerome, R., and Teyssie, P., Molecular design of multicomponent polymer systems. 14. Control of the mechanical-properties of polyethylene polystyrene blends by block copolymers, *J. Polym. Sci. Polym. Phys.*, 27, 775, 1989.
130. Harrats, C., Blacher, S., Fayt, R., Jerome, R., and Teyssie, P., Molecular design of multicomponent polymer systems. 19. Stability of cocontinuous phase morphologies in low-density polyethylene polystyrene blends emulsified by block-copolymers, *J. Polym. Sci. Polym. Phys.*, 33, 801, 1995.

131. Maric, M. and Macosko, C.W., Improving polymer blend dispersions in mini-mixers, *Polym. Eng. Sci.*, 41, 118, 2001.
132. Favis, B.D., Phase size interface relationships in polymer blends — the emulsification curve, *Polymer*, 35, 1552, 1994.
133. Matos, M., Favis, B.D., and Lomellini, P., Interfacial modification of polymer blends — the emulsification curve. 1. Influence of molecular-weight and chemical-composition of the interfacial modifier, *Polymer*, 36, 3899, 1995.
134. Lomellini, P., Matos, M., and Favis, B.D., Interfacial modification of polymer blends — the emulsification curve. 2. Predicting the critical concentration of interfacial modifier from geometrical considerations, *Polymer*, 37, 5689, 1996.
135. Milner, S.T. and Xi, H.W., How copolymers promote mixing of immiscible homopolymers, *J. Rheol.*, 40, 663, 1996.
136. Favis, B.D., and Chalifoux, J.P., Influence of composition on the morphology of polypropylene polycarbonate blends, *Polymer*, 29, 1761, 1988.
137. Cho, Y.G. and Kamal, M.R., Effect of the dispersed phase fraction on particle size in blends with high viscosity ratio, *Polym. Eng. Sci.*, 42, 2005, 2002.
138. Taylor, G.I., The viscosity of a fluid containing small drops of another fluid, *Proc. R. Soc. London*, A138, 41, 1932.
139. Taylor, G.I., The formation of emulsions in definable fields of flow, *Proc. R. Soc. London*, A146, 501, 1934.
140. Evaraert, V., Aerts, L., and Groeninckx, G., Phase morphology development in immiscible PP/(PS/PPE) blends influence of the melt-viscosity ratio and blend composition, *Polymer*, 40, 6627, 1999.
141. Avergopoulos, G.N., Weissert, F.C., Biddison, P.H., and Bohm, G.G.A., Heterogeneous blends of polymers. Rheology and morphology, *Rubber Chem. Technol.*, 49, 93, 1976.
142. Paul, D.R. and Barlow, J.W., Polymer blends (or alloys), *J. Macromol. Sci. — Rev. Macromol. Chem.*, C18, 109, 1980.
143. Metelkin, V.I. and Blekht, V.S., Formation of a continuous phase in heterogeneous polymer mixtures, *Colloid. J. USSR*, 46, 425, 1984.
144. Jordhamo, G.M., Manson, J.A., and Sperling, L.H., Phase continuity and inversion in polymer blends and simultaneous interpenetrating networks, *Polym. Eng. Sci.*, 26, 517, 1986.
145. Chuai, C.Z., Almdal, K., and Lyngaae-Jorgensen, J., Phase continuity and inversion in polystyrene/poly(methyl methacrylate) blends, *Polymer*, 44, 481, 2003.
146. Marin, N. and Favis, B.D., Co-continuous morphology development in partially miscible PMMA/PC blends, *Polymer*, 43, 4723, 2002.
147. Potschke, P. and Paul, D.R., Formation of co-continuous structures in melt-mixed immiscible polymer blends, *J. Macromol. Sci. Polym. Rev.*, C43, 87, 2003.
148. Min, K., White, J.L., and Fellers, J.F., Development of phase morphology in incompatible polymer blends during mixing and its variation in extrusion, *Polym. Eng. Sci.*, 24, 1327, 1984.
149. Walczak, Z.K., Influence of shearing history on properties of polymer melt — 2, *J. Appl. Polym Sci.*, 17, 169, 1973.
150. Huneault, M.A., Shi, Z.H., and Utracki, L.A., Development of polymer blend morphology during compounding in a twin-screw extruder. 4. A new computational model with coalescence, *Polym. Eng. Sci.*, 35, 115, 1995.
151. Martin, P., Devaux, J., Legras, R., Leemans, L., van Gurp, M., and van Duin, M., Complex processing-morphology interrelationships during the reactive compatibilization of blends of poly(butylene terephthalate) with epoxide-containing rubber, *J. Appl. Polym. Sci.*, 91, 703, 2003.

152. Lee, J.K. and Han, C.D., Evolution of a dispersed morphology from a co-continuous morphology immiscible polymer blends, *Polymer*, 40, 6277, 1999.

153. Potente, H., Bastian, M., Gehring, A., Stephan, M., and Potschke, P., Experimental investigation of the morphology development of polyblends in corotating twin-screw extruders, *J. Appl. Polym. Sci.*, 76, 708, 2000.

154. Tjahjadi, M. and Ottino, J.M., Stretching and breakup of droplets in chaotic flows, *J. Fluid Mech.*, 232, 191, 1991.

155. Jana, S.C., Tjahjadi, M., and Ottino, J.M., Chaotic mixing of viscous fluids by periodic changes in geometry — baffled cavity flow, *AIChE J.*, 40, 1769, 1994.

156. Maxwell, B., Miniature injection molder minimizes residence time, *SPE J.*, 28, 24, 1972.

157. Breuer, O., Sundararaj, U., and Toogood, R.W., The design and performance of a new miniature mixer for specialty polymer blends and nanocomposites, *Polym. Eng. Sci.*, 44, 868, 2004.

158. Guido, S., Greco, F., and Villone, M., Experimental determination of drop shape in slow steady shear flow, *J. Coll. Interface Sci.*, 219, 298, 1999.

159. Guido, S. and Greco, F., Drop shape under slow steady shear flow and during relaxation. Experimental results and comparison with theory, *Rheol. Acta*, 40, 176, 2001.

160. Cavallo, R., Guido, S., and Simeone, M., Drop deformation under small-amplitude oscillatory shear flow, *Rheol. Acta*, 42, 1, 2003.

161. Yamane, H, Takahashi, M., Hayaski, R., Okamoto, K., Kahihara H., and Masuda, T., Observation of deformation and recovery of poly(isobutylene) droplet in a poly(isobutylene)/poly(dimethylsiloxane) blend and after application of step shear strain, *J. Rheol.*, 42, 567, 1998.

162. Elkovitch, M.D., Tomasko, D.L., and Lee, L.J., Supercritical carbon dioxide assisted blending of polystyrene and poly(methyl methacrylate), *Polym. Eng. Sci.*, 39, 2075, 1999.

163. Elkovitch, M.D., Lee, L.J., and Tomasko, D.L., Effect of supercritical carbon dioxide on morphology development during polymer blending, *Polym. Eng. Sci.*, 40, 1850, 2000.

164. Elkovitch, M.D., Lee, L.J., and Tomasko, D.L., Effect of supercritical carbon dioxide on PMMA/rubber and polystyrene/rubber blending: viscosity ratio and phase inversion, *Polym. Eng. Sci.*, 41, 2108, 2001.

165. Kwon, O. and Zumbrunnen, D.A., Progressive morphology development to produce multilayer films and interpenetrating blends by chaotic mixing, *J. Appl. Polym. Sci.*, 82, 1569, 2001.

166. Zumbrunnen, D.A. and Chhibber, C., Morphology development in polymer blends produced by chaotic mixing at various compositions, *Polymer*, 43, 3267, 2002.

167. Kwon, O. and Zumbrunnen, D.A., Production of barrier films by chaotic mixing of plastics, *Polym. Eng. Sci.*, 43, 1443, 2003.

168. Jana, S.C. and Sau, M., Effects of viscosity ratio and composition on development of morphology in chaotic mixing of polymers, *Polymer*, 45, 1665, 2004.

169. Sau, M. and Jana, S.C., A study on the effects of chaotic mixer design and operating conditions on morphology development in immiscible polymer systems, *Polym. Eng. Sci.*, 44, 407, 2004.

170. Lebovitz, A.H., Khait, K., and Torkelson, J.M., Stabilization of dispersed phase to static coarsening: polymer blend compatibilization via solid-state shear pulverization, *Macromolecules*, 35, 8672, 2002.

171. Furgiuele, N., Lebovitz, A.H., Khait, K., and Torkelson, J.M., Efficient mixing of polymer blends of extreme viscosity ratio: elimination of phase inversion via solid-state shear pulverization, *Polym. Eng. Sci.*, 40, 1447, 2000.

172. Scott, C.E., *Characterization of the Reactive Polymer Blending Process*, Ph.D. thesis, University of Minnesota, Minneapolis, MN, 1990.

5 The Role of Interfaces and Phase Morphology on Mechanical Properties of Multiphase: Copolymer Systems

Roland Weidisch and Manfred Stamm

CONTENTS

5.1 INTRODUCTION

A vigorously evolving and highly interdisciplinary area of research during the past decades has been the emerging field of block copolymers where much effort has been made to optimize polymer properties to design materials for a specific use. To provide particular mechanical properties, block copolymers, that have different available microphase separated morphologies are being used. The enhancement of toughness in rubber modified materials or polymer blends depends on the morphology of the materials; it is well known that the impact properties of homopolymers can be improved upon by incorporating a dispersed elastomeric phase that results in multiple crazing or multiple cavitation with shear yielding, i.e., the macroscopic phenomenon known as stress whitening (1,2).

In contrast to polymer blends, block copolymers form various ordered morphologies in a size scale of typically 10 to 100 nm and usually reveal a macroscopic grain structure in the size scale of 1 to 10 μm. Such materials exhibit isotropic properties in the absence of macroscopic orientations. In poly(styrene-b-isoprene) (PS-b-PI) diblock copolymers, the following morphologies have been reported: body-center-cubic (BCC) spheres, hexagonally packed cylinders, and "gyroid" and lamellar structures (3–6). In the weak segregation limit, the hexagonally ordered perforated layer (HPL) and the cubic bicontinuous structures (gyroid) have been found (7–9). Recently, several studies (10–12) have reported new morphologies in ABC block copolymers consisting of three different components, demonstrating the complexity of structure formation of block copolymers compared to other polymeric systems. While the influence of morphology on tensile properties and deformation behavior of block copolymers have been studied in detail, the effect of phase behavior and interface formation still remains to be investigated. In Sections 5.3.1 and 5.3.2, we discuss our recent results that reveal a correlation between phase behavior, interface formation, and mechanical properties of block copolymers.

At lower temperatures, most block copolymers undergo a microphase separation to form regular structures. In thin films, there is often a preferential adsorption of one of the two blocks to the films interfaces, leading to a parallel orientation of the lamellae, which allows for the determination of their interfacial width by neutron reflectometry (13). For strongly segregated polystyrene-b-polymethylmethacrylate (PS-b-PMMA) block copolymers, the interfacial thickness could be determined by a hyperbolic tangent profile with a high accuracy to 54 ± 2 Å; similar values have been found for other systems (14). To study the influence of interface formation on mechanical properties, systems with largely varied interfacial widths have to be used. While such a correlation has already been discussed for homopolymer blends (15–20), it has not yet been revealed for block copolymers where two blocks are chemically linked at the interfaces. Because of the advantages in anionic polymerization, a remarkable variety of block copolymer architectures is available, which have mainly been studied by transmission electron microscopy (TEM) and small angle x-ray scattering (SAXS). Recently, there is much interest in the mechanical properties of novel architectures with remarkably enhanced toughness that depend on the molecular architecture (see Section 5.3.3). Another interesting area of study is the blending of block copolymers with homopolymers or other block copolymers resulting in a large variety of morphology and mechanical properties. The use of binary block copolymers allows for the tailoring of morphology and mechanical properties for special applications. However, details concerning the mechanical properties and toughness of binary block copolymer blends have not yet been reported, which is one aspect of our investigations and is reported in the Section 5.4.

5.2 EXPERIMENTAL

Transmission electron microscopic investigations were performed using ultrathin sections (50 nm) cut at −80°C using diamond knives in an Ultramicrotome (Reichert). The polystrene blocks were stained with RuO_4 vapor, and the rubbery blocks were stained by OsO_4. Small angle neutron scattering (SANS) measurements were performed at the small angle scattering facility KWS II located at GKSS Research Centre, Geesthacht, Germany. All the samples were melt pressed into 1 mm thick and 13 mm diameter disks. The instrument configuration was $\lambda = 0.91$ nm and $\Delta\lambda/\lambda = 0.2$ as a result of the velocity selector and sample detector distance of 5.6 m. The scattering data were corrected for detector sensitivity, background scattering, used sample thickness, and transmission, and were placed on an absolute basis using several standards. The SANS profiles at different temperatures were discussed as a function of the scattering vector $q = (4\pi/\lambda) \sin(\theta)$, where 2θ is the scattering angle. Neutron reflectometry (NR) experiments were performed at the neutron reflectometer TOREMA II at GKSS Research Centre, Geesthacht. The wavelength was fixed to $\lambda = 0.43$ nm, and a position sensitive gas detector was used. Samples were prepared by spin coating (21) on 10x10 cm^2 float glass plates as a substrate and were annealed at different temperatures for 2 weeks. They were quenched to room temperature for the NR experiments.

Tensile tests were performed using a universal testing machine (Zwick 1425) at different strain rates and temperature. For each material, at least 10 samples were measured in order to avoid preparation effects. The sample dimensions of tensile specimens had a thickness of 0.5 mm and a total length of 50 mm. The toughness of the diblock copolymers was estimated as absorbed energy from the stress-strain curves.

To investigate the micromechanical deformation behavior, semithin sections with a thickness in the order of 0.5 μm were strained in a high voltage electron microscope (HREM, Jeol 1000, 1 MeV) equipped with an *in situ* tensile device. The advantage of HREM investigations is the possibility to use thicker sections for closer comparison with bulk materials.

To quantify the toughness behavior of block copolymer blends, an instrumented Charpy impact tester with a maximum work capacity of 4 J was used. Single edge notched bend (SENB) specimens with the dimensions 4×10×80 mm^3 according to standard ISO 179 were used with 2 or 4.5 mm deep sharp edgewise notches. Notches were prepared by using a razor blade cutter, depending on the kind of investigation: the determination of the resistance against unstable crack propagation (2 mm) or stable crack initiation and propagation (4.5 mm) (22). To minimize the specimen vibrations, the span is set to be equal to 40 mm; and the pendulum speed was 1 m/sec. The load deflection (F-f) diagrams were analyzed with respect to characteristic loading values, deflection, and energy. Here, the investigated fracture behavior of the materials was characterized by means of fracture mechanics parameters of elastic plastic fracture mechanics such as J integral and crack tip opening displacement (CTOD). The J values allow the quantification of energy dissipation during the crack propagation process because of the energetic definition of the J integral. Due to the predominantly stable crack propagation behavior for blends having LN4 content equal to or higher than 20 wt%, the toughness of these blends could only be characterized by using the crack resistance (R) concept of elastic plastic fracture mechanics. Using crack resistance (R) curves as a function of loading parameter (J or δ_d) versus the stable crack growth Δa, it was possible to calculate fracture mechanics values that describe the resistance against stable crack initiation and propagation. Among different experimental methods to determine R curves (23), the multispecimen method utilizing a stop block technique is found to give the most exact data (22,23). The requirements for geometry independence of fracture mechanics values utilizing R curves may be found in ref. (23).

5.3 INFLUENCE OF PHASE MORPHOLOGY AND INTERFACE FORMATION ON MECHANICAL PROPERTIES OF BLOCK COPOLYMERS

5.3.1 MORPHOLOGY, PHASE DIAGRAM, AND INTERFACE FORMATION OF POLY(STYRENE-B-BUTYL METHACYLATE) DIBLOCK COPOLYMERS

Morphology and phase behavior of diblock copolymers have been intensively studied by many authors during the last few years where different complex, highly ordered structures have been revealed. Based on the mean field theory of Leibler (24), it is possible to give an exact description of the order-disorder transition (ODT) in diblock copolymers by scattering methods and rheology. The Leibler theory predicts that the Flory-Huggins segmental interaction parameter χ depends on temperature according to $\chi = A + B/T$, where A and B are constants. For block copolymers, the product of χN (where N indicates the polymerization degree) governs the degree of segregation of A and B blocks, where with increasing incompatibility for a symmetrical diblock copolymer, the order-disorder transition is predicted by Leibler (24) as:

$$(\chi N)_{ODT} = 10.495 \tag{5.1}$$

Furthermore, the theory of Leiber predicts that the maximum of the reciprocal scattering intensity $(I)^1$ depends linearly on the reciprocal temperature while neglecting fluctuation effects close to ODT. A discontinuous change of peak intensity and $(\Delta q)^2$ at ODT can be observed for many systems due to the thermal force effect. The theory of Fredrickson and Helfand (25) includes fluctuation effects that has lead to a correction of the mean field theory whereby the order-disorder transition is predicted at higher values χN as obtained by mean field theory. In contrast to mean field theory, the theory of Fredrickson predicts direct transitions from a disordered state to hexagonal and lamellar structures for asymmetric compositions. The mean field theory was extended to triblock copolymers, star block copolymers (26), and graft copolymers (27).

While most block copolymer systems show a transition to disordered state as the temperature is raised, which is called an order-disorder transition or upper critical order transition (UCOT), for example poly(styrene-b-isoprene) diblock copolymers

(28), several studies (29,30) reported the existence of both an UCOT and lower critical order transition (LCOT) in poly(styrene-b-butyl methacrylate) (PS-b-PBMA) diblock copolymers. Earlier, Hashimoto et al. (31) already reported a LCOT in poly(vinyl methylether-b-styrene) diblock copolymers (PS-b-PVME) by SANS. In contrast to the classical enthalpically driven UCOT, the LCOT results from a decrease in entropy (32) arising either from strong interactions (e.g., dipole-dipole interactions) or molecular packing differences in weakly interacting systems such as PS-b-PBMA diblock copolymers, where molecular packing differences lead to differences in free volume of the pure components that are reflected in their specific volumes and thermal expansion coefficients. The disorder-order transition (DOT) in PS-b-PBMA diblock copolymers has been explored by Ruzette et al. (33) and Weidisch et al (34) by SANS and rheology. While for poly(styrene-b-isoprene) diblock copolymers a relatively large dependence of χ on temperature was observed, for PS-b-PBMA diblock copolymers the dependence of χ in the LCOT region is quite small (34):

$$\chi = (0.0243 \pm 0.0004) - \frac{(4.56 \pm 0.169)}{T} \tag{5.2}$$

As shown by Ruzette et al. (33) for PS-b-polyalkyl methacrylate diblock copolymers with long alkyl side chain methacylates (n > 5), the entropic and enthalpic contributions are comparable in size with our system, which, however, shows an UCST behavior. While PS-b-alkylmethacrylate diblock copolymers with long side chain methacrylates (n ≥ 6) show UCOT behavior, diblock copolymers with short side chain methacrylates (n < 5) reveal an LCOT behavior (33). Furthermore, recently Pollard et al. (35) have reported that an increase of pressure drives the system into the disordered state. Recently, Jeong et al. (36) have shown that the LCOT and ODT of weakly interacting PS-PBMA changes significantly with only the addition of one end functional group, which has been explained by the relative magnitudes of the three interaction parameters, including the end functional group. Obviously, the decrease in LCOT is more than twice the increase of ODT by the addition of MAH or carboxyl (COOH) end functional groups. Furthermore, close loop phase behaviour composed of LCOT and UCOT for PS-b-penthyl methacrylate (PS-b-PenMA) has been reported by same group (37).

The results of investigation of phase behavior and morphology of PS-b-PBMA have been summarized in a phase diagram (38). The phase diagram displayed in Figure 5.1 is asymmetrical, which arises from differences in the segment length and monomer volume for PS and PBMA. The ratio of segment length and monomer volume of PS and PBMA, as a measure for asymmetry of the phase diagram, has been calculated as 1.608 compared to 1.5 for PS-b-PI and 2.72 for PI-b-poly ethylene oxide (PEO-b-PI), the largest asymmetry known so far for diblock copolymers (39).

Furthermore, only a small molecular weight dependency of the UCOT as compared to the dependency of the LCOT has been found for PS-b-PBMA, which may arise from the fact that the mixing temperatures are located below the glass transition temperature (T_g) of the pure PS block. Therefore, it was difficult to reach the UCOT while cooling the samples. The dynamics of the system will be hindered, and a true

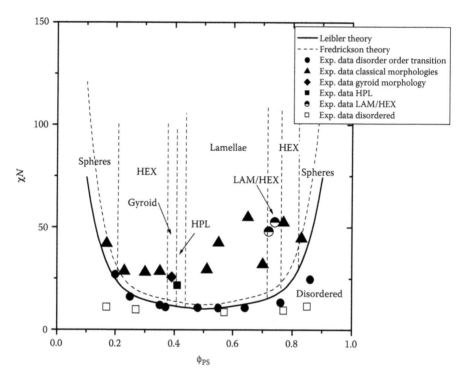

FIGURE 5.1 Phase diagram of the system PS-b-PBMA with LCOT data for molecular weights of about 100, 200, and 400 kg/mol. LCOT of samples with a molecular weight of about 100 kg/mol (deuterated samples); symbols are explained in the inset legend. Classical morphology are spheres, cylinder (HEX), and lamellae (LAM). UCOT is not shown since it is only possible to observe it by heating the samples. Experimental data are compared to corresponding data from the theories of Leibler (*Macromolecules*, 13, 1602, 1980) and Fredrickson (25).

equilibrium or phase separated systems is not reached, forcing us to conclude that a reversibility of the UCOT transition is very difficult to observe. A similar observation with respect to the reversibility and the problems of their detection for the system described has been noticed as well earlier by Russell et al. (29). Our results confirm the UCOT behavior observed by Russell's group using SAXS. However, the UCOT is not reversible due to the proximity of T_g and UCOT, which is associated with very long relaxation times near the T_g of PS.

Being above the LCOT transition temperature, it was possible to quench samples rapidly and to inspect their morphology. The equilibrium morphologies are, in principle, the same morphologies as known so far for block copolymers systems being at temperatures below the UCOT. The phase diagram shows all the principal morphologies that are known from other systems as well as transition morphologies such as the gyroid phase and hexagonally perforated lamellae (40,41) (Figures 5.2a and 5.2b). However, as discussed earlier (40,41), the appearance of morphologies is highly asymmetric. The diversity of the morphologies is less rich at high contents

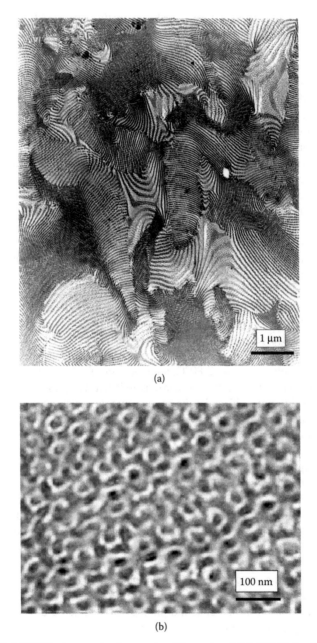

(a)

(b)

FIGURE 5.2 (a) TEM micrograph of a sample of the system PS-b-PBMA (molecular weight 270 kg/mol, 39% PS). The sample shown was not stained; a natural contrast appeared after a short exposure to the electron beam. The picture shows a Ia3d or gyroid structure, together with a low angle electron diffraction pattern obtained from the same sample. (b) TEM micrograph of a sample of the system PS-b-PBMA (molecular weight 212 kg/mol, 40% PS). The sample shown was not stained; a natural contrast appeared after a short exposure to the electron beam. The picture shows a hexagonally perforated lamellar (HPL) structure.

of PS in the block copolymers. Gyroid or hexagonally perforated lamellae structures have not been observed so far in the PS rich area of the phase diagram, a region where lamellae and cylindrical structure coexist instead (38). This structure is most likely a nonequilibrium structure resulting from a very small N and hence a very small driving force for separation value together with a small diffusion coefficient connected with the high molecular weight of the sample. This region appears, however, in the PS-rich part of the phase diagram ($0.72 < f_{PS} < 0.74$). Furthermore, order-order transitions have also not been observed in this system for the same reason. Calculations of the free energy of the system, as a combination of elastic energy and surface energy using the segment length and the molecular volumina of the monomeric units, support the experimentally found dividing lines between the areas of different morphologies.

The existence of the demixed phase at lower temperatures has been discussed as a consequence of a phase separation under the action of the solvent alone (38,40). However, the reported morphologies were all found from annealed samples; thus, an instable system should be mixed to a homogeneous system. Furthermore, a system of binary blends of the homopolymers in a solvent [tetrahydrofuran (THF)] shows a clear tendency to phase separate at lower temperatures (38). The presence of stable phase separated samples has also been reported earlier on the deuterated system by Russell et al (29) where, again, phase separated morphologies were found on samples that have been annealed at 160°C prior to investigation of the UCOT and the LCOT. A reversibility of the UCOT seems to be particularly difficult since the demixing must take place below or around 80°C, a temperature far below the glass transition of one of the pure blocks, i.e., PS. One can conclude that samples with high molecular weights (> 100 kg/mol) will make the phase separation also very difficult, at least to reach an equilibrium morphology, since diffusion times are expected to be rather large. In addition, the driving force will be very small since the interaction parameter is very small and the system is clearly in the weak segregation limit even at larger molecular weights (> 100 kg/mol).

It is well known that miscibility of polymer blends is closely related to interfacial width between the respective components. To study the influence of compatibility even at large molecular weights, we have studied two symmetrical d8PS-b-PBMA (deuterated styrene) diblock copolymers by neutron reflectometry with $M_n > 248300$ g/mol (SBM250) and $M_n = 148800$ g/mol (SBM150). The analysis of specular reflectivity curves (Figure 5.3 and Figure 5.4) yields information on the neutron refractive index and composition profile perpendicular to the interface (42).

The theory by Semenov (discussed in refs. 34 and 28) predicts long period L and interfacial width a_I for lamellae morphology of symmetric diblock copolymers in the strong segregation limit:

$$L = 4\left(\frac{3}{\pi^2}\right)^{\frac{1}{3}} \frac{b}{\sqrt{6}} N^{\frac{2}{3}} \chi^{\frac{1}{6}}$$

(5.3)

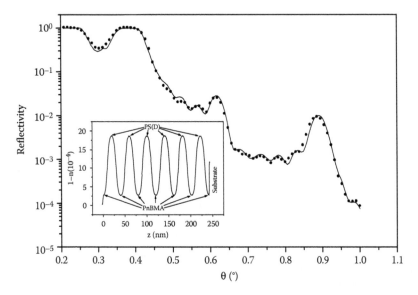

FIGURE 5.3 Neutron reflectivity data (•) and calculated fit curve from a thin film of a dPS-b-PBMA diblock copolymer with M_n=148 kg/mol and 50% PS. The inset shows the corresponding refractive index profile normal to the film surface that was used to calculate the reflectivity.

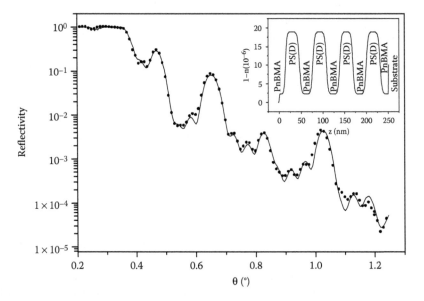

FIGURE 5.4 Neutron reflectivity data (•) (NR) and calculated fit curve from a thin film of a d8PS-b-PBMA diblock copolymer with M_n=248 kg/mol and 51% PS. The inset shows the corresponding refractive index profile normal to the film surface that was used to calculate the reflectivity.

$$a_I = \frac{2b}{\sqrt{6\chi}} \left[1 + \frac{1.34}{(\chi N)^{\frac{1}{3}}} \right] + \text{fluctuation corrections} \qquad (5.4)$$

Calculating the period L from Eq. (5.3) based on the χ values determined by SANS, one finds a good agreement with the NR results. Regarding the interfacial width a_I, the experimentally determined values of 9.0 and 8.4 nm for sample SBM150 and SBM250, respectively, are slightly larger than predicted from Eq. (5.4). The deviation might be due to the fact that Eq. (5.4) was derived in the strong segregation limit, and for the present case, χN is 16.7 for the sample SBM150 and 27.9 for sample SBM250 at 150°C. Fluctuations like capillary waves may not play an important role for the investigated system, which might be a consequence of the thin film preparation. It was shown in one of our previous papers (43) that a TEM investigation of bulk samples yields comparable values for the interfacial width. Fluctuations could be quenched by the influence of the smooth substrate and surface and may also not have fully developed within the annealing time.

The strength of segregation expressed by χN is about 28 for sample SBM250, placing this sample in the intermediately segregation limit. This result puts the measured large value of interface width determined by NR in the right perspective with respect to the smaller values calculated utilizing strong segregation theory [Eq. (5.4)]. For sample SBM150, the strength of segregation is only about 17, which indicates that this block copolymer is almost weakly segregated.

The observed large interfacial width and low χ parameter for the investigated system compared to other block copolymer systems confirms the observed partial miscibility at $M_n > 200$ kg/mol and different compositions observed by dynamic mechanical analysis (DMA) (34,40,41).

5.3.2 INFLUENCE OF PHASE MORPHOLOGY AND INTERFACIAL WIDTH ON TENSILE PROPERTIES OF DIBLOCK COPOLYMERS

Most diblock copolymer studies have been focused on morphology and phase behaviour. Mechanical properties and deformation behaviour of di- and triblock copolymers based on PS and PB (polybutadiene) have been studied in detail, especially the PS-b-PB-b-PS (SBS) triblock copolymers, such as Kraton ®, which is a commercially used thermoplastic elastomers (TPE) (44) and which consists of a hard block (crystalline or glassy) and a rubbery soft block. The TPE Kraton ® (Shell Oil Co.) is such a material consisting of a glassy PS block and a rubbery PB middle block (44,45) that is well known for the unique thermomechanical properties associated with phase morphology of PS domains dispersed in a continuous rubbery PB matrix. Whereas PS-b-PB diblock copolymers show only a very small tensile strength, the presence of bridged midblock conformations in the SBS triblock copolymers provides improved mechanical strength by physical midblock bridging. The deformation behaviour of the PS cylinders in SBS triblock copolymers at higher strains has been intensively investigated by various methods (46–49). Several studies

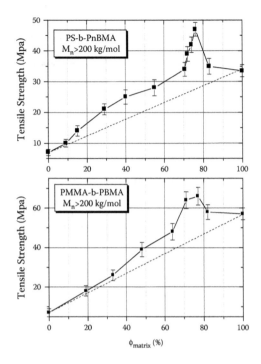

FIGURE 5.5 Dependence of tensile strength on volume fraction for PS-b-PnBMA and PMMA-b-PBMA diblock copolymers with $M_n > 200$ kg/mol at a strain rate of $\varepsilon = 1.6 \times 10^{-4} s^{-1}$.

(50–52) reported deformation behaviour of the gyroid and hexagonal morphology in triblock copolymers by SAXS and TEM. Schwier et al. (53) and Yamaoka (54) have studied the influence of morphology and composition of thermoplastic block copolymers on tensile properties.

We have investigated the morphology, tensile properties, and deformation behaviour of PS-b-PBMA diblock copolymers (40,41,55), where a strong increase in tensile strength with increasing PS content has been observed. In contrast to other diblock copolymers, we have observed a maximum of tensile strength at 76% PS (Figure 5.5), which is about 40% higher than that of pure PS. This synergism of tensile strength was found in the composition range between 70 and 80% PS content. At the transition of a mixed lamellar-hexagonal structure to a pure hexagonal morphology at 76% PS content, the strain at break decreased and the block copolymers became brittle (40). With a further increase of polystyrene content up to 83%, the block copolymers behave almost like polystyrene. Also for PMMA-b-PBMA diblock copolymers a maximum of tensile strength at 77% PMMA was observed that, however, only slightly exceeded that of pure PMMA (Figure 5.5). For PMMA-b-PBMA diblock copolymers, the strain at break is generally smaller than that of PS-b-PBMA diblock copolymers, which can be attributed to the stronger incompatibility of the components in PMMA-b-PBMA (Figure 5.6).

The reasons for the observed different tensile properties of PS-b-PBMA and PMMA-b-PBMA diblock copolymers can be discussed on the basis of the different

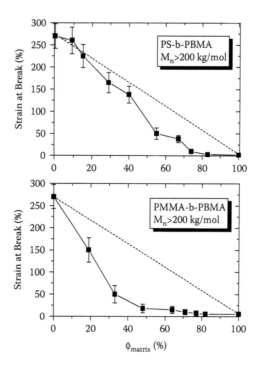

FIGURE 5.6 Dependence of strain at break on volume fraction of PS for PS-b-PnBMA and PMMA-b-PBMA diblock copolymers with $M_n > 200$ kg/mol at a strain rate of $\varepsilon = 1.6 \times 10^{-4} s^{-1}$.

interaction parameters χ between the components. For a symmetrical dPMMA-b-PBMA diblock copolymer with 45.7 kg/mol, the interaction parameter was determined to be $\chi_{PMMA/PBMA} = 0.062$ at $T=140°C$ (34). In contrast, for dPS-b-PBMA diblock copolymers, we determined χ to be equal to 0.0133 at $T=140°C$ by SANS, which is smaller than that of PMMA-b-PBMA diblock copolymers; for both systems, χ is much smaller than that obtained for PS-b-PI diblock copolymers [$\chi=71.4/T - 0.00857$ (34,56)]. While for symmetrical PMMA-b-PBMA diblock copolymer in the composition range between 70 and 80% PMMA content, χN is about 120 ($M_n \approx 200$ kg/mol); PS-b-PBMA diblock copolymers are only intermediately segregated at 120°C (for a sample with $M_n = 278$ kg/mol and 50% PS content, one obtains $\chi N_{120°C} = 29.5$). It should be kept in mind that deuteration also changes the thermodynamics slightly if we compare deuterated and nondeuterated samples, but this effect is believed to be relatively small and is assumed to be unimportant for our discussions. A smaller interfacial width was observed for dPMMA-b-PBMA diblock copolymers [e.g., about 3.5 nm at $T = 140°C$ determined for a sample with $M_n = 45.7$ kg/mol and 43% PMMA (57)] compared to dPS-b-PBMA diblock copolymers with even much larger molecular weight of about 250 kg/mol [interfacial width of 7.7 nm at $T = 40°C$ (28)], clearly providing reasons for the observed differences in tensile properties of both systems. This obviously demonstrates that compatibility and interfacial width have a pronounced influence on tensile properties as earlier found in polymer blends. Obviously, the combination of nanoscaled PBMA cylinder in a PS matrix and enhanced interfacial width are capable of increasing the strain at break, which is consistent with the theoretical predictions by Smith et al. (58–60) who suggested that incorporation of relatively stiff rubbery particles with good adhesion, arising from a larger interfacial width, may stabilize postyield behaviour such as strain hardening in polymers. A brittle-to-tough transition (BTT) has been predicted by the same group (61), suggesting that particles below a critical thickness cause a remarkable increase in strain at break for polymer blends. The enhanced yield stress of PS-b-PBMA diblock copolymers (70 to 74% PS) compared to pure PS is obviously much more difficult to explain. However, the combination of improved postyield behavior of PS-b-PBMA and larger interfacial width (reduced number of defects and larger interfacial adhesion) may also reduce critical triaxial stresses. Furthermore, the glass transition temperature of the PBMA-block increases compared to neat PBMA, resulting in larger stiffness of the PBMA cylinder partially mixed with PS. A larger interfacial width therefore may be associated with a reduction of critical dilative stresses near small flaws or defects primarily responsible for enhanced craze initiation stress, thus improving the defect sensitivity and resulting in a large yield stress observed for PS-b-PBMA with 25 to 30% PBMA (while for PS yielding, this does not occur). The superior properties of PS-b-PBMA diblock copolymers are believed to result from the different effects discussed.

Many studies were attributed to the correlation between deformation behavior and morphology of PS-b-PB diblock copolymers (53,54,62,63). Schwier et al. (53) proposed a model for craze growth in diblock copolymers based on void formation in the rubbery domains under the concentrated stresses of the craze tip. This is followed by drawing and fibril formation in the PS phase. It was shown that block copolymers do not show a crazing mechanism because the microphase separated morphology in the nanoscale is too small to initiate crazes (63). Therefore, deformation in block copolymers occurs via a two-step cavitation mechanism.

In order to establish a unified correlation between phase behavior and deformation mechanisms of block copolymers, samples with different strength of segregation have been used. χN can be varied by changing temperature, resulting in a change of χ due to temperature or molecular weight (expressed by polymerization degree N). We used both ways to confirm the discussed correlation.

At the ODT, a lamellar morphology is formed in the case of symmetrical composition. Within the weak segregation limit (WSL), the microscopic density profile of the components is considered to vary sinusoidal in space and the chains of the components interpenetrate to a high degree. Between the WSL and strong segregation limit (SSL), an intermediate segregation regime (ISR) at $12.5 < \chi N < 95$ has been identified where the chains are more stretched than in the strong segregated regime due to coarsening of the density profile as χN increases from that observed in the WSL (64). The interface is then broadened, and the junction points are not completely localized in the interfacial region. In the strong segregation limit at $\chi N > 100$, the interface is sharp and the phases are completely separated. The interfacial width of strongly segregated block copolymers is about 1 to 2 nm (65).

Figure 5.7 presents a unified scheme of deformation mechanisms for disordered, weakly segregated, intermediately segregated, and strongly segregated diblock

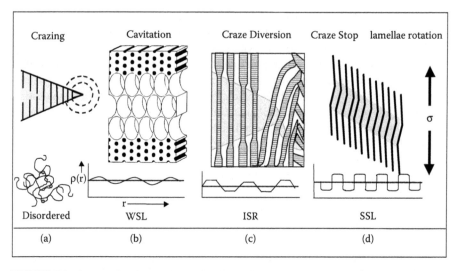

FIGURE 5.7 Schematic presentation of deformation mechanisms of diblock copolymers depending on strength of segregation, χN. (a) Disordered block copolymers are deformed by crazing independent on their molecular architecture. The deformation mechanism is the same as observed in the corresponding glassy homopolymer. (b) Weakly segregated (WSL) diblock copolymers are deformed by the cavitation mechanism. Diblock copolymers undergo a transition from crazing to cavitation mechanism at the order-disorder transition. (c) Intermediately segregated (ISR) diblock copolymers show a cavitation mechanism as well as termination and diversion of crazes. (d) Strongly segregated (SSL) diblock copolymers show the same mechanisms as observed for intermediately diblock copolymers but a much lower craze initiation stress compared to case (c).

copolymers. It shows a schematical representation of deformation mechanisms of diblock copolymers depending on strength of segregation, χN, and reveals the correlation between micromechanical deformation mechanisms and phase behavior. While for disordered block copolymers, the deformation occurs via crazing (Figure 5.7a), the ODT is associated with a transition from crazing to cavitation mechanism (Figure 5.7b). Within the ISR, cavitation can also be observed as was already found for weakly segregated block copolymers. However, for intermediately segregated block copolymers, the mechanisms of diversion and termination of crazes can also be found (Figure 5.7c). For ordered block copolymers, craze propagation mainly depends on local orientation of the grains. Within the SSL, the same deformation mechanisms as was seen for ISR is observed (Figure 5.7d), including the mechanisms of lamellae rotation and kink band formation (66,67). The most striking difference between SSL and ISR is the enhanced craze initiation stress in the ISR due to the larger interfacial width (Figure 5.7). While the ODT can be considered as the reason for the transition from crazing to cavitation in weakly segregated block copolymers, the increasing grain size with increasing χN is associated with the observation of additional deformation mechanisms of termination and diversion of crazes (66).

This correlation should also be reflected in a change in tensile properties, resulting in a correlation between interaction parameter (interfacial width) and mechanical

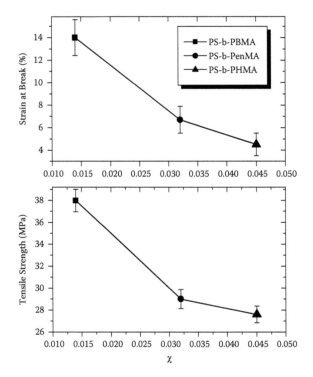

FIGURE 5.8 Dependence of the tensile strength and strain at break on Flory-Huggins interaction parameter χ for poly(styrene-b-alkyl methacrylate) diblock copolymers with about 70% PS at a strain rate of $\varepsilon = 1.6 \times 10^{-4} s^{-1}$.

properties of block copolymers. For this purpose, a series of diblock copolymers with different alkyl methacrylate blocks were investigated with respect to their tensile properties. The glass transition temperature depends of the length of alkyl methacrylate block, therefore, $T - T_g$ has been kept constant, e.g., materials have been tested at slightly different temperatures. The χ parameter for the different block copolymers have been given in recent studies (33,34,57). Figure 5.8 shows the dependence of tensile strength and strain at break on χ for a series of poly(styrene-b-alkyl methacrylate) diblock copolymers where it is obvious that both tensile strength and strain at break decrease with increasing incompatibility. While the tensile strength for PS-b-PBMA diblock copolymers is comparable with PS, significantly lower values were found in the case of PS-b-PenMA and PS-b-hexyl methacrylate (PS-b-PHMA) block copolymers. Figure 5.9 represents a correlation between tensile properties and χN. While PS-b-PHMA samples are strongly segregated, PS-b-PenMA and PS-b-PBMA diblock copolymers are intermediately segregated. For weakly segregated PBMA-b-PS-b-PBMA triblock copolymers, a steep increase of tensile strength was observed (66). In contrast, the adsorbed energy shows a maximum in the ISR (Figure 5.9). For weakly segregated triblock copolymers, the tensile strength increases at the expense of toughness. Furthermore, it has been observed that the ODT is associated with a brittle-to-tough transition (67).

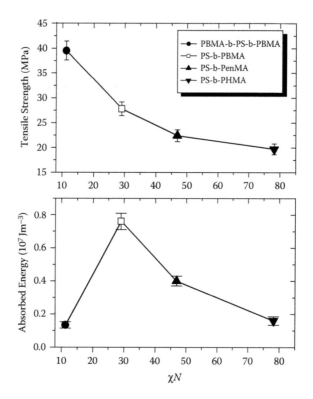

FIGURE 5.9 Dependence of the tensile strength and adsorbed energy on χN for different poly(styrene-b-alkyl methacrylate) diblock copolymers and PBMA-b-PS-b-PBMA triblock copolymers with about 50% PS measured at strain rate of $\varepsilon = 1.6 \times 10^{-4} s^{-1}$.

Figure 5.10 summarizes the property profile of different block copolymers with respect to different strengths of segregation compared to commercial polymer materials. It is shown that classical materials show either a high toughness or a high stiffness and strength. It is usually difficult to obtain materials with a high toughness, a high stiffness, and strength; it is only possible to realize this by utilizing high performance polymeric materials that are usually quite expensive. While the property profile for disordered and strongly segregated diblock copolymers is similar to that of classical materials or corresponding homopolymers, nonclassical property profiles are observed with intermediately and weakly segregated block copolymers. For intermediately segregated block copolymers, a combination between high toughness, high stiffness, and strength is shown in Figure 5.10, as discussed in detail in our previous paper (40). However, for weakly segregated block copolymers, a very high stiffness and strength was observed at the expense of toughness (Figure 5.10) due to the small long range order of weakly segregated block copolymers that makes deformation mechanisms such as termination and diversion of crazes less effective in a larger volume element. For block copolymers, a minimum grain size (about 2 µm) is necessary to achieve a larger toughness (68).

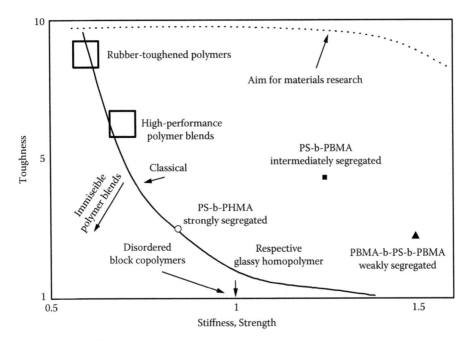

FIGURE 5.10 Qualitative presentation of dependence of toughness on tensile strength and stiffness for different polymeric systems, which demonstrate the influence of strength of segregation on property profile. Nonclassical property profiles exist for intermediately and weakly segregated block copolymers.

5.3.3 INFLUENCE OF MOLECULAR ARCHITECTURE OF BLOCK COPOLYMERS ON MECHANICAL PROPERTIES AND DEFORMATION BEHAVIOR

While in the previous chapter the influence of phase behaviour on deformation of block copolymers has been discussed, an indication of the influence of molecular architecture on deformation of block copolymers has been reported by Drolet and Fredrickson (69) who have calculated enhanced average chain bridging fractions of internal blocks for pentablock copolymers compared to triblock copolymers causing enhanced toughness. Ryu et al. (70) have investigated deformation and fracture behavior of poly(vinyl cyclo-hexane)/polyethylene (PCHE/PE), diblock (CE), triblock (CEC) and pentablock (CECEC) copolymers. A brittle-to-ductile transition was observed for pentablock copolymers that deform by formation of shear zones caused by the formation of bridging molecules.

In order to investigate the influence of molecular architecture on deformation behavior, block copolymers of polystyrene and polybutyl methacrylate with different architectures were studied by *in situ* high-voltage electron microscopy (HVEM). The small interaction parameter between PS and PBMA (34) allows for the study of disordered and weakly segregated star block copolymers (by using samples with different N) while the arm molecular weight remains above the PS entanglement molecular weight (arm molecular weight M_n > 35000 g/mol). From the previously proposed scheme (66), a cavitation mechanism is expected for microphase separated diblock copolymers. This is shown in Figure 5.11a for a PS-b-PBMA diblock copolymer that illustrates the large difference in deformation behavior between di- and star block copolymers. Details of deformation behavior of microphase separated diblock copolymers has been discussed in previous studies (71,72).

Interestingly, weakly segregated samples show a transition to shear deformation as shown in Figure 5.11, clearly in contrast to disordered samples where crazing is the dominant deformation mechanism in the material. While di- and triblock copolymers deform by formation of crazes by cavitation (Figures 5.11a and 5.11b), for star block copolymers shear yielding has been observed arising from the influence of the molecular architecture on deformation behavior.

In contrast to diblock copolymers, which show a transition from crazing to cavitation at the order-disorder-transition, ODT, star block copolymers reveal a *transition from crazing to shear deformation*, suggesting that disordered star block copolymers are still deformed by crazing and ordered star block copolymers are deformed by the formation of shear deformation zones (75).

Shear deformation is well known from homopolymers and several rubber modified polymers. For example, polycarbonate is deformed by shear yielding; this phenomenon was also observed in rubber modified PC, rubber modified PMMA, and modified epoxy (73). Shear deformation zones usually propagate about 45° toward the direction of external stress. In rubber modified polymers, multiple shearing is initiated in the matrix (for example PC) because of the effective distance of the rubber particles, causing an enhanced toughness of the material. In star block copolymers, a high volume fraction of the material is deformed and turned into highly deformed material by shear yielding that causes multiple shear deformation zones as shown in Figure 5.11c. Higher magnifications of shear zones have revealed that grains with different orientations are deformed of about 45° toward the external stress field by shear yielding (75).

While in homopolymers the network density and in polymer blends the effective distance of rubbery particles cause a shear yielding, in star block copolymers, the molecular architecture and grain morphology are associated with a transition to shear deformation. Recently, Ryu et al. (70) have reported shear deformation in pentablock copolymers in contrast to their tri- and diblock analogues. Drolet and Fredrickson (69) have calculated enhanced average bridging fractions of internal blocks for pentablock copolymers compared to triblock copolymers that initiate a transition to shear deformation. For star block copolymers compared to triblock copolymers, an increased fraction of entanglements is present, thus causing a transition to shear deformation. In addition, a reduction of yield stress occurs because of the low molecular weight of the PS arms compared to that of triblock copolymers. Similar to this study, Mori et al. (74) have observed improved strain at break in CECEC pentablock copolymers compared to CEC triblock copolymers. By adding the pentablock to the triblock copolymer, a brittle-to-tough transition has been observed at about 10% pentablock, applying external strain normal to the lamellae, which is consistent with calculations that assume that the pentablock chains bridge the lamellae.

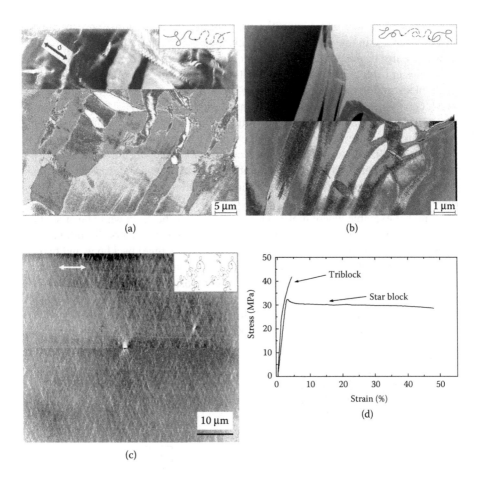

FIGURE 5.11 (a) HVEM micrograph of deformation zones of an intermediately segregated PS-b-PBMA diblock copolymer at $\chi N = 39$ already discussed in detail in our previous study ($\Phi_{PS} = 0.70$, $M_n = 315$ kg/mol, lamellae). The craze propagation is influenced by morphology. (b) HVEM micrograph of craze structure (lower and higher magnification) of a weakly segregated PBMA-b-PS-b-PBMA diblock copolymer at $\chi N = 21.4$ ($\Phi_{PS} = 0.52$, $M_n = 201$ kg/mol, lamellae) that is deformed via cavitation. Strain direction is parallel to craze fibril direction. (c) HVEM micrograph of shear zones of the weakly segregated PS-PBMA star block copolymer ST35 at $\chi N = 50.3$ ($\Phi_{PS} = 0.52$, $M_n = 456$ kg/mol, lamellae) that is deformed by shear deformation. Strain direction is about 45° to propagation direction of shear bands. (d) Stress-strain diagrams of a different weakly segregated PBMA-PS-PBMA triblock copolymer and weakly segregated star block copolymer ST35 measured at a strain rate of $\varepsilon = 1.6 \times 10^{-4}\, s^{-1}$.

Smith et al. (58–60) have shown that enhanced local mobility of polymer segments and reduced yield stress may result in strain hardening and can eliminate strain softening of the material resulting in an enhanced toughness. In star block copolymers, the reduced yield stress and enhanced mobility of polymer chains in the thin PS lamellae (thickness about 10 nm) may support the transition to shear deformation.

It is well known that a transition to multiple shear deformation can cause a marked increase in toughness of rubber modified polymer blends, and therefore, the same may also be expected for block copolymers. Figure 5.11d compares the tensile properties of triblock and star block copolymers consisting of PS and PBMA. While the triblock copolymer shows a strain at break of only about 2% (75), for the star block copolymer, the strain at break is increased to about 50%. The yield stress of the star block copolymer is lower than that of the triblock copolymer, suggesting that a large number of low molecular weight PS arms reduces the yield stress. Furthermore, a brittle–to–tough transition occurs upon changing the architecture from triblock to star block. The increased number of bridging molecules in the star block copolymers initiates a transition to shear deformation, which explains the strong increase in absorbed energy.

Another example where molecular architecture has been shown to be crucial for mechanical properties are tetrafunctional multigraft copolymers. Tetrafunctional multigraft copolymers consisting of a rubbery poly(isoprene) backbone and two PS branches at each branch point are schematically represented in Figures 5.12a and 5.12b. Multigraft copolymers with different compositions and about eight branch points, each with a different microphase separated morphology, were investigated. It has been found that materials between 8 and 37% PS exhibit the behavior of a TPE as indicated by the large increase of stress at higher strains. With increasing

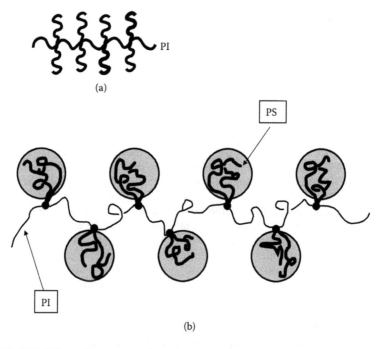

(a)

(b)

FIGURE 5.12 Schematic representation of (a) a tetrafunctional multigraft copolymers with PS domains in a PI matrix and (b) molecular chain architecture of a tetrafunctional multigraft copolymer demonstrating the strong force between branch points at the PI backbone and PS domains resulting in "physically multicrosslinked" domains.

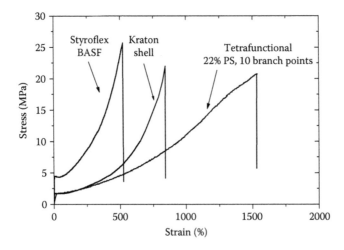

FIGURE 5.13 Mechanical properties of multigraft copolymers (MG 4-7-21 and MG 4-10-21) compared to commercial TPEs Kraton® (20% PS) and Styroflex® (60% PS, BASF), revealing the exceptional large strain at break of tetrafunctional multigraft copolymers.

PS content, the strain at break decreases dramatically without a corresponding increase in strength. This strong decrease in strain at break is accompanied by the change in morphology from cylindrical to lamellar morphology (76). At high PS contents, the material becomes brittle where the strain at break is only about 2% and yielding does not occur. Clearly the most intriguing TPE properties with both high tensile strength and strain at break can be observed at 22% PS where wormlike cylindrical PS domains in a PI matrix has been observed that lack long-range order (77), which is clearly in contrast to common SBS triblock copolymers where an ordered hexagonal morphology is observed.

Figure 5.13 shows a comparison between two multigraft copolymers and two commercial TPEs, a Shell Kraton® and a BASF Styroflex®. Kraton, one of the leading commercial TPE on the market, is a typical SBS triblock copolymer with a hexagonal morphology. Styroflex is a S-SB-S triblock copolymer with a SB random copolymer as the middle block provided by the BASF Company. Styroflex shows a higher yield stress and tensile strength than Kraton but a lower strain at break. Obviously, the tetrafunctional multigraft copolymers show much higher strains at break than observed for the commercial TPEs. For a multigraft copolymer with 22% PS and 10 branch points, the strain at break is almost twice that observed for Kraton with 20% PS. The combination of a huge strain at break and an acceptable tensile strength of about 21 MPa indicate an exceptional property profile for tetrafunctional multigraft copolymers.

The first hysteresis experiments (Figure 5.14) at about 300 to 700% deformation have demonstrated the high elasticity of these multigraft copolymers compared to Kraton and Styroflex. The residual strain of multigraft copolymers is much less than that of linear triblock copolymer analogous, demonstrating superior performance of multigraft copolymers. The residual strain obviously does not change much with increasing applied strain, while for Styroflex, revealing a similar morphology, a

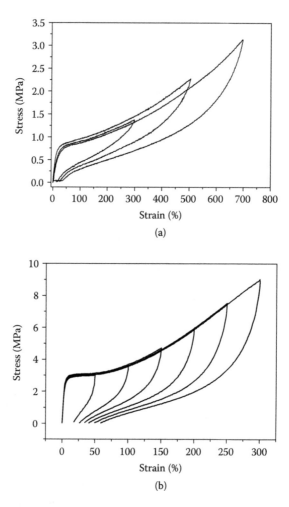

FIGURE 5.14 Hysteresis behavior of (a) a tetrafunctional multigraft copolymer with 25% PS and about seven branch points compared to that of (b) Styroflex (BASF; S-SB-S triblock).

much larger slope is observed arising from its linear architecture. While triblock copolymers have a linear architecture, multigraft copolymers have a rubbery backbone coupled to multiple glassy branches, resulting in a more branched architecture.

The mechanical properties of common TPEs, such as SBS triblock copolymers, arise from their midblock conformation, as a result of their triblock architecture, leading to bridges between the PS blocks. The requirements for this behavior are strongly segregated components with narrow interfaces and PS blocks with sufficiently high molecular weights. Thus, molecular architecture and morphology of SBS triblock copolymers leads to a *physical crosslinking* that is generally reported as the reason for their excellent elastomeric properties. It was shown by Quirk and Legge (78) that the mechanical properties of TPEs only show a molecular weight dependence below 100 kg/mol that is attributed to the limit of phase separation. Higher molecular weights ($M_n > 100$ kg/mol) do not influence the mechanical properties. The multigraft copolymers used in this study have molecular weights far exceeding 100 kg/mol, thus mechanical properties are independent of molecular weight. In contrast to triblock copolymers, graft copolymers have a backbone with chemically different branches, and their morphology and properties can be tailored by changing molecular architecture (79). To achieve acceptable mechanical properties with graft copolymers, it is necessary to use a rubbery backbone with glassy branches at two or more branch points and a fairly high molecular weight of rubbery backbone and glassy branches (79). The tetrafunctional multigraft copolymers used in our study consist of two glassy branches per grafting point that provide enhanced stress transfer between the rubbery matrix and polystyrene domains compared to triblock copolymers, which is schematically illustrated in Figure 5.12b where a rubbery PI backbone has multiple branch points connected to two PS chains. In the microphase separated morphology, the junction points must reside at the interface between the PS cylindrical domains and the PI matrix. Recently, we have shown that the enhanced mechanical properties of the multigraft materials arises from improved coupling of each highly elastic PI backbone to the reinforcing PS domains. This results from having two PS arms at each branch point and a relatively large number of branch points per molecule. Note that in such a microphase separated morphology, the two PS blocks that share a common junction point must both reside in the same PS domain (76).

It is shown in Figure 5.13 that tetrafunctional multigraft copolymers have an exceptional level of strain far exceeding that of commercial TPEs. It is interesting that this exceptional property profile can be achieved with a less ordered morphology. For SBS triblock copolymers, it was reported (80,81) that a well ordered morphology is one requirement that must be met in order to get acceptable TPEs. In the case of multigraft copolymers, however, a large number of branch points are necessary rather than a well ordered morphology. The existence of smaller grain sizes can also be correlated with an improved property profile as recently shown in the case of weakly segregated block copolymers (66,82). Multigraft copolymers with fewer than five branch points reveal a quite low strain at break and strength as shown in Figure 5.15 where it is obvious that the mechanical properties linearly

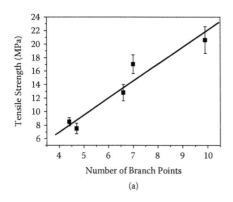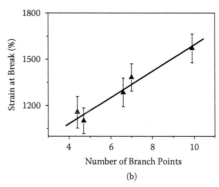

FIGURE 5.15 Influence of number of branch points on (a) tensile strength and (b) strain at break for tetrafunctional multigraft copolymers with about 25% PS.

depend on the number of branch points. This means that an increasing number of branch points connected to the PS domains provide a high strength combined with a smaller grain size. The enhanced number of physical crosslinks in tetrafunctional multigraft copolymers leads to a *physical multicrosslinking*, causing a huge strain at break and elasticity (83).

5.4 MECHANICAL PROPERTIES AND CRACK TOUGHNESS BEHAVIOR OF BINARY BLOCK COPOLYMER BLENDS

5.4.1 MORPHOLOGY DEVELOPMENT IN BINARY BLOCK COPOLYMER BLENDS

Binary blends of diblock copolymers have been intensively studied recently. Most of the work done on the morphology of diblock copolymer blends is summarized in the monograph of Hamley (28). Comprehensive studies have been published by Hashimoto and coworkers (84–86) who have initially started studying lamellae forming diblock copolymer blends but extended their investigations to other morphologies later. Spontak et al. (87) used SI diblock copolymers of different compositions but similar molecular weight to vary the morphology by changing the relative amounts of the two block copolymers. Sakurai et al. (88,89) found an order-order transition between the lamellar and the gyroid phase in a block copolymer blend upon increasing temperature. While in these studies, mixed morphologies were always obtained, there are also studies that demonstrate that the chain topology and the molecular weights of the blocks may also affect the miscibility in block copolymer blends. Spontak et al. (90) studied a blend of a symmetric SI diblock copolymer and a $(SI)_4$ multiblock copolymer with the same overall molecular weight by electron tomography and found an architecturally induced macrophase separation into two different lamellar phases. Furthermore, blends of triblock and diblock copolymers have been studied by the same group where the blends have been found to have enhanced mechanical properties compared to that of the pure triblock, which mainly arise from the increased number of physical crosslinks (91). Hashimoto and coworkers (92) reported a macrophase separation in blends of two symmetric SI diblock copolymers with molecular weights differing by more than a factor of 10 that separated into partly mixed phases of long and short chains. While the morphology and phase behaviour of binary block copolymer blends have been studied by many groups, their mechanical properties have not yet been studied in detail. Only a few studies report on the effect of microphase morphology on strength and stiffness of blends containing amorphous block copolymers (54, 62, 93); however, homopolymers were used in most of these studies. Recently, we have published mechanical and fracture behaviour of partial miscible block copolymer blends of star block and triblock copolymers, resulting in two brittle-to-tough transitions and in yield stress that depended exponentially on the composition (94,95).

We have studied blends of triblock and star block copolymers with a similar composition of about 72% PS. Both materials show different morphologies arising

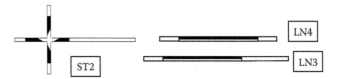

FIGURE 5.16 Molecular architecture of (a) ST2 (tapered transition and asymmetric architecture), (b) LN4 (SB-random copolymer middle block), and (c) LN3 (SB-random copolymer middle block and asymmetric architecture). Black represents polybutadiene (PB) and white represents polystyrene (PS).

from differences in architecture as shown in Figure 5.16. While the star block copolymer shows a lamellae morphology due to asymmetric architecture, the triblock copolymer consists of a SB random copolymer as the middle block, which gives rise to an enhanced compatibility and wormlike morphology. Equilibrium morphologies of samples ST2 and LN4 comprise alternating polybutadiene (PB) and polystyrene (PS) lamellae with additional PS domains inside the PB lamellae and randomly distributed PS cylinders in a styrene-butadiene copolymer matrix (94). Equilibrium morphology (films prepared from toluene) of ST2/LN4 show that the separated macrophase consists of dispersed LN4 domains in an ST2 matrix (95). However, for technical applications of block copolymers, thermodynamic equilibrium is never achieved because of technological constraints like limited processing time (nonequilibrium temperature and deformation regime) and shear stresses. A comparison of the equilibrium morphology of sample ST2 with injection molded morphology clearly demonstrates a substantial influence of processing on the morphology of the block copolymer (Figure 5.17a). High shear stresses lead to preferential orientation and pronounced long-range order of the lamellae along the injection direction. While the influence of processing on the morphology of pure LN4 is less pronounced than that of pure ST2 (Figure 5.17b), the macrophase separation in

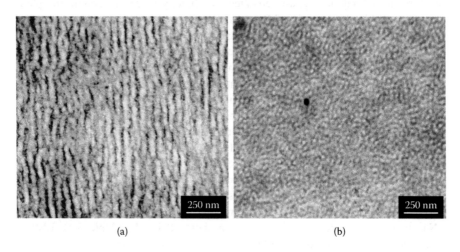

(a)   (b)

FIGURE 5.17 TEM micrographs of injection molded block copolymers: (a) ST2 and (b) LN4.

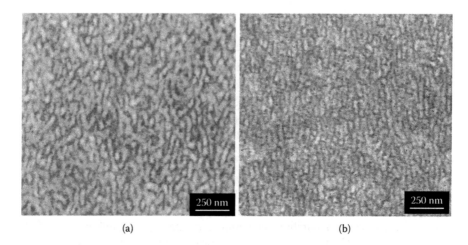

(a)   (b)

FIGURE 5.18 TEM micrographs of injection molded ST2/LN4 block copolymer blends: (a) 20 wt% LN4, and (b) 40 wt% LN4 (stained with OsO$_4$). Dark areas represent PB domains, and light areas represent PS domains.

the injection molded blends of ST2 and LN4 is suppressed by the shear stresses in the melt during injection molding (Figures 5.18a and 5.18b). Basically, two types of morphology were found in the injection molded blends due to the partial miscibility [see results of DMA (94)] of ST2 and LN4. The characteristics of the lamellar morphology found in pure ST2 also predominate at lower LN4 content (0 to 20 wt% LN4). The microphase separated structures at higher LN4 content (40 to 80 wt% LN4) are similar to pure LN4. The existence of partial miscibility of ST2/LN4 blends was proven by shifts in the glass transition temperatures (T_gs), which depended on the composition, leading to weak segregation into PB-rich and PS-rich domains and resulting in a mixed macrophase separated morphology. Our results on the morphology of these blends show that different molecular architectures may induce macrophase separation, confirming the results of Spontak and coworkers (90). Partial miscibility has been observed because of the tapered transition in ST2 and the random middle blocks in LN4, resulting in a coexistence of macrophase separated PS-rich and PB-rich domains over the entire composition range under equilibrium conditions as was recently found by Adhikari et al. (96).

Binary blends of S-SB-S triblock copolymer with different molecular architectures have been used to study the influence of blend composition and morphology on mechanical properties. Both S-SB-S triblock copolymers with similar architecture (LN3, LN4) contain a styrene/butadiene (SB) random copolymer middle block with similar S/B ratios (Figure 5.16). Both block copolymers, LN3 and LN4, used in the blends are linear triblock copolymer based on styrene and butadiene with outer polystyrene blocks and a middle block composed of a SB random copolymer. While LN4 shows a symmetric architecture with short PS blocks (Table 5.1), LN3 is asymmetric with respect to the PS blocks (Table 5.1). Molecular characterization of the materials is summarized in Table 5.1 revealing that both materials have different morphologies despite their similar overall composition. We have used materials with

TABLE 5.1
Characterization of the Blend Components LN3 and LN4

Material	M_n (g/mol)	M_w/M_n	Total PS content (wt%)	PS_1/S-B/PS_2 (wt%)	S/B ratio in the random copolymer block	$T_{g\text{-}PB}$ (°C) (DSC)	$T_{g\text{-}PS}$ (°C) (DSC)	Morphology
LN3[a]	111,000	1.21	75	12/49/39	1:1	−33	100	Lamellae
LN4[b]	119,000	1.30	66	16/68/16	1:1	−24	71	Wormlike

[a] LN3:PS$_1$ PS-co-PB PS$_2$
[b] LN4: PS PS-co-PB PS

small differences in overall composition of about 10 wt% and with similar compositions of the SB middle blocks in LN3 and LN4, thereby leading to better miscibility. Differences in the materials are mainly the following: the molecular architecture (asymmetry in LN3), the different weight fractions of the outer PS blocks of 32 and 50% for LN3 and LN4, respectively, and the different weight fractions of the SB blocks (50 and 68% for LN3 and LN4, respectively).

Furthermore, molecular characterization of the materials, summarized in Table 5.1, shows that both materials have similar total PS composition but different SB middle block compositions, causing very different mechanical behaviour (LN3 is thermoplastic while LN4 is elastomeric). For LN4 exhibiting an overall weight fraction of the glassy phase of 32%, the outer PS blocks are partly miscible with the SB middle block, which results in a weakly segregated wormlike structure (Figure 5.19a) that lacks long-range order as shown by the existence of only one reflection in the SAXS experiments (98). LN3 has an overall weight fraction of the outer glassy PS blocks of about 50% and shows a well ordered lamellae morphology (Figure 5.19b) as confirmed by the existence of three reflections. The interplanar spacing (as seen by SAXS) of LN3 and LN4 are 33.9 and 26.2 nm, respectively, in good agreement with the results obtained by TEM. The differences in morphology of S-SB-S triblock copolymers compared to that of common diblock copolymers with the same overall composition clearly arise from the existence of the SB middle blocks acting as rubbery phase while the PS forms domains or lamellae. Furthermore, the middle SB blocks increases the S-SB interaction, leading to a lower effective interaction parameter that turns the materials in the weak segregation limit as proven by the shift of glass transition temperatures (T_g) of the materials (97). The T_g of the SB blocks in LN3 and LN4 is furthermore governed by the S to B ratio in the SB blocks of about 50%. The T_g can be calculated from the Gordon-Taylor equation (98) and is in well agreement with the T_g determined for the SB phase in LN3 but not for LN4, confirming the discussion of the existence of a PS/PB mixed phase in LN4. For both materials, a lamellae morphology would be expected from the composition of the outer PS blocks of 32% (LN4) and 50% (LN3), with rubbery lamellae formed by the SB phase instead of a pure PB phase in the case of SBS triblock copolymers. While this is true for LN3, for LN4 the weak segregation of the material

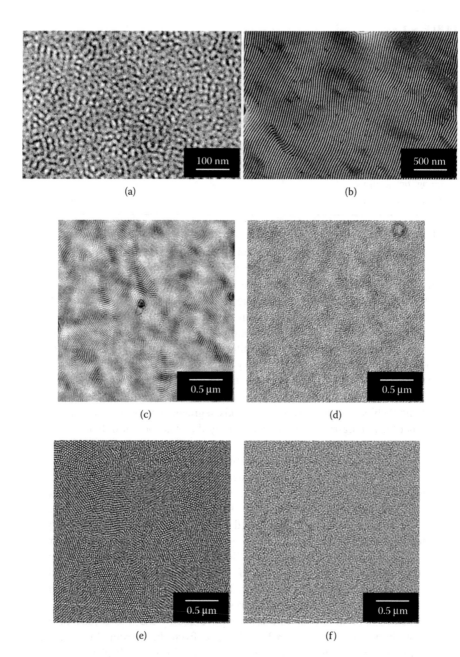

(a)   (b)

(c)   (d)

(e)   (f)

FIGURE 5.19 TEM micrographs of isotropic LN3/LN4-blends stained with OsO_4 (glassy phase: light areas, and soft phase: dark areas). (a) LN4, (b) LN3, (c) LN3/LN4 90:10, (d) LN3/LN4 80:20, (e), LN3/LN4 50:50, and (f) LN3/LN4 20:80.

strongly changes the morphology because of the partially mixed phases. This allows the lamellae morphology to extend to about 75 wt% PS, which is not observed for di- and triblock copolymers with pure PB or PI middle blocks.

The morphological study of these blends has revealed the absence of macrophase separation in a broad range of composition and the existence of different structures not observed in the neat copolymers. Wormlike, hexagonal, bicontinuous, and lamellae morphologies have been found that depend on the composition, which is similar to findings of Spontak and coworkers (87) in SI diblock copolymer blends. As shown in Figures 5.19c to 5.19f, wormlike and hexagonal morphologies have a larger composition range (0 to 60 wt% LN3) than bicontinuous and lamellae morphologies. The bicontinuous morphology resembles the gyroid structure found in block copolymers and their blends (28) but has not yet been confirmed by SAXS. We have only found two reflections in our experiments, which is not sufficient to allow to characterize the lattice; they probably arise from the coexistence of hexagonal and gyroid morphologies. In contrast to diblock copolymers and their blends, bicontinuous morphology seems to extend to a composition range of about 15 to 20 wt%, probably as a result of the compatibility of S-SB-S triblocks as compared to SI diblocks. Obviously, triblock copolymers offer interesting opportunities for tailoring blend morphologies by incorporating random copolymers as middle blocks. The wormlike and hexagonal morphologies show a mixed SB matrix and PS-rich domains where the similar composition of the SB copolymers promotes their miscibility. The following morphological transitions have been observed so far in this system: 0 to 10 wt% LN4: lamellae (Figures 5.19b and 5.19c); 20 to 30 wt% LN4: bicontinuous (Figure 5.19d); 35 wt% LN4: transition to hexagonally ordered PS cylinders; 40 to 50 wt% LN4: hexagonally ordered cylinders (Figure 5.19e); 60 wt% LN4: locally ordered short cylinders; and 70–100 wt% LN4: wormlike morphology (Figures 5.19a and 5.19f). Furthermore, a coexistence of bicontinuous and hexagonal morphology has been locally observed for 35 wt%, indicating a tendency to macrophase separation in a very limited composition range as already observed earlier by Spontak and coworkers (87). SAXS patterns for samples with 50 and 70% LN3 have proved the existence of a hexagonal morphology (97). Similar results have been found for 60% LN3. For 60 to 90% LN4, only one reflection that is very similar to that of pure LN4 has been found, indicating the existence of a wormlike morphology that lacks long-range order arising from partial miscibility

5.4.2 MECHANICAL PROPERTIES AND TOUGHNESS OF BINARY BLOCK COPOLYMER BLENDS

Fracture behavior of injection molded ST2/LN4 blends has been studied by an instrumented charpy tester. Experimental details and method are described in (94,95).

For sharply notched specimens of the blends, only small plastic deformation of the materials was observed up to 20 wt% LN4. Since the maximum stable crack growth in this composition range was very small ($\Delta a \leq 70$ μm, small scale yielding), crack arrest is generally impossible (unstable fracture). Hence, the experimental determination of crack resistance curves (R curves), i.e., the relation between fracture mechanics parameters (J integral J_d as well as crack tip opening displacement, CTOD

δ_d) and stable crack growth Δa was only possible at LN4 weight fractions equal to or larger than 20%. The quantitative description of stable crack propagation behavior is based on fracture mechanics parameters as resistance against stable crack initiation and stable crack propagation. The technical crack initiation values determined at an arbitrary value of Δa (here, $J_{0.05}$ is determined at $\Delta a = 0.05$ mm instead of 0.2 or 0.1 mm, which is usually found in standards and in the literature) are taken into account in order to characterize the stable crack initiation. Furthermore, crack initiation is followed by crack propagation and occurs at larger Δa values. Figure 5.20 shows the fracture mechanics parameters as a function of LN4 content where a strong increase of toughness at 20 and 60% LN4 has been observed to be associated with brittle-to-tough transitions (as discussed later). The indices "0.05" and "i" represent the technical and physical crack initiation values, respectively, as resistance against stable crack propagation determined at $\Delta a = 0.05$ mm. The index "Id" represents the resistance against unstable crack propagation. The physical crack initiation values J_i are not influenced by the morphology of the materials, which is in accordance with previous results on heterophase polymeric materials (94). This means that the crack initiation values are quite insensitive to the change of morphology, which has been shown for both heterophase polymers and metals (94). This is, however, not true in the case of crack propagation values. Thus, usually only J_i and J_{Id} can be considered as the relevant toughness parameters related to well defined fracture behavior. These values, i.e., J_i and J_{Id}, can be clearly correlated to the processes of stable crack initiation and unstable crack propagation, respectively.

The application of heterogeneous polymeric materials is often limited by the brittle-to-tough transition temperature and the critical modifier concentration. Generally, this transition is determined as an average value of high and low plateau values of toughness where, unfortunately, the conventional notched impact strength is mostly used. In contrast to current trends to develop novel polymeric materials based on binary polymer blends such as ethylene-propylene copolymers and other high-impact PP materials, the binary SB block copolymer blends, consisting of a thermoplastic block copolymer (ST2) and a TPE (LN4), investigated in this study, combine high-impact behavior with excellent transparency, thus allowing the stiffness to toughness ratio to be adjusted over a wide range. Two transitions are observed with increasing LN4 content, the conventional brittle-to-tough transition (Figure 5.20), and the tough to high-impact transition. While the former is caused by the transition from unstable to stable crack propagation, the latter transition should be caused only by a transition in deformation mechanism. In contrast to sample ST2, materials with about 60 wt% LN4 reveal a sufficiently high toughness under impact loading applications and still maintain their high level of transparency. According to the present investigation, two different BTTs were also observed in heterophase PP systems (99) and poly(acrylonitrile-butadiene styrene) (ABS) (100): the conventional BTT transition (BTT 1) and the second one (BTT 2). BTT 1 can be used as a measure of the safety against unstable crack propagation while BTT 2 can be described as a measure of the safety against stable crack propagation. While a pronounced BTT 1 can be observed at 20 wt% of LN4, a quite wide BTT 2 is seen at 60 wt% of LN4. As demonstrated recently (94), BTT 1 occurs if the crack growth mechanism changes from unstable to stable. Here, this behavior is combined with a change in the predominant microdeformation

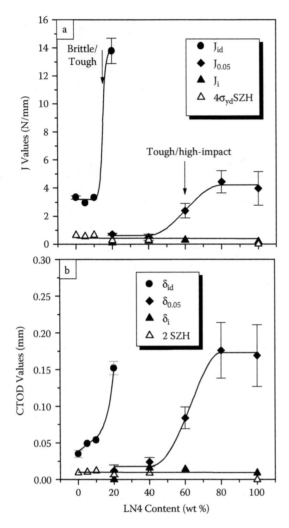

FIGURE 5.20 Technical initiation values, $J_{0.05}$ and $\delta_{0.05}$, and physical initiation values, J_i and d_i, as functions of resistance against stable crack initiation, and values, J_{Id} and d_{Id}, as functions of resistance against unstable crack propagation for ST2/LN4 blends as a function of weight fraction of LN4. (a) J_{Id}, $J_{0.05}$, J_i, and $4s_{yd} \times$ SZH (stretch zone height); (b) d_{Id}, $\delta_{0.05}$, d_i, and $2 \times$SZH.

process in the plastic zone ahead of the crack tip (change from thin layer yielding and coalescence of holes to shear flow). The BTT 2 is correlated with a change in the microdeformation mechanism, the transition from shear flow to rubberlike tearing. For conventionally toughened polymers (19), differences in deformation mechanisms associated with BTT 1 and BTT 2 can be explained by Wu's percolation concept (101) and Margolina's concept (102), respectively.

The correlation between morphology, fracture toughness, and deformation mechanisms as a function of LN4 content can be described by a transition in deformation mechanism that is similar to that of lamellae structured semicrystalline polymers,

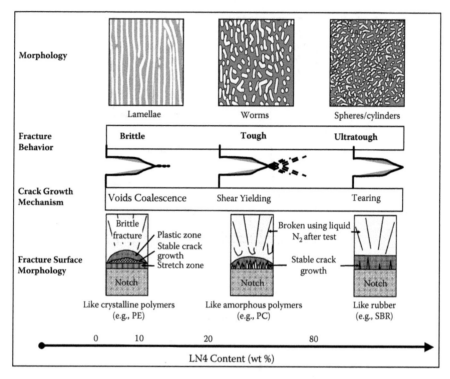

FIGURE 5.21 Schematic of the correlation between morphology, fracture toughness, and deformation mechanisms in ST2/LN4 blends.

amorphous polymers, and elastomers (Figure 5.21). It should be mentioned that the morphology-toughness correlation observed in the block copolymer blends is fundamentally different from that found in conventional impact modified or reinforced polymeric systems with matrix-particle morphologies that are based on their nanometer structured morphologies. Our investigations have demonstrated that utilizing blends of block copolymers can lead to concepts for developing toughened and transparent nanostructured polymer materials.

While two brittle-to-tough transitions have been observed for the partial miscible system star block/triblock copolymer blends, the mechanical properties, such as yield stress and strain at break, depend exponentially or linearly on composition (94,95). The morphologies of these blends resemble their neat copolymers, thereby causing only a simple change in mechanical properties (96). In contrast, hexagonal and bicontinuous morphologies have been observed for the miscible LN3/LN4 blends, which is not the case for the neat copolymers (i.e., they have similar architectures), resulting in a nonmonotonic dependence of mechanical properties on composition. While LN3 is a thermoplastic material, LN4 is a thermoplastic elastomer despite the similar compositions. The elastomeric property profile of LN4 is caused by the larger SB middle block fraction that acts as a rubbery matrix, while the PS block content of only 32% provides a property profile similar to that of SBS thermoplastic elastomers but exhibiting slightly larger tensile strength.

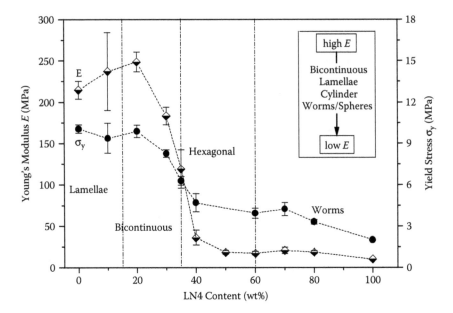

FIGURE 5.22 Correlation between morphology (and LN4 content) and mechanical properties (Young's modulus E and yield stress s_y) for LN3/LN4 blends.

The idea behind combining two miscible triblock copolymers was to create a TPE with increased tensile strength while maintaining the elasticity of each. Furthermore, the question arose as to how the different morphologies may affect the individual mechanical properties.

Figure 5.22 shows that both the Young's modulus and yield stress display a strong increase at the transition from hexagonal bicontinuous morphology. The bicontinuous morphology is known to possess a larger Young's modulus and yield stress compared to lamellae morphology as reported by Dair and Thomas (50). For the LN3/LN4 blends, the Young's modulus shows a maximum in the bicontinuous composition range, obviously increasing despite a volume fraction of the elastomeric triblock copolymer LN4 of about 20 wt%. The yield stress does not show a maximum, but it is approximately constant in the composition range of 0 to 20 wt% LN4 where lamellae and bicontinuous morphologies are apparent. Figure 5.23 shows stress-strain curves for LN3/LN4 blends in the composition range of 0 to 35 wt% LN4. While lamellae forming samples LN3 and LN3/LN4 90:10 show similar properties, materials in the bicontinuous composition range obviously show about the same yield stress but larger strain at break compared to LN3. If one compares materials property in both composition ranges, we can conclude that bicontinuous forming materials have about the same yield stress, associated with a pronounced yield point, enhanced Young's modulus, and improved strain at break, demonstrating superior property profile of bicontinuous (gyroid) morphology compared to lamellae forming materials. With increasing LN4 content, the yield point diminishes due to the transition from gyroid to hexagonal morphology at about 35 wt% LN4, assigning the transition from thermoplastic to elastomeric behavior and extending the elasto-

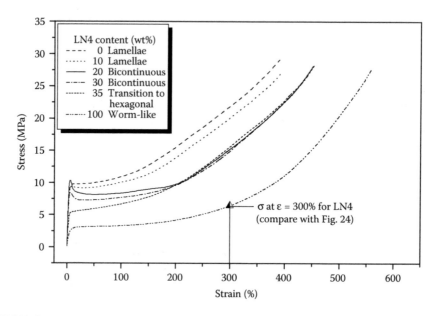

FIGURE 5.23 Stress-strain curves of LN3/LN4-blends as a function of LN4 content.

meric property profile up to a thermoplastic (TP) content of 65%. Bicontinuous structures such as the gyroid are obviously capable of providing improved properties compared to other morphologies as observed earlier by Dair and Thomas (50), apparently strengthening the thermoplastic property profile of the material due to the absence of a rubbery matrix; the change in properties at 35 wt% TPE (LN4) arises from the presence of a rubbery SB-matrix.

Another interesting result is that both the Young's modulus and yield stress do not vary much in the composition range of 50 to 100 wt% LN4 where hexagonal and wormlike morphologies have been observed. In this composition range, as shown in Figure 5.22, the PB-rich SB matrix is associated with elastomeric behavior, while an increase of LN3 content leads only to a slight increase in yield stress. To characterize elastomeric properties in this composition range, the stress at different given strains have been compared. The stress at 300% strain (as seen in Figure 5.24) increases from 5.5 MPa for LN4 to about 10 MPa for LN3/LN4 50:50 while still maintaining its elastomeric property profile, which has also been proved by hysteresis tests for up to 300% strain (103). Hysteresis tests of these materials revealed a decrease in elasticity as the content of LN3 is increased, but the elasticity still remained at a low level up to 40 wt% of the thermoplastic block copolymer LN3.

Similarly, the Young's modulus in Figure 5.22 linearly increases from 10 to only 36 MPa in the composition range 0 to 50 wt% LN3 but strongly increases in the gyroid morphology composition range. A low Young's modulus and a high elasticity are two of the requirements for successful TPE materials. Such materials combine the high strength of elastomers and the stress values usually provided by thermo-

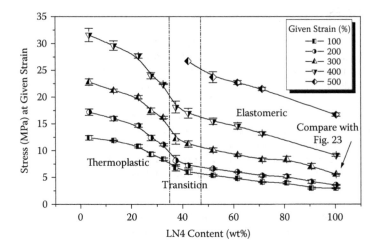

FIGURE 5.24 Stress at given strain ($\varepsilon = 100, 200, 300, 400,$ and 500%) for LN3/LN4-blends as a function of LN4 content.

plastic materials, demonstrating the potential of combining highly miscible thermoplastic and elastomeric block copolymers with similar architectures (98,103).

Highly miscible binary triblock copolymer blends may be an interesting tool to tailor mechanical properties and to provide interesting property profiles as unique morphologies not observed in the pure block copolymers are created, leading to new materials engineering design of either toughened thermoplastic materials or high strength elastomers. In addition, these materials reveal a remarkable level of transparency, which is usually more difficult to obtain by using only polymer blends.

ACKNOWLEDGMENTS

The authors would like to thank for their fruitful cooperation in the areas of morphology, deformation, and fracture of block and graft copolymers the research groups of Prof. S. P. Gido (Amherst, USA), Prof. J. Mays (Knoxville, USA), Prof. N. Hadjichristidis (Athens, Greece), Prof. G. H. Michler (Halle, Germany), and Prof. W. Grellmann (Halle). R.W. acknowledges Heisenberg-professorship from Deutsche Forschungsgemeinschaft (DFG). We acknowledge the cooperation and helpful discussions with Dr. R. Lach (Halle), Dr. K. Schneider (IPF Dresden, Germany), Dipl.-Ing U. Reuter, H. Scheibner (IPF Dresden), Dr. R. Adhikari (Halle), Prof. V. Abetz (GKSS Geesthacht), Dr. D. W. Schubert (Freudenberg), and Dr. H. Fischer (TU Eindhoven, The Netherlands). We would like to acknowledge the synthesis of the used block copolymers from the groups of Doz. Dr. M. Arnold (Halle), Doz. Dr. S. Höring (Halle, Germany), Prof. N. Hadjichristidis (Athens), Prof. J. Mays (Knoxville), and Prof. R. Jerome (Liege, Belgium). Furthermore, we would like to acknowledge financial support and supply of PS-b-PB block copolymers from BASF, especially Dr. K. Knoll for a very fruitful cooperation.

REFERENCES

1. Bucknall, C.B. *Toughened Plastics*, Appl. Sci. Publ. Ltd., London, 1977.
2. Michler, G. H. *Kunststoffmikromechanik*, Carl Hanser Verlag, München Wien, 1992.
3. Hashimoto, T, Yamasaki, K., Koizumi, S., and Hasegawa, H., Ordered structure in blends of block copolymers. 2. Self-assembly for immiscible lamella-forming copolymers, *Macromolecules*, 26, 1562, 1994.
4. Hasegawa, H., Tanaka, H., Yamasaki, K., and Hashimoto, T., Bicontinuous microdomain morphology of block copolymers. 1. Tetrapod-network structure of polystyrene-polyisoprene diblock polymers, *Macromolecules*, 20, 1651, 1987.
5. Winey, K.I., Gobran, D.A., Xu, Z., Fetters, L.J., and Thomas, E.L., Compositional dependence of the order-disorder transition in diblock copolymers, *Macromolecules*, 27, 2392, 1994.
6. Khandpur, A. K, Förster, S., Bates, F. S., Hamley, I. W., Ryan, A. J., Bras, W., Almdal, K., and Mortensen, K., Polyisoprene-polystyrene diblock copolymer phase diagram near the order-disorder transition, *Macromolecules*, 26, 8796, 1995.
7. Hamley, I.W., Koppi, K.A., Rosedale, J.H., Bates, F.S., Almdal, K., and Mortensen, K., Hexagonal mesophases between lamellae and cylinders in a diblock copolymer melt, *Macromolecules*, 26, 5959, 1993.
8. Hadjuk, D.A., Harper, P.E., Gruner, S.M., Honeker, C.C., Kim, G., Thomas, E.L., and Fetters, L.J., The gyroid: a new equilibrium morphology in weakly segregated diblock copolymers, *Macromolecules*, 27, 4063, 1994.
9. Förster, S., Khanpur, A.K., Zhao, J., Bates, F.S., Hamley, I.W., Ryan, A.J., and Bras, W., Complex phase behavior of polyisoprene-polystyrene diblock copolymers near the order-disorder transition, *Macromolecules*, 27, 6922, 1994.
10. Beckmann, J, Auschra, C., and Stadler, R., Ball at the wall — a new lamellar multiphase morphology in a polystyrene-block-polybutadiene-block-poly(methyl methacrylate) triblock copolymer, *Macromol. Rapid Commun.*, 15, 67, 1994.
11. Krappe, U., Stadler, R., and Voigt-Martin, I., Chiral assembly in amorphous abc triblock copolymers. formation of a helical morphology in polystyrene-block-polybutadiene-block-poly(methyl methacrylate) block copolymers, *Macromolecules*, 28, 4558, 1995.
12. Stadler, R., Auschra, C., Beckmann, J., Krappe, U., Voigt-Martin, I., and Leibler., L., Morphology and thermodynamics of symmetric poly(A-block-B-block-C) triblock copolymers, *Macromolecules*, 28, 3080, 1995.
13. Higgins, J.S. and Benoit, H.C., *Polymers and Neutron Scattering*, Clarendon Press, Oxford, 1994.
14. Anastasiadis, S.H., Russell, T.P., Satija, S.K., and Majkrzak, C.F., The morphology of symmetric diblock copolymers as revealed by neutron reflectivity, *J. Chem. Phys.*, 92, 5677, 1990.
15. Schubert, D.W. and Stamm, M., The influence of chain length on the interface width of an immiscible polymer blend, *Europhys. Lett.*, 35: 419, 1996.
16. Stamm, M., Reflection of neutrons for the investigation of polymer interdiffusion at interfaces, in *Physics of Polymer Surfaces and Interfaces*, Sanchez, I.C. and Fitzpatrick, L.E., Eds., Butterworth-Heinemann, Boston, 1992, p.163.
17. Broseta, D., Fredrickson, G.H., Helfand, E., and Leibler, L., Molecular weight and polydisparity effects at polymer-polymer interfaces, *Macromolecules*, 23, 132, 1990.
18. Russell, T.P., Hjelm, R.P., and Seeger, P.A., Temperature dependence of the interaction parameter of polystyrene and poly(methylmethacrylate), *Macromolecules*, 23, 890, 1990.

19. Tcherkasskaya, O.S. and Winnik, M.A., Direct energy transfer studies of the domain-boundary interface in polyisoprene-poly(methylmethacrylate) block copolymer films, *Macromolecules*, 29, 610, 1996.

20. Schnell, R., Stamm, M., and Creton, C., Mechanical properties of homopolymer interfaces: Transition from simple pullout to crazing with increasing interfacial width, *Macromolecules*, 32, 3420, 1999.

21. Schubert, D.W., Spin coating as a method for polymer molecular weight determination, *Polymer Bulletin*, 38, 177, 1997.

22. Grellmann, W., Seidler, S., and Hesse, W., in *Deformation and Fracture Behaviour of Polymers*, Grellmann, W. and Seidler, S., Eds., Springer, Berlin-Heidelberg, 2001, p. 71.

23. Grellmann, W., Lach, R., and Seidler, S., in *From Charpy to Present Impact Testing (ESIS Publication 30)*, Pineau, A. and Francois, D., Eds., Elsevier Science, Amsterdam, 2002, p. 145.

24. Leibler, L., Theory of microphase separation in block copolymers, *Macromolecules*, 13, 1602, 1980.

25. Fredrickson, G.H. and Helfand, E., Fluctuation effects in the theory of microphase separation in block copolymers, *J. Chem. Phys.*, 87, 697, 1987.

26. Mayes, A.M. and Olvera de la Cruz, M., Microphase separation in multiblock copolymer melts, *J. Chem. Phys.*, 91, 7228, 1989.

27. Dobrynin, A.V. and Erukhimovich, I.Y., Computer-aided comparative investigation of architecture influence on block copolymer phase diagrams, *Macromolecules*, 26, 276, 1993.

28. Hamley, I.W., *The Physics of Block Copolymers*, University Press, Oxford, 1998.

29. Russell, T.P., Karis, T.E., Gallot, Y., and Mayes, A.M, A lower critical ordering transition in a diblock copolymer melt, *Nature*, 368, 729, 1994.

30. Karis, T.E., Russell, T.P., Gallot, Y., and Mayes, A.M., Rheology of the lower critical ordering transition, *Macromolecules*, 28, 1129, 1995.

31. Hashimoto, T., Hasegawa, H., Hashimoto, T., Katayama, H., Kamigaito, M., Sawamoto, M., and Imai, M., Synthesis and SANS characterization of poly(vinyl methyl ether)-block-polystyrene, *Macromolecules*, 30, 6819, 1997.

32. Olabisi, O., *Polymer-Polymer Miscibility*, Academic Press, New York, 1978.

33. Ruzette, A., Banerjee, P., Mayes, A. M., Pollard, M., Russell, T. P., Jerome, R., Slawecki, T, Hjelm, R., and Thiyagarajan, P., Phase behavior of diblock copolymers between styrene and n-alkyl methacrylates, *Macromolecules*, 31, 8509, 1998.

34. Weidisch, R., Stamm, M., Schubert, D.W., Arnold, M., Budde, H., and Höring, S., Correlation between phase behavior and tensile properties of diblock copolymers, *Macromolecules*, 32, 3405, 1999.

35. Pollard, M., Russell, T.P., Ruzette, A.V., Mayes, A.M., and Gollot, Y., The effect of hydrostatic pressure on the lower critical ordering transition in diblock copolymers, *Macromolecules*, 31, 6493, 1998.

36. Jeong, U., Ryu, D.Y., and Kim, J.K., Phase behavior of polystyrene-*block*-poly(*n*-butyl methacrylate) copolymers with various end-functional groups, *Macromolecules*, 36, 8913, 2003.

37. Ryu, D.Y., Jeong, U., Lee, D.H., Kim, J., Shik Youn, H., and Kim, J. K., Phase behavior of deuterated polystyrene-*block*-poly(*n*-pentyl methacrylate) copolymers, *Macromolecules*, 36, 2894, 2003.

38. Fischer, H., Weidisch, R., Stamm, M., Budde, H., and Höring, S. The phase diagram of the system PS-b-PBMA, *Polymer Colloid Sci.*, 287, 1019, 2000.

39. Floudas, G., Vazaiou, B., Schipper, F., Ulrich, R., Wiesner, U., Iatrou, H., and Hadjichristidis, N., Poly(ethylene oxide-b-isoprene) diblock copolymer phase diagram, *Macromolecules*, 34, 2947, 2001.

40. Weidisch, R., Stamm, M., Michler, G.H., Fischer, H., and Jerome, R., Mechanical properties of weakly segregated block copolymers 3. Influence of strain rate and temperature on tensile properties of poly(styrol-b-butylmethacrylate) diblock copolymers with different morphologies, *Macromolecules*, 32, 742, 1999.

41. Weidisch, R., Michler, G.H., Fischer, H., Hofmann, S., Arnold, M., and Stamm, M., Mechanical properties of weakly segregated block copolymers. 1. Synergism on tensile properties of poly(styrene-b-butylmethacrylate) diblock copolymers, *Polymer*, 40, 1191, 1999.

42. Stamm, M. and Schubert, D.W., Interfaces between incompatible polymers, *Annu. Rev. Mater. Sci.*, 25, 325, 1995.

43. Schubert, D.W., Weidisch, R., Stamm, M., and Michler, G.H., Interface width of poly(styrene-b-butylmethacrylate) diblock copolymers, *Macromolecules*, 31, 3743, 1998.

44. Holden, G., in *Thermoplastic Elastomers*, Legge, N.R., Holden, G., and Schroeder, H.E., Eds., Hanser-Verlag, Munich, 1987, p. 481.

45. Morton, M., in *Encyclopedia of Polymer Science and Technology*, Vol.15, Bikales, M., Ed., Wiley-Interscience, New York, 1971, p. 508.

46. Segula, R. and Prud'homme, J., Affinity of grain deformation in mesomorphic block polymers submitted to simple elongation, *Macromolecules*, 21, 635, 1988.

47. Odell, J.A. and Keller, A., Orientation of block copolymer melts, *Polym. Eng. Sci.*, 17, 544, 1977.

48. Pakula, T., Saijo, K., Kawai, H., and Hashimoto, T., Deformation behavior of styrene-butadiene-styrene triblock copolymer with cylindrical morphology, *Macromolecules*, 18, 1294, 1985.

49. Sakamoto, J., Sakurai, Doi, K., and Nomura, S., Effects of microdomain structures on the molecular orientation of poly(styrene-block-butadiene-block-styrene) triblock copolymer, *Macromolecules*, 26, 3351, 1993.

50. Dair, B.J. and Thomas, E.L., Mechanical properties of the double gyroid phase in oriented thermoplastic elastomers, *J. Mater. Sci.*, 35, 5207, 2000.

51. Honeker, C.C., Thomas, E.L., Albalak, R.J., Hadjuk, D.A., Gruner, S.M., and Capel, M.C., Perpendicular deformation of a near-single-crystal triblock copolymer with a cylindrical morphology. 1. Synchrotron SAXS, *Macromolecules*, 33, 9395, 2000.

52. Honeker, C.C., Thomas, E.L., Albalak, R.J., Hadjuk, D.A., Gruner, S.M., and Capel, M. C., Perpendicular deformation of a near-single-crystal triblock copolymer with a cylindrical morphology. 2. TEM, *Macromolecules*, 33, 9407, 2000.

53. Schwier, C.E., Argon, A.S., and Cohen, R.E., Crazing in polystyrene-polybutadiene diblock copolymers containing cylindrical polybutadiene domains, *Polymer*, 26, 1985, 1985.

54. Yamaoka, I., Effects of morphology on the toughness of styrene — butadiene — styrene triblock copolymer/methyl methacrylate — styrene copolymer blends, *Polymer*, 39, 1765, 1998.

55. Weidisch, R., Michler, G.H., Arnold, M, Hofmann, S., Stamm, M., and Jérôme, R., Synergism for the improvement of tensile strength in block copolymers of polystyrene and poly(n-butylmethacrylate), *Macromolecules*, 30, 8078, 1997.

56] Khandpur, A.K., Foerster, S., Bates, F.S., Hamley, I.W., Ryan, A.J., Bras, W., Almdal, K., and Mortensen, K., Polyisoprene-polystyrene diblock copolymer phase diagram near the order-disorder transition, *Macromolecules*, 26, 8796, 1995.

57. Scherble, J., Stark, B., Stuehn, B., Kressler, J., Schubert, D.W., Budde, H., Hoering, S., Simon, P., and Stamm, M., Comparison of interfacial width of block copolymers of d_8-poly(methyl methacrylate) with various poly(n-alkyl methacrylate)s and the respective homopolymer pairs as measured by neutron reflection, *Macromolecules*, 32, 1859, 1999.

58. Smith, R.J.M., Brekelmans, W.A.M., and Meijer, H.E.H., Predictive modelling of the properties and toughness of polymeric materials. Part I Why is polystyrene brittle and polycarbonate tough? *J. Mater. Sci.*, 35, 2855, 2000.

59. Smith, R.J.M., Brekelmans, W.A.M., and Meijer, H.E.H., Predictive modelling of the properties and toughness of polymeric materials. Part II Effect of microstructural properties on the macroscopic response of rubber-modified polymers, *J. Mater. Sci.*, 35, 2869, 2000.

60. Smith, R.J.M., Brekelmans, W.A.M., and Meijer, H.E.H., Predictive modelling of the properties and toughness of polymeric materials. Part III Simultaneous prediction of micro- and macrostructural deformation of rubber-modified polymers, *J. Mater. Sci.*, 35, 2881, 2000.

61. van Melick, H.G.H., Govaert, L.E., and Meijer, H.E.H., Prediction of brittle-to-ductile transitions in polystyrene, *Polymer*, 44, 457, 2003.

62. Argon, A.S. and Cohen, R.E., in *Crazing in Polymers*, Vol. 1, Kausch, H.H., Ed., Springer Verlag, 1983, p..

63. Argon, A.S. and Cohen, R. E., in *Crazing in Polymers*, Vol. 2, Kausch, H.H., Ed., Springer Verlag, 1990, p..

64. Matsen, M.W. and Bates, F., Unifying weak- and strong-segregation block copolymer theories, *Macromolecules*, 29, 1091, 1996.

65. Anastasiadis, S.H., Russell, T.P., Satija, S.K. and Majkrzak, C.F., The morphology of symmetric diblock copolymers as revealed by neutron reflectivity, *J. Chem. Phys.*, 92, 5677, 1990.

66. Weidisch, R., Enßlen, M., Michler, G.H., Arnold, M., Budde, H., and Höring, S., A novel scheme for prediction of deformation mechanisms of block copolymers based on phase behavior, *Macromolecules*, 34, 2528, 2001.

67. Weidisch, R., Michler, G.H., and Arnold, M., Mechanical properties of weakly segregated block copolymers. 2. Influence of phase behaviour on tensile properties of poly(styrene-b-butyl methacrylate) diblock copolymers, *Polymer*, 41, 2231, 2000.

68. Weidisch, R., Michler, G.H., Arnold, M., and Fischer, H., Mechanical properties of weakly segregated block copolymers. 4. Influence of chain architecture and miscibility on tensile properties of block copolymers, *J. Mat. Sci.*, 35, 1257, 2000.

69. Drolet, F. and Fredrickson, G.H., Optimizing chain bridging in complex block copolymers, *Macromolecules*, 34, 5317, 2001.

70. Ryu, C.Y., Ruokolainen, J., Fredrickson, G.H., and Kramer, E.J., Chain architecture effects on deformation and fracture of block copolymers with unentangled matrices, *Macromolecules*, 35, 2157, 2002.

71. Weidisch, R., Ensslen, M., Michler, G.H., and Fischer, H, Deformation mechanisms of weakly segregated block copolymers. 1. Influence of morphology of poly(styrene-b-butyl methacrylate) diblock copolymers, *Macromolecules*, 32, 5375, 1999.

72. Weidisch, R., Schreyeck, G., Ensslen, M., Michler, G.H., Stamm, M., Schubert, D.W., Arnold, M., Budde, H., Höring, S., and Jerome, R., Deformation mechanisms of weakly segregated block copolymers. 2. Correlation between phase behavior and deformation mechanisms in diblock copolymers, *Macromolecules*, 33, 5495, 2000.

73. Haward, R.N., *The Physics of Glassy Polymers*, Applied Publishers Ltd., London, 1997.

74. Mori, Y., Lim, L.S., and Bates, F.S., Consequences of molecular bridging in lamellae-forming triblock/pentablock copolymer blends, *Macromolecules*, 36, 9879, 2003.

75. Weidisch, R., Laatsch, J., Michler, G.H:, Schade, B., Arnold, M., and Fischer, H., Transition from crazing to shear deformation in star block copolymers, *Macromolecules*, 35, 6585, 2002.

76. Weidisch, R., Gido, S.P., Uhrig, D., Iatrou, H., Mays, J., and Hadjichristidis, N., Tetrafunctional multigraft copolymers as novel thermoplastic elastomers, *Macromolecules*, 34, 6333, 2001.

77. Beyer, F., Gido, S.P., Büschl, C., Iatrou, H., Uhrig, D., Mays, J.W., Chang, M.Y., Garetz, B.A., Balsara, N.P., Tan, N.B., and Hadjichristidis, N., Graft copolymers with regularly spaced, tetrafunctional branch points: morphology and grain structure, *Macromolecules*, 33, 2039, 2000.

78. Quirk, R.P. and Legge, N.R., in *Thermoplastic Elastomers*, Legge, N.R., Holden, G., Quirk, R., and Schroeder, H.E., Eds., Hanser, Muinch, 1987, p. 71.

79. Kennedy, J.P., in *Thermoplastic Elastomers*, Legge, N.R. Holden, G., Quirk, R., and Schroeder, H.E., Eds., Hanser, Muinch, 1996, p. 365.

80. Holden, G., in *Thermoplastic Elastomers*, Legge, N.R. Holden, G., Quirk, R., and Schroeder, H.E., Eds., Hanser, Muinch, 1987, 481–506

81. Rader, C.P., in *Modern Plastics: Encyclopedia 1996*, Vol. 2, Kaplan, W.L., Ed., McGraw-Hill Co., New York, 1996, p..

82. Weidisch, R. and Michler, G.H., in *Block Copolymers*, Balta Calleja, F.J. and Roslaniec, Z., Eds., Marcel Dekker, New York, 2000, p. 215.

83. Zhu, Y., Weidisch, R., Hong, S., Gido, S.P:, Iatrou, H., and Hadjichristidis, N., Morphology and mechanical properties of a series of block double-graft copolymers, *Macromolecules*, 35, 5903, 2002.

84. Hashimoto, T., Koizumi, S., and Hasegawa, H., Ordered structure in blends of block copolymers. 2. Self-assembly for immiscible lamella-forming copolymers, *Macromolecules*, 27 1562, 1994.

85. Yamaguchi, D., Shiratake, S., and Hashimoto, T., Ordered structure in blends of block copolymers. 5. Blends of lamella-forming block copolymers showing both microphase separation involving unique morphological transitions and macrophase separation, *Macromolecules*, 33, 8258, 2000.

86. Koizumi, S., Hasegawa, H., and Hashimoto, T., Ordered structure in blends of block copolymers. 3. Self-assembly in blends of sphere- or cylinder-forming copolymers, *Macromolecules*, 27, 4371, 1994.

87. Spontak, R.J., Fung, J.C., Braunfeld, M.B., Sedat, J.W., Agard, D.A., Kane, L., Smith, S.D., Satkowski, M.M., Ashraf, A., Hajduk, D.A., and Gruner, S.M., Phase behavior of ordered diblock copolymer blends: effect of compositional heterogeneity, *Macromolecules*, 29, 4494, 1996.

88. Sakurai, S., Irie, H., Umeda, H., Nomura, S., Lee, H.H., and Kim, J.K., Gyroid structures and morphological control in binary blends of polystyrene-block-polyisoprene diblock copolymers, *Macromolecules*, 31, 336, 1998.

89. Sakurai, S., Umeda, H., Furukawa, C., Irie, H., Nomura, S., Lee, H.H., and Kim, J.K., Thermally induced morphological transition from lamella to gyroid in a binary blend of diblock copolymers, *J. Chem Phys.*, 108, 4333, 1998.

90. Spontak, R.J., Fung, C.J., Braunfeld, M.B., Sedat, J.W., Agard, D.A., Ashraf, A., and Smith, S.D., Architecture-induced phase immiscibility in a diblock/multiblock copolymer blend, *Macromolecules*, 29, 2850, 1996.

91. Kane, L., Norman, D.A., White, S.A., Matsen, M.W., Satkowski, M.M., Smith, S.D., and Spontak, R.J., Molecular, nanostructural and mechanical characteristics of lamellar triblock copolymer blends: effects of molecular weight and constraint, *Macromol. Rapid Commun.*, 22, 281, 2001.
92. Hashimoto, T., Yamasaki, K., Koizumi, S., and Hasegawa, H., Ordered structure in blends of block copolymers. 1. Miscibility criterion for lamellar block copolymers, *Macromolecules*, 26, 2895, 1993.
93. Ivankova, E., Adhikari, R., Michler, G.H., Weidisch, R., and Knoll, K., Investigation of the micromechanical deformation behavior of styrene-butadiene star block copolymer/polystyrene blends with high-voltage electron microscopy, *J. Polym. Sci. Part B: Polymer Phys.*, 41, 1157, 2003.
94. Lach, R., Adhikari, R., Weidisch, R., Huy, T.A., Michler, G.H., Grellmann, W., and Knoll K., Crack toughness behavior of binary styrene-butadiene block copolymer blends, *J. Mater. Sci.*, 39, 1283, 2004.
95. Adhikari, R., Lach, R., Grellmann, W., Weidisch, R., Michler, G.H., and Knoll, K., Morphology and crack resistance behavior of binary block copolymer blends, *Polymer*, 43, 1943, 2002.
96. Adhikari, R., Michler, G.H., Henning, S., Godehardt, R., Huy, T.A., and Goerlitz, S., and Knoll, K., Morphology and micromechanical deformation behavior of styrene–butadiene block copolymers. IV. Structure–property correlation in binary block copolymer blends, *J. Appl. Polym. Sci.*, 92, 1219, 2004.
97. Lach, R., Weidisch, R., and Knoll, K., Morphology and mechanical properties of binary triblock copolymer blends, *J. Polym. Sci., Polym. Phys.*, in press.
98. Lach, R., Staudinger, U., Weidisch, R., and Knoll, K., in preparation.
99. Grellmann, W. and Seidler, S., Eds., *Deformation and Fracture Behaviour of Polymers*, Springer-Verlag, Berlin-Heidelberg, 2001.
100 Han, Y., Lach, R., and Grellmann, W., Effects of rubber content and temperature on unstable fracture behavior in ABS materials with different particle sizes, *J. Appl. Polym. Sci.*, 79, 9, 2001.
101. Wu, S., Phase structure and adhesion in polymer blends: a criterion for rubber toughening, *Polymer*, 26, 1855, 1985.
102. Margolina, A., Toughening mechanisms for nylon-rubber blends — the effect of temperature, *Polym. Commun.*, 31, 95, 1990.
103. Weidisch, R. and Knoll, K., *Blends of PS-PB Block Copolymers*, German Patent, DE 10 2004 014 585.7, 23.03.2004.

6 Polymer-Polymer Interfaces: Theoretical, Experimental, and Adhesion Aspects

Hideaki Yokoyama

CONTENTS

6.1 INTRODUCTION

There are a variety of polymers available in the market. However, a single polymer often cannot achieve the desired properties for industrial applications. As developing a new polymer with a new chemical structure that fits the needs of the market is becoming increasingly difficult, blending polymers is an indispensable method for developing new polymeric materials. Two or more polymers are often blended together to achieve the desired properties; however, such attempts often lead to materials with poor mechanical properties (1–2). Polymer-polymer interfaces play an important role in the properties of polymer blends. Most polymer pairs are immiscible, and the interfaces dividing polymer pairs are thin and weak (3). The presence of such weak interfaces between immiscible phases in blends makes the material very brittle. It is essential to have a fundamental understanding of the structures and properties of polymer–polymer interfaces in addition to knowing many practical methods for compatibilizing polymer blends to achieve the desired properties. While the polymer–polymer interfaces of materials of interest to industry are often very complicated, it is nevertheless helpful to understand the physics of polymer–polymer interfaces of simple model systems. We believe that the understanding of simple model polymer–polymer interfaces can help us to understand and develop new polymeric materials. In this chapter, the theoretical aspects of interfaces, reinforcement of interfaces by copolymers, and interfacial reaction of polymers are discussed. First, we introduce theories of polymer–polymer interfaces that are often used to analyze the interface of polymers. Second, the typical segregation of block copolymers and model compatibilizers of polymer blends will be analyzed. The segregation process often leads to unstable interfacial structures and properties. We discuss the diffusion of block copolymers, which tends to be so slow the interfacial segregation may be kinetically limited. Based on our understanding of the segregation of block copolymers, we will present what is known about the adhesion aspect of polymer–polymer interfaces reinforced by block copolymers. Lastly, the reaction of functional polymers at interfaces will be discussed. We especially focus on the analogy and difference between the interfacial reaction of end-functionalized polymers and the segregation of premade block copolymers.

6.2 THEORETICAL APSECTS OF INTERFACES

6.2.1 HELFAND-TAGAMI'S SELF-CONSISTENT FIELD

Very few pairs of polymers show some degree of miscibility. The heat of mixing between the segments of polymer pairs is usually positive. Unlike small molecules, polymers have a very small entropy of mixing that is proportional to the number of polymer molecules because of the connectivity of the segments (4). Therefore, the entropy of mixing is proportional to $1/N$, where N is the number of segments and is usually large, making most polymer pairs immiscible.

In this section, we discuss only the essence of the theoretical treatment of polymer-polymer interface using the self-consistent-field (SCF) theory (5–8), which is frequently used to analyze interfacial structures and properties. The analytical solutions for the set of SCF equations are not obtained except for very simple cases. However, analytical solutions are valuable for capturing the physics of the interfacial structures and properties, and are widely used to interpret experimental results. Although this section is not intended to go into the mathematical detail of the theory, an understanding of the concept of SCF mean field theory is still helpful. In addition, understanding SCF theories can help us to solve SCF equations numerically in more complex situations of experimental interest to us, such as polymer-polymer interfaces reinforced with block copolymers, which is discussed in Section 6.2.2.

Let us consider the interface between a pair of polymers A and B. The degree of repelling forces between segments is the defined nondimensional χ parameter in a scale of $k_B T$, which is a measure of the chemical potential change of A when its neighbors are changed from A to B; k_B and T are the Boltzman constant and temperature, respectively. When two nearly pure polymer phases meet at the interface, the unlike segments repel each other so that the interface becomes sharp to

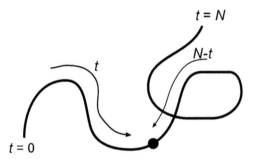

FIGURE 6.1 Schematic representation of probability distribution function $q_k(x, t)$. A segment t of a polymer chain is expressed as a function of time t on a diffusion path.

avoid undesired contacts. Let us introduce a partition function, $q_k(x,t)$, of a polymer molecule k with number of segments, N_k, and segment length, b_k, where k is either A or B in this case. For simplicity, we consider a one-dimensional problem in the direction x normal to the interface. $q_k(x,t)$ is the partition function of finding t-th segment of polymer k at position x. In the absence of an external field, $U(x)$, the chain conformation has a Gaussian distribution, and the partition function $q_k(x,t)$ follows the diffusion equation with the diffusion coefficient of $b_k^2/6$ (9)

$$\frac{q_k}{\partial t} = \frac{b_k^{\,2}}{6} \nabla^2 q_k \; . \tag{6.1}$$

The diffusion equation includes the connectivity of the polymer chain in the partition function by way of the discrete segment number t and continuous time t as schematically shown in Figure 6.1 since the Gaussian random walk process is used to describe both the polymer chain and diffusion process. The solution to this diffusion equation with the initial condition of $q_k(x,0) = 1$ is a Gaussian distribution. When an external field $U_k(x)$ is applied, the diffusion equation is more generally expressed as

$$\frac{q_k}{\partial t} = \left(\frac{b_k^{\,2}}{6} \nabla^2 - \frac{U_k(x)}{k_B T} \right) q_k \tag{6.2}$$

with an initial condition $q_k(x,0) = 1$. The probability distribution of any segment of polymer k at position x is given by summing the probability of all segments from the chain ends on both sides as

$$\rho_k = \int_0^N q_k(x,t) q_k(x, N_k - t) dt \; . \tag{6.3}$$

The free energy density f of a uniform system can be written in the general Flory-Huggins form as

$$\frac{f}{k_B T} = \frac{\phi_A \ln \phi_A}{N} + \frac{\phi_B \ln \phi_B}{N} + \chi \phi_A \phi_B \qquad (6.4)$$

It is more useful to introduce the chemical potential form, that is,

$$\frac{U_k(x)}{k_B T} = \chi \left[\rho_{k'}(x) + w(x) \right], \qquad (6.5)$$

where $\rho_{k'}(x)$ is the probability of the segment of the polymer "not k" ($k = $ A, B and $k' = $ B, A) at x. The first term in brackets represents the Flory-Huggins interaction chemical potential. The excess energy of the undesired contacts of the segments is included in this term. The second term is the compressibility function at x, and, in the Helfand-Tagami formula (5–7), it is expressed as

$$w(x) = \frac{\zeta}{\chi} \left[\sum \rho_k(x) - 1 \right], \qquad (6.6)$$

where ζ is the incompressibility parameter. This term gives additional potential energy so that the sum of the probability of polymer A and B does not deviate from unity for the incompressible system. The partition function is obtained by solving the modified diffusion equation [Eq. (6.2)] if the external potential field or the probability distribution of the segments is known. However, the probability distribution of the segments is obtained from the partition functions by Eq. (6.3). We have to find a "self-consistent" solution to the set of recurring equations. Although, in most cases, this self-consistent-field approach requires numerical calculations, Helfand and Tagami found an analytical solution for the case that the number of segments in the polymer A and B are infinite. By assuming $q_k(x,t)$ to be time independent, that is, $q_k(x)$, for polymers with an infinite number of segments except for the chain ends, they found the steady state solution

$$\rho_k(x) = q_k^2(x) \qquad (6.7)$$

$$0 = \frac{b_k^2}{6} \frac{\partial^2}{\partial x^2} q_k - \chi q_{k'}^2 q_k - \zeta(q_k^2 + q_{k'}^2 - 1) q_k \qquad (6.8)$$

Solving Eqs. (6.7) and (6.8) and assuming χ/ζ is small and negligible, then the interfacial width d [defined as $d = (\partial \rho / \partial x)_{x=0}^{-1}$], the interfacial tension γ, and the density distribution $\rho_k(x)$ are

$$d = \frac{2b_k}{\sqrt{6\chi}} \qquad (6.9)$$

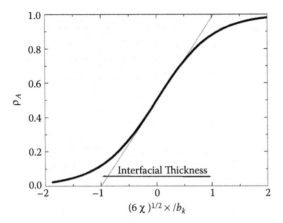

FIGURE 6.2 Probability distribution ρ_A of polymer A at the interface with polymer B computed using Eq. 6.11. The coordinate x, which is in the direction normal to the interface, is normalized with the interaction parameter χ and b_k, the segment lengths of polymers A and B.

$$\gamma = b_k \rho_0 k_B T \sqrt{\chi/6} \tag{6.10}$$

$$\rho_k(x) = \frac{1}{2} + \frac{1}{2}\tanh\left[\frac{\sqrt{6\chi}x}{b_k}\right], \tag{6.11}$$

respectively. The steady state solution is very informative and provides the following conclusions. The interfacial width is scaled as approximately $\chi^{-1/2}$ for small c, and the density profile is given by Eq. (6.11). The probability distribution is expressed by the hyperbolic tangent function as shown in Figure 6.2. Equations (6.9) to (6.11) indicate that the surface structures and properties are asymptotically molecular weight independent for large N. The interfacial tension goes as approximately $\chi^{1/2}$. The excess free energy density of order c over the length of order $\chi^{-1/2}$ leads to the scaling of the interfacial tension g equal to approximately $\chi^{1/2}$. As a consequence of this theory, the typical interfacial width is on the order of nanometers, given that χ has an order of 0.01. The density at the interface varies over a distance much greater than the segment lengths of polymers, which are typically less than 1 nm. Under such gradual change of density and the potential field, replacing the polymer chain by the Gaussian random walk path near the interface is self-consistent. Broseta et al. (10) extended the Helfand-Tagami result for the interface of polymers with finite molecular weights

$$d = \frac{2b_k}{\sqrt{6\chi}}\left[1 + \frac{\ln 2}{\chi}\left(\frac{1}{N_A} + \frac{1}{N_B}\right)\right], \tag{6.12}$$

where N_A and N_B are the number of segments of polymers A and B, respectively. The interfacial width becomes broader for low molecular weight polymers and approaches the Helfand-Tagami result for the high molecular weight limit. The results of Helfand and Tagami (5–7), and Broseta et al. (10) reasonably agree with experiments (11–16); the presence of a capillary wave, however, needs to be taken into account.

6.2.2 NUMERICAL SELF-CONSISTENT FIELD

As the analytical SCF by Helfand-Tagami (5–7) revealed, the properties of simple polymer–polymer interfaces are successfully predicted. For instance, the interfacial width of the pair of polymers with χ of 0.05 and b_k of 0.6 nm is computed to be only 2.2 nm. The interfacial width of the polymer pairs with a large value of χ is so small that the mechanical properties of the blends tend to be severely deteriorated by the poor adhesion of the interface. In order to use such polymer blends for practical use, the interface must be reinforced. In addition, the interfacial tension, which is often large for a large value of χ, impedes good dispersion of the phases during mixing and drives coalescence. The interfacial tension must be somehow reduced to achieve the finer dispersion of phases. Such a process is often called *compatibilization*. One classic example is adding A–B block copolymers to the blends of polymers A and B to reduce the interfacial tension and to reinforce the interface. The added A–B block copolymers segregate to the interface, reduce the interfacial tension, and reinforce the adhesion. The details of the experimental findings concerning the segregation, diffusion, and adhesion of block copolymers are discussed later in this chapter. While compatibilization is quite important in practice, SCF theory cannot find analytical solutions when compatibilizers are added to the interface. In SCF theories, adding block copolymers in the set of SCF equations is easily achieved, but finding the analytical solution to the set of equations is, in general, impossible. We can, however, find the numerical solutions for the modified SCF equations for ternary blends and extract a variety of physical properties from the equations. This method was extensively used by Hong and Noolandi (17–21) for blends, block copolymers, and solutions. The interface of homopolymers A and B in the presence of A–B block copolymers has been investigated numerically using SCF equations by Shull and Kramer (22), and their results are compared with the equivalent experiments (23). In addition to the linear architectures that can be computed with ease, nonlinear architectures can be integrated into the SCF equations (24–28). The SCF theory illuminates the potential of the theory to simulate the complicated experimental problems we face. Adding a block copolymer (3,29–31), random copolymer (32,33), graft copolymer (34), and third homopolymer (35) to the blends improves the interfacial properties as the experiments revealed. Such an experimental system can be adopted easily in the numerical SCF theories, if necessary, and directly compared with the corresponding experiment as Shull and coworkers (22,23) presented. Therefore, numerical SCF is a very powerful technique to "fit" experimental results with the molecular models. Unlike analytical theories based on simple model polymer system, the numerical SCF theory provides the precise structures and properties of interfaces, for example, density distribution, excess free

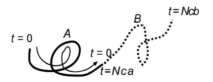

FIGURE 6.3 Schematic presentation of the partition functions for a block copolymer. Two partition functions q_{ca} and q_{cb} are introduced that depend on t, which is the position in a curvilinear coordinate of a chain.

energy, and distribution of chain ends. We briefly review numerical SCF theory of block copolymers at homopolymer interfaces in order to extend SCF theory to systems that interest us when necessary.

As an introduction to the numerical SCF analysis of polymer-polymer interfaces, we show how we describe block copolymers in the set of SCF equations and what we obtain by solving the equations. Let us consider the interface of two immiscible homopolymers A and B in the presence of an A–B diblock copolymer. The partition function $q_k(x, t)$ of the homopolymer k (k is either A or B) in the system can be described as a solution of the modified diffusion equations. $q_k(x, t)$ for the homopolymers are the same as those in the original Helfand-Tagami form in the previous section. As shown in Figure 6.3, the partition function can be separated into two functions $q_{ca}(x, t)$ and $q_{cb}(x, t)$ of A and B blocks, respectively, depending on t so that the block structure is captured in the equations:

$$q_{ca}(x, t) \ (0 < t < N_{ca}) \tag{6.13}$$

$$q_{cb}(x, t) \ (0 < t < N_{cb}) \tag{6.14}$$

$$N_c = N_{ca} + N_{cb}. \tag{6.15}$$

The connectivity of A and B blocks at a joint requires

$$q_{ca}(x, N_{ca}) = q_{cb}(x, 0). \tag{6.16}$$

The initial conditions for the chain ends are $q_{ca}(x, 0) = q_{cb}(x, N_{cb}) = 1$. The density of the A block of the block copolymer is given by

$$\rho_{ca} = A_{ca} \int_0^{N_{ca}} q_{ca}(x,t) q_{ca}(x, N_{ca} - t) dt \ . \tag{6.17}$$

Similarly, the B block density is given by

$$\rho_{cb} = A_{cb} \int_0^{N_{cb}} q_{cb}(x,t) q_{cb}(x, N_{cb} - t) dt \tag{6.18}$$

where the parameters A_k are the normalization factors. The modified diffusion equations to be solved are expressed as Eq. (6.19):

$$\frac{\partial q_k(x,t)}{\partial t} = \frac{b_k^2}{6} \frac{\partial^2 q_k(x,t)}{\partial x^2} - \frac{w_k(x)}{k_B T} q_k(x,t) \qquad (6.19)$$

$$w_{kc}(x) = \frac{\mu_{kc}(x)}{N_{kc}} - \frac{k_B T \ln \rho_k(x)}{N_k} - \Delta w(x) - c_k . \qquad (6.20)$$

The chemical potentials are given by the mean field, Eq. (6.20). The second term of Eq. (6.20) is the translational entropy, and c_k is a constant. Hong and Noolandi used solvent molecules instead of introducing $\Delta w(x)$ to satisfy the incompressibility requirement. In this nonsolvent system,

$$\Delta w(x) = \zeta(1 - \sum_k \rho_k) \qquad (6.21)$$

where ζ is the incompressibility parameter introduced by Helfand and Tagami. For $\chi/\zeta = 0$, the system is incompressible. For numerical analysis, we chose negligibly small but nonzero values of c/z. Dw(x) is the excess free energy at the position x. The recursive relations of Eqs. (6.17) to (6.21) and the equations for the homopolymers, which are the same as those in the previous section, are solved by iterating the loop of equations from an initial guess until we obtain a satisfying set of values. The interested readers can find the details of the calculations in the original paper (22).

Here we see what we obtain from the calculations. Using the numerical SCF calculations, we are able to obtain important interfacial structures and properties, such as the interfacial width and interfacial tension. Moreover, the density distributions of the polymers A and B and the A-B block copolymer are precisely calculated. The volume fractions (densities) of the monomers A and B of the homopolymers (f_{ha} and f_{hb}) and the block copolymer (f_{ca} and f_{cb}) are plotted in Figure 6.4. The interfacial width for the segments A and B increases only slightly as shown in Figure 6.4c while the interface is totally replaced by the added block copolymer. Almost no overlap between the homopolymers is found. The interface is occupied and bound together by the block copolymer. The interfacial tension is calculated by the SCF theory; moreover, the distribution of the excess energy, $\Delta w(x)$, can be calculated as shown in Figure 6.5. The homopolymer interface (solid line) shows a sharp peak of $\Delta w(x)$, which is the source of interfacial tension, at the interface. On the other hand, by adding the block copolymer, the peak at the interface is significantly reduced. Two minima next to the interface arise from the stretching elastic energy of A and B blocks of the A-B block copolymer, giving a negative contribution to the interfacial tension. The small peaks next to the minima are the interface between A (B) block of the A-B block copolymer and the A (B) homopolymer. Since the A (B) homopolymer cannot penetrate into the A (B) block domain of the A-B block copolymer,

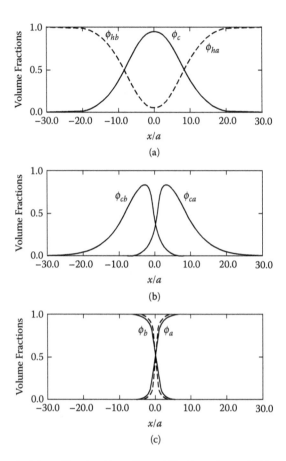

FIGURE 6.4 Interfacial profiles for $N_{ca} = N_{cb} = 200$, $N_{ha} = N_{hb} = 1000$, $\chi_{ab} = 0.1$, and the chemical potential $\mu_c = 4k_BT$. (a) Homopolymer volume fractions (– – –)and the copolymer volume fractions (——), (b) volume fractions of the individual block copolymer blocks, and (c) overall volume fractions of A and B segments in the presence of block copolymer (——) and in the absence of block copolymer (– – –) (Reproduced from Shull and Kramer, *Macromolecules*, 23, 4769, 1990 by permission of the American Chemical Society.)

$\Delta w(x)$ becomes positive at the interface although the interface is between the same segments. The overlap between A (B) homopolymer and the A (B) block of the block copolymer is so small for large χ that the interface can be the weakest link for the interfacial failure. The failure mechanism of the interface reinforced by block copolymers are discussed in a later section.

The calculations reveal the segregation of A-B diblock copolymer to the interface of polymer A and B. It is straightforward to extend the calculation scheme to the other copolymers with different monomer sequence, e.g., triblock copolymers, by modifying the partition functions. Predicting and analyzing the structures and properties of the interface between polymers in the presence of compatibilizers other than simple block copolymers is feasible with the numerical SCF theories and will be helpful for designing and developing new polymeric materials.

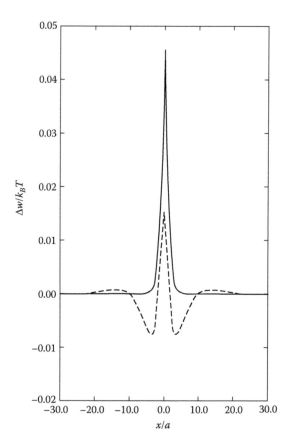

FIGURE 6.5 Excess free energy, $\Delta w(x)$, as a function of x/a, where a is the segment length for the interface of homopolymers (1000) (——) and homopolymers (1000) with block copolymer (200–200) (– – –) for $\chi_{ab} = 0.1$. (Reproduced from Shull and Kramer, *Macromolecules*, 23, 4769, 1990 by permission of the American Chemical Society.)

6.3 REINFORCEMENTS OF INTERFACES BY BLOCK COPOLYMERS

Because of the small change in the entropy of mixing of polymers, the interface becomes very thin for a pair of dissimilar polymers as predicted by the Helfand-Tagami equation, Eq. (6.9), (5–7) and confirmed by the experiments (11–16). The interface between such a pair of polymers is mechanically weak because the interpenetration and entanglement in the interfacial region are limited. In addition, the interfacial tension between dissimilar polymers is large in general as predicted by Eq. (6.10). When two dissimilar polymers are blended together in a mixer, such large interfacial tension impedes dispersion and promotes coalescence. In order to reach good dispersion in polymer blends, we need to reduce the interfacial tension between polymers. One classic idea of overcoming the above two problems is adding a compatibilizer such as a block copolymer. In the following subsection, we discuss how block copolymers segregate to the interface spontaneously and the effect of architecture on the segregation and adhesion of the interface.

6.3.1 SEGREGATION OF COPOLYMERS

A classical example of interfacial reinforcement is the use of block copolymers as compatibilizers (1,2). For industrial applications, however, a variety of copolymers other than block copolymers are used as compatibilizers. Moreover, reactive processing, in which block or graft copolymers are produced during processing, are also widely used by industry. While a variety of compatibilizers are being used, it nevertheless behooves us to understand simple block copolymers as compatibilizer because the interfacial reinforcement by block copolymers can be analyzed precisely with the current theories. Here we discuss the theories and experiments of the segregation, diffusion, and adhesion of block copolymers in homopolymers.

Let us assume an ideal system of polymers A and B separated by an interface. When an A-B block copolymer is added to the blends of A and B, the block copolymer is in one of the following three states as schematically shown in Figure 6.6. (a) The block copolymer is dissolved in either A or B homopolymer phase. The A-B block copolymer chains in A (B) phase experience unfavorable contacts of the B (A) block with the surrounding A (B) homopolymer. (b) The block copolymer chain segregates to the interface and partitions its A and B block to the A and B phases, respectively. (c) The A-B block copolymer chains form micelles in either the A or B phase. When the concentration of added block copolymer is lower than the critical micelle concentration (CMC), the block copolymers reside either in the bulk or at the interface as shown in Figures 6.6a or 6.6b. In this case, the segregation of the block copolymers to the interface is expressed as a function of the concentration of the block copolymer in bulk, which determines the chemical potential of the block copolymer. The interfacial excess z^* can be evaluated as a function of concentration in bulk as shown in Figure 6.7. Figure 6.7 is produced based on the experimental data of the segregation study by Shull and Kramer (23). In Figure 6.7, the linear portion of the curve at low concentration represents the segregation of the block copolymer chains to the interface that is not fully covered by the block copolymer as in Figure 6.6a. When the concentration of block copolymer is low, block copolymers can easily find the bare interface and segregate. The interfacial excess is linearly dependant on the concentration of the copolymer f_c. Even below the CMC, the rate of increment of the interfacial excess becomes very small as seen in Figure 6.7. As the interfacial excess or the number of block copolymer chain at the interface increases, the block copolymers are stretched in the direction perpendicular to the interface and impede the block copolymer chains approaching the interface as schematically shown in Figure 6.6b. Leibler (36) introduced the idea of dry brush in his theory for segregated block copolymers at the polymer-polymer interface. By introducing the elastic energy of the brush, the equilibrium segregation isotherm, i.e., the relation of the interfacial excess and concentration of copolymer in bulk, was derived. The unfavorable interaction of A (B) block of A-B block copolymer in B (A) homopolymer phase is simply minimized by placing the block copolymer at the interface at the expense of a loss of translational entropy due to

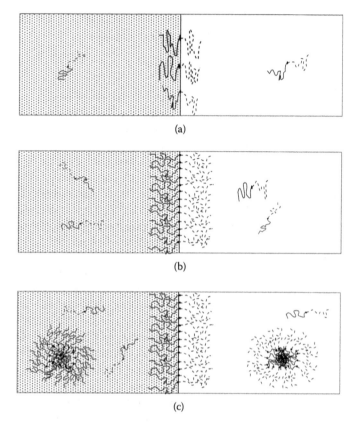

FIGURE 6.6 Schematic picture of the state of the block copolymer near the homopolymer interface. The block copolymer segregates to the interface and reduces the interfacial energy. (a) No interaction between segregated block copolymer chains. (b) Segregated block copolymer chains are stretched. (c) Above the critical micelle concentration.

chain localization and of configurational entropy (or an increase of elastic energy) due to stretching. The concept of a polymer brush appears in many problems of polymers at interface and surface (36). Leibler's theory (36) successfully provides the segregation isotherm such as that seen in Figure 6.7 if the interaction parameter is used as a fitting parameter. While Leibler's theory captures the physics of the segregation problem, the fitted interaction parameter does not always agree quantitatively with the value determined by other methods (38–40). Semenov (41) proposed a modified model in which the block copolymer joints are not strictly placed at the interface. Semenov's prediction showed better agreement with the corresponding experiments and provided a reasonable value for the interaction parameter; however, the model failed to predict the dependence of the segregation isotherm on homopolymer molecular weights (39). To predict the experimental results quantitatively, the numerical self-consistent field theory, which is described in the previous section, is necessary (38).

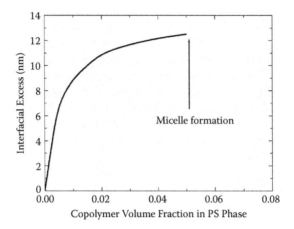

FIGURE 6.7 Interfacial excess is plotted against the block copolymer fraction in polystyrene phase. The plot is produced based on the data in Shull and Kramer, *Macromolecules*, 23, 4769, 1990.

Segregation of block copolymers reduces interfacial tension as predicted by the theories. Experimentally, linear reduction of interfacial tension with increasing diblock copolymer concentration is observed in the concentration regime well below the critical micelle concentration (CMS) (42). The interfacial tension reduction is linear in the copolymer concentration and agrees very well with the theoretical predictions of Leibler (36) and self-consistent field calculations by Noolandi and Hong (17,18). With increasing concentration of the block copolymer in the blends, the interfacial tension shows a leveling off, which indicates saturation of block copolymer at the interface followed by micellization of the block copolymer. This trend is consistent with the segregation isotherms discussed above. As an indirect observation of interfacial tension reduction, the morphology change of polymer blends with a block copolymer was investigated (43). The dispersed domain diameter, which is proportional to the interfacial tension, is reduced linearly with block copolymer concentration and levels off at higher concentration. Consequently, the linear reduction of interfacial tension is predicted by the theories and confirmed by the experiments while a leveling-off of interfacial tension, which is the result of oversaturation and stretching of block copolymer chains at the interface followed by micellization, is observed at higher concentrations.

Once the micelles are formed in homopolymer phases (44,45), as in Figure 6.6c, the added copolymers do not segregate to the surface but form additional micelles in homopolymer phases (23,46). The chemical potential to drive the segregation of block copolymer seldom increases with the concentration above CMC (36). In the numerical SCF theory, the chemical potential must be given by the mean field theory; however, the chemical potential cannot be computed for the concentration above CMC.

Most of the theories and the analyses of experiments discuss the equilibrium structures and properties of the interfaces reinforced with block copolymers whereas the equilibrium properties are seldom attained in practice due to the slow diffusion process of copolymers relative to the time scale of processing operations. There have been a very limited number of investigations concerning the dynamics of the segregation processes (41,47–52). The studies suggest that the diffusion of copolymers in homopolymer influences the kinetics of segregation; the diffusion of copolymers in a matrix of homopolymers is not, however, completely understood. Block copolymers form micelles in homopolymer when the block copolymer concentration is above its CMC. The micellization significantly alters the diffusion and the segregation to the interface of block copolymers. In the following section, the diffusion of block copolymers in homopolymers, which is often overlooked in the study of segregation, is discussed.

6.3.2 DIFFUSION OF BLOCK COPOLYMERS IN HOMOPOLYMERS

In order to reinforce the polymer-polymer interface, only a small amount of block copolymers is necessary in theory. However, in practice, a greater amount of block copolymers is often required to achieve the required properties. Although the block copolymers added to the blend eventually reach and reinforce the interface, it may take a very long time to reach the equilibrium distribution, especially when micelles are formed in the blends. While it has been hypothesized that the slow diffusion of block copolymers may influence the segregation process, very few experiments have shown this (47–52). In this section, the diffusion mechanisms of diblock copolymers in neat block copolymers and in homopolymers are discussed. The diffusion mechanisms are closely related to the segregation of diblock copolymers to the interface; the extremely slow diffusion of block copolymers in homopolymers will be quantitatively discussed to show that reaching equilibrium is sometimes impossible or at least difficult when the interaction between the polymer pair is strong.

Let us think of an A-B block copolymer that is mixed with a homopolymer A. When the concentration of A-B block copolymer is above its CMC, micelles are formed in the A homopolymer matrix (44–46). What is the diffusion coefficient of this A-B block copolymer in this mixture? There are many parameters influencing the diffusion coefficient. The monomeric friction coefficients of monomer A and B influence the diffusivity. Friction coefficients and miscibility are closely related, but the effect of the difference in friction coefficients on the diffusion coefficient is not trivial (53–54). When one of the blocks is glassy at the temperature of interest, the diffusion of single chains is completely prohibited (55). When the temperature is above the glass transition temperatures of both polymers and the interaction between the polymers is strong, the effect of the difference in the monomeric friction on the diffusion is only minor. By choosing a system such as polystyrene (PS) and poly-2-vinylpyridine (PVP) with approximately the same glass transition temperatures and monomeric frictions (56), only the thermodynamic effect on the diffusivity can be extracted to simplify the problem.

While we want to discuss the diffusion of block copolymers in homopolymers, we first consider the self-diffusion of block copolymers in the neat diblock copolymers, in which the diblock copolymer forms a variety of ordered structures depending on the block fraction. A reasonable guess would be that the diffusion in neat diblock copolymers is not related to the segregation of block copolymers to the interface. However, the mechanism of the diffusion in neat diblock copolymers gives us an insight into the segregation dynamics of block copolymers. The diffusion mechanism of diblock copolymer in the ordered spherical domain structures provides insight into the basic principles behind the diffusion of block copolymers micelles in homopolymers since there are no shortcuts for the diffusion. On the contrary, the diffusion of lamellar and cylindrical structures is anisotropic (57). The slowest diffusion path in such ordered structures is perpendicular to the interface dividing the domains, e.g., perpendicular to the lamellae or cylinders (58,59). Diffusion along the interface, which is faster than diffusion in the direction perpendicular to the interface in some degree, is not important either for sphere forming block copolymers, including diffusion to the interface from micelles.

The reduction in diffusion in the direction perpendicular to the interface is a result of the thermodynamic penalty that ensues when diblock copolymer chains detach (i.e., become activated) unless the spherical domains themselves cooperatively diffuse. The possibility of cooperative diffusion is discussed later in this section. The self-diffusion of block copolymers in sphere forming diblock copolymer has been studied (60–63) and provides useful insight into the segregation process of diblock copolymer to the interface. Let us assume that the process consists of a single A-B diblock copolymer chain with shorter B block hopping from one domain to another. The hopping frequency is governed by the probability that the B block of the block copolymer in continuous A domain will face unfavorable contacts. We call this state "activated." The excess energy of B block in A domain is χN_B, and thus the characteristic time for this state is expressed by

$$\tau \sim \exp(\chi N_B) \qquad (6.22)$$

where τ is the characteristic activated hopping time scale (61).

The diffusion coefficient controlled by the activated hopping process is given by

$$D/D_0 = \exp(-\chi N_B) \qquad (6.23)$$

where D_0 is the hypothetical diffusion coefficient of the diblock copolymer when no ordered structure is present (60,62). The studies of poly(styrene-*b*-2-vinylpyridine) (60,61) and poly(ethylene-*alt*-propylene-*b*-dimethylsiloxane) (62) proved that the normalized diffusion coefficient D/D_0 is exponentially reduced by χN_B, and hence the activated hopping diffusion is the dominant diffusion process in spherically ordered block copolymer melt. The result of the activated hopping diffusion of poly(styrene-*b*-2-vinylpyridine) is shown in Figure 6.8. The normalized diffusion coefficient D/D_0 is exponentially dependent on χN_{PVP}, where N_{PVP} is the number of segments of the shorter block and corresponds to N_B in our notation. Langevin dynamics simulation of single block copolymer chain in a periodic potential field also supports the experimental results and the simple theoretical argument of the activated hopping mechanism (63). The activated hopping diffusion mechanism has been found to be the dominant diffusion process even in the case of a concentrated block copolymer forming packed spherical micelles in a mixture with a small amount of homopolymer (64).

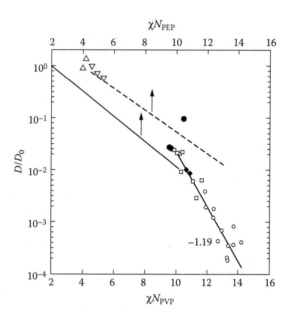

FIGURE 6.8 Normalized diffusion coefficients, D/D_0, of poly(styrene-b-vinylpyridine) block copolymers as a function of the product of the interaction parameter, χ, and the number of monomers of the minor block, N_{PVP}. The normalized diffusion coefficients are reduced exponentially. (Reproduced from Yokoyama et al., *Macromolecules*, 31, 7871, 1998 by permission of the American Chemical Society.)

Below the CMC, the rate of diffusion of single chains of the diblock copolymer and of homopolymers are similar, and thus the segregation process is rapid. When the micelles are interacting and forming ordered structures in the mixture, the diffusion mechanism is that of "activated hopping" of single block copolymer chains (64). Let us consider the case where a block copolymer is mixed with a homopolymer at a concentration above its critical micelle concentration but far below the ordering concentration of micelles. Under those circumstances, we need to consider the possibility of both activated hopping diffusion and cooperative diffusion of micelles. This problem has been experimentally studied by measuring the diffusion coefficients of poly(styrene-b-2-vinylpyridine) in mixtures with polystyrene with various molecular weights (64). When the molecular weight of the homopolymer is sufficiently low, the Stokes-Einstein diffusion of spherical micelles in a matrix of viscosity η is observed as shown in Figure 6.9, where η depends on the molecular weight of homopolymer. The diffusion coefficient is expressed as

$$D \sim \eta^{-1} \sim P^{-3} \tag{6.24}$$

$$D = \frac{k_B T}{6\pi\eta R_H} \tag{6.25}$$

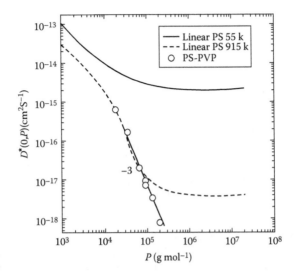

FIGURE 6.9 Diffusion coefficient, $D(0, P)$, of the dilute block copolymer () with M_w of 100 kDa in homopolymer as a function of the number of segments P. $D(0, P)$ shows the power law $D \sim P^{-3} \sim \eta^{-1}$, indicating that it is following Einstein-Stokes diffusion of spherical micelles. (Reproduced from Yokoyama et al., *Macromolecules*, 32, 3353, 1999 by permission of the American Chemical Society.)

where η, P, and R_H are the viscosity, homopolymer molecular weight, and hydro-dynamic radius, respectively. The diffusivity is controlled by the viscosity of the homopolymer and therefore by the molecular weight to approximately a power of three. Such a strong power law slows the diffusion down tremendously as the molecular weight increases. Therefore, when the molecular weight is in the range of compounds of interest to industry, the diffusion of micelles is too slow to be observed. In addition, when such micelles approach the interface, there is no guarantee that the micelles will release the block copolymers to the interface. Therefore, it is unclear as to why the cooperative diffusion of micelles contributes to the segregation of block copolymers to the interface.

When the molecular weight of the matrix homopolymer is sufficiently high, the diffusion of micelles or cooperative diffusion of diblock copolymer becomes too slow as a result of the power law [Eq. (6.24)] and cannot be observed under typical experimental time frames. Nevertheless, the diffusion by activated hopping mechanism, which is the dominant diffusion mechanism in neat diblock copolymers, is possible even in a very high molecular weight homopolymer matrix. The concentration profile after annealing of a thin layer of PS-PVP diblock copolymer sandwiched between PS layers with molecular weight of 100 kDa was measured by secondary ion mass spectrometry (SIMS) as shown in Figure 6.10 (64). The broadening of the PS-PVP middle layer, which is indicative of the diffusion of the micelles, is seldom observed. The peak of the block copolymer simply decreases, and the new peaks appear at the interface of the silicon substrate and surface. The segregation of the block copolymer at the silicon substrate interface is due to the interaction of

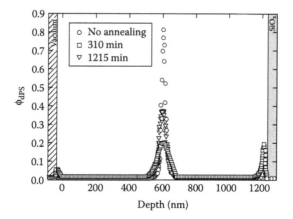

FIGURE 6.10 $\phi_{dPS\text{-}PVP}$ of the dPS-PVP block copolymer in homopolymer of the molecular weight of 100 kDa. A 20 nm layer of dPS-PVP is sandwiched by layers of PS (100 kDa). After annealing at 178°C, the dPS-PVP micelles release the dPS-PVP block copolymer single chains that eventually segregate to the silicon wafer interface and the surface. Almost no diffusion of micelles is observed. (Reproduced from Yokoyama et al., *Macromolecules*, 32, 3353, 1999 by permission of the American Chemical Society.)

the PVP block with silicon oxide. The same principles can be applied for polymer-polymer interfaces, which similarly attract block copolymers instead of silicon substrates. The micelles hardly move in the experimental time frame; nonetheless, jumping (long-distance hopping) of single diblock copolymer chains from the micelles in the middle layer to the silicon interface is clearly seen. Once the diblock copolymer chain is activated, the diblock copolymer chain diffuses rapidly, resulting in the non-Fickian concentration profile. The rate determining step is the activation of the diblock copolymer from the micelles, which is similar to the self-diffusion of block copolymer in neat diblock copolymer to form spherical domains. While we cannot define the diffusion coefficient D in this problem, we can define the characteristic time for hopping t as

$$\tau \sim \exp(\chi N_B) \qquad (6.26)$$

It should be noted that the activation frequency, $1/\tau$, which should be approximately the same as that in the neat diblock copolymer (60), becomes extremely small for a large value of χN_B; therefore, the segregation isotherm should be difficult to achieve, if not impossible, in practice under experimental conditions where strong segregation is the norm. As already mentioned, even when cooperative diffusion dominates (e.g., micelles in low molecular weight homopolymers), such diffusion may not contribute anything to the segregation of the diblock copolymer to the interface. It seems reasonable to think that the activation frequency of the block copolymer chains from the micelles is the rate-determining step of the segregation process, yielding exponential dependence on χN_B [Eq. (26)].

6.3.3 Mechanical Strength of Polymer/Polymer Interfaces

In polymer blends, reducing the interfacial tension decreases the size of polymer phases and makes coalescence less likely. When block copolymers are not present at the interface, the interfacial width tends to be too small for the polymer pairs to entangle across the interface. The lack of entanglements results in mechanical failure at the interface. Reducing interfacial tension (65) by adding block copolymers promotes finer dispersion of the phases and an increase in the interface width (38,66–67), providing better adhesion. In addition, the segregated block copolymer chains reinforce the interface by connecting two immiscible polymer phases. The block copolymer chains segregated to the interface actually bridge the two immiscible phases and reinforce the interface mechanically; therefore, it helps to understand on a molecular level the adhesive qualities of block copolymers.

First, we discuss the direct relation between interfacial width and fracture toughness of the interface. The correlation between the interfacial width and the adhesion of the interface has been investigated using neutron reflectivity and fracture toughness measurements on the same interface as shown in Figure 6.11 (68). A pair of weakly interacting polymers is chosen for this study and is annealed at a variety of temperatures, or equivalently c, to provide a wide range of interfacial widths. As seen in Figure 6.11, the fracture toughness, G_c, increases with increasing interfacial width, a_i, until it reaches approximately 11.5 nm. The interfacial width controls the adhesion of the interface provided that the polymers have molecular weights greater than that required for entanglement. A separate experiment on the interface of two sheets of the same homopolymer that were welded after thermal annealing also revealed that a sufficient level of entanglement is needed to reach the interfacial bulk toughness (68). Promoting entanglements at the interface is key to the success of good interfacial adhesion.

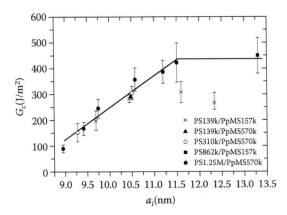

FIGURE 6.11 The fracture toughness, G_c, of the polystyrene/poly-p-methylsyrene interface plotted against the measured interfacial width, a_i. (Reproduced from Schnell et al., *Macromolecules*, 31, 2284, 1998 by permission of the American Chemical Society.)

When block copolymers are added to blends as compatibilizers, it is believed that the added block copolymers segregate to the interface and make bridges between two immiscible polymer phases. In such a case, the reinforcement is controlled by the areal density of strands, which is equivalent to the interfacial excess of the block copolymer.

The presence of a critical molecular weight for block copolymer reinforcement of interfaces has been demonstrated by Creton et al. (69). Blends of polystyrene (PS) and polyvinylpyridine (PVP) are known to deform plastically by crazing under shear deformation. Thin films of the blends of PS and PVP with and without PS-PVP block copolymers were deformed on copper grids for subsequent study by transmission electron microscopy (TEM). When the crazing propagating in PS found the PVP particles, the PVP particles, which were reinforced by the PS-PVP block copolymer, were deformed and became part of the drawn fibrils in the crazing as shown in the TEM micrograph in Figure 6.12a. The bright, intermediate, and dark areas are the crazing zone, the PS phase, and the PVP phases, respectively. In an absence of the block copolymer, the dark PVP phases do not deform so that the crack initiates at the interface between PVP and crazed PS phase as shown in Figure 6.12b. Effective reinforcement requires that each block of the diblock copolymer be larger than the molecular weight between entanglements (69). The block copolymer must have entanglements with the homopolymers in both phases to reinforce the interface effectively.

The efficiency of the reinforcement by the block copolymer depends on the areal chain density of the block copolymer at the interface and the molecular weight of the block copolymers. The failure mechanism at the interface for crack propagation can be either chain scission (70,71) or chain pull-out (72–74) from the interface. Chain scission likely occurs when the block copolymer has the blocks with the high molecular weights that entangle with the homopolymers in both phases, and has a low areal chain density. In contrast, chain pull-out is caused by the lack of entanglements with the homopolymer of the relatively short block. Such failure mechanisms were successfully identified by surface analysis, e.g., secondary ion mass spectrometry and forward recoil spectrometry (FRES), of the fractured surface reinforced with deuterated block copolymers. Deuterium sensitive surface analysis techniques such as SIMS and FRES can actually determine where the deuterium labeled PS (dPS) block resides after interfacial failure.

It has been pointed out that an optimum areal chain density exists for fracture toughness of the interface reinforced by block copolymers (75,76). When the thickness of the block copolymer layer at the homopolymer interface is greater than a single layer of brush (half of a lamella), the interface is covered by multiple lamellae, which have the additional interfaces between the block copolymer lamellae. In strong segregation, the brush of the diblock copolymers is stretched and the interface between the brushes or the interface between the block copolymer brush and homopolymer becomes the weakest link (75). This is a good example that shows the importance of interfacial structure on adhesion of the interfaces.

We have attempted to show that an understanding of interfaces at the molecular scale is crucial if we want to understand adhesion of polymer-polymer interfaces.

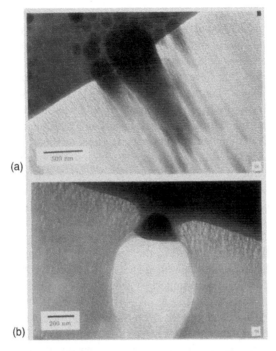

FIGURE 6.12 The interfacial failure of the polystyrene/poly-2-vinylpyridine interface (a) with and (b) without the PS-PVP block copolymer. The brighter areas are crazed PS. The darkest areas are PVP phases stained by iodine. The intermediate areas are the undeformed PS phases. In (a), the dark PVP phases that stained by iodine are deformed in the crazed zone. In (b), the brightest area is the crack, the dark PVP phases are not deformed, and the crack initiates at the PS/PVP interface. (Reproduced from Creton et al., *Macromolecules*, 24, 1846, 1991 by permission of the American Chemical Society.)

We only discussed the simplest case of the interface reinforced by diblock copolymers. In practice, we encounter complex interfaces in polymer blends where we can apply the same strategies and analytical techniques.

6.4 INTERFACIAL REACTION OF POLYMERS

Reactive processing is widely used in industry as an alternative to adding premade copolymers to reinforce the interface of polymer blends (77). For most commercial polymer blends, the block copolymer of a particular pair of polymers is not always available even though they are suitable for investigating the physics of reinforcement of the interface. However, by reacting polymers A and B at their interface, a copolymer of A and B can be formed at the interface. When polymers A and B are end-functionalized, an A-B block copolymer is formed. When one of the polymers has multiple functional groups, which is often the case in industrial polymer blends, the copolymers that form at the interface are graft-copolymers and have a varied distribution in their chemical composition and architecture. Such compositional and architectural inhomogeneities make further analysis difficult. Although reaction between end-functionalized polymer pairs may not be employed in industrial applications, it is useful to compare such pairs with the large numbers of block copolymers that have been studied. There are two methods that are in use in this field. One of the methods involves analyzing the products of reactive polymers after processing in a mixer. Reactive polymers are mixed in the molten state, and then the blends after processing are analyzed. The advantage of this style of analysis is that many conventional analytical methods can be applied, e.g., TEM for morphological studies and gel permeation chromatography (GPC) for extracting the extent of reaction are typical examples. On the other hand, the disadvantage of this method is that no direct information on the interface is extracted. The second method utilizes bilayers of polymers. The interface of the two polymer layers is parallel to the surface so that surface analysis techniques are used to analyze the reaction. As opposed to the studies using mixers, the surface area can be well-defined; therefore, the result can be discussed in detail and compared with the results of theoretical studies. The disadvantage of this method is that shear, which may play an important role in reactivity at the interface, cannot be applied to the interface.

To simplify the problem, we focus on the reaction of a pair of one-end-functionalized polymers at the interface. The product produced at the interface is a well-defined block copolymer. In this case, the mechanism of the reinforcement by reactive blending is essentially the same as that of added block copolymers. The difference is, however, that in the case of premade block copolymers, the block copolymer is not added to the interface directly but to the bulk of the homopolymer phases as opposed to the case of reactive processing where the block copolymer is created directly at the interface. When polymers A and B are immiscible, the reaction between end-functionalized A and B occurs only at the interface since almost no A (B) chains are in the B (A) phase, the probability of which is proportional to $e^{-\chi N_A}$ ($e^{-\chi N_B}$). When the extent of reaction and the amount of excess block copolymer at the interface are low, the end-functionalized polymers can find their counterpart at the interface without any difficulty. In this case, the reaction rate is controlled by the combination of the reactivity of the pair of functional groups and the probability of the counterparts finding each other. The mechanisms of the interfacial reaction have been studied experimentally (78–80) and theoretically (81–86). When the reaction between functional groups is fast and the functional groups are dilute, the reaction kinetics may be controlled by the translational diffusion of the end-functionalized polymer chains. This diffusion limited reaction mechanism is predicted by the theories (82,84); the reactivity of functional groups is usually so slow that the reaction tends to be controlled by the reactivity of the functional groups (78,79,85).

As the extent of reaction increases, the amount of block copolymer at the interface increases and thus the block copolymer begins to form a "dry brush" structure (37), preventing homopolymer with a functional group from approaching the interface. This same problem has been addressed for the segregation of block copolymers to the interface in the previous sections. Since the homopolymer with functional group is excluded from the interface, the rate of reaction is significantly reduced (78–80,82,84). Experimental studies often show a slow rate of reaction even before the dry brush is formed at the interface (77,78). Even a strongly reactive pair such as maleic anhydride and primary amine is often considered to have a "weak" reaction, leading to an interfacial reaction controlled by the reaction of the functional groups (85,86). Thus, even in the case that the interaction between the polymer pair is strong, the interface remains very thin, giving the reactive pair a lower chance for meeting at the interface, with the result being a very slow rate of reaction at the interface. This problem is recently addressed in experiments that studied several different functional polymer pairs with different interaction parameter χ (87) as well as different molecular weights (80,87). The studies found that the miscibility of the polymers as well as the interfacial width may play important roles in ensuring that there is sufficient volume for the reaction to actually occur at the interface.

Although the reaction is often controlled by the reactivity of the functional groups and eventually by the exclusion effect of the block copolymer brush, the extent of reaction gradually increases and the interfacial tension decreases with time. It has been reported that the reduced interfacial tension makes the interface rough as the areal chain density increases (80,87–91). It is still unclear whether the interfacial tension becomes negative or not with the increasing areal chain density of copolymers produced by the reaction at the interface. While aspects of the roughening and broadening of the interface caused by the interfacial reactions are still controversial and require further study, it is important to consider the possible mechanism of these scientifically interesting and industrially useful processes. If the block copolymer chains produced by the reaction leave the interface and dissolve or form micelles in the homopolymer phases, the roughening of the interface may not occur, which is the opposite of what occurs when block copolymer are added to blends. In the latter, when the concentration of the premade block copolymer in the homopolymer phases exceeds the critical micelle concentration, micelles are formed in the homopolymer phases. The chemical potential, therefore, stays approximately constant with increasing block copolymer concentration so that further segregation to the interface is prevented. Once the micelles are formed, however, the premade block copolymers may remain trapped in the micelles and never reach equilibrium. Also, the addition of premade block copolymers is not reported to lead to interfacial roughening but is more often seen in the case of reactive polymer blends. For reactive polymers, the block copolymer is produced at the interface; therefore, the situation is the opposite of that found when adding premade block copolymer to homopolymer phases to increase the concentration of the block copolymer at the interface. For reacted copolymers, micellization may be preferred to staying at the interface; however, such a process may be limited by the diffusion of the block copolymer produced at the interface leaving the interface. Moreover, micellization requires oversaturation of the block copolymer concentration in the homopolymer phase (36,41). Such a kinetic barrier for micellization may prevent the block copolymer from leaving the interface, thus boosting the interfacial roughening that is driven by the reduced interfacial tension.

6.5 CONCLUSION AND FUTURE OUTLOOK

In this chapter, several fundamental problems of the interfaces between polymers have been discussed. Special attention was paid to the segregation, diffusion, and reinforcement of simple block copolymers. While such simple block copolymers are seldom used in industrial applications where complicated polymer blends and composites are the norm, the fundamental physics of block copolymers at the interface of homopolymers is still valuable to understand. Interfacial reactions, which are widely used in industrial applications, are not fully understood. Simple model experiments of polymer–polymer bilayers do not quantitatively agree with bulk experiments using mixers. While polymer blends prepared by mixers are closer to those produced by industry, we have no direct access to the interfaces, unlike the case of bilayers. Because of this difference, the effect of shear on the interfacial reaction is one of the unsolved problems in reactive processing.

Polymers have been replacing metals and ceramics in a wide range of applications because of their low density and good processabilities. It is no doubt that blends and composites play an essential role in achieving the desired properties, especially since it is increasingly difficult to develop new polymers. Polymer blends and composites will be used even for micro- and nanodevices where the size of the devices is comparable to the domains of polymer blends. The surface and interface dominate the properties of such small devices, and so an understanding of polymer-polymer interfaces will be of increasing importance for developing and characterizing such small devices made of polymer blends and composites.

REFERENCES

1. Paul, D.R. and Newman, S., Eds., *Polymer Blends*, Academic Press, San Diego, CA, 1978.
2. Utracki, L.A., *Polymer Alloys and Blends — Thermodynamics and Rheology*, Hanser Publishers, Munchen, Germany, 1989.
3. Brown, H.R., The adhesion between polymers, *Ann. Rev. Mater. Sci.*, 21, 463, 1991.
4. Flory, P.J., *Principles of Polymer Chemistry*, Cornell University Press, Ithaca, NY, 1953.
5. Helfand E. and Tagami, Y., Theory of the interface between immiscible polymers, *J. Polym. Sci.*, 9, 741, 1971.
6. Helfand E. and Tagami, Y., Theory of the interface between immiscible polymers. II, *J. Chem. Phys.*, 56, 3592, 1971.
7. Helfand E. and Tagami, Y., Theory of the interface between immiscible polymers, *J. Chem. Phys.*, 57, 1812, 1972.

8. Helfand, E. and Sapse, A.M., Theory of unsymmetric polymer-polymer interfaces, *J. Chem. Phys.*, 62, 1327, 1975.
9. Edwards, S.F., The statistical mechanics of polymers with excluded volume, *Proc. Phys. Soc. London*, 85, 613, 1965.
10. Broseta, D., Fredrickson, G.H., Helfand E., et al., Molecular weight and polydispersity effects at polymer-polymer interfaces, *Macromolecules*, 23, 132, 1990.
11. Sferrazza, M., Xiao, C., Bucknall D.G., et al., Interface width of low-molecular-weight immiscible polymers, *J. Phys. Cond. Matt.*, 13, 10269, 2001.
12. Sferrazza, M., Xiao, C., Jones, R.A.L., et al., Evidence for capillary waves at immiscible polymer/polymer interfaces, *Phys. Rev. Lett.*, 78, 3693, 1997.
13. Fernandez, M.L., Higgins, J.S., Penfold, J., et al., Neutron reflection investigation of the interface between an immiscible polymer pair, *Polymer*, 29, 1923, 1998.
14. Anastasiadis, S.H., Russell, T.P., Satija S.K., et al., The morphology of symmetric diblock copolymers as revealed by neutron reflectivity, *J. Chem. Phys.*, 92, 5677, 1990.
15. Schubert, D.W., Abetz, V., Stamm, M., et al., Composition and temperature dependence of the segmental interaction parameter in statistical copolymer/homopolymer blends, *Macromolecules*, 28, 2519, 1995.
16. Guckenbiehl, B., Stamm M., and Springer, T., Interface properties of blends of incompatible polymers, *Physica B*, 198, 127, 1994.
17. Noolandi, J. and Hong, K.M., Effect of block copolymers at a demixed homopolymer interface, *Macromolecules*, 17, 1531, 1984.
18. Noolandi, J. and Hong, K.M., Interfacial properties of immiscible homopolymer blends in the presence of block copolymers, *Macromolecule*, 15, 482, 1982.
19. Noolandi, J. and Hong, K.M., Theory of interfacial tension in ternary homopolymer-solvent systems, *Macromolecule*, 14, 736, 1981.
20. Hong, K.M. and Noolandi, J., Theory of inhomogeneous multicomponent polymer systems, *Macromolecule*, 14, 727, 1981.
21. Hong, K.M. and Noolandi, J., Theory of unsymmetric polymer-polymer interfaces in the presence of solvent, *Macromolecule*, 13, 964, 1980.
22. Shull, K.R. and Kramer, E.J., Mean-field theory of polymer interfaces in the presence of block copolymers, *Macromolecules*, 23, 4769, 1990.
23. Shull, K.R., Kramer, E.J., Hadziioannou, G., et al., Segregation of block copolymers to interfaces between immiscible homopolymers, *Macromolecule*, 23, 4780, 1990.
24. van de Grampel, R.D., Ming, W., Laven, J., et al., A self-consistent-field analysis of the surface structure and surface tension of partially fluorinated copolymers: the influence of polymer architecture, *Macromolecule*, 35, 5670, 2002.
25. van der Linden, C.C., Leermakers, F.A.M., and Fleer, G.J., Adsorption of comb polymers, *Macromolecules*, 29, 1000, 1996.
26. van der Linden, C.C., Leermakers, F.A.M., and Fleer, G.J., Adsorption of semiflexible polymers, *Macromolecules*, 29, 1172, 1996.
27. Scheutjens, J.M.H.M. and Fleer, G.J., Statistical theory of the adsorption of interacting chain molecules. 2. Train, loop, and tail size distribution, *J. Phys. Chem.*, 84, 178, 1980.
28. Scheutjens, J.M.H.M. and Fleer, G.J., Statistical theory of the adsorption of interacting chain molecules. 1. Partition function, segment density distribution, and adsorption isotherms, *Phys. Chem.*, 83, 1619, 1979.

29. Fayt, R., Jérôme, R., and Teyssié, P., Molecular design of multicomponent polymer systems. XIV. Control of the mechanical properties of polyethylene-polystyrene blends by block copolymers, *J. Polym. Sci. Part B: Polym. Phys.*, 27, 775, 1989.

30. Creton, C., Kramer, E.J., and Hadziioannou, G., Critical molecular weight for block copolymer reinforcement of interfaces in a two-phase polymer blend, *Macromolecules*, 24, 1846, 1991.

31. Dai, C.-A., Jandt, K.D., Iyengar, D.R., et al., Strengthening polymer interfaces with triblock copolymers, *Macromolecules*, 30, 549.

32. Dai, C.-A., Osuji, C.O., Jandt, K.D., et al., Effect of the monomer ratio on the strengthening of polymer phase boundaries by random copolymers, *Macromolecules*, 30, 6727, 1997.

33. Dai, C.-A., Dair, B.J., Dai, K.H., et al., Reinforcement of polymer interfaces with random copolymers, *Phys. Rev. Lett.*, 73, 2472, 1994.

34. Gersappe, D., Irvine, D., Balazs, A.C., et al., The use of graft-copolymers to bind immiscible blends. *Science*, 265, 1072, 1994.

35. Hobbs, S.Y., Dekkers, M.E.J., and Watkins, V.H., Effect of interfacial forces on polymer blend morphologies, *Polymer*, 29, 1598, 1988.

36. Leibler, L., Emulsifying effects of block copolymers in incompatible polymer blends, *Makromol. Chem. Macromol. Symp.*, 16, 1, 1988.

37. de Gennes, P.G., Conformations of polymers attached to an interface, *Macromolecules*, 13, 1069, 1980.

38. Dai, K.H., Norton, L.J., and Kramer, E.J., Equilibrium segment density distribution of a diblock copolymer segregated to the polymer/polymer interface, *Macromolecules*, 27, 1949, 1994.

39. Dai, K.H., Kramer, E.J., and Shull, K.R., Interfacial segregation in two-phase polymer blends with diblock copolymer additives: the effect of homopolymer molecular weight, *Macromolecules*, 25, 220, 1992.

40. Green, P.F. and Russell, T.P., Segregation of low molecular weight symmetric diblock copolymers at the interface of high molecular weight homopolymers, *Macromolecules*, 24, 2931, 1991.

41. Semenov, N., Theory of diblock-copolymer segregation to the interface and free surface of a homopolymer layer, *Macromolecules*, 25, 4967, 1992.

42. Anastasiadis, S.H., Gancarz, I., and Koberstein, J.T., Compatibilizing effect of block copolymers added to the polymer/polymer interface, *Macromolecules*, 22, 1449, 1989.

43. Thomas, S. and Prud'homme, R.E., Compatibilizing effect of block copolymers in heterogeneous polystyrene/poly(methyl methacrylate) blends, *Polymer*, 33, 4260, 1992.

44. Rigby D. and Roe, R.J., Small-angle x-ray scattering study of micelle formation in mixtures of butadiene homopolymer and styrene-butadiene block copolymer, *Macromolecules*, 17, 1778, 1984.

45. Selb, J., Marie, P., Rameau, A., et al., Study of the structure of block copolymer-homopolymer blends using small-angle neutron-scattering, *Polym. Bull.*, 10, 444, 1983.

46. Shull, K.R., Winey, K.I., Thomas, E.L., et al., Segregation of block copolymer micelles to surfaces and interfaces, *Macromolecules*, 24, 2748, 1991.

47. Clarke, C.J., Jones, R.A.L., Edwards, J.L., et al., Kinetics of formation of physically end-adsorbed polystyrene layers from the melt, *Polymer*, 35, 4065, 1994.

48. Jones, R.A.L. and Kramer, E.J., Kinetics of formation of a surface enriched layer in an isotopic polymer blend, *Philos. Mag. B*, 62, 129, 1990.

49. Kim, E., Kramer, E.J., Garrett, P.D., et al., Surface segregation in blends of styrene-acrylonitrile copolymers, *Polymer*, 36, 2427, 1995.
50. Budkowski, A., Losch, A., and Klein, J., Diffusion-limited segregation of diblock copolymers to a homopolymer surface, *Isr. J. Chem.*, 35, 55, 1995.
51. Geoghegan, M., Nicolai, T., Penfold, J., et al., Kinetics of surface segregation and the approach to wetting in an isotopic polymer blend, *Macromolecules*, 30, 4220, 1997.
52. Cho, D., Hu, W., Koberstein, J.T., et al., Segregation dynamics of block copolymers to immiscible polymer blend interfaces, *Macromolecules*, 33, 5245, 2000.
53. Chapman, B.R., Hamersky, M.W., Milhaupt, J.M., et al., Structure and dynamics of disordered tetrablock copolymers: composition and temperature dependence of local friction, *Macromolecules*, 31, 4562, 1998.
54. Lodge, T.P. and McLeish, T.C.B., Self-concentrations and effective glass transition temperatures in polymer blends, *Macromolecules*, 33, 5278, 2000; and references therein.
55. Schaertl, W., Tsutsumi, K., Kimishima, K., et al., FRS study of diffusional processes in block copolymer/homopolymer blends containing glassy spherical micelles, *Macromolecules*, 29, 5297, 1996.
56. Eastman, C.E. and Lodge, T.P., Self-diffusion and tracer diffusion in styrene/2-vinylpyridine block copolymer melts, *Macromolecules*, 27, 5591, 1994.
57. Lodge, T.P. and Dalvi, M.C., Mechanisms of chain diffusion in lamellar block copolymers, *Phys. Rev. Lett.*, 75, 657, 1995.
58. Hamersky, M.W., Hillmyer, M.A., Tirrell, M., et al., Block copolymer self-diffusion in the gyroid and cylinder morphologies, *Macromolecules*, 31, 5363, 1998.
59. Rittig, F., Fleischer, G., Kärger, J., et al., Anisotropic self-diffusion in a hexagonally ordered asymmetric PEP-PDMS diblock copolymer studied by pulsed field gradient NMR, *Macromolecules*, 32, 5872, 1999.
60. Yokoyama, H. and Kramer, E.J., Self-diffusion of asymmetric diblock copolymers with a spherical domain structure, *Macromolecules*, 31, 7871, 1998.
61. Yokoyama, H., Kramer, E.J., Rafailovich, M.H., et al., Structure and diffusion of asymmetric diblock copolymers in thin films: a dynamic secondary ion mass spectrometry study, *Macromolecules*, 31, 8826, 1998.
62. Cavicchi, K.A. and Lodge, T.P., Self-diffusion and tracer diffusion in sphere-forming block copolymers, *Macromolecules*, 36, 7158, 2003.
63. Yokoyama, H., Kramer, E.J., and Fredrickson, G.H., Simulation of diffusion of asymmetric diblock and triblock copolymers in a spherical domain structure, *Macromolecules*, 33, 2249, 2000.
64. Yokoyama, H., Kramer, E.J., Hajduk, D.A., et al., Diffusion in mixtures of asymmetric diblock copolymers with homopolymers, *Macromolecules*, 32, 3353, 1999.
65. Hu, W., Koberstein, J.T., Lingelser, J.P., et al., Interfacial tension reduction in polystyrene/poly(dimethylsiloxane) blends by the addition of poly(styrene-b-dimethylsiloxane), *Macromolecules*, 28, 5209, 1995.
66. Russell, T.P., Anastasiadis, S.H., Menelle, A., et al., Segment density distribution of symmetric diblock copolymers at the interface between two homopolymers as revealed by neutron reflectivity, *Macromolecules*, 24, 1575, 1991.
67. Russell, T.P., Menelle, A., Hamilton, W.A., et al., Width of homopolymer interfaces in the presence of symmetric diblock copolymers, *Macromolecules*, 24, 5721, 1991.
68. Schnell, R., Stamm, M., and Creton, C., Direct correlation between interfacial width and adhesion in glassy polymers, *Macromolecules*, 31, 2284, 1998.

69. Creton, C., Kramer, E.J., and Hadziioannou, G., Critical molecular weight for block copolymer reinforcement of interfaces in a two-phase polymer blend, *Macromolecules*, 24, 1846, 1991.

70. Brown, H.R., Deline, V.R., and Green, P.F., Evidence for cleavage of polymer-chains by crack-propagation. *Nature*, 341, 221, 1989.

71. Char, K., Brown, H.R., and Deline, V.R., Effects of a diblock copolymer on adhesion between immiscible polymers. 2. Polystyrene (PS)-PMMA copolymer between poly(phenylene oxide) (PPO) and PMMA, *Macromolecules*, 26, 4164, 1993.

72. Creton, C., Kramer, E.J., Hui, C.Y., et al., Failure mechanisms of polymer interfaces reinforced with block copolymers, *Macromolecules*, 25, 3075, 1992.

73. Creton, C., Brown, H.R., and Shull, K.R., Molecular weight effects in chain pullout, *Macromolecules*, 27, 3174, 1994.

74. Washiyama, J., Kramer, E.J., Creton, C.F., et al., Chain pullout fracture of polymer interfaces, *Macromolecules*, 27, 2019, 1994.

75. Washiyama, J., Creton, C., Kramer, E.J., et al., Optimum toughening of homopolymer interfaces with block copolymers, *Macromolecules*, 26, 6011, 1993.

76. Creton, C., Brown, H.R., and Deline, V.R., Influence of chain entanglement on the failure modes in block copolymer toughened interfaces, *Macromolecules*, 27, 1774, 1994.

77. For example, Lavengood, R.E., Eur. Patent, No. 202214, Monsanto Company, 1986.

78. Schulze, J.S., Cernohous, J.J., Hirao, A., et al., Reaction kinetics of end-functionalized chains at a polystyrene/poly(methyl methacrylate) interface, *Macromolecules*, 33, 1191, 2000.

79. Schulze, J.S., Moon, B., Lodge, T.P., et al., Measuring copolymer formation from end-functionalized chains at a PS/PMMA interface using FRES and SEC, *Macromolecules*, 34, 200, 2001.

80. Yin, Z., Koulic, C., Pagnoulle, C., et al., Probing of the reaction progress at a PMMA/PS interface by using anthracene-labeled reactive PS chains. *Langmuir*, 19, 453, 2003.

81. Fredrickson, G.H. and Milner, S.T., Time-dependent reactive coupling at polymer-polymer interfaces, *Macromolecules*, 29, 7386, 1996.

82. Fredrickson, G.H., Diffusion-controlled reactions at polymer-polymer interfaces, *Phys. Rev. Lett.*, 76, 3440, 1996.

83. O'Shaughnessy, B. and Sawhney, U., Reaction kinetics at polymer-polymer interfaces, *Macromolecules*, 29, 7230, 1996.

84. O'Shaughnessy, B. and Sawhney, U., Polymer reaction kinetics at interfaces, *Phys. Rev. Lett.*, 76, 3444, 1996.

85. O'Shaughnessy, B. and Vavylonis, D., Reactive polymer interfaces: how reaction kinetics depend on reactivity and density of chemical groups, *Macromolecules*, 32, 1785, 1999.

86. O'Shaughnessy, B. and Vavylonis, D., Reactions at polymer interfaces: transitions from chemical to diffusion-control and mixed order kinetics, *Europhys. Lett.*, 45, 638, 1999.

87. Jones, T.D., Schulze, J.S., Macosko, C.W., et al., Effect of thermodynamic interactions on reactions at polymer/polymer interfaces, *Macromolecules*, 36, 7212, 2003.

88. Jiao, J., Kramer, E.J., de Vos, S., et al., Morphological changes of a molten polymer/polymer interface driven by grafting, *Macromolecules*, 32, 6261, 1999.

89. Jiao, J., Kramer, E.J., de Vos, S., et al., Polymer interface instability caused by a grafting reaction, *Polymer*, 40, 3585, 1999.

90. Lyu, S.-P., Cernohous, J.J., Bates, F.S., et al., Interfacial reaction induced roughening in polymer blends, *Macromolecules*, 32, 106, 1999.
91. Koriyama, H., Oyama, H.T., Ougizawa, T., et al., Studies on the reactive polysulfone–polyamide interface: interfacial thickness and adhesion, *Polymer*, 40, 6381, 1999.

7 Phase Morphology and Solidification under Shear in Immiscible Polymer Blends

Philippe Cassagnau, Yves Deyrail, and René Fulchiron

CONTENTS

7.1 INTRODUCTION

Generally speaking, the morphology of immiscible polymer blends develops during
melt processing in a very complex way. Phenomenalike deformation, breakup, and/or
coalescence of the dispersed phase can occur when an immiscible blend system is
submitted to an external mechanical treatment. These processes have been extensively
studied in the last few decades. The deformation of a melt polymer droplet suspended
in another polymer, whose study was pioneered by the works of Taylor (1) and Grace
(2), has been considered by many subsequent authors from both theoretical and
experimental points of view. Two dimensionless parameters that enable morphology
prediction in the melt state were defined, i.e., (a) the Capillary number:

$$Ca = \frac{\sigma R}{\alpha} \tag{7.1}$$

where σ represents the shear stress, R the droplet radius, and α the interfacial tension
between polymer phases, and (b) the viscosity ratio:

$$p = \frac{\eta_d}{\eta_m} \tag{7.2}$$

where η_d and η_m are the viscosity of the dispersed phase and the viscosity of the
matrix, respectively. At the critical value of the capillary number Ca_{crit}, the deformed
droplets are expected to break up in the dispersed phase. This critical value Ca_{crit}
depends on the viscosity ratio p. However, it has been demonstrated that, in shear
flow, no breakup of the deformed droplets is noticed for blends with $p > 4$. In this
case, the capillary number remains constant. Droplets with a capillary number below
the critical value Ca_{crit}, develop a stable ellipsoidal shape. In contrast, viscous drops
subjected to an extensional flow field can break up at any value of p, provided that
$Ca > Ca_{crit}$. Above Ca_{crit}, the particles break up into smaller droplets. Grace (2) and
many other authors (3,4,5) have shown that Ca_{crit} depends not only on p but also
on the nature of the flow. The critical value of Ca can be calculated by the following
expression derived from de Bruijn (6) as a function of the viscosity ratio p. For
shear flow:

$$\log(Ca_{crit}) = -0.506 - 0.0994 \log(p) + 0.124(\log(p))^2 - \frac{0.115}{\log(p) - 0.6107} \tag{7.3}$$

Many authors have been interested in droplet deformation under well-defined
flow conditions (7–16). A new interesting result was reported by Janssen et al. (17)
who observed that the critical capillary number becomes concentration dependent
for the most concentrated emulsions, producing a decrease of more than one order
of magnitude in very concentrated emulsions. Furthermore, drops with viscosity
ratio $p > 4$, which are known not to break up in single drop experiments, did show

breakup at elevated concentrations of the dispersed phase. However, in most of the studies discussed in literature, the blend components used are model fluids, generally Newtonian fluids, whereas in the usual processing conditions, polymers are vis-coelastic fluids. Therefore, some authors have become interested in the effects of elasticity on morphology development. Mighri et al. (18,19) demonstrated the influence of elasticity on the critical capillary number using Boger model fluids. They are syruplike fluids that exhibit elastic properties without shear thinning behavior. The elastic properties of the blends are defined by the ratio ($k' = \tau_d/\tau_m$) between the relaxation times of the droplet phase and matrix. These relaxation times, namely τ_d and τ_m, were calculated from the first normal stress difference of the components:

$$\tau_d = \frac{N_{1d}}{2\eta_d\gamma^2} \tag{7.4}$$

$$\tau_m = \frac{N_{1m}}{2\eta_m\gamma^2} \tag{7.5}$$

Thus, these authors demonstrated that matrix elasticity helps to deform the droplets whereas droplet elasticity resists droplet deformation. They also demonstrated that the value of the critical capillary number increases with k' for $k' < 4$ and remains constant for $k' > 4$. Most of the studies on the influence of the elastic properties of the components of the blends on the droplet deformation led to the same conclusions. Lerdwijitjarud et al. (20) observed the same influence of elasticity on droplet deformation with polystyrene/high-density polyethylene (HDPE) blends. Indeed, vis-coelastic droplets in a viscoelastic matrix are more difficult to break up compared to Newtonian droplets in a Newtonian matrix because of the contribution of droplet elasticity and matrix shear thinning. The difficulty often encountered in studies of the influence of the elastic properties of the components is to manage to increase the elasticity ratio of the blend without increasing its viscosity ratio. Lerdwijitjarud et al. (21) succeeded in decoupling these parameters; however, the components of the blends in their study are model fluids with weak elastic properties. They came to the same conclusions as those previously drawn by other authors but could explicitly attribute them to the elasticity of the droplet phase. In the work of Deyrail et al. (22) on blends of copolymers of ethylene-methyl acrylate (EMA) and poly-carbonate (PC), strong elastic properties were created in the dispersed phase through solidification of an amorphous PC phase under shear flow. It was shown that despite the fact that elastic properties were quite strong in this blend, the shear viscosity ratio still drove the morphology. In order to avoid such an occurrence in the more recent study, Deyrail and Cassagnau (23) proposed to cross-link the dispersed phase in order to generate permanent elastic properties independent of shear rate and temperature. However, they have shown that it was not possible to clearly demonstrate the dominance of the elasticity ratio over the viscosity ratio since their individual influences are similar with respect to droplet deformation. Indeed, increasing the viscosity ratio or the elasticity ratio tends to limit droplet deformation.

Nevertheless, drop deformation and breakup are somehow transient phenomena that require a specific time, which must be defined. As was done with the capillary number, a time scale can be defined (for shear flow for instance) in a dimensionless form:

$$t_{crit} = \frac{t_b \gamma}{Ca} \qquad (7.6)$$

where t_b is the breakup time under quiescent conditions via Rayleigh instabilities

$$t_b = \frac{2\eta_m R_0}{\alpha \Omega(\chi_m, p)} Ln\left(\frac{\alpha_r}{\alpha_i}\right) \qquad (7.7)$$

with

$$\alpha_i = \left(\frac{21kT}{8\pi^{3/2}\alpha}\right)^{1/2} \qquad (7.8)$$

$$\alpha_r = 1.64 R_0 \qquad (7.9)$$

where $\Omega(\chi_m, p)$ is a complex function of the viscosity ratio and the instability wavenumber, tabulated by Tomotika (24), α_i and α_r are the initial and breakup amplitude of the Rayleigh instability, respectively, and R_0 represents the initial radius of the thread. Consequently, the drop can break up only if $p < 4$ and $t > t_b$.

In addition, Huneault et al. (25) gave the breakup time for a melt fiber in quiescent state or submitted to shear flow as:

$$t_b = \left[\frac{\eta_m Ln\left(\alpha_r / \alpha_i\right)}{\alpha \Omega(\chi_m, p) \gamma^{1/2}} R_0\right]^{2/3} \qquad (7.10)$$

However, the quenching time t_q, which is the time for freezing the morphology by crystallization or solidification of the dispersed phase, must be taken into account. Usually, the morphologies and/or the mechanical properties are considered at room temperature. Consequently, these properties depend not only on the shear and elongation of the dispersed phase in the melt state but also on the quenching processing conditions.

Three typical polymer blend systems are used in this chapter to illustrate the dispersed phase deformation and solidification under shear flow:

1. Deformation of polycarbonate droplets imbedded in a molten ethylene-methyl acrylate copolymer matrix at different temperatures. In this way,

it is possible to study the deformation of PC droplets in different viscoelastic states, from nearly Newtonian at high temperature to solid viscoelastic at temperatures near the glass transition temperature (T_g).

2. Deformation of polybutylene terephthalate (PBT) droplets in a molten
   matrix of EMA. The crystallization under shear flow of the PBT droplet
   phase in the molten EMA matrix is considered.

3. Deformation of a cross-linked EVA droplet in a polydimethylsiloxane
   (PDMS) matrix. The aim is to show the influence of the cross-linking
   density (i.e., elasticity) of the dispersed phase on its droplet deformation.

Furthermore, the control of the morphology development of these blends in shear
flow is also discussed. Thus, a rodlike amorphous dispersed phase and fibers of a
semicrystalline or cross-linked dispersed phase imbedded in a molten polymeric
matrix are generated. Finally, processing applications are discussed.

7.2 DROPLET DEFORMATION

If we denote γ as the macroscopic strain applied by a shear device, Elemans et al.
(26) defined the strain λ really applied to the droplet as:

$$\lambda = \sqrt{1 + \frac{\gamma^2}{2} + \frac{\gamma}{2}\sqrt{4 + \gamma^2}} \qquad (7.11)$$

The droplet deformation is denoted by λ_d:

$$\lambda_d = \frac{a}{R_0} \qquad (7.12)$$

where a is the major axis of the ellipsoid and R_0 is the initial radius of the droplet.
Delaby et al. (8,27,28) derived the expression based on Palierne's theory (29) of the
time-dependent drop deformation during a start-up uniaxial extensional flow in the
case of viscoelastic phases and constant interfacial tension.

Dynamic droplet deformation during shear dynamic deformation $(\lambda^* = 1 + \gamma_0 e^{j\omega t})$
satisfies the following equation:

$$\frac{\lambda_d^* - 1}{\lambda^* - 1} = \frac{5G_m^*(19G_d^* + 16G_m^*)}{(2G_d^* + 3G_m^*)(19G_d^* + 16G_m^*) + 40\alpha / R(G_d^* + G_m^*)} \qquad (7.13)$$

where G_m^* and G_d^* are the complex shear moduli of the matrix and droplet, respectively, at the frequency of the dynamic strain, α is the interfacial tension, and R is
the droplet radius.

Actually, we need the relation between λ and λ_d during a start-up flow at a
constant shear rate γ for which $\gamma(t) = \lambda(t) - 1 = \gamma t$ at t > 0. By analogy with the

work of Delaby et al. (8) who studied elongation strain, Deyrail et al. (22) derived different expressions of droplet deformation under shearing, as explained in the forthcoming section.

7.2.1 NEWTONIAN PHASES WITHOUT INTERFACIAL TENSION

Obviously, in this case, $G_m^* = j\omega\eta_m$ and $G_d^* = j\omega\eta_d$. Furthermore, $\alpha = 0$ and Eq. (7.13) reduces to the Taylor equation:

$$\frac{\lambda_d - 1}{\lambda - 1} = \frac{5}{2p + 3} \tag{7.14}$$

Actually, Taylor's result is obtained under conditions of constant strain rate flow.

7.2.2 NEWTONIAN PHASES WITH INTERFACIAL TENSION

In this case, $G_m^* = j\omega\eta_m$, $G_d^* = j\omega\eta_d$, and $\alpha \neq 0$:

$$\lambda_d(t) - 1 = \frac{19p + 16}{16p + 16}\frac{\eta_m \gamma R_0}{\alpha}[1 - e^{-t/\tau}] \tag{7.15}$$

where τ is a shape characteristic relaxation time for the drop deformation process:

$$\tau = \frac{R\eta_m}{40\sigma}\frac{(2p + 3)(19p + 16)}{(p + 1)} \tag{7.16}$$

As a consequence, drop deformation will depend on the matrix viscosity, drop radius, and the shear rate. At long times, the equilibrium droplet deformation is identical to the result derived by Taylor:

$$\lambda_d - 1 = \frac{19p + 16}{16p + 16}\frac{\eta_m \gamma R_0}{\alpha} \tag{7.17}$$

7.2.3 VISCOELASTIC PHASES WITH $\alpha = 0$

Assuming that both phases are Maxwell fluids with single relaxation times, Delaby et al. (8) derived the following limiting behavior at long times:

$$\lambda_d(t) - 1 = \frac{5}{2p + 3}\gamma(t) + \frac{10p\gamma}{(2p + 3)^2}(\tau_d - \tau_m) \tag{7.18}$$

where τ_d and τ_m correspond to the reputation relaxation times [which can be calculated from Eqs. (7.4) and (7.5)] of the inclusion and the matrix, respectively. This model takes into account the viscosity ratio and the relaxation times of both the matrix and the inclusion. The model is also derived from Taylor's equation and predicts, for a start-up flow, that it is easier to deform a viscoelastic drop than a purely viscous one. However, a drop is more deformed in a viscous matrix than in a viscoelastic one, with the same zero shear viscosity. At long times, phases can behave as Newtonian fluids, leading to the same deformation predicted by the Taylor model [Eq. (7.14)].

7.2.4 SOLID–VISCOELASTIC PHASE DISPERSED IN A NEWTONIAN MATRIX

For a chemically cross-linked droplet in a Newtonian matrix, it can be assumed:

$$\alpha \approx 0; \; G_d^* \gg G_m^* \; ; \text{ and } G_d^* = Ge \tag{7.19}$$

where Ge is the equilibrium shear modulus of the cross-linked phase. Therefore, Eq. (7.13) can be simplified to:

$$\frac{\lambda_d^* - 1}{\lambda^* - 1} = \frac{G_m^*}{2G_e} \tag{7.20}$$

where $G_m^* = j\omega\eta_m$ is for a Newtonian matrix.

The calculation of the droplet deformation for a start-up flow at a constant shear rate can be made from the Laplace-Carson transform of Eq. (7.20). A straightforward expression is then obtained:

$$(\lambda_d - 1) = \frac{5\gamma\eta_m}{2G_e} \tag{7.21}$$

As the deformation experiments are carried out in steady shear flow, the viscous stress remains constant. Therefore, a constant deformation of the pure elastic droplet is expected that does not depend on the applied strain.

One should keep in mind that these models are restricted to small deformations as they are derived from a linear theory. Consequently, nonlinear effects can be expected in the deformation range that has been used. However, Delaby et al. (27) showed for nearly Newtonian phases at high capillarity numbers that experimental data were found to be in good agreement with the predictions of the Newtonian theory up to relatively high drop deformation ($\lambda < 5$). Furthermore, the linear theory from Palierne can be generalized to viscoelastic phases although some slight discrepancies remain due to nonlinear effects.

7.3 DEFORMATION OF A DISPERSED PHASE

All the results reported here come from the work of Deyrail developed during his Ph.D. thesis (30) at the University of Lyon. The experiments were carried out using a Leitz Orthoplan microscope (transmission) with 20x lens coupled with a Linkam CSS-450 high temperature shearing stage. Readers can find more experimental details in the references of Deyrail et al. (22,23,31). Furthermore, measurements of the interfacial tension were carried out using the drop retraction method according to Xing et al. (32). In a few words, a step deformation was imposed to a melt dispersed droplet imbedded in molten matrix. The relaxation process was then optically recorded until the drop returned from an ellipsoidal to an equilibrium spherical shape. Knowing the viscosity of both components, the interfacial tension could be inferred from the dimensional variations of the deformed drop as a function of time. The following interfacial tensions were obtained for the different blend systems reported in the forthcoming discussion:

$$\alpha(EMA \, / \, PC)_{240°C} = 2.7 \text{ mN/m} \tag{7.22}$$

$$\alpha(EMA \, / \, PBT)_{250°C} = 3.6 \text{ mN/m} \tag{7.23}$$

$$\alpha(EVA \, / \, PDMS)_{140°C} = 3.1 \text{ mN/m} \tag{7.24}$$

7.3.1 DEFORMATION OF AN AMORPHOUS PHASE

Deformation of a polycarbonate ($T_g \approx 150°C$) droplet dispersed in an EMA matrix ($T_m \approx 80°C$) is used to illustrate the present work. For this blend system, the viscosity ratio increases considerably with decreasing temperature from $T = 240°C$ to the glass-transition temperature of the PC phase. Furthermore, the viscoelasticity of the PC phase gradually changes from Newtonian viscous behavior at high temperature to an elastic behavior with decreasing temperature up to T_g. Consequently, we can expect that the deformation and breakup of an elastic PC droplet will be quite different near and far above its T_g.

Therefore, the deformation of PC droplets at different viscoelastic states in a molten EMA matrix can be investigated by changing the temperature only. Figure 7.1 shows both the variation of the zero shear viscosity ratio p and elasticity ratio k' or τ_d/τ_m as a function of temperature. Note that the viscosity ratio was calculated at $\gamma = 0.1 \text{ s}^{-1}$ because the deformation of a PC droplet was performed at this shear rate. It can be observed that both the viscosity and elasticity ratio drastically increases with decreasing temperature.

Following this, step strain experiments were carried out at various temperatures. The selected temperatures were 240, 220, 200, and 180°C. The results are presented in Figure 7.2. The droplet deformation decreases with decreasing temperature because the viscosity ratio increases with decreasing temperature. Furthermore, a droplet deformation can be calculated from the models described by Eqs. (7.14),

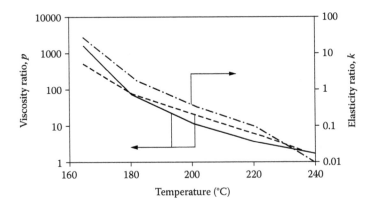

FIGURE 7.1 Viscosity ratio and elasticity ratio as a function of the temperature. For $\gamma = 0.1$ s^{-1}, solid line = zero shear viscosity, dotted line = viscosity ratio, dashed line = elasticity ratio. (Reprinted with permission from Deyrail et al., *Can. J. Chem. Eng.*, 80, 1017–1027, 2002.)

(7.15), and (7.18). The average relaxation times have been determined from dynamic frequency sweep tests at the maximum of the Cole-Cole plot of the complex viscosity. Furthermore, we assumed that the viscosity ratio p is identical to the viscosity ratio obtained from the shear rate $p(\gamma)$ data.

Figure 7.3 shows the experimental results compared to the predictions of each model. Taylor's model response to matrix deformation is found to be quite linear in the considered deformation range. As expected, the droplet deformation is clearly

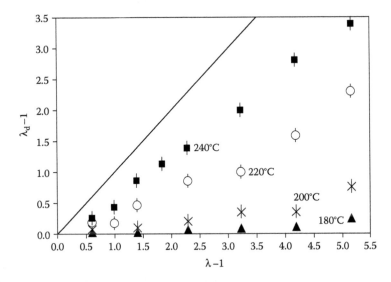

FIGURE 7.2 Drop deformation vs. matrix deformation for several temperatures. Step strain experiment at $\gamma = 0.1$ s^{-1}. The solid line represents the affine deformation. (Reprinted with permission from Deyrail et al., *Can. J. Chem. Eng.*, 80, 1017–1027, 2002.)

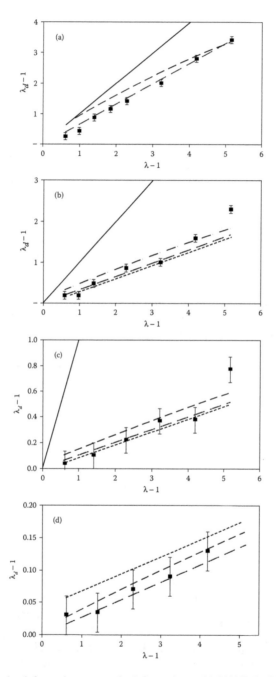

FIGURE 7.3 Droplet deformation vs. matrix deformation at (a) 240°C, (b) 220°C, (c) 200°C, and (d) 180°C at $\gamma = 0.1\ s^{-1}$. Comparison to model predictions: (-- -- -- --) Taylor [Eq. (7.14)], Newtonian phases with interfacial tension (-----) [Eq. (7.15)], and viscoelastic phase without interfacial tension (·····) [Eq. (7.18)]. Solid line represents the affine model. (Reprinted with permission from Deyrail et al., *Can. J. Chem. Eng.*, 80, 1017–1027, 2002.)

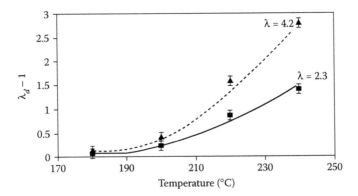

FIGURE 7.4 Drop deformation as a function of temperature. Step strain occurred at 0.1sec^{-1}. The curves are based on the prediction of the Taylor's model [Eq. (7.14)]. (Reprinted with permission from Deyrail et al., *Can. J. Chem. Eng.*, 80, 1017–1027, 2002.)

not affine. Although an overall agreement is observed, some discrepancies remain. Indeed, for the smallest deformations, the interfacial model overestimates the droplet deformation. Furthermore, we can observe at low temperature (180°C) that Taylor's model fits the experimental data better than the other two. In the temperature range that was used, elastic behavior was stronger at 180°C. As the Taylor model does not take into account elastic effects, it can be concluded that the shear viscosity ratio is the most influential parameter and consequently that the hydrodynamic effects are more important than the interfacial ones.

According to model deformations and the time-temperature superposition principle, we can predict the deformation of the PC dispersed phase on a large temperature scale. For example, it can be seen in Figure 7.4, that Taylor's model agrees well with the experimental data. The most interesting result of this study was the major influence of the viscosity ratio on droplet deformation. Indeed, Taylor's model, which is based on deformation of Newtonian liquids, can be successfully applied to the predictions of the deformation of viscoelastic phases, assuming that the viscosity ratio can be expressed by the shear viscosity p_γ. Furthermore, the results indicate that droplet deformation for high deformations of the matrix can be predicted using a linear theory developed for small deformations. This type of successful extrapolation from linear to nonlinear behavior reminds us of the well-known Cox Merz rule.

7.3.2 Deformation of a Cross-Linked Phase

To illustrate the deformation of a cross-linked phase imbedded in a molten matrix, we studied the deformation of a cross-linked EVA phase dispersed in PDMS matrix. It was quite a difficult task to choose between the various cross-linking chemistries. Indeed, the mixing of the cross-linking agent with the polymer melt must be separated from the cross-linking itself. Hence, the temperature should be high enough to mix the cross-linking agent in the EVA melt without starting the cross-linking reaction. However, the characteristic cross-linking temperature should also be lower than EVA's degradation temperature (above 200°C). Finally, the chemistry involved

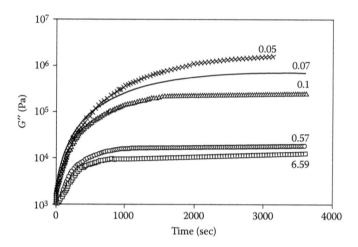

FIGURE 7.5 Dynamic time sweep tests on reactive EVA. Evolution of the storage modulus as a function of time of cross-linking for various levels of cross-linking. The values of tan δ at the end of the cross-linking process ($\Delta t = 1$ h) are reported on the corresponding curves. $T = 160°C$ and $\omega=1$ rad/sec. (Reprinted with permission from Deyrail and Cassagnau, *J. Rheol.*, 48(3), 505–524, 2004.)

has to be inert with respect to the matrix regardless of the temperature. As far as the PDMS matrix is concerned, radical chemistry should be avoided. For these reasons, Deyrail and Cassagnau (23) studied the cross-linking of EVA based on an ester-silane interchange reaction (33) between the vinyl acetate (VA) group of EVA and the silane (OR) groups of tetrapropoxysilane (TPOS). The extent of cross-linking is controlled through the VA/OR ratio, which corresponds to the molar ratio of acetate groups of the EVA chains to alkoxy groups of the TPOS. Thus, the lower the VA to OR ratio, the higher the extent of cross-linking.

As the cross-linking is taking place, the storage and loss moduli increase with time. For example, Figure 7.5 shows the variation of the elastic shear modulus for different VA to OR ratios. This figure shows that the extent of cross-linking can be controlled by varying the VA to OR ratio. It can also be observed that the cross-linking reaction is complete within an hour. The values of tan δ are given for different VA to OR ratios in Figure 7.5. Because of the shear thinning behaviour of cross-linked EVA, these parameters have been measured at the shear rate used during the deformation experiments, i.e., 0.1 sec^{-1}, using the Cox-Merz rule. Beyond the gel point, when the EVA phase is cross-linked, the Cox-Merz rule fails. In this study, we have chosen the tan δ at $\omega = 0.1$ rad/sec of the blend components as the relevant parameter to characterize the extent of cross-linking, which obviously corresponds to a decrease of tan δ. Thus, values of tan δ are very low for high levels of cross-linking. The elasticity ratio of the system was defined as the ratio of the storage moduli of the components of the blend ($G_r' = G_d'/G_m'$). Therefore, before the gel point, which is defined by tan δ = 1 for this system (34), the elasticity ratio of the reactive system is defined as the ratio of the storage modulus of the reactive EVA phase to the storage modulus of the PDMS at the shear rate used:

$$k' = \frac{G'_d(\omega)}{G'_m(\omega)} \text{ with } \omega \equiv \gamma = 0.1 \text{ s}^{-1} \tag{7.25}$$

Beyond the gel point, G'_d becomes G_e, the equilibrium modulus, determined at low frequencies, i.e., $G_e = \lim_{\omega \to 0} G'(\omega)$. $G'_m(\omega)$ is not affected by the gel point. As a matter of fact, the ratio k' will be characteristic of the elasticity of the dispersed phase whereas tan δ is a relevant parameter of the reaction progress.

The "deformation" experiments conducted for every sample consists of an applied strain, the amplitude of which varies from 1 to 4 at a shear rate of 0.1 sec^{-1}. As one might expect, Figure 7.6 shows that the droplet deformation decreases with increasing extent of cross-linking since both the elasticity and viscosity of the dispersed phase increase with the level of cross-linking. In addition, it can be observed that droplet deformation increases linearly as a function of the applied deformation (λ–1 < 3.5) regardless of the extent of the cross-linking. Hence, the resulting normalized droplet deformation (λ_d – 1)/(λ – 1) is the same regardless of the strain amplitude. If one considers the normalized droplet deformation (λ_d – 1)/(λ – 1) as a function of tan δ, as shown in Figure 7.7, no characteristic effect at the gel point can be seen on droplet deformation. Furthermore, deformation of cross-linked droplets is even possible when tan δ is held between 0.57 and 0.1. Finally, a drastic decrease in droplet deformation is observed for tan $\delta \leq 0.07$, which corresponds to an insoluble polymer fraction greater than 60%. Beyond this value, droplet deformation cannot be measured optically regardless of the applied strain. It should be pointed out that the shortest distance measurable with the optical shear device is about 3 μm, which corresponds to an optical accuracy for the deformation measurement of approximately 10%. As a consequence, the fact that the droplet deformation cannot be measured does not absolutely imply that the droplet does not deform.

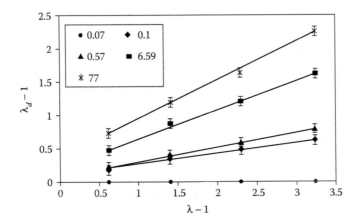

FIGURE 7.6 Droplet deformation vs. matrix deformation as a function of the cross-linking extent. Pure EVA deformation has been added for comprehension purpose. Values related to symbols correspond to the value of tan δ of the cross-linked phase. (Reprinted with permission from Deyrail and Cassagnau, *J. Rheol.*, 48(3), 505–524, 2004.)

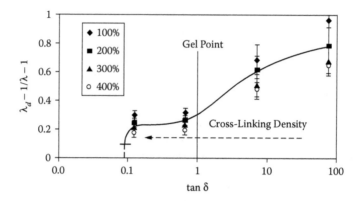

FIGURE 7.7 Normalized droplet deformation as a function of tan δ. Pure EVA deformation has been added for comprehension purposes. The vertical line at tan δ = 1 corresponds to the gel point position (Cassagnau in *Handbook of Advanced Material Testing*, 1995, pp. 925–932). The dashed vertical bar signifies the optical measure of accuracy. (Reprinted with permission from Deyrail and Cassagnau, *J. Rheol.*, 48(3), 505–524, 2004.)

Thus, a vertical dashed bar with a horizontal limit represents the droplet deformations that are not measurable in Figure 7.7. The major difficulty with modeling this system is that the dispersed phase is a viscoelastic fluid before the gel point and becomes solid afterwards.

7.3.2.1 Before the Gel Point

Normalized droplet deformation as a function of the shear viscosity ratio and a comparison with the Taylor's model predictions before the gel point are presented in Figure 7.8. Once again, no discontinuity is noted near the gel point. In the area considered, meaning before the gel point, predictions using Taylor's model agree well with the experimental results. The viscosity ratio was calculated from the modulus of the complex viscosity $|\eta^*_{\omega=0.1\ \text{sec}^{-1}}|$ of the EVA branched phase using the Cox Merz rule. Furthermore, it can be clearly seen in Figure 7.8 that beyond tan δ = 6.59, the predicted droplet deformation drastically decreases with increasing branching level (and so increasing the viscosity ratio p up to the gel point). For instance, at tan δ = 6.59, the experimental droplet deformation is around 60% and at tan δ = 1, the predicted droplet deformation is 26%.

7.3.2.2 Beyond the Gel Point

An increase in elasticity of the cross-linked droplets should reduce their deformability under shear. From the droplet deformation experiments that have been carried out at various extents of cross-linking, the importance of the elasticity ratio for the viscosity ratio cannot be demonstrated since the increase in the viscosity ratio due to the cross-linking also results in a decrease of the deformability of the dispersed phase. Indeed, from the literature it is known that elastic droplets deform less than

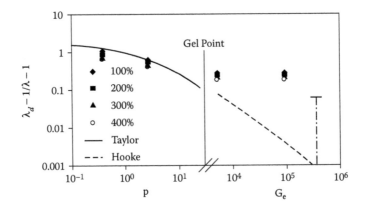

FIGURE 7.8 Normalized droplet deformation. Before the gel point: as a function of the shear viscosity ratio. Beyond the gel point: as a function of the equilibrium modulus G_e. Comparison with Taylor's model before the gel point [Eq.1(7.4)] and theoretical elastic deformation [Eq. (7.21)]. (Reprinted with permission from Deyrail and Cassagnau, *J. Rheol.*, 48(3), 505–524, 2004.)

Newtonian ones in a Newtonian matrix. The same problem was encountered with EMA/PC blends as is discussed below. Decreasing the experimental temperature to the glass transition temperature of PC was expected to generate higher elasticity properties for these blends. However, it was demonstrated that the shear viscosity ratio still drove the morphology of the blend. In the present case, since cross-linked droplets still deform beyond the gel point and since the viscosity of the droplets no longer drive the droplet deformation, we may infer the influence of droplet elasticity. Nevertheless, as shown in Figure 7.6, droplet deformation increases with applied strain regardless of the level of cross-linking. Since the viscous forces applied to the droplets were constant, the resulting deformation of pure elastic droplets should have remained constant as well. In Figure 7.8, the deformation after the gel point is plotted as a function of the elastic equilibrium modulus G_e. It can be seen that the theoretical elastic deformation given by Eq. 21, is always lower than the optical accuracy. However, the experimental droplet deformations are much greater for tan $\delta = 0.57$ and 0.1. Such observations can be explained by the viscous contribution, which is not negligible in our blends. Indeed, the values of tan δ, even for a highly cross-linked phase, remain quite high. Consequently, the cross-linked EVA samples are not purely elastic.

In conclusion, the deformation of cross-linked droplets cannot be anticipated by classical modeling, both from an elastic and a viscous point of view, notably because of the ambiguity concerning the gel point. The most interesting result is that the deformation of a cross-linked dispersed phase beyond the gel point is possible, and thus the influence of the permanent nature of the elastic properties can be demonstrated. Furthermore, such a behavior was also qualitatively reported in the literature (35) on the deformation of a thermoset phase in a thermoplastic matrix under elongational flow at the die exit of an extruder.

7.4 SOLIDIFICATION OF A DISPERSED PHASE

Two parameters are of importance for the blend morphology control: the breakup time (t_b) of a fiber and the quenching time (t_q). Fibers (or filaments) can be generated by applying a shear treatment with a Ca greater than the critical value; however the breakup of the fiber is expected afterwards. In addition, Testa et al. (36) have shown that for strong drop deformation, generating slender filaments, a sudden decrease in shear rate leads to breakup. Breakup times could be predicted from the work of Huneault et al. (25) and Elemans et al. (37) based on the theory of Tomotika (24). According to Eqs. (7.7) and (7.10), the breakup time of a fiber depends on the size of the initial droplet. From a calculation point of view, the breakup time of fibers must be taken as the average value of the theoretical breakup times of initial fibers, considering an average diameter of the fibers after a preshear treatment. The quenching time can be defined as the necessary time to change the viscoelasticity of the dispersed phase so that it cannot deform anymore under shear flow. Three typical cases can be differentiated:

1. The dispersed phase is an amorphous phase. The quenching time can be defined as the time necessary to decrease the temperature (T ≈ 240°C) at which PC fibers have been generated to below 180°C, the temperature at which PC droplets do not deform anymore.
2. The dispersed phase is a semicrystalline phase. The quenching time is the time for a fiber to crystallize so that the morphology is frozen by a strong variation of the viscoelastic properties from liquid to solid.
3. The dispersed phase is a reactive phase leading to cross-linking. The quenching time here is a characteristic reaction time leading to a cross-linking density corresponding to an insoluble fraction higher than 60%.

In addition, a dimensionless number can be defined, called λ_{DC}, corresponding to the ratio of the quenching time and the breakup time of fibers:

$$\lambda_{DC} = \frac{\text{quenching time}}{\text{breakup time}} \qquad (7.26)$$

Starting from an initial fiber morphology of the dispersed phase, the three following cases can be considered:

$\lambda_{DC} \ll 1$, fibrillar morphology is expected.
$\lambda_{DC} \gg 1$, nodular morphology is expected.
$\lambda_{DC} \approx 1$, heterogeneous morphology resulting from the fact that thin fibers lead to nodules and thick fibers adopt a more or less pronounced wavy shape.

This concept is illustrated in the forthcoming section.

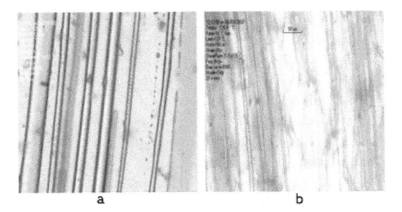

FIGURE 7.9 Morphology control of PC fiber dispersed in EMA matrix. The two images have the same scale and the pictures were taken at 130°C. Static solidification occurred under 30°C/min. (a) Preshear at 1.5 sec^{-1} for 60 sec at $T = 240$°C. (b) Preshear at 1.5 sec^{-1} for 120 sec at $T = 240$°C. (Reprinted with permission from Deyrail et al., *Can. J. Chem. Eng.*, 80, 1017–1027, 2002.)

7.4.1 SOLIDIFICATION OF AN AMORPHOUS PHASE

Let us consider the previous case of the deformation of a PC phase dispersed in a molten EMA matrix. As previously discussed and reported by Deyrail et al. (22), it is possible to obtain thin regular filaments of PC in an EMA matrix after shear treatment. Figure 7.9 presents two ways for obtaining a stable defined fibrillar morphology: a preshear treatment at 1.5 sec^{-1} for 60 sec followed by a static solidification (without shear flow) and a preshear at 1.5 sec^{-1} for 60 sec followed by a dynamic solidification (under shear flow with the same rate as that of preshear) under 30°C/min cooling rate. Several conclusions can be drawn from these experiments. Longer preshear enables fiber diameter reduction. Following this, solidification does not modify the shape of the fiber. Lastly, rapid cooling prevents the fibers from developing Rayleigh instabilities. From a deformation point of view, we previously showed that PC fibers can be significantly deformed until the temperature is decreased below 180°C. Consequently, fibers in Figure 7.9b are more deformed than those in Figure 7.9a, as shown by their respective diameters.

Since a substantial decrease in temperature can prevent fibers from breaking up and since low temperatures can limit deformation of PC droplets, we can assume that deformation control of the PC fibers is possible. Furthermore, short PC fibers should be obtained by using a preshear treatment that enables progressive deformation of PC droplets followed by a rapid cooling of the blend. For example, let us consider an experiment where a preshear of 0.5 sec^{-1} is applied during 120 sec at 240°C followed by a temperature decrease with a cooling rate of 30°C/min without stopping the shear flow. Figure 7.10 shows that the morphology obtained consists of small rods of PC.

FIGURE 7.10 Morphology control of PC fiber dispersed in an EMA matrix. Preshear at 1.5 sec^{-1} during 60 sec (T = 240°C) following by quenching under cooling rate at T = 30°C/min with shear flow reduced at 0.17 sec^{-1}. (Reprinted with permission from Deyrail et al., *Can. J. Chem. Eng.*, 80, 1017–1027, 2002.)

7.4.2 CRYSTALLIZATION OF A SEMICRYSTALLINE PHASE

Deyrail et al. (31) studied the crystallization under shear flow of polybutylene terephthalate morphologies (droplets and/or fibers) dispersed in a melt matrix of ethylene-methyl acrylate copolymer. Shear rates between 0.05 and 0.2 sec^{-1} were used to demonstrate the influence of shear rate on the crystallization of PBT fibers.

The influence of flow (shear or extension) during the crystallization process is nowadays widely investigated (38–44). Thus, the ability of the shear to accelerate crystallization has been linked to the presence of long chain molecules in the polymer (43). Hence, it appears that when the elastic behavior of the polymer melt is influenced by long chain molecules, the molecular orientation due to the shear will produce numerous nuclei, leading to a higher rate of crystallization (40, 44).

On the other hand, when the polymer is dispersed as small droplets in a matrix, its crystallization may be delayed because of the fractionated crystallization phenomenon (see, for example, Chapter 12). Such observations were made on PE (polyethylene)/PBT blends (45) for which it was shown that molten PBT inclusions still exist at high supercooling. This behavior is attributed to the low probability of finding a potential crystallization nucleus in a small droplet. So, if the droplet size distribution is broad, the crystallization times of the inclusions will vary over a large range. Therefore, for the sheared blends, the opposite trends can appear for the crystallization of the dispersed phase, i.e., an increase or decrease of crystallization kinetics because of deformation or fractionation, respectively.

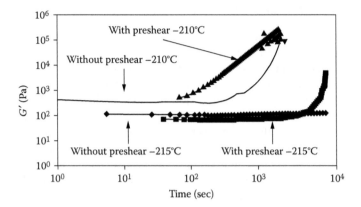

FIGURE 7.11 Variation of the storage modulus vs. time. Influence of a preshear treatment on the isothermal crystallization time of pure PBT at 210 and 215°C. (Reprinted with permission from Deyrail et al., *Polymer*, 43(11), 3311–3321, 2002.)

Although PBT chains are relatively short, as is the case generally for polycondensates, which leads to a low elastic behavior of the melt, an eventual increase of the crystallization kinetics as a result of the shear should be expected. In this way, dynamic time sweep tests, using cone-plate geometry, were carried out at $\omega = 1$ rad/sec on samples of pure PBT that have undergone, or not, a preshear treatment at $\gamma = 1$ sec^{-1} for 2 min. A substantial decrease is observed (Figure 7.11) on the crystallization time of shear-induced crystallized PBT during an isothermal treatment at both 210 and 215°C. Conversely, shear enhanced crystallization of PBT seems to be less effective in dispersed conditions, depending on the process. Considering an EMA/PBT 70:30 blend submitted to preshear, the PBT phase crystallizes between 220 and 190°C during a rapid cooling process. A non-presheared blend crystallizes between 160 and 115°C. Under isothermal conditions, dispersed PBT does not crystallize, with or without preshear treatment, even for temperatures much lower than $T = 210$°C. A corresponding experiment carried out in differential scanning colorimetry (DSC) leads to the same conclusion. This result is obviously linked to the fractionated crystallization phenomenon.

In addition, it must be pointed out that samples used for these tests were obtained from blends prepared in an internal mixer. As a consequence, the phase morphology of the blends (isolated droplets) used with the optical shear device is much coarser ($\phi \approx 100$ μm) than the phase morphology of EMA/PBT samples used for the rheological tests ($\phi \approx 2$μm). Actually, the crystallization behavior of samples used in the optical and rheological experiments is expected to be different. In other words, the effect of shear on the crystallization of PBT domains observed in the shear optical device is the dominant phenomenon since fractionated crystallization can be neglected for domains larger than 10μm. From an experimental point of view, the crystallization of PBT domains have been investigated under static conditions without flowing (static crystallization) and under dynamic conditions by applying a shear flow.

7.4.2.1 Static Crystallization

Let us consider the following experiment: after a preshear is applied for 3 min, the shear is stopped as soon as the temperature is reached or as soon as the cooling ramp starts for isothermal or nonisothermal experiments, respectively. As previously discussed, the breakup time of molten fibers and the time of crystallization must be compared. If the crystallization is too slow, the molten fibers are allowed to relax and recover their spherical shape. Since they were generated under shear treatment with a Ca greater than the critical value, breakup of the fibers is expected. The increase in the Rayleigh instabilities generated by stopping the flow make the fibers break up sooner. A nodular morphology has been obtained for two isothermal experiments, at 210 and 215°C. Referring to the DSC analysis, we could predict the start of the crystallization after about 200 sec at 210°C and 1100 sec at 215°C. Consequently, the increase in fiber viscosity because of crystallization, does not slow down efficiently the relaxation process and thereby does not prevent the fibers from breaking. A nodular morphology is therefore expected. Regarding the crystallization times involved, a shear enhancement phenomenon following the fiber formation by shear flow should be negligible in this case. Such a phenomenon is demonstrated in an isothermal experiment carried out at 209°C. The rate of crystallization is too slow to stop the relaxation of the fiber by the crystallization process. The thinnest fibers having short relaxation times will break up into several drops. The biggest ones crystallized during the relaxation process, and this leads to two types of structures. The former consists of fibers that have been broken into smaller domains as they were partially crystallized. The latter consists of wavy fibers, the shape of which comes from Rayleigh instabilities that have been fixed by the crystallization. The competition between relaxation of the molten PBT fibers and their crystallization is shown in Figure 7.12.

From that point of view, a fibrillar morphology should be obtained if the crystallization time is shorter than the breakup time. This is demonstrated in an isothermal experiment carried out at 206°C, which results mainly in a fibrillar morphology with a few particles forming from the partial relaxation of the broken fibers. The morphological change that occurred during this experiment is shown in Figure 7.13.

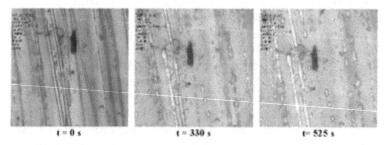

t = 0 s   t = 330 s   t = 525 s

FIGURE 7.12 Static crystallization of PBT fibers in molten EMA at 209°C; competition between relaxation of molten PBT fibers and their crystallization. (Reprinted with permission from Deyrail et al., *Polymer*, 43(11), 3311–3321, 2002.)

FIGURE 7.13 Static crystallization of PBT fibers in molten EMA at 206°C. The small white line in the pictures corresponds to 50 µm. The scale is the same for all pictures. (Reprinted with permission from Deyrail et al., *Polymer*, 43(11), 3311–3321, 2002.)

Note that the shear enhancement of crystallization of molten PBT fibers is quite important. Indeed, Figures 7.12 and 7.13 show that the crystallization is completed before the half time of crystallization as measured in a DSC. In order to drastically reduce the crystallization time, a controlled cooling ramp could be applied to the molten PBT fibers as soon as the shear flow is stopped. Based on the DSC analysis, we could predict the crystallization onset temperatures at each cooling rate. At 2, 10, and 30°C/min, the crystallization onset temperature is about 201, 198, and 190°C, repectively. For the two fastest cooling rates, the crystallization is so fast that the shear enhancement could be neglected. This assumption is justified since the experimental start temperatures are 206, 198, and 190°C at 2, 10, and 30°C/min, respectively. The shear effect on crystallization is only effective for long crystallization times. As can be seen in Figure 7.14, decreasing drastically the crystallization time (or increasing the cooling rate from 10 to 30°C/min) prevents the fibers from developing Rayleigh instabilities. Thus, the PBT fibers that have been crystallized at 10°C/min are slightly wavy, whereas those crystallized at 30°C/min are regular. The crystallization times involved during such experiments are much shorter (71 sec at 10°C/min and 42 sec at 30°C/min) than those observed for isothermal experiments.

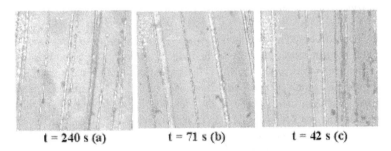

FIGURE 7.14 Static crystallization of PBT fibers in melt EMA under different cooling rate conditions: (a) $T = 2$°C/min, (b) $T = 10$°C/min, and (c) $T = 30$°C/min. The small white line on the pictures correspond to 50 µm. (Reprinted with permission from Deyrail et al., *Polymer*, 43(11), 3311–3321, 2002.)

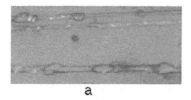
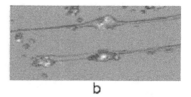

a   b

FIGURE 7.15 Dynamic crystallization. (a) Isotherm at 206°C and (b) isotherm at 210°C. Zoom on the development of irregularities in a PBT fiber caused by heterogeneous crystallization along the fiber. (Reprinted with permission from Deyrail et al., *Polymer*, 43(11), 3311–3321, 2002.)

7.4.2.2 Dynamic Crystallization

In this experiment, a continuous shear flow is applied during the crystallization of the PBT fibers generated by a preshear of 3 min, i.e., dynamic crystallization. As previously demonstrated, the crystallization time must be short enough in order to retain the fibrillar morphology. During dynamic experiments, other parameters have to be taken into account, for instance, the heterogenous nature of the crystallization along the fiber under shear flow. Such a phenomenon was already observed during static experiments and was responsible for the wavy shape of partially crystallized fibers. In dynamic experiments, the same effect appears and is responsible for the premature breaking of partially crystallized fibers. Indeed, as the PBT crystallized, its viscosity increases drastically until it becomes infinite, when no more chain movements are possible. Consequently, there is a major viscosity difference between the molten parts of the fiber and the already crystallized parts. Under shear flow, the molten parts are still able to deform whereas partially crystallized ones cannot deform anymore. Such a breaking up process is shown in Figure 7.15. In these pictures, it can be seen that a molten part of the fiber, located between two crystallized parts, is able to deform. This process leads to a fast diameter reduction of the molten areas that could eventually break up. This explains the irregularities in fiber width as well as the number of short broken fibers observed at the end of the crystallization. Final morphologies obtained under isothermal conditions at 206 and 210°C are shown in Figure 7.16.

It has been observed that decreasing the crystallization time limits the development of the irregularities along the fiber. Indeed, although the crystallization starts in a heterogeneous way along the fiber, the remaining molten parts crystallize in a rather short time, which prevents the fiber from further deformation. This complex mechanism drives the morphology of the blend during the dynamic crystallization. Therefore, in order to keep the fibrillar morphology through the dynamic crystallization process, an "instantaneous crystallization" is required. Even by using the fastest cooling rate of the shear device (30°C/min), it was not possible to reduce efficiently the crystallization time. Therefore, it was not possible to obtain regular fibers with the dynamic crystallization process. Figure 7.17 shows the final morphology of PBT domains that have undergone a dynamic crystallization at 0.1 sec^{-1} and cooling rate of 30°C/min. Even though the remaining fibers are rather long, their shapes confirm the heterogeneous development of the crystallization along the fiber.

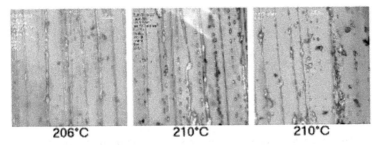

FIGURE 7.16 Dynamic crystallization: isothermal experiments. Breaking of PBT filaments due to heterogeneous crystallization along the fiber. The small white line on the pictures correspond to 50 μm. (Reprinted with permission from Deyrail et al., *Polymer*, 43(11), 3311–3321, 2002.)

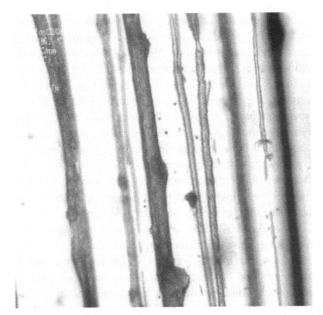

FIGURE 7.17 PBT fibers crystallized during dynamic cooling at 0.1 sec^{-1} with a cooling rate of 30°C/min. (Reprinted with permission from Deyrail et al., *Polymer*, 43(11), 3311–3321, 2002.)

7.4.3 CROSS-LINKING OF A REACTIVE PHASE

One can also imagine creating a desired morphology, either nodular or fibrillar, by controlling the cross-linking of the dispersed phase. This concept is developed in the present section where an EVA dispersed phase is cross-linked in order to prevent the fibers from breaking.

From a theoretical point of view, the EVA fibers that have been cross-linked beyond the gel point cannot break up. However, deformation of cross-linked droplets at tanδ ≈ 1 is possible as previously shown. Consequently, we can assume that under steady shear treatment, substantial deformation of the EVA particles is expected until

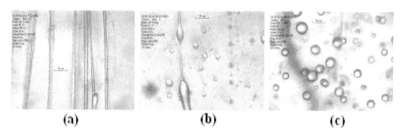

(a)   (b)   (c)

FIGURE 7.18 Morphology control through cross-linking of the dispersed phase: (a) tan δ = 0.05; (b) tan δ = 0.1; (c) nonrestrictive EVA after tanδ = 0.1. (Reprinted with permission from Deyrail and Cassagnau, *J. Rheol.*, 48(3), 505–524, 2004.)

the level of cross-linking is lower than that at the gel point. According to Huneault et al. (25), theoretical breakup times of EVA fibers under shear and under quiescent conditions can be calculated. For a steady shear treatment at 0.1 sec^{-1} for 120 sec generating fibers, the average diameter of which is 10 μm, the theoretical breakup time in quiescent conditions is about 7 min. Such a breakup time is incompatible with cross-linking kinetics. Indeed, as can be seen in Figure 7.5, a curing time of an hour is required to obtain a high degree of cross-linking. Therefore, to generate a stable fibrillar morphology, it is necessary to create the EVA fibers during the curing process. The shear rate has to be sufficiently low in order to prevent the fibers from breaking up during the shearing process. However, the extent of deformation applied to the EVA phase must be significant since cross-linking gradually limits it with curing time. For instance, a shear treatment of 0.05 sec^{-1} for a 60-min curing period at 160°C leads to a stable fibrillar morphology (see Figure 7.18a). It is important to note that the rate of cross-linking level be sufficiently high to fix the fibrillar morphology, which means that tan δ must be under 0.1. Actually, it was experimentally observed that cross-linked EVA fibers can break up into rather spherical droplets in spite of a cross-linking density corresponding to tan δ = 0.57 (see Figure 7.18b).

Finally, a stable nodular morphology can be obtained in the same way. Actually, a nodular morphology can be generated thanks to an appropriate shear rate that allow for short breakup times of the EVA fibers. High shear rates lead to a fine nodular morphology, while low shear rates lead to a coarse nodular morphology. Knowing that the cross-linking kinetics are much slower than the breakup of the EVA fibers under shear flow, a fixed nodular morphology is easily generated. Then, depending on the cross-linking density, it is possible to observe a relaxation of the nodules. For instance, a high level of cross-linking will prevent the cross-linked nodules from further deformation. In Figure 7.18c, a steady shear rate of 0.05 sec^{-1} for one hour leads to a coarse nodular morphology for pure EVA dispersed in PDMS.

7.5 PROCESSING APPLICATIONS

7.5.1 MICROFIBRILLAR COMPOSITES

The concept of microfibrillar reinforced composites (MFCs) consisting of an isotropic matrix from a lower melting polymer reinforced by microfibrils of a higher melting polymer was recently developed (46–51). First, the two polymers are melt-blended in order to obtain a good dispersion. Then, to create the microfibrils, two different routes can be used: either draw the blend material in a tensile machine at a temperature held between the melting and glass-transition temperature (46–48,50,51) or deform the dispersed phase under adequate shear or elongation flows followed by a subsequent quenching process (49). Concerning the latter case, crystallization (or solidification for an amorphous dispersed phase) of the fibrillated higher melting (or glass transition temperature) polymer leads to a more or less fibrillar morphology of the blend. Following this, this blend can be reprocessed in order to obtain the MFC final product. Consequently, the process temperature is set above the melting temperature of the matrix but below the melting temperature (or solidification) of the dispersed phase (49,52). However, this approach can only be applied to very specific polymer blends. Indeed, the blending temperature of the components must be low enough to avoid thermal degradation of the low melting temperature component (matrix). Furthermore, the matrix must have a lower melting temperature than the reinforced phase in order to offer a large enough temperature window for processing.

Let us imagine extrusion processing conditions for generating a fibrillar morphology in immiscible polymer blends. From an experimental and theoretical point of view, elongational flow is more effective than shear flow in order to generate a fibrillar morphology with controlled size. Indeed, the influence of the viscosity ratio is less important in elongational flow than in shear flow, and the critical value of the capillary number is lower for elongational flow than for simple shear flow. The evolution of the morphology along the screw channel of a twin screw extruder is a complex problem as the nature and intensity of flow can vary considerably along the screws. However, it is generally observed (53–55) that a fine dispersion of the dispersed phase is obtained during the melting process of the two components. Its further evolution along the screw profile remains limited.

Pesneau et al. (56) studied the morphology development of polyamide (PA) in polypropylene (PP) in a capillary die of an extruder. They calculated that for the usual processing conditions, it is quite difficult to generate a stable fibrillar morphology under shearing. In the die, the blends are submitted to a nonhomogeneous shear flow. The apparent shear rate decreases from $\gamma = 4Q / \pi R^3$ at the wall (Q is the volumetric flow rate and R is the capillary radius) to zero at the center. Then, the shear rate gradient in the capillary die results in a gradient of the deformation of the dispersed phase. Furthermore, the time for breaking up for such fibers is in the same order of magnitude as the residence time in the capillary (about 1 sec). Thus, the formed threads disintegrate into drops before their crystallization. For these reasons, a droplet morphology is usually observed at the die exit of an extruder and a fiber morphology is generally unexpected.

Nevertheless, the extrudate can be easily deformed under an elongation flow at the die exit. Hence, two rolls draw it while the two polymers are still in the molten state. The fibrillation, caused by phase deformation, takes place during this drawing operation. After this operation, the extrudate is water-quenched in order to lock in the morphology. For example, Monticciolo et al. (49), Pesneau et al. (56), and Boyaud et al. (57) developed this processing approach for different blend systems.

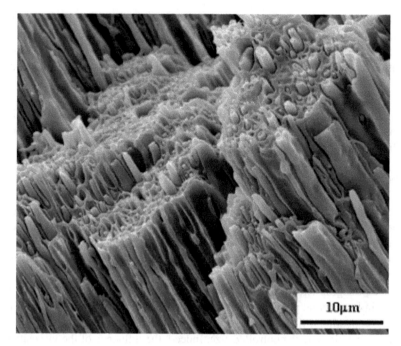

10μm

FIGURE 7.19 Fibrillar morphology of PET in a PE matrix.

Furthermore Pesneau et al. (56) showed that the breakup time in extensional flow is much higher than in shear flow. In addition, it is less dependent on the initial droplet radius. For example, Figure 7.19 shows polyethylene terephthalate (PET) fibers in a PE matrix generated under extensional flow at the die exit of the extruder.

7.5.2 Dynamic Quenching Process

In the same way, Cassagnau and Michel (58–60) developed the concept of the dynamic quenching process. The idea of the dynamic quenching process is to lock in the morphology of a minor phase dispersed in a viscoelastic matrix under flow; the process takes place in a batch mixer or in an extruder. Furthermore, *in situ* reinforcing organic composites can be directly generated in one step. In order to perform a dynamic quenching process in a twin screw extruder, for example, the configuration of the screws and the temperature profile must be designed, taking into account the different aspects of morphology development. Polymers are melted in the first left-handed element. Through this negative conveying element, charac-terized by a high back pressure, a drastic reduction of the domain size of the minor phase occurs in a few seconds during the melting (or softening for amorphous polymers). Going down the screws, the molten blend is being conveyed at a tem-perature higher than the melting (or softening) temperature of the dispersed phase. The check set point temperature of the last barrels and the die are lower ($\Delta T \approx 50°C$) than the melting (or softening) temperature of the dispersed phase. As a consequence, the dispersed phase undergoes a dynamic quenching process under flow of the molten

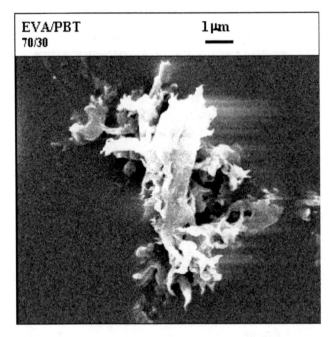

FIGURE 7.20 New type of morphology of EVA/PBT blends elaborated from the dynamic quenching process. SEM micrograph after extraction of the EVA matrix. 30% PBT. (Reprinted with permission from Cassagnau and Michel, *Polymer*, 42, 3139–3152 2001.)

matrix. Assuming a nodular morphology of the dispersed phase at the exit of the left screw element, droplets of the dispersed phase are going to crystallize (or solidify) in molten viscoelastic matrix under complex flow. As the transition between liquid and solid occurs in a dynamic process where shear and elongational flow can be superposed, unexpected types of morphologies are observed. For example, Figure 7.20 shows an example of PBT particles generated in an EVA matrix from the dynamic quenching process. This picture describes a complex morphology without any well-defined shape as in a coral structure. This morphology can be described as a microporous body full of ramifications, and its global size is rather rough. This morphology has a mean critical size of about 5 μm whereas some filaments linked to this morphology have a thickness lower than 0.5 μm. Actually, the crystallization of the PBT dispersed phase simultaneously involves two different physical mechanisms in a transient state imposed by a temperature gradient and a superposed shear and elongation flow during a finite processing time: breakup and coalescence equilibrium as well as deformation and crystallization of fine morphology in the dispersed media, as explained previously.

A quite different morphology was observed for EVA/PC blends. For example, Figures 7.21a and 7.21b show the morphology of the EVA/PC blends obtained from the dynamic quenching process for 30 and 50 wt% of PC, respectively. A fine dispersion of nanoscale PC rods can be observed. The mean length of the rods is about 300 nm with a mean diameter of 100 nm. This particular morphology is well defined at a magnification of 20,000 in Figure 7.22.

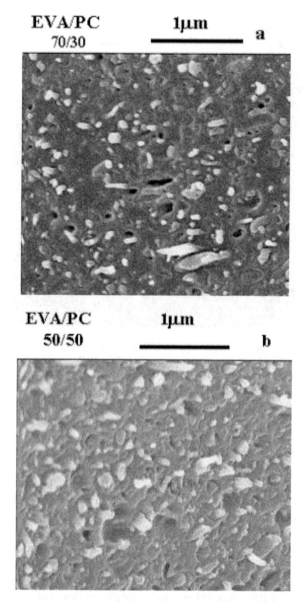

FIGURE 7.21 Morphology of EVA/PC blends elaborated from the dynamic quenching process. SEM pictures. (a) 30% PC and (b) 50% PC. (Reprinted with permission from Cassagnau and Michel, *Polymer*, 42, 3139–3152 2001.)

As previously explained, the mechanism of the morphology development of an amorphous phase is different from the one described for a semicrystalline phase. As PC is an amorphous polymer, the viscosity ratio and the elasticity of the PC phase considerably increase with decreasing temperature to the PC glass transition temperature. Consequently, the mechanisms of deformation, breakup and coalescence

EVA/PC   0.5μm

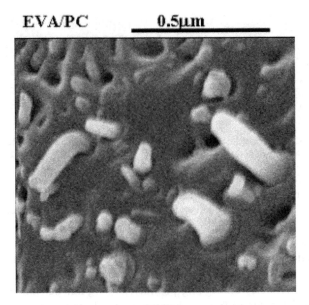

FIGURE 7.22 Rod particles of 30%PC; magnification 20,000. (Reprinted with permission from Cassagnau and Michel, *Polymer*, 42, 3139–3152 2001.)

of PC particles change considerably under shear or elongation flow. When the melt temperature reaches the glass transition temperature of PC, the shape of the PC particles then "freezes" under flow.

Furthermore, using the usual processing conditions at high concentration of the PC phase ($\phi > 30\%$), the PC morphology is not stable as the phenomenon of coalescence becomes predominant. Consequently, the dynamic quenching process is a way to control and to form a rodlike morphology at high concentration of PC. This process can be then used as a way to reinforce an elastomer matrix by finely dispersed amorphous polymers.

7.5.3 CROSS-LINKED BLEND SYSTEM

Countless papers on thermoplastic vulcanized (TPV) processing have been reported in the literature these last few decades. By definition, the process of dynamic vulcanization consists of vulcanizing an elastomer (in major phase) during its melt-mixing with a thermoplastic resin (in minor phase), which results more often in a final morphology of the TPV material where the elastomer phase is fully vulcanized and finely dispersed in the thermoplastic matrix. The processing key factor is to control the phase inversion between the two phases in order to obtain material with good elastic properties at long times. Chapter 9 is devoted to the phase morphology of dynamic vulcanized thermoplastics. Apart from the literature on TPV processing, only a few studies have discussed the morphology development of a reactive minor phase.

Deloor et al. (61,62) studied the morphology of a reactive blend consisting of an EVA/EMA phase dispersed in a polypropylene matrix. The EVA/EMA dispersed

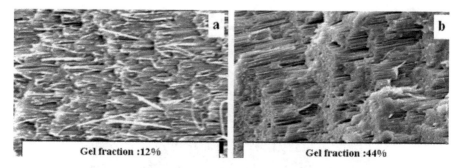

FIGURE 7.23 SEM micrographs of a PS/thermoset (80:20) blend at the die exit of the extruder. Draw ratio $\lambda = 3$. (a) Gel fraction is 12% and (b) gel fraction is 45%. (Reprinted with permission from Fenouillot and Perier-Camby, *Polym. Eng. Sci.*, 44(4), 625–637, 2004.)

phase (20 wt%) was cross-linked *in situ* during extrusion by means of a transesterification reaction. In addition, Meynié et al. (63) studied in a batch mixer the morphology development of a blend based on a thermoplastic matrix with a thermoset system (40 wt%) undergoing polymerisation. Although these two reactive systems are quite different from a chemical point of view, their general behavior with respect to the extent of the cross-linking reaction can be analyzed. Indeed, the local morphology is strongly dependent on the extent of the reaction. Near the gel point (gel fraction < 60%), the equilibrium between breakup and coalescence shifts toward coalescence as the breakup process becomes impossible. At higher gel fraction levels the classical mechanisms are largely modified and are replaced by a "coagulation" mechanism. It was shown that, for the fully cross-linked dispersed phase, the final morphology remained remarkably stable during successive extrusion steps. These results, described in a qualitative form, agree well with our quantitative findings previously described on the deformation of a cross-linked phase under controlled shearing.

Furthermore, Fenouillot et al. (35) recently investigated the formation at the die exit of an extruder of a fibrillar morphology of a thermoset phase dispersed in a thermoplastic matrix. Indeed, the elongational flow imposed by drawing the extrudate at the die exit permitted the generation of a fibrillar morphology of the reactive thermoset phase. A fibrillar morphology (Figure 7.23) was obtained when the gel fraction did not exceed a value that the authors estimated to be between 45 and 70%. For upper values of the gel fraction, the morphology coarsened drastically because of the lower deformability of the droplets and also probably because of an agglomeration phenomenon. However, increasing the gel fraction to 100% is necessary in order to stabilize the structure for a second processing step, for example, injection molding.

7.6 CONCLUSIONS

The most interesting feature of this study refers to the major influence of the viscosity ratio on droplet deformation. Taylor's model based on deformation of Newtonian liquids can be successfully applied to the prediction of the deformation of viscoelastic phases assuming that the viscosity ratio can be expressed by the ratio of the shear viscosity p_γ. Furthermore, the droplet deformation for high deformations of the matrix can be predicted using a linear theory developed for small deformations. However, the deformation of cross-linked droplets cannot be predicted by classical modeling, both from an elastic and a viscous point of view, notably because of the ambiguity concerning the gel point. Nevertheless, the most interesting result is that the deformation of a cross-linked dispersed phase beyond the gel point is possible for the insoluble fraction lower than 60% approximately.

On the other hand, the mechanical properties of polymer blend systems are generally observed at room temperature. Consequently, these properties depend not only on the shear and elongation of the dispersed phase in the molten state, but also on the quenching and processing conditions. This morphology development of a dispersed phase under well-defined processing conditions (flows and cooling conditions) can be predicted using a dimensionless number, λ_{DC}, corresponding to the ratio of the quenching time on the breakup time of fibers:

$$\lambda_{DC} = \frac{\text{quenching time}}{\text{breakup time}}$$

Starting from an initial fibrillar morphology of the dispersed phase, the three following cases can be considered:

$\lambda_{DC} \ll 1$, fibrillar morphology is expected.

$\lambda_{DC} \gg 1$, nodular morphology is expected.

$\lambda_{DC} \approx 1$, heterogeneous morphology resulting from the fact that thin fibers lead to nodules and thick fibers adopt a more or less pronounced wavy shape.

Three typical cases have been studied:

1. The dispersed phase is an amorphous phase. The quenching time is the time necessary to decrease the temperature at which fibers have been generated to a temperature close to the glass transition temperature at which the amorphous phase does not deform anymore.
2. The dispersed phase is a semicrystalline phase. The quenching time is the time for a fiber to crystallize so that the morphology is locked in by a strong variation of the viscoelastic properties from liquid to solid.
3. The dispersed phase is a reactive phase leading to cross-linking. The quenching time is a characteristic reaction time leading to a cross-linking density that corresponds to an insoluble fraction higher than 60%.

Lastly, we discussed the fact that fiber morphologies (or other types of morphology) can be well controlled under the usual processing application conditions.

REFERENCES

1. Taylor, G.L., The deformation of emulsions in definable fields of flow, *Proc. R. Soc. London*, A138, 41–48, 1932.
2. Grace, H.P., Dispersion phenomena in high viscosity immiscible fluid systems and application of static mixers as dispersion devices in such systems, *Chem. Eng. Sci.*, 14, 225–277 1982.
3. Acrivos, A., The breakup of small drops or bubbles in shear flows, *Annals New York Acad. Sci.*, 404, 1–11, 1983.
4. Rallison, J.M., A numerical study of the deformation and burst of a viscous drop in general shear flows, *J. Fluid. Mech.*, 109, 465–482, 1981.
5. Bentley, B.J. and Leal, L.G., An experimental investigation of drop deformation and breakup in steady, two dimensional linear flows, *J. Fluid Mech.*, 167, 241–283, 1986.
6. De Bruijn, R.A., *Deformation and Breakup of Drops in Simple Shear Flows*, Ph.D. thesis, University of Technology, Eindhoven, The Netherlands, 1989.
7. Lyngaae-Joergensen, J., Soendergaard, K., and Utracki, L.A., Formation of ellipsoidal drop in simple shear flow, *Poly. Net. Blends*, 3(4), 167–181, 1993.
8. Delaby, I., Ernst, B., and Muller, R., Droplet deformation during elongation flow in blends of viscoelastic fluids. Small deformation theory and comparison with experimental results, *Rheol. Acta*, 34, 525–533, 1995.
9. Levitt, L., Macosko, C.W., and Pearson, S.D., Influence of normal stress difference on polymer droplet deformation, *Polym. Eng. Sci.*, 36, 1647–1655, 1996.
10. Vinckier, I., Moldenaers, P., and Mewis, J., Transient rheological response and morphology evolution of immiscible polymer blends, *J. Rheol.*, 41, 705–718, 1997.
11. Guido, S. and Villone, M., Three-dimensional shape of a droplet under simple shear flow, *J. Rheol.*, 42, 395–415, 1998.
12. Guido, S., Minale, M., and Maffettone, P.L., Droplet shape dynamics under shear-flow reversal, *J. Rheol.*, 44, 1385–1399, 2000.
13. Khayat, R.E., Luciani, A., Utracki, L.A., et al., Influence of shear and elongation on droplet deformation in convergent-divergent flows, *Int. J. Multiphase Flow*, 26, 17–44, 2000.
14. Velankar, S., Puyvelde, P.V., Mewis, J., et al., Effect of compatibilization on the breakup of polymeric droplets in shear flow, *J. Rheol.*, 45, 1007–1019, 2001.
15. Hayashi, R.M., Takahashi, H., Yamane, H., et al., Dynamic interfacial properties of polymer blends under large deformations: shape recovery of a single droplet, *Polymer*, 42, 757–764, 2001.
16. Hayashi, R., Takahashi, M., Kajihara, T., et al., Application of a large double-step shear strains to analyze deformation and shape recovery of a polymer droplet in an immiscible polymer matrix, *J. Rheol.*, 45, 627–636, 2001.
17. Jansen, K.M.B., Agterof, W.G.M., and Mellema, J., Droplet breakup in concentrated emulsions, *J. Rheol.*, 45, 227–236, 2001.
18. Mighri, F., Ajji, A., and Carreau, P.J., Influence of elastic properties on droplet deformation in elongational flow, *J. Rheol.*, 41, 1183–1201, 1997.
19. Mighri, F., Carreau, P.J., and Ajji, A., Influence of elastic properties on droplet deformation and breakup in shear flow, *J. Rheol.*, 42, 1477–1490, 1998.
20. Lerdwijitjarud, W., Sirivat, A., and Larson R.G., Influence of elasticity on dispersed-phase droplet size in immiscible polymer blends in simple shearing flow, *Polym. Eng. Sci.*, 42 798–809, 2002.

21. Lerdwijitjarud, W., Larson, R.G., Sirivat, A., et al., Influence of weak elasticity of dispersed phase on droplet beahavior in sheared polybutadiene/poly(dimethylsiloxane) blends, *J. Rheol.*, 47, 37–58, 2003.

22. Deyrail, Y., Michel, A., and Cassagnau, P., Morphology in immiscible polymer blends during solidification of an amorphous dispersed phase under shearing. *Can. J. Chem. Eng.*, 80, 1017–1027, 2002.

23. Deyrail, Y. and Cassagnau, P., Phase deformation under shear in an immiscible polymer blend: influence of strong permanent elastic properties, *J. Rheol.*, 48(3), 505–524, 2004.

24. Tomotika, S., On the instability of a cylindrical thread of a viscous liquid surrounded by another viscous fluid, *Proc. R. Soc. London*, A150, 322–337, 1935.

25. Huneault, M.A., Champagne, M.F., and Luciani, A., Polymer blend mixing and dispersion in the kneading section of a twin-screw extruder, *Polym. Eng. Sci.*, 36, 1694–1706, 1996.

26. Elemans, E.H.M., Bos, H.L., Janssen, J.M.H., and Meijer, H.E.H., Transient phenomena in dispersive mixing, *Chem. Eng. Sci.*, 48, 267–276, 1993.

27. Delaby, I., Ernst, B., and Muller, R., Droplet deformation in polymer blends during elongation flow, *J. Macromol. Sci. Phys.*, B35, 547–561, 1996.

28. Delaby, I., Ernst, B., Froelich D., et al., Droplet deformation in immiscible polymer blends during transient uniaxial elongational flow, *Polym. Eng. and Sci.*, 36, 1627–1635, 1996.

29. Palierne, J.F., Linear rheology of viscoelastic emulsions with interfacial tension, *Rheol. Acta*, 29, 204–214, 1991.

30. Deyrail, Y., *Evolution de la Morphologie sous Cisaillement dans les Mélanges de Polymers Non Miscibles: Cristallisation, Solidification et Réticulation de la Phase Dispersée*, Ph.D. thesis, University of Lyon, France, 2003.

31. Deyrail, Y., Fulchiron, R., and Cassagnau, P., Morphology development in immiscible polymer blends during crystallization of the dispersed phase under shear flow, *Polymer*, 43(11), 3311–3321, 2002.

32. Xing, P., Bousmina, M., Rodrigue D., et al., Critical experimental comparison between five techniques for the determination of interfacial tension in polymer blends: model system of polystyrene/polyamide-6, *Macromolecules*, 33, 8020–8034, 2000.

33. Bounor-Legaré V., Ferreira, I., Verbois, A., et al., New transesterification between ester and alkoxysilane groups: application to ethylene-co-vinyl acetate copolymer crosslinking, *Polymer*, 43, 6085–6092, 2002.

34. Cassagnau, P., Rheological characterization of crosslinking, in *Handbook of Advanced Material Testing*, 1st Ed., Cheremisinoff, N.P. and Cheremisinoff, P.N., Eds., Marcel Dekker Inc., New York, 1995, pp. 925–932.

35. Fenouillot, F. and Perier-Camby, H., Formation of a fibrillar morphology of crosslinked epoxy in a polystyrene continuous phase by reactive extrusion, *Polym. Eng. Sci.*, 44(4), 625–637, 2004.

36. Testa, C., Sigillo, I., and Grizzuti N., Morphology evolution of immiscible polymer blends in complex flow fields, *Polymer*, 42, 5651–5629, 2001.

37. Elemans, P.H.M., Janssen, J.M.H., and Meijer, H.E.H., The measurement of interfacial tension in polymer/polymer systems: the breaking thread method, *J. Rheol.*, 34, 1311–1325, 1990.

38. Janeschitz-Kriegl, H., Ratajski, E., and Stadlbauer, M., Flow as an effective promotor of nucleation in polymer melts: a quantitative evaluation, *Rheol. Acta*, 42(4), 355–364, 2003.

39. Duplay, C., Monasse, B., Haudin, J.M., et al., Shear-induced crystallization of polypropylene: influence of molecular weight, *J. Mater. Sci.*, 35(24), 6093–6103, 2000.

40. Zuidema, H., Peters, G.W.M., and Meijer, H.E.H., Development and validation of a recoverable strain-based model for flow-induced crystallization of polymers, *Macromol. Theory Simul.*, 10(5), 447–460, 2001.

41. Pogodina, N.V., Lavrenko, V.P., Srinivas, S., et al., Rheology and structure of isotactic polypropylene near the gel point: quiescent and shear-induced crystallization, *Polymer*, 42(21), 9031–9043, 2001.

42. G. Kumaraswamy, G., Kornfield, J.A., Yeh, F., et al., Shear-enhanced crystallization in isotactic polypropylene. 3. Evidence for a kinetic pathway to nucleation, *Macromolecules*, 35(5), 1762–1769, 2002.

43. Doufas, A.K. and McHugh, A.J., Two-dimensional simulation of melt spinning with a microstructural model for flow-induced crystallization, J. Rheol., 2001, 45(4), 855–879.

44. Koscher, E. and Fulchiron, R., Influence of Shear on Polypropylene Crystallization: Morphology Development and Kinetics, Polymer, 2002, 43(25), 6931–6942.

45. Pesneau, I., Cassagnau, P., Fulchiron R., et al., Crystallization from the melt at high supercooling in finely dispersed polymer blends: DSC and rheological analysis, *J. Polym. Sci.*, 36, 2573–2585, 199p.

46. Evstatiev, M. and Fakirov, S., Microfibrillar reinforcement of polymer blends, *Polymer*, 33, 877–880, 1992.

47. Fakirov, S., Evstatiev, M., and Schultz, J.M., Microfibrillar reinforced composites from drawn poly(ethylene terephthalate)/nylon-6 blend, *Polymer*, 34, 4669–4679, 1993.

48. Fakirov, S., Evstatiev, M., and Petrovich, S., Microfibrillar reinforced composites from binary and ternary blends of polyesters and nylon-6, *Macromolecules*, 26, 5219–5226, 1993.

49. Monticciolo, A., Cassagnau, P., and Michel, A., Fibrillar morphology development of PE/PBT blends: rheology and solvent permeability, *Polym. Eng. Sci.*, 38(11), 1882–1889, 1998.

50. Fakirov, S., Evstatiev, M., and Friedrich, K., From polymer blends to microfibrillar reinforced composites, in *Polymer Blends, Vol. 2, Performance*, Paul, D.R. and Bucknall, C.B., Eds., John Wiley & Sons, New York, 1999, pp. 455–475.

51. Sarkissova, M., Harrats, C., Groeninckx, G., et al., Design and characterization of microfibrillar reinforced composite materials based on PET/PA12 blends, *Compos. Part A: Appl. Sci. Manuf.*, 35A(4), 489–499, 2004.

52. Sabol, E.A., Handlos, A.A., and Bavid, D.G., Composites based on drawn strands of thermotropic liquid crystalline polymer reinforced polypropylene, *Polym. Compos.*, 16(4), 330–345, 1995.

53. Willis, J.M., Favis B.V., and Lunt, J., Reactive processing of polystyrene co-maleic anhydride/elastomer blends: Processing-morphology property relationships, *Polym. Eng. Sci.*, 30, 1073–1084, 1990.

54. Favi, B.D. and Therrien, D., Factors influencing structure formation and phase size in an immiscible polymer blend of polycarbonate and polypropylene prepared by twin screw extrusion, *Polymer*, 32, 1474–1481, 1991.

55. De Loor, A., Cassagnau, P., Michel, A., et al., Morphological changes of a polymer blend into a twin screw extruder, *Int. Polym. Process*, 9, 211–218, 1994.

56. Pesneau, I., Aït Kadi, A., Bousmina M., et al., From polymer blends to in situ polymer/polymer composites: morphology control and mechanical properties, *Polym. Eng. Sci.*, 42(10), 1990–2004, 2002.
57. Boyaud, M.-F., Ait-Kadi, A., Bousmina, M., et al., Organic short fiber/thermoplastic composites: morphology and thermorheological analysis, *Polymer*, 42 (15), 6515–6526, 2001.
58. Cassagnau, P. and Michel, A., Composite polymer/polymer material with high content in amorphous dispersed phase and preparation method, *PCT Int. Appl.*, 1, 23, 2001.
59. Cassagnau, P. and Michel, A., Microcomposite polymer/polymer materials with semi-crystalline dispersed phase and preparation method, *PCT Int. Appl.*, 1, 33, 2001.
60. Cassagnau, P. and Michel, A., New morphologies in immiscible polymer blends generated by a dynamic quenching process, *Polymer*, 42, 3139–3152, 2001.
61. De Loor, A., Cassagnau, P., Michel, A., et al., Development and control of a blend morphology by in situ crosslinking of the dispersed phase, *J. Appl. Polym. Sci.*, 53, 1675–1686, 1994.
62. De Loor, A., Cassagnau, P., Michel, A., et al., Reactive blending in a twin screw extruder: Experimental and theoretical approaches, *Int. Polym. Process*, 11, 211–218, 1996.
63. Meynié, L., Fenouillot, F., and Pascault, J.P., Polymerisation of a thermoset system into a thermoplastic matrix. Effect of shear, *Polymer*, 45(6), 1867–1877, 2004.

8 Simultaneous Interpenetrating Network Structured Vinylester/Epoxy Hybrids and Their Use in Composites

J. Karger-Kocsis

CONTENTS

8.1 INTRODUCTION

Nowadays, there is an increasing interest in the use of hybrid thermosetting resins for composite applications. This interest is fueled by the potential of property improvements (e.g., stiffness, strength, toughness, fiber to matrix adhesion, hydrolytic resistance) that can be achieved via a combination of resins that often show an interpenetrating network structure (IPN). The traditional way of optimizing the properties of thermosets is to tailor the density of their cross-linked network as well as the chain flexibility between the network points. When the possibilities related to this kind of network architecturing are exhausted, micrometer scaled modification strategies are then employed. For example, toughness upgrade is reached by a fine dispersion (in the micrometer range) of reactive rubbers in the corresponding thermoset matrix. To enhance the stiffness and strength, usually various fibrous reinforcements (e.g., discontinuous or continuous fibers, two- and three-dimensional textile assemblies) are incorporated in the resins. On the other hand, the IPN structuring of thermosets represents an entirely different strategy that opens a bright horizon for the development and use of thermosets and their composites. To highlight this issue, attention should be paid to the IPN structure itself. The IPN is held together exclusively by mutually entangled bands of nanometer scale dimension (in the range of a tenth of a nanometer). This kind of "supermolecular" structure favors shear yielding, which is the major energy dissipation mechanism in thermosets. Note that the capability for shear yielding controls the toughness in thermosets. The IPN structuring offers unique possibilities for composites with respect to fiber to matrix adhesion. Provided that the IPN structure is not influenced by the composite production, an intermittent bonding of the reinforcement can be guaranteed via this peculiar matrix structure. Intermittent here means that the fiber experiences repeatedly good and poor bonding toward the matrix along its length (1) owing to the IPN structure as schematically depicted in Figure 8.1.

Intermittent bonding is believed to improve the durability of composites. The mechanisms behind the bonding (often termed Cook-Gordon type of failure) are crack deflections accompanied with long range debonding of the fibers aligned transverse to or inclined with the plane of the crack front. In the past, attempts were made to achieve intermittent bonding by intermittent sizing of the fibers, which could be done only on a large scale (millimeter range) (2,3). To solve this problem, however, from the matrix side (Figure 8.1) via self-assembly is far more promising and reproducible than from the fiber side (intermittent sizing).

8.2 INTERPENETRATED THERMOSETS

As this book contains no other chapter on IPN structured thermosets, a brief description on their grouping, manufacturing routes, and structure-property relationships is given prior to tackling the actual topic by way of recent achievements with vinylester/epoxy (VE/EP) hybrid resins.

According to Sperling (4), IPNs are defined as a combination of two or more polymers in network form that are synthesized in juxtaposition. For thermoset IPNs, the above definition should be further modified by emphasizing that the related networks are chemically cross-linked. The term IPN suggests some "interpenetration" on a molecular scale, which is, however, never the case. The thickness of the network strands varies from 10 to 1000 nm according to Utracki (5). However, the reader may find other (usually smaller) data for the upper threshold value in the literature.

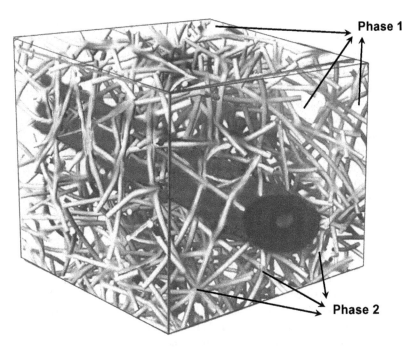

FIGURE 8.1 Scheme of the intermittent fiber bonding achieved via matrix organization (i.e., IPN structure).

Thermoset IPNs are prepared either sequentially or simultaneously (the latter is also called a one-shot process). In the sequential case, the first cross-linked polymer network is swollen by the monomer of the second polymer that is polymerized and/or cross-linked afterward. By contrast, simultaneous IPNs are prepared by cross-linking of two (or more) monomer systems at the same time. The related reactions should be noninterfering, which can be achieved by a straightforward combination of chain growth– and step growth–polymerization reactions (4). "Chain" polymerization reactions cover free radical–induced (co)polymerizations and anionic and cationic homopolymerization processes. "Step" polymerizations may involve polyaddition and polycondensation reactions. In case of coreactions (cross-reactions) between the cross-linked networks, the resulting IPN is termed "grafted."

The morphology of IPNs is determined by mechanisms of the phase separation. This occurs via nucleation and growth and/or spinodal decomposition [e.g., see refs. (4–6) and references therein]. Nucleation and growth kinetics produce spheres, droplets of the second phase within the first phase (matrix), followed by their growth. During spinodal decomposition, interconnected cylinders of the second phase appear in the matrix that grow by increasing their wave amplitude (4). In the following stage, coarsening and coalescence that are superimposed to cross-linking–related phenomena (e.g., gelation) occur. If gelation precedes phase separation during polymerization and/or cross-linking, then the extent of phase separation is restricted. As a consequence, the apparent miscibility of the components increases. The macroscopic appearance of the latter is an inward shift in the glass transition temperature

(T_g) or α-relaxation. As a consequence, the T_g of the IPN lies between those of the plain components. It is, however, still a topic of dispute whether this inward T_g shift during IPN development can be simply interpreted as improved miscibility.

Note that IPNs often do not possess a cocontinuous structure that we would assume to happen based on its designation as IPN. In order to receive a cocontinuous IPN structure, the volume fraction ratio of the components should be equal to their viscosity ratio. As both viscosity and volume fraction may change during cross-linking, the resulting morphology will depend on the reaction conditions (6).

Thermoset IPNs have been mostly used as tough, impact modified polymers, matrices in composites, and as sound and vibration damping materials. As the latter aspect is beyond the scope of this chapter, the interested reader is referred to several monographs (4,7,8). It is intuitive that all the beneficial properties of IPN structured thermosets are controlled by their morphology, which is either cocontinuous or a finely dispersed one [e.g., see ref. (5)].

A large body of works is available on thermoset IPNs with various resin combinations. To produce simultaneous thermoset IPNs, usually chain and step polymerization and/or cross-linking methods are combined. As mentioned before, chain polymerization covers free radical–induced cross-linking of polymers and monomers bearing two or more double bonds along their chains, which is the case for unsaturated polyester (UPs) and VE resins, and cross-linking of vinyl (acryl) compounds possessing one double bond with two or multifunctional monomers of similar structure. For step polymerizations, the polyaddition reactions of the polyurethane (PU) chemistry, polyaddition type curing of EP, and polycondensation cross-linking of phenolic resins are frequently exploited. The resin combinations, listed below, have been the favorite topics of scientific investigations:

UP/PU [see refs. (9,10) and references therein]
VE/PU [see refs. (11–14) and references therein]
Acrylate/PU [e.g., see ref. (15)]
Acrylate/EP [e.g., see ref. (16)]
Acrylate/phenolic [e.g., see ref. (17)]
EP/PU [e.g., see ref. (18)]

The hybridization of UP and VE with EP is the basic topic of this chapter, and thus it is treated separately.

8.3 SELECTION CRITERIA FOR VINYLESTER/EPOXY HYBRIDS

What is the driving force to combine vinylester and epoxy resins? There are numerous reasons for combining them. Both VE and EP resins are widely used in structural composites and thus the related know-how is available. VE and EP resins are compatible. The corresponding networks are formed (quasi)simultaneously via non-interfering reactions. This is an important prerequisite of simultaneous IPN formation [e.g., see refs. (4,19)]. Note that VE cross-links via free radical–induced copolymerization with suitable unsaturated monomers (usually styrene). On the other hand, polyaddition and homopolymerization reactions may serve for EP curing. A great variety of VE/EP hybrids can be produced that combine various VE and EP resins. However, it is emphasized later that the versatility of VE/EP hybrid resins is given by the EP part instead of VE. The viscosity of these hybrid resins is very low, rendering them especially suited for resin transfer molding (RTM) applications. In RTM applications, low viscosity of the resins promotes the wet-out of the reinforcement that is placed in the mold before its infiltration starts with the resin. A VE/EP combination is promising also with respect to fiber to matrix adhesion as this issue is best solved for EP resins in the present praxis.

Works performed on hybrid resins composed of unsaturated polyester and EP resins may serve to guide R&D with VE/EP systems, albeit they have appeared also recently (20–24). On the other hand, ref. (20) treats some common aspects of both UP/EP and VE/EP hybrid resin systems.

8.4 IPN STRUCTURED VINYLESTER/EPOXY HYBRIDS

8.4.1 MORPHOLOGY

Information on the morphology of simultaneous IPNs can be derived from differential scanning calorimetry (DSC), dynamic mechanical thermal analysis (DMTA), scanning and transmission electron microscopy (SEM and TEM, respectively), and nuclear magnetic resonance (NMR) proton spin diffusion studies [e.g., see refs. (6,19)]. At present, one of the most powerful techniques to resolve the IPN structure is atomic force microscopy (AFM). Laser ablation (25) and ion bombardment (26) proved to be suitable techniques for selective "physical etching" of the samples for AFM inspection. Needless to say, the ablation behavior of one of the constituting phases should differ considerably from the other in order to get the necessary surface relief. Unfortunately, no guidelines are available on this issue at present. In addition, microheterogeneity in thermosets, by whatever method detected, is a steady issue of scientific dispute.

Figure 8.2 displays the morphology of a VE and a VE/EP hybrid as assessed by AFM. According to Figure 8.2a, VE possesses a microgel-like structure. In this two-phase structure, more or less spherical VE nodules are dispersed in a polystyrene-rich matrix phase. The latter is less resistant to ablation than the VE-rich one. Consequently, the nodules, the size of which is in the range of approximately 70 nm, pertain to the VE phase. On the other hand, the presence of IPN composed of mutually entangled bands is obvious in Figure 8.2b. Note that in this VE/EP combination, the EP phase was far less resistant to Ar^+ bombardment than the VE (26). The apparent width of the entangled strands in Figure 8.2b is approximately 90 nm.

The IPN morphology becomes more compact upon postcuring (26,27) as shown by AFM micrographs in Figure 8.3. Increasing compactness means that the width of the bands and their heights are reduced. It is not yet clear whether compactness is controlled by increasing conversion in the corresponding phases (more likely but not yet studied in detail) or grafting reactions between them (less likely). Note that the secondary hydroxyl groups of the VE may react with the epoxy groups of the

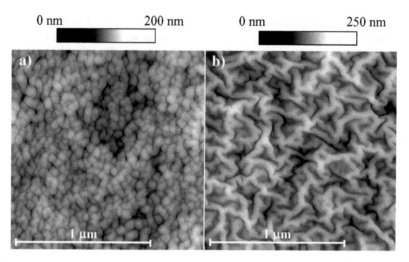

FIGURE 8.2 AFM micrographs (height modulated) taken from the ion-etched polished surface of (a) VE and (b) VE/Cal-EP+Al-Am. *Notes:* Physical etching occurred by Ar⁺ ion bombardment. Cal-EP = cycloaliphatic EP resin (1,4-cyclohexanedimethanol-diglycidylether) that was hardened by an aliphatic diamine [1,2-bis(2-aminoethoxy)ethane (Al-Am)]. The maximum temperature of curing was $T = 150°C$. The VE/EP ratio was 1:1. The conversion of the resins was not assessed.

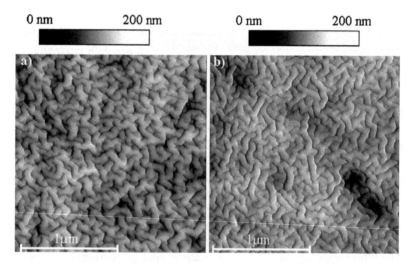

FIGURE 8.3 AFM micrographs (height images) taken from ion etched surfaces of VE/Al-EP+Cal-Am cured at maximum T = (a) 150°C and (b) 200°C. *Notes:* Al-EP = aliphatic EP resin (1,4-butanediol-diglycidylether) that was hardened by a cycloaliphatic diamine [2,2′-dimethyl-4,4′-methylene-bis(cyclohexylamine) (Cal-Am)]. For further notes, see those in Figure 8.2.

TABLE 8.1
Tensile Mechanical Properties, Fracture Energy (G_c), and Glass Transition Temperature (T_g) of VE/EP (1:1) Systems the EP Component of Which Was Cured by Polyaddition Reactions with Diamines

Composition	Maximum Curing Temperature (°C)	Tensile Strength (Mpa)	Ultimate Tensile Strain (%)	Young's Modulus (MPa)	G_c (kJ/m²)	T_g (°C)
VE/Al-EP+Cal-Am	150	50	5.5	2600	3.7	81
	200	60	4.5	2700	1.1	128
VE/Cal-EP+Al-Am	150	14	38.5	700	7.3	57
	200	46	5.1	2900	5.2	87
VE	150	48	2.8	3200	0.54	157
	200	52	2.1	3400	0.45	160

Notes: The data represents the highest measured values; T_g was read as the maximum temperature of the α-relaxation peak.

EP. In addition, Michael's type grafting reactions may occur between the amine hardener and VE. It is intuitively obvious that the morphology change shown in Figure 8.3 would have a strong affect on the mechanical and thermal behavior. This is demonstrated later. How to describe the IPN morphology? In a first approximation, the apparent band width and some surface roughness parameters, all of them derived from the AFM scans, can be considered (28). On the other hand, the fractal geometry may be a useful tool to describe the IPN structure accordingly.

8.4.2 Properties and Their Modification

Table 8.1 lists the basic mechanical and fracture mechanical properties of VE/EP hybrids, the EP phase of which was amine cured. As expected, all IPN systems are very sensitive to postcuring, which results in a substantial increase in the T_g. This is associated, however, with some decrease in the fracture energy (G_c). It was found that VE/EP hybrids, in which the idealized network of the EP phase is composed of aliphatic and cycloaliphatic units, show outstanding toughness values (especially high G_c values) (29). This finding was traced to the possibility of chair to boat conformational changes of the cyclohexylene moieties. G_c is a key parameter of the resin, especially when the goal is the development of a matrix material for composites since the G_c of the matrix and composite are highly correlated. The related G_c values can be determined using fracture mechanical approaches on suitable specimens in mode I (crack opening — G_c or G_{Ic}), mode II (in-plane shear — G_{IIc}), and their mixed-type (mode I and mode II) loading. According to the related relationship (Figure 8.4), a small change in the G_c of a very brittle resin (having very low G_c value) yields a relatively larger improvement for the composite than a large change in G_c of a very ductile resin. This is a result of the volume constraint between the aligned endless fibers in advanced composites (note that their fiber volume content may reach greater than 60 vol%) that hampers the development of

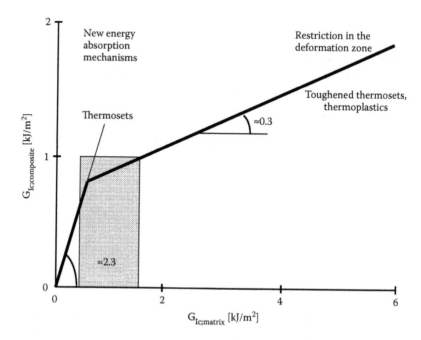

FIGURE 8.4 Fracture energy of the composite ($G_{c,composite}$) as a function of that of the matrix ($G_{c,matrix}$). *Notes:* G_c was determined in the same loading (mode I or crack tip opening). *Source:* Hunston, *Compos. Technol. Rev.*, 6, 176–180, 1984.

the crack tip damage zone. Further information on the transfer of matrix fracture energy to the composite interlaminar fracture energy can be obtained from the cited works in refs. (30,31).

The data in Table 8.1 indicate that VE resins are very brittle, whereas the IPN structured VE/EP hybrids are very tough (their G_c values are comparable to those of tough thermoplastics). However, this outstanding toughness, which is prominent also in fatigue crack propagation tests (32), was achieved at the cost of other properties. Note that stiffness, strength, and T_g were reduced as a result of IPN formation. In addition, the IPN structure likely results in higher water uptake and a decrease in hygrothermal resistance (33). Therefore, an important question is: What strategies should be developed to modify the properties of VE/EP resins? These systems show usually one single-relaxation peak (T_g), albeit a very broad one (20,26,29,34,35). So the T_g of the IPN is in between those of the pure resin components. In order to increase the T_g of the IPN, VE and/or EP resins with enhanced T_g values have to be selected. This can hardly be solved from the VE side as VE resins of both bisphenol-A and novolak types have very similar characteristics (the latter gives slightly higher T_g than the bisphenol-A type VE). Consequently, the properties of VE/EP hybrids can be tailored by the selection of the EP characteristics (e.g., type, hardening, relative amount).

Changing the amount, type, and hardener of the EP should result in different IPN structured VE/EP resins. Their properties should vary over a broad range, which appears to be the case (Figure 8.5). In this figure, traces of the mechanical loss factor (tanδ) as a function of temperature (T) are displayed for VE/EP systems containing aliphatic, cycloaliphatic, and aromatic EP resins (Al-EP, Cal-EP, and Ar-EP, respectively) but cured with the same aliphatic amine (Al-Am). Note that the T_g of the neat EP hardened by Al-Am increases as expected according to the ranking: Al-EP < Cal-EP < Ar-EP. A very broad mechanical relaxation peak, located between the T_g values of the pure resins, appears in all the systems, at least in a given composition range (VE/EP = 75:25... 25:75) (34). It is intuitive that by the proper choice of the EP, not only the T_g of the resulting IPN, but its other characteristics (mechanical properties, etc.) can be influenced as well.

EP can be modified further to influence IPN formation in VE/EP hybrids. EP can be cured catalytically (termed homopolymerization), for example, by using a borontrifluoride complex (BF3) [e.g., see ref. (28)], an imidazole compound [e.g., see ref. (36)], or anhydrides. Selecting an unsaturated anhydride, such as maleic anhydride (MAH), provides us with the opportunity to work with a dual phase cross-linking agent (28).

Figure 8.6a shows the effect of EP hardening on an example of a VE/Cal-EP hybrid at a composition ratio of 1:1. Note that the dual phase cross-linker MAH produced the highest stiffness value for the IPN, whereas the lowest one was produced by the aliphatic diamine. AFM images taken from the etched surfaces of VE/Cal-EP+Al-Am and VE/Cal-EP+MAH are given in Figure 8.6b. It is clear that increasing compactness (low apparent band width, low surface roughness) was accompanied with a steep increase in stiffness and also in T_g (see Figure 8.6a); however, the toughness was jeopardized. G_c was found to decrease with decreasing width of the entangled bands and surface roughness (mean or maximum roughness depth values were considered) (28). Curing with MAH and BF3 was accompanied by other beneficial properties, such as lower water uptake and increased resistance to thermooxidative degradation, compared to amine curing (28).

8.4.3 Cross-Linking and Chemorheology

The effect on cross-linking on chemorheology is markedly less studied for simultaneous VE/EP IPNs than their mechanical performance. Based on DSC and rheological measurements, it was supposed that IPN generation in these systems is governed by VE cross-linking and gelling, albeit no direct evidence was found that the cross-linking reaction of EP was delayed when compared to VE (34). Indirect evidence for this reaction sequence was obtained when VE/EP systems were modified by amine intercalated organoclays. The organoclay did not exfoliate and/or intercalate but changed the VE/EP composition ratio in the interphase (resulting in EP enrichment). The poor intercalation of the organoclay was attributed to the fast curing of VE, causing space confinement during IPN formation (37).

Dean and coworkers (20,38) reported a reduction in the reaction rate in the IPN phases and attributed this effect to the dilution of each reacting system by the other resin component. The authors also showed that grafting between the amines (hardener for EP) and the methacrylate group of VE took place via Michael's type addition reaction. Full cure for most IPN systems studied could be obtained by postcuring

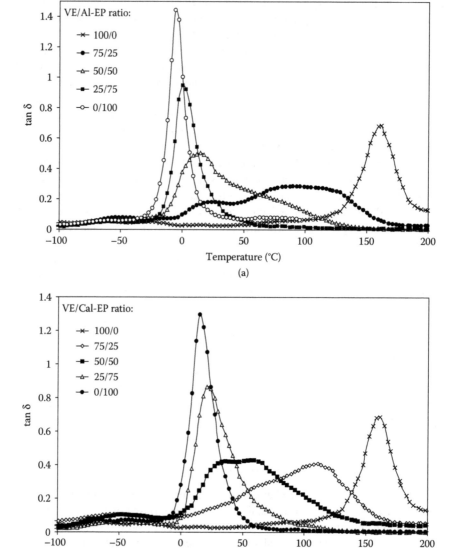

FIGURE 8.5 Mechanical loss factor (tanδ) as a function of temperature (*T*) for VE/EP+Al-Am hybrids at varying compositions using different types of EP (Karger-Kocsis et al., *SPE-ANTEC*, 60, 751–754, 2002). (a) Aliphatic EP (Al-EP; cf. notes to Figure 8.3); (b) cycloaliphatic EP (Cal-EP; cf. notes to Figure 8.2); and (c) aromatic EP [bisphenol-A-diglycidylether (Ar-EP)] *Note:* for Al-Am, see notes to Figure 8.2.

(Continued.)

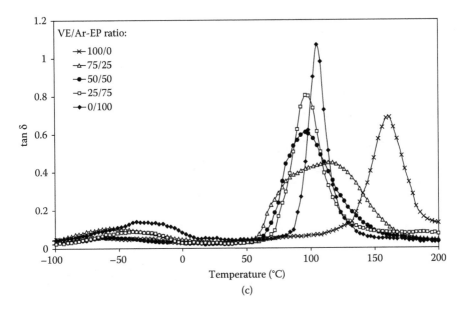

FIGURE 8.5 *Continued.*

at elevated temperatures. This finding is of paramount importance for the practical use of such systems. It was also demonstrated by infrared spectroscopic and rheological measurements that the VE cured more quickly than the EP component (hardened either by imidazole or diamine compounds) and that the gelation of VE controls the overall gel behavior of the IPN (20,39). This was corroborated also in our work (40). In a recent work, Dean and Cook (41) demonstrated the feasibility of interchanging the cure order of the components in VE/EP hybrids. This has been achieved by a proper selection of azo initiators (the criterion being the half-life temperature) for the curing of VE. Because of these modifications, the cure kinetics were also markedly influenced.

8.5 USE OF IPN STRUCTURED VINYLESTER/EPOXY IN FIBER REINFORCED COMPOSITES

As mentioned before, VE/EP hybrid resins may be promising matrix materials in composites. This prediction is based on the following facts: The reactivity of the epoxy groups may enhance the interfacial adhesion between fiber and matrix when the fiber surface contains suitable reactive groups; excellent wet-out of the fibrous reinforcement is given by the low viscosity of these hybrids; and the fast cycle time in RTM processing is guaranteed by the rapid cross-linking of the VE phase.

8.5.1 FLAX FIBER MAT REINFORCEMENT

Before we could produce flax fiber reinforced composites, we needed to verify that the reaction of the hydroxyl groups of the cellulose with the epoxy groups of the

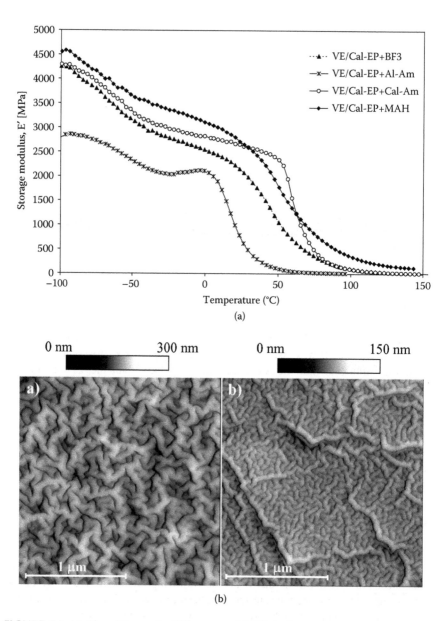

FIGURE 8.6 (a) E' vs. T traces for IPN-structued VE/Cal-EP hybrids, the EP component of which was cured by different methods. (b) AFM height images of the VE/Cal-EP hybrids showing the highest (using MAH as dual hardener, left) and lowest stiffness values (hardening by Al-Am, right) at ambient temperature. *Notes:* BF3 = trifluoroboron(4-chlorobenzeneamine). For Cal-EP, Al-Am, and Cal-Am, see the notes to Figures 8.2 and 8.3. The EP/hardener ratio was stoichiometric for all EP formulations except EP/MAH where it was set to 1:0.5.

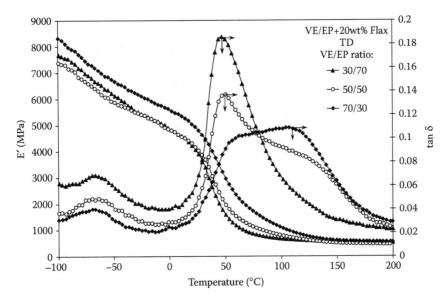

FIGURE 8.7 E' vs. T and tanδ vs T traces for carded flax fiber mat (20 wt%) reinforced VE/EP hybrids as a function of the VE/EP ratio. *Notes:* Specimens were taken from the transverse direction (TD) to that of the carding. The EP resin was an Al-EP+Cal-Am combination (see the notes to Figure 8.3)

EP resin could be triggered to create a good interface between the flax and matrix *in situ*, i.e., without any pretreatment of the flax fiber. This was not the case as the tensile and flexural properties of the composites containing 20 wt% carded flax fiber mat were only slightly better than those of the plain matrix (42). It was also observed that the stiffness (and also the strength) parameters did not vary much as a function of the VE/EP composition (changed between 30:70 and 70:30) (Figure 8.7). SEM inspection of the fracture surface of specimens in which the flax fibers were aligned transverse to the loading direction showed some mechanical anchorage between the flax and matrix. This was attributed to the good wet-out performance of this low viscosity hybrid system (42).

8.5.2 CERAMIC AND BASALT FIBER MAT REINFORCEMENTS

These composites were manufactured to check whether sizings reactive with either the VE or the EP phase were efficient and yielded the same improvements in the mechanical response. Based on the IPN structuring, it was expected that the outcome with a VE or EP compatible (i.e., coreactive) coatings of the fibers would be the same. The tensile and flexural characteristics listed in Table 8.2 clearly show the property improvement achieved by epoxy- and vinylsilane sizings. The corresponding data corroborate, in fact, that the efficacy of epoxy- and vinylsilane coatings is practically identical (1). Recall that this was expected based on the IPN structure of these VE/EP hybrid resins. SEM micrographs taken from the fracture

TABLE 8.2

Tensile and Flexural Properties of the Neat VE/EP Hybrid and Its Composites with Ceramic Fiber Mat Reinforcement (30 wt%), the Fibers of Which Were Sized Differently

Properties	Unit	Matrix VE/EP (1:1)	Fiber Sizing in Composites (30 wt%)		
			As-Received	Epoxysilane	Vinylsilane
Tensile					
E-modulus	GPa	2.4	2.7	6.6	7.6
Strength	MPa	40.8	56.8	86.1	96.6
Elongation	%	4.5	4.5	3.4	3.8
Flexural					
E-modulus	GPa	2.2	2.6	6.4	6.5
Strength	MPa	75.9	99.3	145.3	133.1
Elongation	%	5.2	5.9	6.7	6.3

Source: Szabó et al., *Compos. Sci. Technol.*, 64, 1717–1723, 2004.

surfaces suggested a good adhesion between the matrix and sized ceramic (1) and basalt fibers.

Next, our interest was focused on clarifying whether the structure of the interphase differs from that of the bulk. To shed light on this issue, the earlier described physical etching technique was adopted also in this case. Figure 8.8 shows an AFM image taken from the interphase region of an epoxysilane coated basalt fiber embed-

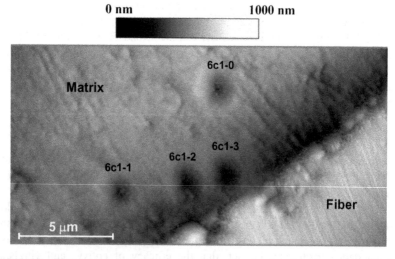

FIGURE 8.8 AFM micrograph taken from the ion-etched polished surface in the vicinity of a basalt fiber treated with epoxysilane and embedded in a VE/Al-EP+Cal-Am based IPN structured resin. *Notes:* AFM micrographs were captured by a Hysitron-TriboScope-Indenter, the imprints of which — used for stiffness mapping — are well resolved.

ded in VE/Al-EP+Cal-Am. The interphase region extended to approximately 3 μm. Nanoindentation tests, performed in the AFM, showed that the stiffness of the interphase was considerably less than the bulk. So here an EP-rich phase was formed owing to the reaction with the epoxy sizing. The observed EP enrichment is in agreement with the etching results. Recall that the EP was far more sensitive to ablation than VE (26). On the contrary, with vinylsilane coating of the fibers, the interphase became stiffer than the bulk because in that case a VE-rich phase (however, markedly less thin compared to the EP-rich one) was formed. Accordingly, the morphology of the interphase differs from that of the bulk. Assuming that this alteration in the morphology was caused by the difference in the cure rates of the components, the strategy proposed by Dean and Cook (41) could be adopted to overcome this problem. Restoring the nanoscale IPN in the interphase would guarantee the intermittent bonding outlined in Section 8.1 (see Figure 8.1).

8.5.3 UNIDIRECTIONAL CARBON FIBER REINFORCEMENT

Unidirectional (UD) carbon fiber (CF) reinforced laminates were produced in the wet filament winding procedure [described in ref. (43)] using epoxy sized high tenacity CF (Tenax® HTA 5131). The CF content of the laminates with negligible offset in their alignment (less than 1°) was between 50 and 60 vol%. For mode I and mode II fracture mechanical tests of these composites, the double cantilever beam (DCB, mode I) and end-notched flexural configurations (ENF, mode II) were selected [e.g., see ref. (44)]. To serve as a starter crack, a 25 μm thick PTFE film was used that was built in the UD laminates during filament winding. The G_c values under mode I and mode II type loading are listed in Table 8.3 for both matrices and composites. The vinylester/urethane hybrid (VEUH) resin, used in this case for reference purposes, is even more brittle than VE as additional cross-links were

TABLE 8.3
G_{Ic} and G_{IIc} Data Determined on Different Matrices (VEUH, VE/Al-EP+Cal-Am, VE/Cal-EP+Al-Am) and on Their UD CF-Reinforced Composite Laminates (Fiber Volume Content: 50 to 60 vol%)

| | G_{Ic} (kJ/m²) | | G_{IIc} (kJ/m²) |
Resin Combination	Matrix	Composite (UD Laminate)	Composite (UD Laminate)
VEUH	0.13	0.22	0.83
VE/Al-EP+ Cal-Am (T=150°C)	4.30	1.50	0.54
VE/Cal-EP+Al-Am (T=200°C)	5.23	1.02	0.88

Notes: VEUH = vinylester/urethane hybrid resin. For its curing, both styrene and a polyisocyanate compound were used. The composition ratio of the VE/EP hybrids was 1:1. The VE/Al-EP+Cal-Am and VE/Cal-EP+Al-Am hybrids are identical to those reported in Table 8.1 and Figures 8.2 and 7.3.

created therein via incorporation of polyisocyanates (45,46). The latter reacted with the secondary hydroxyl groups of the VE, resulting in a tightly cross-linked network exhibiting very high T_g and stiffness. Note that the related G_{Ic} of the composite is higher than that of the VEUH matrix, and it is likely also the case with respect to G_{IIc} (which was not determined for the matrices). This improvement can be assigned to the beneficial effect of the polyisocyanate, which improved the adhesion between the CFs and the matrix. A further hint for this is given by the G_{IIc} value of the VEUH based composite. Recall that under mode II loading, shear deformation dominates, and the resulting hackle pattern on the fracture surface contains valuable information on the fiber to matrix adhesion. The strong improvement in the G_{Ic} values of the IPN structured VE/EP hybrids are not fully transferred to the composites. This is in harmony with the prediction schematically depicted in Figure 8.4 (30,31).

SEM micrographs taken from the fracture surface of DCB specimens (interlaminar failure under mode I) suggested very good adhesion between the IPN matrix and CF compared to the VEUH based composite (see Figure 8.9) (47). On the other hand, the G_{IIc} values of the composites with VE/EP hybrid matrices are unexpectedly low. This suggests that these IPN structured resins are prone to shear deformation (as argued before) and their inherent strength is likely too low. Fortunately, the most important failure in composites always occurs under mode I type loading, and thus low G_{IIc} values are less problematic.

Preliminary results achieved on model laminates (cross-ply, UD reinforcement along the loading direction) corroborate that the IPN structuring of the matrix support the onset of the Cook-Gordon mechanism as speculated in Section 8.1.

8.6 SUMMARY AND FUTURE OUTLOOK

The major task of this survey is to show the potential of resin hybridization by taking VE and EP combinations as an example for composite applications. IPN structuring of thermosets is the right tool for property modifications, and the related research shows some analogies with the blending of thermoplastics. Note that the blending of thermoplastics has become a success story with respect to both theory and praxis. We might face a similar development with some hybrid thermosets in the future.

There are several problems that have to be addressed for VE/EP hybrids in order for them to reach their commercial breakthrough. The highest priority is their durability (e.g., effects of physical aging, hygrothermal resistance, thermal spiking behavior).

IPN structuring may be a suitable "carrier" to set some desired or unique properties (e.g., flame resistance, electric and magnetic conductivity) in a reproducible manner. Note that a preferred enrichment of the related modifiers in one of the IPN forming phases may be enough to obtain the required property. IPN structuring may also be a helpful tool to promote the adhesion between various thermosets for which even new processing techniques, like coinjection RTM (48), may be developed. It should be borne in mind that more sophisticated IPNs, composed of more than two phases, may also be created and used in composites [e.g., see ref. (49)].

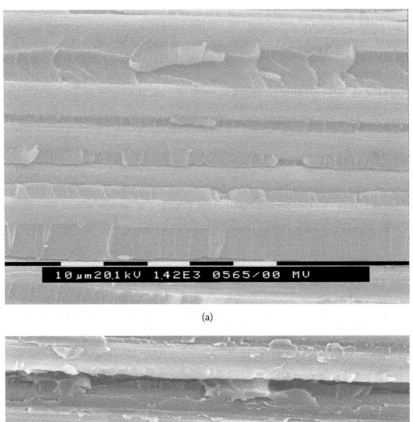

(a)

(b)

FIGURE 8.9 SEM micrographs showing the difference in the failure mode between UD CF-reinforced composite laminates with VEUH (a) and VE/Al-EP+Cal-Am (b) matrices. *Notes:* Images were taken from the interlaminar fracture surfaces of DCB specimens (mode I loading). The IPN resin is similar to the one used in Figure 8.3.

The strategy of producing IPNs is very straightforward when thermoset resins, derived from natural resources [e.g., functionalized plant oils (50), animal oils, sugar derivates], are used. Thermoset resins from such natural products show low T_g, low stiffness, and strength when cross-linked alone. The related products usually show rubbery characteristics. Therefore, they have to be combined with some "high end" petrochemical–based resins in order to meet the requirements of composite resins for structural applications. This resin hybridization can be considered as a straightforward combination of "green chemistry" and petrochemistry.

ACKNOWLEDGMENTS

This work was supported by grants of the German Science Foundation (DFG; Ka 1202) and Fonds der Chemischen Industrie (FCI). The author would like to thank Dr. O. Gryshchuk for his help with the artwork.

REFERENCES

1. Szabó, J.S., Karger-Kocsis, J., Gryshchuk, O., and Czigány, T., Effect of fibre surface treatment on the mechanical response of ceramic fibre mat-reinforced interpenetrating vinylester/epoxy resins, *Compos. Sci. Technol.*, 64, 1717–1723, 2004.
2. Atkins, A.G., Intermittent bonding for high toughness/high strength composites, *J. Mater. Sci.*, 10, 819–832, 1975.
3. Mai, Y.W. and Castino, F., Fracture toughness of Kevlar-epoxy composites with controlled interfacial bonding, *J. Mater. Sci.*, 19, 1638–1655, 1984.
4. Sperling, L.H., Interpenetrating polymer networks: an overview, in *Interpenetrating Polymer Networks*, (Adv. Chem. Ser. 239), Klempner, D., Sperling, L.H., and Utracki, L.A., Eds., Amer. Chem. Soc., Washington, 1994, pp. 3–38.
5. Utracki, L.A., Thermodynamics and kinetics of phase separation, in *Interpenetrating Polymer Networks*, (Adv. Chem. Ser. 239), Klempner, D., Sperling, L.H., and Utracki, L.A., Eds., Amer. Chem. Soc., Washington, 1994, pp. 77–123.
6. Lipatov, Y.S., *Phase Separated Interpenetrating Polymer Networks*, USChTU, Dnepropetrovsk, Russia, 2001.
7. Sophiea, D., Klempner, D., Sendijarevic, V., Suthar, B., and Frisch, K.C., Interpenetrating polymer networks as energy-absorbing materials, in *Interpenetrating Polymer Networks*, (Adv. Chem. Ser. 239), Klempner, D., Sperling, L.H., and Utracki, L.A., Eds., Amer. Chem. Soc., Washington, 1994, pp. 39–75.
8. Klempner, D. and Sophiea, D., Interpenetrating polymer networks, in *Elastomer Technology Handbook,* Cheremisinoff, N.P., Ed., CRC Press, Boca Raton, FL, 1993, pp. 421–444.
9. Chou, Y.C. and Lee, L.J., Kinetic, rheological, and morphological changes of polyurethane-unsaturated polyester interpenetrating polymer networks, in *Interpenetrating Polymer Networks*, (Adv. Chem. Ser. 239), Klempner, D., Sperling, L.H., and Utracki, L.A., Eds., Amer. Chem. Soc., Washington, D.C., 1994, pp. 305–331.
10. Hsu, T.J. and Lee, L.J., Reaction kinetics of polyurethane-polyester interpenetrating polymer network in the bulk polymerization, *Polym. Compos.*, 25, 951–958, 1985.

11. Wang, G.Y., Zhu, M.Q., and Hu C.P., Interpenetrating polymer networks of polyurethane and graft vinyl ester resin: polyurethane formed with diphenylmethane diisocyanate, *J. Polym. Sci. Part A:Chem.*, 38, 136–144, 2000.
12. Chen, C.-H., Chen, W.-J., Chen, M.-H., and Li, Y.-M. Simultaneous full-interpenetrating polymer networks of blocked polyurethane and vinyl ester. Part I. Synthesis, swelling ratio, thermal properties and morphology, *Polymer*, 41, 7961–7967, 2000.
13. Chen, C.-H., Chen, W.-J., Chen, M.-H., and Li, Y.-M., Simultaneous full-interpenetrating polymer networks of blocked polyurethane and vinyl ester. II. Static and dynamic mechanical properties, *J. Appl. Polym. Sci.*, 71, 1977–1985, 1999.
14. Qin, C.-L., Cai, W.-M., Cai, J., Tang, D.-Y., Zhang, J.-S., and Qin, M., Damping properties and morphology of polyurethane/vinyl ester resin interpenetrating polymer network, *Mater. Chem. Phys.*, 85, 402–409, 2004.
15. Cano, F.B. and Visconte, L.L.Y., UV-polymerization of 2-ethylhexyl acrylate in interpenetrating polymer networks — some mechanical properties, *Polym. Bull.*, 40, 83–87, 1998.
16. Ma, S. and Tang, X., Morphology of acrylic-epoxy bichain simultaneous interpenetrating polymer networks, in *Interpenetrating Polymer Networks*, (Adv. Chem. Ser. 239), Klempner, D., Sperling, L.H., and Utracki, L.A., Eds., Amer. Chem. Soc., Washington, 1994, pp. 405–426.
17. Yamamoto, K., Phenolic interpenetrating polymer networks for laminate applications, in *Interpenetrating Polymer Networks*, (Adv. Chem. Ser. 239), Klempner, D., Sperling, L.H., and Utracki, L.A., Eds., Amer. Chem. Soc., Washington, 1994, pp. 233–243.
18. Chern, Y.C., Tseng, S.M., and Hsieh, K.H., Damping properties of interpenetrating polymer networks of polyurethane-modified epoxy and polyurethanes, *J. Appl. Polym. Sci.*, 74, 328–335, 1999.
19. Athawale, V.D., Kolekar, S.L., and Raut, S.S., Recent developments in polyurethanes and poly(acrylates) interpenetrating polymer networks, *J. Macromol. Sci. Part C Polym. Rev.*, 43, 1–26, 2003.
20. Cook, W.D., Dean, K., and Forsythe, J.S., Cure, rheology and properties of IPN thermosets for composite applications, *Materials Forum*, 25, 30–59, 2001.
21. Lin, M.-S., Liu, C.-C., and Lee, C.-T., Toughened interpenetrating polymer network materials based on unsaturated polyester and epoxy, *J. Appl. Polym. Sci.*, 72, 585–592, 1999.
22. Shih, Y.-F. and Jeng, R.-J., Carbon black containing IPNs based on unsaturated polyester/epoxy. I. Dynamic mechanical properties, thermal analysis, and morphology, *J. Appl. Polym. Sci.*, 86, 1904–1910, 2002.
23. Shaker, Z.G., Browne, R.M., Stretz, H.A., Cassidy, P.E., and Blanda, M.T., Epoxy-toughened, unsaturated polyester interpenetrating networks, *J. Appl. Polym. Sci.*, 84, 2283–2286, 2002.
24. Ivanković, M., Džodan, N., Brnardi, I., and Mencer, H.J., DSC study on simultaneous interpenetrating polymer network formation of epoxy resin and unsaturated polyester, *J. Appl. Polym. Sci.*, 83, 2689–2698, 2002.
25. Mortaigne, B., Feltz, B., and Laurens, P., Study of unsaturated polyester and vinylester morphologies using excimer laser surface treatment, *J. Appl. Polym. Sci.*, 66, 1703–1714, 1997.
26. Karger-Kocsis, J., Gryshchuk, O., and Schmitt, S., Vinylester/epoxy-based thermosets of interpenetrating network structure: An atomic force microscopic study, *J. Mater. Sci.*, 38, 413–420, 2003.

27. Karger-Kocsis, J. and Gryshchuk, O., Toughness behavior of vinylester/epoxy thermosets with interpenetrating network structure, *Macromol. Symp.*, 217, 317–328, 2004.
28. Gryshchuk, O. and Karger-Kocsis, J., Influence of the type of epoxy hardener on the structure and properties of interpenetrated vinylester/epoxy resins, *J. Polym. Sci. Part A Chem.*, 42, 5471–5481, 2004.
29. Karger-Kocsis, J., Gryshchuk, O., and Jost, N., Toughness response of vinyl-ester/epoxy-based thermosets of interpenetrating network structure as a function of the epoxy resin formulation: Effects of the cyclohexylene linkage, *J. Appl. Polym. Sci.*, 88, 2124–2131, 2003.
30. Hunston, D.L., Composite interlaminar fracture: effect of matrix fracture energy, *Compos. Technol. Rev.*, 6, 176–180, 1984.
31. Compston, P., Jar, P.-Y.B., Burchill, P.J., and Takahashi, K., The transfer of matrix toughness to composite mode I interlaminar fracture toughness in glass-fibre/vinyl ester composites, *Appl. Compos. Mater.*, 9, 291–314, 2002.
32. Szabó, J.S., Gryshchuk, O., and Karger-Kocsis, J., Fatigue crack propagation behavior of interpenrating vinylester/epoxy resins, *J. Mater. Sci. Letters*, 22, 1141–1145, 2003.
33. Merdas, I., Tcharkhtchi, A., Thominette, F., Verdu, J., Dean, K., and Cook, W., Water absorption by uncrosslinked polymers, networks and IPNs having medium to high polarity, *Polymer*, 43, 4619–4625, 2002.
34. Karger-Kocsis, J., Gryshchuk, O., and Jost, N., High toughness vinylester/epoxy-based thermosets of interpenetrating network structure, *SPE-ANTEC*, 60, 751–754, 2002.
35. Zhang, M., Wang, C.Z., Wu, D.Z., and Yang, X.P., Studies on ipns formed from model VERs and epoxy sytems, in *Proceedings of the International Symposium on Advanced Materials and Their Related Science*, Bejing University of Chemical Technology, Beijing, PR China, Oct. 21–24, 2003, pp. 842–846.
36. Dean, K., Cook, W.D., Burchill, P., and Zipper, M., Curing behaviour of IPNs formed from model VERs and epoxy systems. Part II. Imidazole-cured epoxy, *Polymer*, 42, 3589–3601, 2001.
37. Karger-Kocsis, J., Gryshchuk, O., Fröhlich, J., and Mülhaupt, R., Interpenetratring vinylester/epoxy resins modified with organophilic layered silicates, *Compos. Sci. Technol.*, 63, 2045–2054, 2003.
38. Dean, K., Cook, W.D., Zipper, M.D., and Burchill, P., Curing behaviour of IPNs formed from model VERs and epoxy systems. I. Amine cured epoxy, *Polymer*, 42, 1345–1359, 2001.
39. Dean, K., Cook, W.D., Rey, L., Galy, J., and Sautereau, H., Near-infrared and rheological investigations of epoxy-vinyl ester interpenetrating polymer networks, *Macromolecules*, 34, 6623–6630, 2001.
40. Jost, N., *Vernetzung und Chemorheologie von Duromeren mit hybrider und interpenetrierender Struktur*, Ph.D. thesis, IVW Bericht 43, Schlarb, A.K., Ed., Institut für Verbundwerkstoffe GmbH, Kaiserslautern, 2004.
41. Dean, K.M. and Cook, W.D., Azo initiator selection to control the curing order in dimethacrylate/epoxy interpenetrating polymer networks, *Polym. Int.*, 53, 1305–1313, 2004.
42. Szabó, J.S., Romhány, G., Czigány, T., and Karger-Kocsis, J., Interpenetrating vinylester/epoxy resins reinforced by flax fibre mat, *Adv. Compos. Letters*, 12, 115–120, 2003.

43. Hoecker, F., *Grenzflächeneffekte in Hochleistungsfaserverbundwerkstoffen mit Polymeren Matrizes — VDI Fortschrittberichte*, Reihe 5, Nr. 439, VDI Verlag:Düsseldorf, 1996.

44. Moore, D.R., (Ed.). *The Application of Fracture Mechanics to Polymers, Adhesives and Composites*, ESIS Publ. 33, Elsevier: Oxford, 2004.

45. Jost, N. and Karger-Kocsis, J., On the curing of a vinylester-urethane hybrid resin, *Polymer*, 43, 1383–1389, 2002.

46. Klumperman, G., Carbon fibre reinforced vinyl-ester urethane hybrid resins. The alternative to epoxy-carbon systems, in *Proceedings of the Tenth European Conference on Composite Materials — ECCM-10*, Brugge, Belgium,, June 3–7, 2002, paper 248.

47. Rosso, P. and Friedrich, K., *Einfluss der Faser/Matrix-Haftung auf das mechanische eigenschaftsprofil modifizierter kohlenstofffaserverstärkter Vinylester-Harze*, In IVW Bericht 48, Schlarb, A.K., Ed., Institut für Verbundwerkstoffe GmbH:Kaiserslautern, in press, 2004; Rosso, P., Friedrich, K., Lenz, C., and Ye, L., Effects of vinylester-modification on the mechanical and interfacial properties of their carbon fiber reinforced-composites, *Composites A*, submitted, 2005.

48. Singh, K.P. and Palmese, G.R., Network formation during cocure of phenolic resins with vinyl-ester and epoxy-amine systems for use in multifunctional composites, *J. Appl. Polym. Sci.*, 91, 3107–3119, 2004.

49. Parmar, J.S., Patel, G.R., and Patel, A.K., Vinyl ester-based three-component IPNs for glass fibre reinforced composite applications, *Adv. Polym. Technol.*, 23, 71–75, 2004.

50. Sperling, L.H. and Mishra, V. The current status of interpenetrating polymer networks, in *IPNs around the World Science and Engineering*, Kim, S.C. and Sperling, L.H., Eds., Wiley, Chichester, 1997, pp. 1–25.

9 Phase Morphology of Dynamically Vulcanized Thermoplastic Vulcanizates

Hans-Joachim Radusch

CONTENTS

9.1 INTRODUCTION

The production of polymer blends by means of melt-mixing technologies presents itself today as a well established way for the development of polymeric materials. Numerous examples of commercially produced blends have been made — by reactive and nonreactive mixing — using this flexible and successful method for modifying polymer materials. Besides conventional polymer blends, dynamic vulcanizates, examples of which include reactive blended polymer systems based on a rubber and a thermoplastic component, have become increasingly important in the last two decades. Dynamic vulcanization is a superimposed process consisting of melt-mixing and selective rubber cross-linking that enables the production of thermoplastic elastomers (TPE) of the thermoplastic vulcanizate (TPE-V) type, many of which have already found commercial applications, especially in the automotive sector.

For the successful development of a blend, the control of the complex correlations between the thermodynamic interactions and rheological behavior of the polymer components as well as the temperature and forces affecting the material system is of fundamental importance. An important area of focus is the control of the morphology in heterogeneous polymer blends.

Figure 9.1 is a schematic of the complex relationship between the polymer material, the mixing and processing steps, and the polymer's properties. It is clear that the thermodynamic and rheological properties of the components influence the morphology-properties relationship in an essential way.

The realization of a definite phase morphology in polymer blends made via melt-mixing is a complex process during which the type of the resulting blend morphology (heterogeneous or molecular, mixed or homogeneous) is initially determined by the miscibility of the components, which depends on the temperature, pressure, and concentration. The major part of blends contains polymer components that are thermodynamically immiscible, i.e., they consist of separate phases, which can be present either in the pure form of the specific components or in mixed phases too.

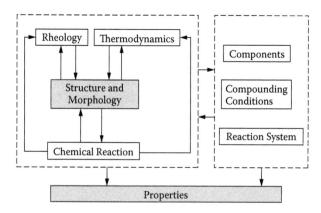

FIGURE 9.1 Complex relationship between processing, morphology, and properties during the production of reactive polymer blends.

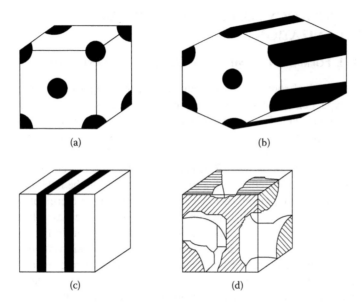

(a)   (b)

(c)   (d)

FIGURE 9.2 Basic types of phase morphology: (a) drop, (b) cylinder, (c) layers, and (d) cocontinuous.

The formation of phase morphologies that can be found in multicomponent polymer systems can differ widely (Figure 9.2). On the one hand, morphology formation depends on the properties of each component, such as the melt viscosity, melt elasticity, and surface tension. On the other hand, the processing conditions, such as temperature, pressure, shear rate, residence time, and component ratios play an important role in the morphology formation process in the melt.

The aim of property enhancement by manipulating the combination of different polymeric components can be achieved in most cases only by compatibilisation of the mostly incompatible polymeric components by means of block or graft copolymers or by means of reactive substances inserted during the melt-mixing process. As a result of all this, the existing thermodynamic interactions between the components, i.e., the processes on the phase boundaries, can be influenced very effectively. The original phase morphology will change as a result of the influences acting on the interfacial tensions between the phases, finally resulting in a change in the complex property spectrum.

The morphology-property relationship can be highly influenced by the use of reactive compounding, in particular, dynamic vulcanization where the chemical reactions on the phase boundary initiated during the melt-mixing process in one phase or two phases induce drastic changes in the rheological behavior of the system and finally in the properties. It is well known that, e.g., dynamic vulcanizates exhibit high stress-strain properties only if their phase morphology is characterized by rubber particles in the submicrometer region (1). Thus, the morphology formation processes during melt-mixing, especially reactive blending, play an essential part in regard to both the properties and the range of applications of the products.

9.2 MORPHOLOGY FORMATION AT DYNAMIC VULCANIZATION

9.2.1 TPE FORMATION BY MEANS OF DYNAMIC VULCANIZATION

Beside the chemical routes for the production of thermoplastic elastomers of the copolymer type, i.e., TPE-S (styrene containing TPE, e.g., styrene-butadiene-styrene block copolymer), TPE-E (ether containing TPE, e.g., polyether block amides) or TPE-U (thermoplastic polyurethane), thermoplastic elastomers can be produced by (a) nonreactive blending of TPE-O [olefin containing TPE, e.g., polypropylene/etylene-propylene-diene copolymer (PP/EPDM)] or (b) reactive compounding of a thermoplastic polymer together with a rubber component. In the latter case, a major particle-like rubber component has to be distributed in a minor thermoplastic matrix. Mixing, i.e., dispersion and distribution of the rubber, as well as selective cross-linking of the rubber are superimposed processes that happen in the melt-mixing process. This reactive compounding process is called *dynamic vulcanization* and, accordingly, the product is called a *dynamic vulcanizate*. Dynamic vulcanizates need rubber particles in the micrometer size range, and the rubber particles have to be cross-linked in a defined degree and embedded in a minor, low viscous thermoplastic component. Therefore, only a controlled and optimized morphology formation process during reactive mixing will guarantee such a specific phase morphology with the desired high mechanical property level.

Two stages have to be realized during dynamic vulcanization: first, a blending step without a chemical reaction, and second, a superimposed cross-linking and mixing process step. The thermodynamic and rheological mechanisms occur in both stages and form the basis for the complex microscale processes that are found in the melt; these are discussed in detail below.

9.2.2 THERMODYNAMIC AND RHEOLOGICAL BASICS

9.2.2.1 Thermodynamic Principles of Polymer Blending

The homogeneity and, as a result, the morphology of polymer blends are basically determined by the thermodynamic interactions of the components. Therefore, the term "polymer blends" shall be defined as a substance resulting from a mixing process of a minimum of two chemically different polymers regardless of the mixing state of the resulting substance. The existing thermodynamic interactions and the intensity of the mixing process determine the homogeneity or heterogeneity of the polymer blends, which can be in a thermodynamically balanced or unbalanced state. The term "homogeneous" describes the mixing by chance on a molecular basis, while the term "inhomogeneous" describes the existence of two phases divided from each other.

The formation of a homogeneous and thermodynamically stable blend under given temperature and pressure conditions — namely the "miscibility" of the components — is controlled by the change in the free mixing enthalpy G_M, which is calculated by using the values of the enthalpies and entropies of the components and the enthalpy and entropy of the blend (i.e., the Gibbs-Helmholtz equation)

$$\Delta G_M = \Delta H_M - \Delta S_M \tag{9.1}$$

where ΔG is the free enthalpy, ΔH is the enthalpy, and ΔS is the entropy.

A system is thermodynamically stable if the change of free enthalpy is negative and the second derivative of the free mixing enthalpy with respect to the concentration of one of the compounds under given temperature T and pressure p is positive:

$$\Delta G_M < 0 \tag{9.2}$$

$$\left. \frac{\partial^2 \Delta G_M}{\partial \varphi^2} \right|_{p,T} > 0 \tag{9.3}$$

Equations (9.2) and (9.3) provide a sharp criterion for the miscibility of polymeric components and the stability of the resulting homogeneous blends.

The free mixing enthalpy G_M can be calculated based on well-known models, such as, e.g., the simple Flory-Huggins model (2,3) according to

$$\frac{\Delta G_M}{RT} = \sum_{i=1}^{n} \frac{\varphi_i \ln \varphi_i}{N_i} + \sum_{j=2}^{n} \sum_{i=1}^{j-1} \varphi_i \varphi_j \chi_{ij} \tag{9.4}$$

where N_i is the number of molecular units, R is the general gas constant, and T is the temperature, as well as the theories of state, taking into consideration the compressibility of polymer melts. Examples include the Flory-Orwoll-Vrij model and the Dee-Walsh model, which depend on the state parameters of pressure, temperature, and volume (4,5). Because the reactions here involve macromolecular components in polymer blends and because the contribution of entropy to the free mixing enthalpy is small in most cases, the components are found to be immiscible. Miscibility is an exception and exists only for very few polymer combinations. The interaction parameter χ_{ij}, which can be determined by using solubility parameters (6), sheds light on the miscibility of polymeric components:

$$\chi_{ij} = \frac{V_O}{RT}(\delta_i - \delta_j) \qquad (V_o = \text{volume}) \tag{9.5}$$

Painter et al. (7) showed that the difference in the solubility parameters δ_{ij} must be $\Delta\delta < 0.2$ $(\text{J/cm}^3)^{1/2}$ for mixing to take place on a molecular level. With the increased likelihood of immiscibility on the part of the components, i.e., with increasing ΔG_M values, the tendency is toward phase separation during mixing, which will result in large morphological units, e.g., dispersed particles as shown schematically in Figure 9.3 (8).

In real melt-mixing processes, miscible components go through a mixing state that is characterized by heterogeneity, i.e., a molecular homogeneous polymer system is not inevitable but does depend on the energy introduced during the mixing as

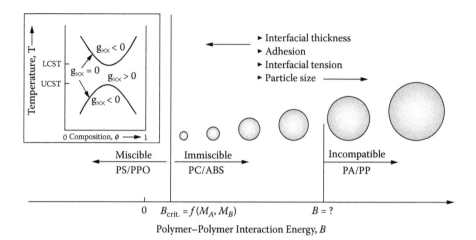

FIGURE 9.3 Schematic of the changes in the phase character of polymer blends as a function of their polymer-polymer interaction energies. (From Paul and Bucknall in *Polymer Blends, Vol. 1, Formulation*, pp. 1–14.)

well as the time of the mixing process; these are important factors for the course of the mixing and the final state of the blend. For the blends discussed here, the interactions resulting from the rheological properties of the components together with the thermodynamic miscibility of the components will determine the final morphological state.

9.2.2.2 Rheological Principles and Microrheology of Blends

Beside the thermodynamic interactions of the polymer components, their rheological behavior also determines to a great extent the morphology that results from a melt-mixing process and therefore the complex property range of the blend. Therefore, large differences must be found between those polymer systems in which the components do not change their rheological behavior, or only change to a small extent, and those in which one or two components undergo a large change of rheological behavior during melt-mixing. Examples of the latter include thermally and mechanically induced molecular degradation processes as well as reactive compounding that produce reactive compatibilised polymer blends or thermodynamic elastomers of the type TPE-V (dynamic vulcanizates). In these cases, the strong chemical reactions affect the thermodynamic conditions and the interactions at the phase boundary layer, producing large changes in the rheological behavior and therefore changes in the phase morphology as well.

9.2.2.2.1 Formation of Morphology in Nonreactive Polymer Systems

The formation of morphology in heterogeneous polymer blends is determined by the mixing mechanisms that take place during melt-mixing, which can be characterized as dispersion and distribution processes. Here, deformation processes such as mass convection and laminar convection are responsible for dislocation and distribution of volume parts while the existing stress conditions cause the dispersion of volume parts.

Mechanisms of Morphology Formation
The formation of morphology in polymer blends, in which one phase forms the matrix (index *c*) and a second component forms the dispersed phase (index *d*), can be described under microrheological aspects by the following ratios and numbers (9–12):

Viscosity ratio of the components

$$\lambda_{\eta} = \frac{\eta_d(T, \gamma)}{\eta_c(T, \gamma)} \tag{9.6}$$

where η_d is the viscosity of the dispersed phase and η_c is the viscosity of the matrix phase

Elasticity ratio of the components

$$\lambda_{\theta} = \frac{\theta_d(T, \gamma)}{\theta_c(T, \gamma)} \tag{9.7}$$

where θ_d is the coefficient of elasticity of the dispersed phase and θ_c is the coefficient of elasticity of the matrix phase

Stress ratio (capillary number or Weber number)

$$\kappa = We = \frac{\eta \cdot \gamma \cdot R}{\gamma_{dc}} \tag{9.8}$$

where γ_{dc} is the interfacial tension and R is the particle radius

Relative duration of the deformation processes as a ratio of deformation and capillary number κ

$$\frac{\gamma}{\kappa} = \frac{\gamma \cdot t_v}{(\eta \cdot \gamma \cdot R) / \gamma_{dc}} = \frac{t_v \cdot \gamma_{dc}}{\eta \cdot R} \tag{9.9}$$

where t_v is the residence time

Coalescence (probability of particle collision P_{coll})

$$P_{coll} = \frac{6 \cdot \gamma \cdot \varphi_d^{2}}{\pi^2 \cdot R^3} \tag{9.10}$$

To determine the interfacial tension, we must take into account the fact that the interfacial tension in the flow γ_{dc} is not equal to the interfacial tension under static conditions γ^0_{dc} since in a viscoelastic medium, the elasticity of the melt, expressed, e.g., by the first normal stress difference $(\sigma_{ii} - \sigma_{jj})_{d,c}$, influences the interfacial tension to a great extent (10,11):

$$\gamma_{dc} = \gamma^0_{dc} + 1/6 \cdot R\left[(\sigma_{11} - \sigma_{22})_d - (\sigma_{11} - \sigma_{22})_c\right] \qquad (9.11)$$

If the capillary number is higher than a critical value κ_{crit}, droplets fragment. If $\kappa <\kappa_{crit}$, only deformation of the dispersed particles takes place.

In order to investigate droplet deformation, two basic experiments were conducted by Taylor (13,14). He investigated the droplet deformation of an oil-water system in simple shear flow by means of the Couette apparatus and in shear and elongational flow by means of a four-roll apparatus. Taylor found that the dispersed particles were stretched into fibrils under the influence of the acting stresses and broke apart into small particles when the stress exceeded a critical point. He pointed out that an elongational flow can cause droplet breakup more effectively than a simple shear flow even in the case of unfavorable viscosity ratios between the dispersed phase and matrix.

Applying the results of Taylor, Cox (15) investigated shear and elongational flow using one Newtonian fluid distributed in another Newtonian fluid. Cox stated that droplet deformation is a function of the viscosity ratio and the capillary number.

Grace (16) investigated droplet breakup in shear and elongational flows and found that droplet destruction in a simple shear field takes place most easily if the viscosity ratio, λ_η, is approximately 1. At a viscosity ratio of $\lambda_\eta \geq 4$ to 5, droplet breakup is possible only in a coupled shear-elongational flow, whereas in a simple shear flow, no droplet breakup is noticed. The deformation conditions of fluid droplets of an emulsion in a shear-field have been classified by Rumscheid and Mason (17) (Figure 9.4).

At a viscosity ratio of $\lambda_\eta \approx 6$, no droplet breakup occurs while for $\lambda_\eta \approx 1$, droplet breakup into two identical smaller droplets is observed. If λ_η is lowered to approximately 0.7, the droplet is deformed into a long cylindrical fibril that finally falls apart into single droplets. The breakup of the fibrils is found to be connected to the distortion of the elongated thread (18) that runs sinusoidally and that depends on the distortion wavelength L, the thread radius R_0, and the matrix viscosity η_m (see Figure 9.8). Tomotika (18) described the stability of the threads by a growth rate parameter of the sinusoidal distortion q

$$q = \gamma_{12}\Omega(L, \lambda_\eta)/\eta_m R_0 \qquad (9.12)$$

where Ω is a function of the wavelength and L the viscosity ratio. Stone et al. (19) found that the breakup of highly deformed liquid threads happens if the flow is stopped.

For very small values of the viscosity ratio $\lambda_\eta > 2 \times 10^{-4}$ and a capillary number $\kappa < \kappa_c$, the strongly deformed droplet remains stable, i.e., droplets are torn off erosion-

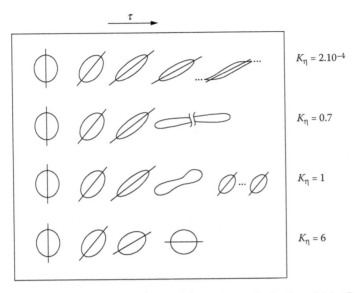

FIGURE 9.4 States of droplet deformation of Newtonian media in shear fields. (Rumscheid and Mason, *J. Colloid Sci.*, 16, 238–261, 1961.)

like (tip streaming) only at the tip of the deformed structure satellite. In this case, the real dispersion process is hindered. This different rheological behavior will affect morphology formation.

Droplet Deformation and Dispersion in Newtonian Media
According to Taylor (13,14), a Newtonian droplet with the viscosity η_d dispersed in a Newtonian matrix with the viscosity η_m under shear deformation is first elongated into 45° with respect to shear direction and deformed into an ellipsoid. Its long axis turns near to the flow direction. With increasing shear deformation of the matrix, the droplet is elongated further, finally falling apart into smaller droplets under the influence of interfacial tension; this process will repeat itself. The process is controlled by two opposing influencing factors: λ_η and γ_{dm}. The ratio of the viscosities of the droplets and matrix, λ_η, determines how easily the droplet can be deformed. At low droplet viscosity, deformation happens easily and the inner circulation is strong. On the other hand, at high viscosity, the droplet only turns with a constant angular speed $\gamma/2$. The interfacial tension between the droplet and matrix, γ_{dm}, tries to keep the droplet surface as small as possible.

In order to achieve droplet dispersion under the influence of a shear rate γ, the viscous forces elongating the droplet must be higher or equal to the interfacial tension. This results in a stress ratio, i.e., Eq. (9.8). κ is the dimensionless capillary number and is the result of the ratio of the viscous stress of the matrix $\eta_c \cdot \gamma$ on a droplet with radius R and of the interfacial tension γ_{dc}. The critical capillary number κ_{crit} is a function of the viscosity ratio, λ_η (Figure 9.5). Droplet dispersion occurs in shear flow only in the area above the curves where $\kappa > \kappa_{crit}$. The droplet radius R and, therefore, the capillary number decreases until a lower limit κ_c is reached. No droplet destruction can be found at $\lambda_\eta > 4$ (droplet is too viscous)

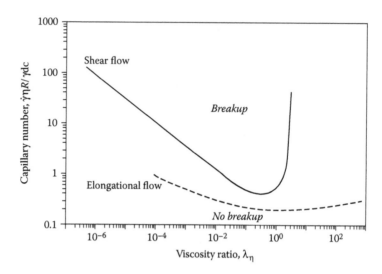

FIGURE 9.5 The dependence of droplet dispersion on the viscosity ratio in a simple shear flow field and under hyperbolic flow. (Grace, *Chem. Eng. Commun.*, 14, 225–277, 1982.)

and very low λ_η values (droplet is drawn into fibrils). The droplet diameter attains its lowest value at a viscosity ratio of approximately 1. Droplet dispersion in shear takes place most easily for viscosity ratios in the range of 0.1 to 1. Out of this range, κ_c becomes very large and droplet dispersion requires very high shear rates, which possibly are above practical reachable values. In an elongational flow, the function κ_{crit} shows a different devolution with respect to shear (see Figure 9.5). According to Wu (20) and Han (21), the minimum is much broader and κ_{crit} reaches values up to 0.2, i.e., the elongational flow allows, when compared to shear flow, for droplet dispersion in a much broader range of viscosity ratios, especially for $\lambda_\eta \gg 1$. Elongational flow components are found in static mixers and twin-screw extruders.

Droplet Deformation and Dispersion in Viscoelastic Fluids

Viscoelastic media under shear experience normal stresses as well as shear stresses. The first difference of normal stresses N_1 $(\sigma_{11}-\sigma_{22})$ is directly measurable. During droplet dispersion, the contribution of normal stresses can be accounted for according to Van Oene (10) using the effective viscoelastic interfacial tension, Eq. (9.11). Here, the normal stress values for the effective shear rates acting near the interface should be inserted because, although Eq. (9.11) was derived for elongational deformations, the equation is valid for shear as well.

An investigation of the deformation of droplets under typical viscoelastic flow conditions has been undertaken by several authors (20–23). Gauthier et al. (23) found a higher critical capillary number in viscoelastic media with respect to Newtonian media. This indicates that the capillary number is also a function of the melt elasticity. Utracki and Shi (12) reported that droplets of a viscoelastic medium in a shear field at $\lambda_\eta > 0.5$ become strongly deformed and break apart as soon as the flow is stopped. The droplets were found to be stable at $\lambda_\eta <$

0.5. De Bruijn (24) found in his experimental investigation that the elasticity of the particles is a stabilizing factor regardless of the viscosity ratios, hindering their dispersion. This fact shows that the dispersion of viscoelastic droplets in a viscoelastic fluid is more difficult to achieve than that of a Newtonian droplet in a Newtonian fluid.

Han and Funatsu (21) emphasized that viscoelastic droplets in shear as well as in elongational flow are much more stable than Newtonian droplets not only in Newtonian media but also in viscoelastic media. For the dispersion of such droplets, higher shear stresses are necessary.

If both phases, the disperse component and the continuous matrix, are viscoelastic, a breakup of dispersed particles is still possible during an extrusion process for $\lambda_\eta > 4$, according to Wu (20). To explain this phenomenon, Wu used the elastic effect, i.e., the presence of elongational flow and a complex viscosity-temperature profile along the screw channel. Van Oene (10) assumed that the "elastic free energy" is changed in each phase with the deformation of the system. From this assumption, Eq. (9.11) was formulated. This relationship, which describes the relation between melt elasticity and interfacial stress during morphology formation, can be used to explain also the difference in morphology and dispersion of submicroscopic dispersed particles that undergo a transformation from a lamellar initial dispersion to droplet dispersion.

Figure 9.6 shows an example of an investigation of the dispersion behavior of real polymer blends. Favis and Chalifoux (25) investigated blends of polypropylene (PP) and polycarbonate (PC) with respect to the correlation between the average droplet size and the viscosity ratio. The torque conditions measured in a laboratory internal mixer were taken as the viscosity ratios and were compared to the particle size of the blends produced under the specific conditions. It became obvious that in a mixing system (real conditions) shear as well as elongational flow must be present since despite the relative high viscosity ratios, dispersion took place. The increase in the content of dispersed phase resulted also in bigger droplets, which imply that coalescence processes take place parallel to dispersion.

Another example of the formation of phase morphology in heterogeneous polymer blends is shown in Figure 9.7. The poly(butylene terephthalate)/polycarbonate (PBT/PC) blend in the experiment contains a lower volume PBT content that have $\lambda_\eta < 1$ (26). Fiberlike structures can be observed clearly that result from the high deformability of the low viscous PBT phase within the more highly viscous PC matrix.

Under appropriately high deformation conditions, the fiberlike structures fall apart into small droplets that are distributed in the higher viscous PC matrix under the influence of shear fields. The result is a heterogeneous phase morphology of the island-matrix type with very small PBT particles embedded in the PC matrix.

The mechanism of destruction of strongly elongated dispersed structures was described theoretically by Taylor and Tomotika, and was experimentally demonstrated, e.g., by Elmendorp (9) as shown in Figure 9.8. The example is that of a dispersed PA phase in a polyethylene (PE) matrix. In this example, the polymer blend systems have $\lambda_\eta > 1$. If a polystyrene (PS) droplet of nearly the same viscosity as PE is deformed under shear above a critical shear rate, it can fall apart into two

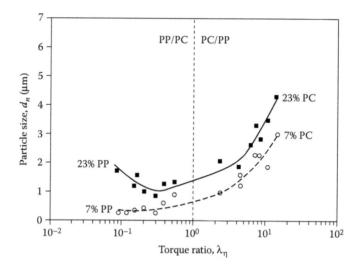

FIGURE 9.6 Average droplet diameter as a function of viscosity ratios of PP/PC-blends. (Favis and Califoux, *Polym. Engin. Sci.*, 27, 1591–1600, 1987.)

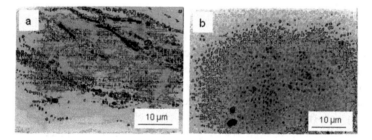

FIGURE 9.7 Optical micrograph (polarized light) showing morphology formation in PBT/PC-blends; $\eta_{BPT} < \eta_{PC}$. (a) Part of a quenched PBT/PC-(10:90) blend; sample was taken during mixing. (b) Phase morphology of a PBT/PC (20:80) blend after mixing. (Radusch in *Frontiers in the Science and Technology of Polymer Recycling*, 1998, pp. 191–211.)

particles. The bigger the droplet and the lower the interfacial tension, the lower the shear stress needed for droplet breakup (Figures 9.9a and 9.9b). In the PP/PS blend with PS as the dispersed phase, it is impossible to obtain a fine dispersion of the PS as the PS has a higher viscosity than the PP matrix (Figure 9.9c). The PS particles show an average size of approximately 10 μm. Obviously, the high interfacial tension of the dispersed phase results in the formation of ball shaped PS particles. Because of the high viscosity of the PS, coalescence cannot progress too far. Furthermore, it can be seen clearly that because of the strong immiscibility of the components, no interface is formed: The PS particles lay detached in the matrix.

Coalescence

Coalescence of dispersed particles results from hydrodynamic and surface effects and is directly opposed to the mixing of the components. Dispersion and coalescence

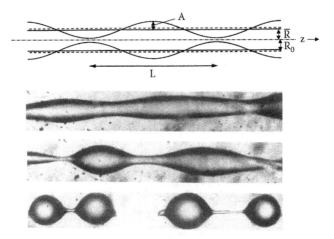

FIGURE 9.8 (Top) Schematic of wave formation in deformed fibres and their breaking off into droplets. (Bottom) Droplet formation in an experiment that consists of PA-6 fibers in a PE matrix at 200°C. (Elmendorp, *A Study on Polymer Blending Microrheology*, 1986.)

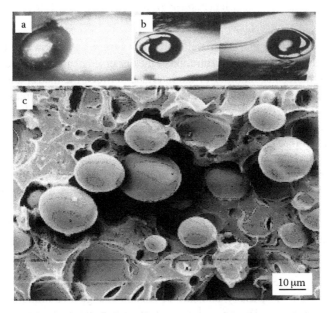

FIGURE 9.9 Examples of melt droplets under low deformation. PS droplets in a PE matrix under shear at 200°C. (a) Below a critical shear rate, (b) just above the critical shear rate, and (c) SEM micrograph of the phase morphology of a PP/PS blend, 30% PS, $\eta_{PS} > \eta_{PP}$ (a) and (b) reproduced with permission from Elmendorp, *A Study on Polymer Blending Microrheology*, 1986; (c) reproduced with permission from Radusch in *Frontiers in the Science and Technology of Polymer Recycling*, 1998, pp. 191–211.

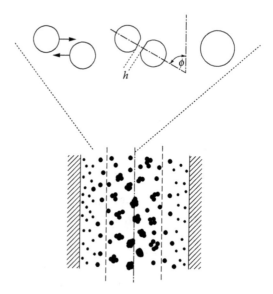

FIGURE 9.10 Coalescence in polymer blends. (Top) Parameters influencing the coalescence. (Bottom) Radial droplet size distribution after extrusion due to superposition of coalescence and dispersion. (Laun and Schuch in *Aufbereiten von Polymerblends*, 1989, pp. 105–128.)

are processes that take place in parallel in these systems. As shown in Eq. (9.10), coalescence depends on the concentration of the dispersed phase, the average particle size, and the molecular mobility of the interface between the matrix and dispersed phase. Therefore, the viscosity is vital for coalescence. Thus, an increase in matrix viscosity will result in a higher dispersion of melt droplets; simultaneously, the viscosity increase will hinder droplet coalescence. The phenomenon of coalescence has to be taken into consideration, especially for long residence times, e.g., in the reservoir of injection molding machines.

Dispersed particles during streaming in capillaries that face the direction of the capillary axis due to normal stress effects, e.g., in extrusion dies, experience an increase in the degree of coalescence of the particles (Figure 9.10b). The results are a concentration gradient over the cross-section, especially for long flow distances, and a strong tendency for coalescence to occur in the middle of the strand (27,28).

Cocontinuous Phases and Phase Inversion
Phase inversion is defined as a phenomenon whereby macroscopic conditions like mixing composition, viscosity, and surface tension change a continuous phase into a discontinuous phase. The phase inversion often happens in a zone containing a cocontinuous phase morphology, i.e., an area where both components of a binary blend interpenetrate and form an interpenetrating network–like phase morphology (dual continuity). Different authors found that phase inversion was dependent on the ratio of the blend composition (component fraction) and the ratio of the viscosities (29,30):

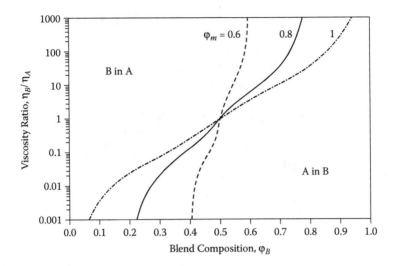

FIGURE 9.11 Phase inversion in binary polymer blends. The theoretical scheme is based on the Krieger-Dougherty equation [Eq. (9.14)]. (Janssen in *Materials Science and Technology, A Comprehensive Treatment, Vol. 18, Processing of Polymers*, 1997, pp. 115–189.)

$$\frac{\varphi_1}{\varphi_2} \cdot \frac{\eta_2}{\eta_1} = X \qquad (9.13)$$

resulting in the following morphology variants:

X > 1: Phase 1 is cocontinuous and phase 2 is dispersed.
X < 1: Phase 2 is cocontinuous and phase 1 is dispersed.
X ≈ 1: Cocontinuous phase morphology (dual continuity) or phase inversion occurs.

Equation (9.13) implies that only the viscosity ratio of the components determines the volume fraction at which phase inversion takes place, i.e., the exchange of the dispersed phase with the matrix.

Figure 9.11 shows schematically the progress of the zone of cocontinuity in a theoretical polymer system. In a zone with $\lambda_\eta \approx 1$, cocontinuous phases can be formed in a more or less broad concentration range of the mixture. Theoretically, cocontinuous phases can be found even for strong concentration differences if certain viscosity ratios are provided. Janssen (31) showed that a drastic change of the line of cocontinuity resulted even for dispersed phases with different packaging densities. The curves in Figure 9.11 were obtained by the application of a modified Krieger-Dougherty equation that was developed for the determination of the viscosity of compounds containing polymer droplets in a polymer matrix (31,32), i.e.,

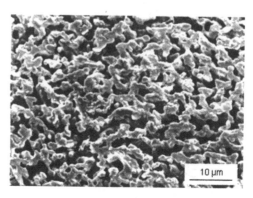

FIGURE 9.12 SEM micrograph (etched) of the cocontinuous phase morphology of a non-cross-linked PP/EPDM 40:60 blend.

$$\eta_{c,eff} / \eta_c = (1 - \varphi / \varphi_m) - 2.5\varphi_m(\lambda_\eta + 0.4) / (\lambda_\eta + 1) \qquad (9.14)$$

where $\eta_{c,eff}$ is the effective matrix viscosity and η_c is the viscosity of the pure continuous matrix phase. Equation (9.14) also contains a maximum packaging fraction φ_m that depends on the viscosity ratio λ_η and the interfacial tension γ_{dc} because of the deformability of the dispersed melt droplets. Equation (9.14) is of special interest for high rubber particle–filled TPE-Vs. Typical cocontinuous phase morphology for a nonreactive PP/EPDM blend is shown in Figure 9.12.

9.2.2.2.2 Formation of Morphology in Reactive Polymer Systems

Many papers describe the phase morphology formation of nonreactive polymer systems consisting of a thermoplastic and a rubber component, where most of the studies are of mainly a phenomenological nature [e.g., refs. (30,33,34)]. Several authors have also studied morphology formation in reactive polymer systems, especially thermoplastic-rubber blend systems with cross-linked rubber phases [e.g., refs. (35–41)].

The formation of the phase morphology in a reactive polymer blend system requires that either the molecular, rheological, and thermodynamic parameters of all the participating components change or only one component changes drastically; the latter is observed, e.g., in the case of selective cross-linking of a rubber phase during the dynamic vulcanization of thermoplastics-rubber blends. Furthermore, chemical reactions at the interface of heterogeneous polymer blend systems can change the phase morphology too, especially because of their influence on the surface tension of the components, thereby affecting the interfacial tension and the associated dispersion effects.

Presuming that only one phase of the blends, i.e., the elastomer phase, has to be selectively cross-linked, the changes in the morphology formation criteria can be characterized as follows (superscript r marks the reactively changed phase):

Viscosity ratio

$$\lambda = \frac{\eta_d(T, \gamma, M_d^{\,\circ})}{\eta_c(T, \gamma, M_c^{\,\circ})} \Rightarrow \lambda^r = \frac{\eta_d^{\,r}(T, \gamma, M_d^{\,r})}{\eta_c(T, \gamma, M_c^{\,\circ})} \tag{9.15}$$

Interfacial tensions and melt elasticities

$$\gamma_{dc} = \gamma_{dc}^{\,\circ} + 1/6R[(\sigma_{11} - \sigma_{22})_d - (\sigma_{11} - \sigma_{22})_c] \Rightarrow$$

$$\gamma_{dc}^{\,r} = \gamma_{dc}^{\,\circ r} + 1/6R[(\sigma_{11} - \sigma_{22})_d^{\,r} - (\sigma_{11} - \sigma_{22})_c] \tag{9.16}$$

Capillary number

$$\kappa = \frac{\eta \cdot \gamma \cdot R}{\gamma_{dc}} \Rightarrow \kappa^r = \frac{\eta^* \cdot \gamma \cdot R^r}{\gamma_{dc}^{\,r}} \tag{9.17}$$

A change in the morphology formation criteria allows for completely different mechanisms of morphology formation. This is the case, e.g., during the dynamic vulcanization of thermoplastic-rubber blends.

9.2.3 MORPHOLOGY FORMATION MODEL FOR DYNAMIC VULCANIZATES

Dynamic vulcanization is characterized by the fact, that a large amount of rubber has to be dispersed into a small amount of a thermoplastic component. The rubber has to be dispersed into the thermoplastic matrix as fine particles in a particle size range of about 1 μm. Such an extraordinary morphology guarantees the thermoplastic processing behavior as well as rubberlike application properties. A superimposed process has to consist of both mixing and dispersion as well as cross-linking of the rubber.

To describe morphology formation in reactive polymer systems, e.g., dynamic vulcanizates or reactive compounded polymer blends, the above discussed rheologically based criteria, i.e., Eqs. (9.6) to (9.12), are no longer applicable. In the systems of interest here, a chemical reaction in a phase or an interface, e.g., a cross-linking reaction in the dispersed rubber, induces a drastic increase in the viscosity of the phase as well as a significant change in the viscosity ratio, thereby affecting also the mixing conditions. However, during the production of thermoplastic elastomers of the TPE-V type (dynamic vulcanizates), the previously discussed morphology formation mechanisms are applicable in the steps prior to the dynamic vulcanization process (cross-linking) since a cocontinuous phase morphology occurs in this period.

The observed viscosity ratio increase during the selective cross-linking reaches a value that is much greater than the critical value for dispersion in heterogeneous polymer blends. The mechanism of the dispersion of the highly viscous rubber component in the low viscosity thermoplastic matrix must take place, therefore, as discussed above, including changes in the materials parameters during cross-linking of the rubber component.

The formation of the typical TPE-V morphology, being characterized by a volume ratio of $\varphi_{rub} > \varphi_{tp}$ and $\eta_{rub} > \eta_{tp}$, plays a significant role only if a cocontinuous initial blend morphology exists before the onset of the selective cross-linking of the rubber phase (42,43).

In order to produce the necessary deformations in a highly viscous but cocontinuous rubber phase, elongational stresses within an elongational flow must act on the continuous phase. These deformations are essential on the one hand for the refinement of the cocontinuous strands and on the other hand for the initiation of breakup of these strands into dispersed particles during the cross-linking process in this highly deformed state, as is shown schematically in Figure 9.13. This riping off of the highly deformed cocontinuous rubber strands happens after the mix exceeds a critical stress, which is reached after the quick viscosity increase initiated by the addition of the cross-linking agent.

The degree of dispersion D_r can be thought of as the ratio of the particle areas before, A_0, and after, $A(t)$, the dispersion process, i.e.,

$$D_r = \frac{A(t)}{A_0} \tag{9.18}$$

The correlation between the viscosity and the degree of dispersion with regards to time is shown schematically in Figure 9.14. In zone I, a cocontinuous phase morphology exists that is characterized by a constant degree of dispersion and unchanged viscosity. In this zone, the cocontinuous phases will be deformed by shear and elongational stresses. Here, the strands of the components become thinner, but the blend morphology (cocontinuous) remains stable. The viscosity of the system is that of a typical blend. After the addition of the cross-linking agent at a definite time t^*, the selective cross-linking reaction in the rubber phase is initiated (zone II). The viscosity of the rubber, and thus the viscosity of the whole system, is strongly increased. Simultaneous with the jumplike increase in the shear viscosity η and the elongational viscosity η_e, there is an increase in the shear and elongational stresses $\tau(t)$ and $\sigma(t)$ acting in the system, resulting in:

$$\tau(t) = \eta^{react}(t) \cdot \gamma \tag{9.19a}$$

$$\sigma(t) = \eta_e^{react}(t) \cdot \varepsilon \tag{9.19b}$$

The greater the number of cross-links in the rubber phase, the greater the load to the rubber phase. If a critical stress is reached at t_c, the rubber phase will break up into small particles and the stress is jumplike reduced again, resulting in a finely dispersed particle-matrix morphology. This is the moment where the cocontinuous

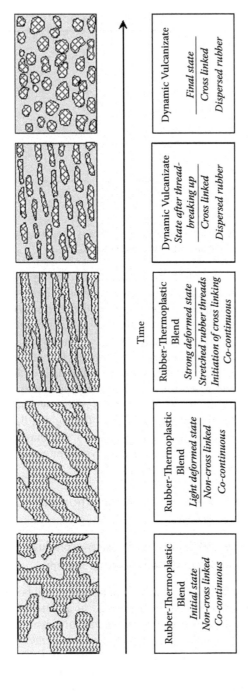

FIGURE 9.13 Schematic of morphology transformation during the dynamic vulcanization of polymer blends.

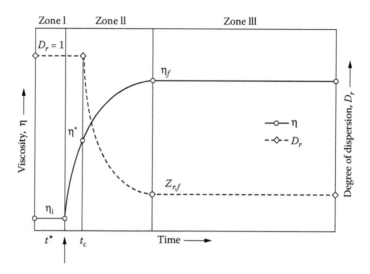

FIGURE 9.14 Schematic of the course of the viscosity η and degree of dispersion D as functions of time t for a dynamic vulcanization process.

morphology is transformed into an island-matrix morphology type. That means that the immediate process of morphology transformation is rather quick.

In zone III, the cross-linked rubber particles now have a very high viscosity, and further dispersion of these particles by the stresses transferred by the low viscous thermoplastic matrix will be impossible. Only a distribution of the particles that improves the macroscopic homogeneity will happen. It should be pointed out once more that the dispersion process takes place instantaneously with the chemical cross-linking reaction and both processes are influencing each other.

The quickly increasing cross-linking of the rubber will lead also to a higher surface tension of the rubber phase. Both cross-linking and the increasing surface tension will lessen the driving force for coalescence, and thus the particle character of the cross-linked major rubber phase is preserved even upon further mixing after the conclusion of the cross-linking process.

9.2.4 VISUALIZATION OF MORPHOLOGY FORMATION DURING DYNAMIC VULCANIZATION

9.2.4.1 Visualization and Control of Phase Morphology Formation in Discontinuous Internal Mixers

The typical stages of the dynamic vulcanization process will be most obvious in the case of a discontinuous procedure realized, e.g., in an internal mixer. Figure 9.15 shows the torque-time dependence of the production of a dynamic vulcanizate based on polypropylene as the thermoplastic component and ethylene-propylene-rubber (EPR) as the elastomer component. The blend composition was 40 wt% PP and 60 wt% EPR, and the viscosity ratio, η_{EPR}/η_{PP}, was about 1.5. A laboratory internal mixer Brabender Plasticoder PL 2000 was used (42,43).

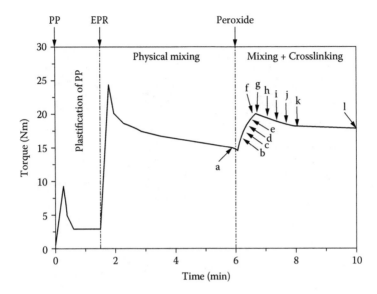

FIGURE 9.15 Torque-time characteristics of a dynamic vulcanization process in an internal mixer. PP/EPR 40:60, peroxidic cross-linked. Points a to l: times of sampling.

After the plastification of the thermoplastic PP (first peak in the torque-time curve) and the addition of the uncross-linked EPR rubber (second peak), a polymer blend is produced that is characterized by a slightly decreasing torque (representing the blend viscosity). The slight decrease in the viscosity is caused by thermomechanically induced molecular destruction processes. In this stage, a cocontinuous phase morphology is needed to ensure an optimal initial phase morphology state for the morphology transformation process. After a definite time, the cross-linking agent is added. Here a peroxidic cross-linking agent (Luperox L 101 from Elf Atochem) and styrene for the inhibition of PP chain scission was used. After the addition of the cross-linking system, a strong increase in the torque ensues that is caused by the rapid cross-linking process (resulting as discussed above in a strong increase in the viscosity). As explained previously, this is followed by the breakup of the highly deformed rubber strands in the cocontinuous phase. Cross-linking and dispersion of the rubber component occur simultaneously and very quickly. The duration of this process depends on the concentration of the cross-linking agent. After finishing the cross-linking process, which is expressed by the third peak, only a distribution of the rubber particles, but not a dispersion, is registered because the cross-linked, very viscous rubber particles "swim" in the low viscosity thermoplastic matrix.

To visualize this characteristic morphology transformation, specimens were taken out of the internal mixer after definite mixing times. These times are marked with small letters, a to l, in Figure 9.15. The microstructure of the samples characterized by scanning electron microscopy (SEM) is shown in Figure 9.16 where the label of the SEM micrographs (a to l) correspond to the sampling points in Figure 9.15.

The non-cross-linked sample taken before the reaction, i.e., the blend, shows a phase morphology that consists of a continuous PP and a continuous EPR phase

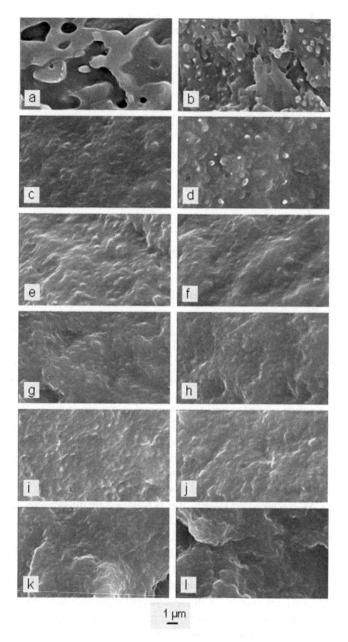

1 μm

FIGURE 9.16 SEM micrographs after edging of dynamic vulcanization monitoring. Samples were taken during melt-mixing in the batch mixer. Labeled according to Figure 9.15.

(cocontinuous) (Figure 9.16a). The cross-linking reaction was initiated with the injection of the peroxide/styrene mixtures into the polymer melt after 6 min of blend mixing. The process of cross-linking can be observed clearly by the increasing torque in the batch mixer (see Figure 9.15). Sample b, which was taken 20 sec after the addition of the peroxide/styrene mixture, still shows a cocontinuous phase morphology, but both phases became finer compared to the starting morphology (Figures 9.16a vs. 9.16b). At point c, 30 sec after the addition of the curing agent, a fine particle-matrix morphology is detected (Figure 9.16c). A comparison between the SEM micrographs of samples d to l shows that after the formation of the particle-matrix structure (after point c), there was no further change in the blend morphology. It seems that the critical stress was reached between points b and c, and the very fast morphology transformation took place here. With the increase in the viscosity due to the cross-linking reaction, the continuous rubber phase was stretched more and more under shear and elongational stresses. Note that the elongational flow is more efficient with respect to phase deformation. When the critical stress is reached, the continuous rubber phase cannot deform efficiently and breaks up into small particles. In this way, the rubber phase is dispersed until a mean particle size of about 1 μm is reached. At the same time, the PP phase keeps a thermoplastic flow behavior because no cross-linking reaction took place in the PP phase. Thus, the PP remains as a continuous phase and forms the matrix in the final blend. The morphology formation process is completed about 30 sec after the initiation of the cross-linking reaction. The torque increases until the cross-linking reaction is finished and reaches a maximum level at about 50 sec after peroxide addition (Figure 9.16g).

The experiments make clear that the conditions for the formation of a fine particle-matrix morphology are not only a suitable volume fraction and viscosity ratio of the components but also, and more decisively, the initiated cross-linking reaction of the rubber phase. The volume fraction and viscosity ratio are important parameters for the formation of a cocontinuous phase morphology before the reaction starts. This cocontinuous structure is an essential condition for the successful transformation into the fine particle-matrix morphology. Only the initial cocontinuous phase morphology allows for the effective loading of shear and elongational stresses on both the thermoplastic and rubber phase, and this is the necessary condition for the breakup of the rubber phase during the reactive melt-mixing or dynamic vulcanization. After the cross-linking reaction is completed, only a distribution of the cured rubber particles in the matrix takes place (Figures 9.16d to 9.16l), but the particle size cannot change anymore because of the rheological properties of the components [see Eq. (9.6)].

9.2.4.2 Visualization and Control of Phase Morphology Formation in Continuous Twin-Screw Kneaders

Using the same polymer blend composition and cross-linking system as in the batch mixer (see Section 9.2.4.1.), the dynamic vulcanization process was attempted in a twin-screw extruder. A twin-screw kneader ZE 25 Ultra-Glide (Berstorff) with a *L/D* ratio of 30 was used (44). The screw geometry is shown in Figure 9.17. The formation of a cocontinuous blend morphology as well as the control of the initiation of the

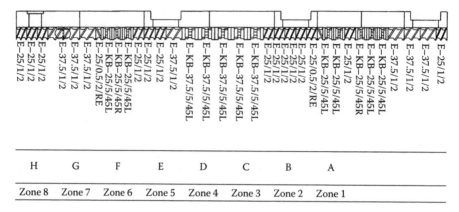

FIGURE 9.17 Screw geometry and positions of sampling (A to H) in the twin-screw kneader ZE 25.

cross-linking reaction was followed by a temperature profile along the extruder barrel. Accordingly, in zones 1 and 2 of the extruder, only melting and mechanical mixing of the polymer components took place. The cross-linking reaction was started in zone 3 where a cocontinuous blend morphology was already formed. Intensive mixing was realized by putting kneading blocks in zones 4 and 6 followed by a left-handed element. Using these kneading blocks, an intensive elongational flow could be achieved, which is sufficient to break up the highly deformed rubber phase.

Figure 9.18 shows the differences in the phase morphology between the mechanical blend (left column) and the dynamic vulcanizate (right column), revealing the serious changes in the course of the processes. In accordance with the results of the internal mixer study, the mechanical blend shows a typical cocontinuous phase morphology over the whole screw range (Figure 9.18 left). It is interesting to see that the general character of the phase morphology in the nonreactive blend system remains unchanged. Also, after passing the screw zones with relatively high mixing intensity, there is no change in the cocontinuous morphology type. Because of the existing viscosity ratio in this blend system and the present blend composition, the cocontinuous morphology is the dominant feature. The non-cross-linked elastomer phase can be deformed further by the application of shear stresses, but the cocontinuous morphology persists (Figure 9.18 left).

A totally different situation occurs if the compounding is realized as dynamic vulcanization, i.e., under addition of a cross-linking agent that selectively acts on the rubber phase (Figure 9.18 right). The cross-linking reaction was initiated at suitable temperatures in zone 3 (Figure 9.17), and a successful dynamic vulcanization associated with a transformation of the phase morphology took place. The initial phase morphology characterized by cocontinuous phases with large PP and EPR domains was changed into a particle-matrix morphology. While the cocontinuous phase morphology still exists in zones 1 to 4 (Figure 9.17), transformation into a particle-matrix morphology type begins in zone 5 immediately after the initiation of the cross-linking reaction. As the micrographs from the specimens taken at $L/D=16$ and 22 show (Figure 9.18), the transformation of morphology runs over a certain

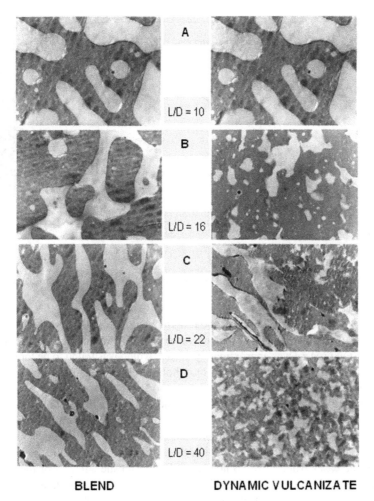

BLEND   DYNAMIC VULCANIZATE

FIGURE 9.18 TEM micrographs (OsO$_4$ stained) of blending (left) and dynamic vulcanization (right) monitoring in a continuous twin-screw kneader (Figure 9.17). Samples were taken at the marked positions after pulling out the screw. The pictures are labeled according to Figure 9.17. Dynamic vulcanization was performed with peroxidic cross-linking agent; PP/EPR 40:60.

length of the extruder. Because of the selective cross-linking reaction, the viscosity of the elastomer phase and therefore the viscosity of the whole material system rises, which is the precondition for the realization of the specific breakup mechanism as described in Section 9.2.2.

The increasing interfacial tensions as a result of the cross-linking and the high elasticity of the particles produce a relaxation of the deformed structures; in an ideal case, spherical particles are formed. The final result is dispersed elastomer particles with higher volume content in a low viscous thermoplastic matrix. Because of the high viscosity ratio, see Eqs. (9.7) and (9.15), the cross-linked elastomer particles can be only distributed but not dispersed anymore in the further course of mixing.

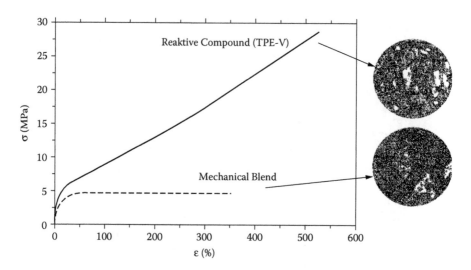

FIGURE 9.19 Stress-strain behavior of a nonreactive and a dynamic vulcanized PP-EPR blend: PP/EPR 40:60.

A mean EPR particle size of about 1μm could be generated under suitable processing conditions (44). The residence time measured for the continuous dynamic vulcanization process was approximately 30 sec, which was found to be long enough for the cross-linking process and the morphology transformation in the individual blend system.

Because of the differences in the morphology, the blends also show significant differences in their mechanical properties. As can be seen from a stress-strain behavior plot (Figure 9.19), an essential improvement in both tensile strength and elongation at break is achieved by the dynamic vulcanization process. While the elongation at break increases from 350% to nearly 600%, the tensile strength of the dynamic vulcanizate reached a value of 24 MPa, which is six times higher than the value of the nonreactive compounded blend. Furthermore, dynamically cross-linked blends based on PP-EPR mixtures are highly transparent and possess excellent processability, features that are desirable in the production of films or thin wall products (45).

9.2.5 MORPHOLOGY REFINEMENT OF INCOMPATIBLE DYNAMIC VULCANIZATES

In dynamic vulcanizates that are based on EPDM rubber and PP, a rather good affinity exists between the components, resulting in products with excellent properties. However, in blend systems with very incompatible components, only poor mechanical properties result from dynamic vulcanization. Thus, the use of polar rubber and nonpolar thermoplastics, e.g., nitrite-butadiene-rubber (NBR) and PP, which is potentially a very interesting polymer system for the production of TPE-V by dynamic vulcanization with good oil resistance, results in only a coarse phase morphology because of the thermodynamic incompatibility of the components (46). The high coarseness of the phase morphology and the low interaction at the interface produces a product with undesirable mechanical properties. In such cases, compatibilizing substances can be used to intensify the phase interactions (47). Block or graft copolymers, functionalized polymers, or low molecular reactive substances can be used for the compatibilization. In addition, compatibilizing copolymers can be formed *in situ* during the melt-mixing process, especially by functionalized polymers or reactive agents.

9.2.5.1 Dynamic Vulcanization and Reactive Compatibilization

Basing their work on the general ideas of Coran and Patel (48,49) of using reactive polymers as compatibilizers in PP/NBR blends, Corley and Radusch (46,50) used maleic anhydride functionalized PP and other polymers and an amine terminated NBR for compatibilization. Two different ways were proposed for achieving component compatibilization. In the first, a copolymer was produced from the functionalized components and added at different concentrations to the incompatible polymers during dynamic vulcanization. In the second, the functionalized polymers were added directly into the mixture during dynamic vulcanization. In the *in situ* reaction during the mixing process, a copolymer should be generated that can act directly as a compatibilizer.

The methods are a reflection of the importance of surface tension during melt-mixing of polymers. It is assumed that the mean thickness of the strands of the necessary cocontinuous phase morphology will be reduced in the presence of a suitable surfactant, and therefore the particles during morphology transformation will become smaller than they would be in the absence of the compatibilizer.

Several maleic anhydride grafted polymers as well as amine terminated NBR were used (Table 9.1) and tested for their compatibilizing effect and their ability to improve the mechanical properties of the products. In each case, a partial substitution of the basis polymer components by the reactive polymers at dynamic vulcanization was realized.

Figure 9.20 shows the torque-time characteristics of a dynamic vulcanization process of PP/NBR 30:70 with and without addition of the reactive components

TABLE 9.1
Compatibilizers for the TPE-V System PP/NBR

Polymers	Compatibilizers
Polypropylene PP	Maleic anhydride functionalized PP: MDEX 94-6-8, EXXELOR® PO 2011, BYNEL® CXA 50E561
	Acrylic acid functionalized PP: SCONA® TPPP 2110 FA
	Maleic anhydride functionalized EPM: EXXELOR® VA 1803
	Maleic anhydride functionalized SEBS: KRATON® FG 1901 X
Acrylonitrile butadiene rubber NBR	Amine terminated NBR: HYCAR® ATNB

Source: Corley and Radusch, *J. Macromol. Sci. Physics B*, 37(2), 265–273, 1998.

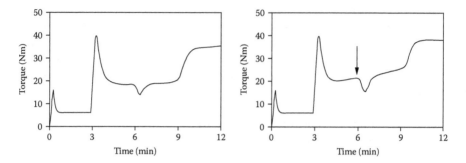

FIGURE 9.20 Torque-time dependence of dynamic vulcanization of PP/NBR 30:70. Non-compatibilized (left) and compatibilized (right) with MDEX and ANTB.

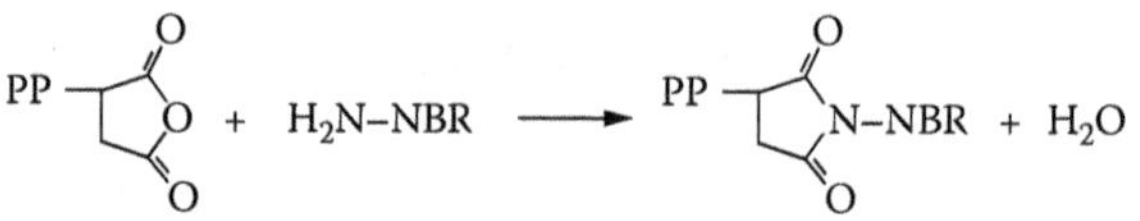

FIGURE 9.21 Chemical reaction of maleic anhydride grafted PP with amine terminated NBR.

MDEX (maleic anhydride functionalized PP) and ATNB (amine terminated NBR) for *in situ* compatibilization of the mixture. A comparison of the mixing characteristics of the dynamic vulcanization process of the incompatible PP/NBR mixture and a PP/NBR mixture compatibilized by MDEX and ATNB shows a typical reaction peak only in the case of the compatibilized mixture caused by the *in situ* formation of copolymer [see arrow in Figure 9.20 (right)]. There is also a rise in the viscosity during melt-mixing before the start of the dynamic vulcanization process because of the rising molecular weight of the *in situ* formed copolymers.

Fourier transform infrared (FTIR) spectroscopy investigations have shown that coupling reactions between, e.g., maleic anhydride grafted polypropylene and amine terminated acrylonitrile butadiene rubber occur during the melt-mixing and dynamic vulcanization process in the presence of the reactive polymers. The reaction can lead to copolymers according to Figure 9.21.

This *in situ* generated copolymer is acting as a compatibilizer as measured by the reduction in the interfacial tension and by the dispersation processes that are running much more effectively than in noncompatibilized blends. As a result, a finer and more equalized phase morphology is found in the compatibilized dynamic vulcanizates. Furthermore, a marked concentration effect exists. As shown in Figure 9.22, the mean rubber particle diameter clearly decreases with increasing amount of the functionalized PP.

In addition to *in situ* compatibilization, compatibilization by the addition of compatibilizing copolymers was also successful (46,50). Special copolymers produced in a prestep before dynamic vulcanization were inserted during dynamic vulcanization in stepped concentrations. Generally, the same compatibilizing effect as expressed by a significant reduction in the rubber particle size could be observed. Figure 9.23 shows an example of dynamic vulcanization of PP/NBR in

FIGURE 9.22 TEM micrographs (OsO₄ stained) of dynamic vulcanizates of PP/NBR 30:70. (a) Noncompatibilized, (b) compatibilized with 1.5% MDEX + 3.5% ATNB, and (c) compatibilized with 6% MDEX + 3.5% ATNB.

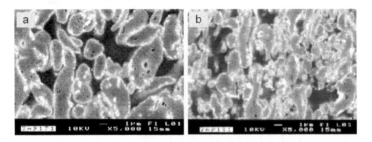

FIGURE 9.23 SEM micrographs (surface staining with OsO₄) of dynamic vulcanizates of PP/NBR 30:70. (a) Noncompatibilized and (b) compatibilized with 5% copolymer.

the presence of a copolymer compatibilizer. The rubber particle size reduction can be seen clearly.

9.2.5.2 Morphology-Property Relationship of Compatibilized Dynamic Vulcanizates

Morphology refinement and coupling reactions at the interface lead to improved phase adhesion. As a result of these compatibilization effects, the mechanical stress-strain properties are markedly improved in comparison to unmodified PP/NBR dynamic vulcanizate. Figure 9.24 (left) shows that the mean particle size is dependent on PP/NBR graft copolymer content, and Figure 9.24 (right) shows the correlation between the mean particle size and the stress-strain properties of these compatibilized dynamic vulcanizates. It is clear that compatibilizers, whether added as copolymers or generated *in situ* by adding reactive polymers, influence the phase morphology and the morphology based mechanical properties.

The investigation of the effect of the type of reactive polymer on the compatibilizing activity of these products showed that the maleic anhydride functionalized polypropylene is a very effective compatibilizer (Figure 9.25). Certainly, there is a strong dependence of the concentration of the maleic anhydride containing component. Although lower amounts of the maleic anhydride functionalized polypropylene (5 and 10% MDEX) improved the properties, only higher concentrations (20%) delivered essentially higher stress-strain properties. In contrast, the acrylic acid–modified polypropylene (Scona® TPPP) and also the maleic anhydride–functionalized EPR (Exxelor® VA) as well as the maleic anhydride–grafted polypropylene

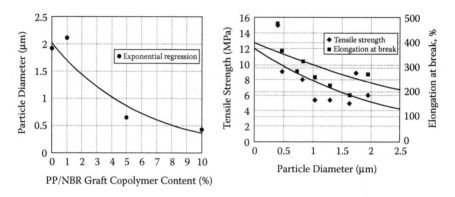

FIGURE 9.24 (Left) Mean particle size as a function of PP/NBR graft copolymer content and (right) tensile strength and elongation at break as functions of mean particle size of compatibilized dynamic vulcanizates.

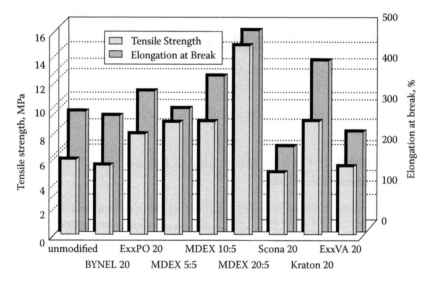

FIGURE 9.25 Comparison of the effectiveness of different compatibilizers on the *in situ* compatibilization of PP/NBR by ATNB.

with only a low degree of grafted maleic anhydride (Bynel® CXA) did not show a sufficient compatibilizing effect.

Another aspect of the compatibilization and the dynamic vulcanization of a rubbery component such as EPDM and a thermoplastic polymer such as nylon-6 is the very high strain recovery exhibited by the obtained blend with more than 60 wt% dispersed rubber phase. In this system, it has been demonstrated using atomic force microscopy and transmission electron microscopy that the matrix exhibits an inhomogeneous plastic deformation in those zones where the nylon matrix between the rubber particles is the thinnest. Even at high strains, the thick ligaments of the nylon matrix remain almost undeformed and act as adhesion points holding the rubber particles together. When the external force is removed, the elastic force of the stretched, dispersed rubber phase pulls back the plastically deformed nylon parts by either buckling or bending (51–53).

9.2.6 DILUTION OF DYNAMIC VULCANIZATES

One of the main advantages of dynamic vulcanizates as thermoplastic elastomers is thermoplastic processability and connected to this is good recycleability. In this context, it is easy to mix, e.g., polypropylene based dynamic vulcanizates with neat polypropylene because PP forms the matrix. In practice, this can happen during the recycling of TPE-V, or the good mixing behavior can be used for the generation of modified dynamic vulcanizates by a dilution procedure.

Dilution of dynamic vulcanizates was investigated to determine the effect on the mechanical stress-strain behavior (54,55). Normally, because of the factors controlling morphology formation, see Eqs. (9.6) to (9.12), the type of resulting morphology in a heterogeneous blend system after melt-mixing is determined by the composition of the mixture, i.e., the volume fractions of the components, and the rheological behavior of the components. Accordingly, under normal conditions, the major component forms the matrix and the minor component forms the dispersed phase. Realizing a concentration row in a binary system phase inversion occurs at a definite composition. This is not the case if dilution of dynamic vulcanizates is performed. An ideal dynamic vulcanizate is characterized by a major cross-linked rubber phase dispersed as fine particles in a minor continuous thermoplastic matrix. Because of the cross-linked character of the dispersed phase, it is relatively stable against deformation in the melt and in the solid state, too. Dilution of a dynamic vulcanizate by melt-mixing with a matrix-like polymer component does not lead to a change of the morphology type. The mean particle size remains constant; only the mean particle distance is increased by the addition of thermoplastics to the dynamic vulcanizate. Figure 9.26 shows a dynamic vulcanizate made from PP and EPR with 70% EPR and a diluted product with a remaining rubber content of 18%. PP was used as the dilution polymer. From Figure 9.26, it is clear that the particle size has remained constant, i.e., the cross-linked rubber particles are not influenced very much by the dilution procedure.

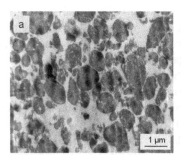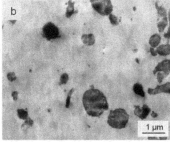

FIGURE 9.26 (a) Phase morphology of a dynamic vulcanizate PP/EPR 30:70 and (b) a diluted product of this with 18 wt% rubber.

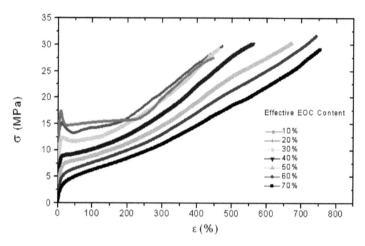

FIGURE 9.27 Stress-strain behavior of a dilution row based on the dynamic vulcanizate PP/EOC 30:70.

The dilution procedure enables a definite control of the blend morphology, and the properties of the blend system are changed incrementally. As shown in Figure 9.27, the character of the stress-strain curves is changed with the composition of the diluted blend. An increasing content of polypropylene changes the typical elastomeric character of the dynamic vulcanizate more and more to that of the semicrystalline thermoplastic. Whereas a first yield point appears at a rubber content below 50%, a pronounced yield point maximum is visible only at a rubber content below 30%.

It is notable that while the tensile stress values of all the compositions are similar, only the elongation at break values are reduced with decreasing rubber content. This indicates that the thermoplastic semicrystalline matrix of the PP component determines the strength behavior, but the rubber content is responsible for the elongational behavior of the blend.

9.3 SUMMARY

The control of the morphology formation processes during dynamic vulcanization guarantees an optimal phase morphology and property level of this specific class of thermoplastic elastomers. Knowledge about the microrheological and thermodynamic basics makes it possible to understand the complicated mechanisms that occur during the superimposed processes of dispersion and selective cross-linking during dynamic vulcanization, and to regulate the technological parameters like mixing temperature, energy input, and geometric parameters of the mixing apparatus as well. For dynamic vulcanization, the general factors that influence morphology formation in polymer blends, such as the ratio of rheological parameters, stress ratio, and interfacial tension, as well as the compatibility of the components, need to be kept in mind as well as the reactive processes that occur more or less exclusively in the rubber phase and that influence the morphology formation processes drastically.

Elongational flow is especially needed to sufficiently deform the high viscous rubber phase that results in a morphology transformation from a cocontinuous morphology of the non-cross-linked rubber thermoplastic blend into the island-matrix morphology of the dynamic vulcanizate. Compatibilization and/or dilution are specific methods for influencing the phase morphology of dynamic vulcanizates and for modifying the applicational properties of these TPE-V.

REFERENCES

1. Coran, A.Y. and Patel, R., Rubber-thermoplastic compositions. Part I. EPDM-polypropylene thermoplastic vulcanizates, *Rubber Chem. Tech.*, 53, 141–150, 1980.
2. Scott, R.L., The thermodynamics of high polymer solutions. V. Phase equilibria in the ternary system: polymer 1-polymer 2-solvent, *J. Chem. Phys.*, 17, 279–284, 1949.
3. Tompa, H., Phase relationships in polymer solutions, *Trans. Faraday Soc.*, 45, 1142–1152, 1949.
4. Flory, P.J., Orwoll, R.A., and Vrij, A., Statistical thermodynamics of chain molecule liquids. I. An equation of state for normal paraffin hydrocarbons, *J. Am. Chem. Soc.*, 86, 3507–3514, 1964.
5. Dee, J.T. and Walsh, D.J., Equation of state for polymer liquids, *Macromolecules*, 21, 811–815, 1988.
6. Hildebrand, J.H. and Scott, R.L., *The Solubility of Electrolytes*, Reinhold, New York, 1950.
7. Painter, P.C., Park, Y., and Coleman, M.M., Thermodynamics of hydrogen bonding in polymer blends. 1. Application of association models, *Macromolecules, 22*, 570–579, 1989.
8. Paul, D.R. and Bucknall, C.B., Introduction, in *Polymer Blends, Vol. 1, Formulation*, John Wiley & Sons, Inc., New York, 1999, pp. 1–14.
9. Elmendorp, J.J., *A Study on Polymer Blending Microrheology*, Ph.D. thesis, Technical University, Delft, The Netherlands, 1986.
10. Van Oene, H.J., Modes of dispersion of viscoelastic fluids in flow, *J. Colloid Interf. Sci.*, 40, 448–467, 1972.
11. Lyngaae-Jørgensen, J., Domain stability during capillary flow of well dispersed two phase polymer blends, *Org. Coat. Plast. Chem.*, 15, 174–179, 1981.
12. Utracki, L.A. and Shi, Z.H., Development of polymer blend morphology during compounding in a twin-screw extruder, *Polym. Eng. Sci.*, 32(24), 1824–1845, 1992.
13. Taylor, G.I., The viscosity of a fluid containing small drops of another fluid, *Proc. R. Soc. London*, A138, 41–48,1932.
14. Taylor, G.I., The deformation of emulsion in definable fields of flow, *Proc. R. Soc. London*, A146, 501–523, 1934.
15. Cox, R.G., The deformation of a drop in a general time-dependent fluid flow, *J. Fluid Mech.*, 37, 601–623, 1969.
16. Grace, H.P., Dispersion phenomena in high viscosity immiscible fluid systems and application of static mixers as dispersion devices in such systems, *Chem. Eng. Commun.*, 14, 225–277, 1982.
17. Rumscheid, F.D. and Mason, S.G., Particle motions in sheared suspensions. XII. Deformation and burst of fluid drops in shear and hyperbolic flow, *J. Colloid. Sci.*, 16, 238–261, 1961.

18. Tomotika, S., On the instability of a cylindtical thread of a viscous liquid surrounded by another viscous fluid, *Proc. R. Soc. London*, A150, 322–337, 1935.

19. Stone, H.A., Bentley, B.J., and Leal, L.G.J., An experimental study of transient effects in the break-up of viscous drops, *J. of Fluid Mech.*, 173, 131–158, 1986.

20. Wu, S., Formation of dispersed phase in incompatible polymer blends: interfacial and rheological effects, *Polym. Eng. Sci.*, 27, 335–343, 1987.

21. Han, C.D. and Funatsu, K., An experimental study of droplet deformation and breakup in pressure-driven flows through converging and uniform channels, *J. Rheol.*, 22, 113–133, 1978.

22. Flumerfelt, R.W., Drop breakup in simple shear fields of viscoelastic fluids, *Ind. Eng. Chem. Fundam.*, 11(3), 312–318, 1972.

23. Gauthier, F., Goldsmith, H.L., and Mason, S.G., Particle motions in non-Newtonian media. I. Couette flow, *Rheol. Acta*, 10, 344–364, 1971.

24. De Bruijn, R.A., *Deformation and Breakup of Drops in Simple Shear Flows*, Ph.D. thesis, University of Technology, Eindhoven, The Netherlands, 1989.

25. Favis, B.D. and Califoux, J.P., The effect of viscosity ratio on the morphology of propylene/polycarbonate blends during processing, *Polym. Engin. Sci.*, 27, 1591–1600, 1987.

26. Radusch, H.-J., Morphology development during processing of recycled polymers, in *Frontiers in the Science and Technology of Polymer Recycling*, Akovali, G., Ed., Kluwer Acad. Publ., Dortrecht, 1998, pp. 191–211.

27. Radusch, H.-J., Hendrich, R., Michler, G., et al. Morphologiebildungs und Dispergiermechanismen bei der Herstellung und Verarbeitung von Polymerkombinationen aus Acrylnitril-Butadien-Styrol bzw. Styrol-Acrylinitril-Copolymerisaten mit thermoplastischem Polyurethan, *Die Angew. Makromol. Chem.*, 194, 159–178, 1992.

28. Laun, H.M. and Schuch, H., Rheologische aspekte der blendaufbereitung, in *Aufbereiten von Polymerblends*, VDI Verlag, Düsseldorf, Germany, 1989, pp. 105–128.

29. Paul, D.R. and Barlow, J.W., Polymer blends (or alloys), *J. Macromol. Sci.-Rev. Macromol. Chem.*, C18, 109–168, 1980.

30. Jordhamo, G.M., Mason, J.A., and Sperling, L.H., Phase continuity and inversion in polymer blends and simultaneous interpenetrating networks, *Polym. Eng. Sci.*, 26(8), 517–524, 1986.

31. Janssen, J.M.H., Emulsions: the dynamics of liquid-liquid mixing, in *Materials Science and Technology, A Comprehensive Treatment, Vol. 18, Processing of Polymers*, Cahn, R.W., Haasen, P., Kramer, E.J., and Meijer, H.E.H., Eds., Wiley, New York, 1997, pp. 115–189.

32. Krieger, I.M. and Dougherty, T.J., A mechanism for non-Newtonian flow in suspensions of rigid spheres, *Trans. Soc. Rheol.*, 3, 137–152, 1959.

33. Thomas, S and Groeninckx, G., Nylon 6/ethylene propylene rubber (EPM) blends: phase morphology development during processing and comparison with literature data, *J. Appl. Polym. Sci.*, 71, 1405–1429, 1999.

34. Ito, E.N., Hage, E., Jr., Mantovani, G.L., et al., Development of phase morphology for extruded and moulded PBT/ABS blends, *Proceedings of the 19th Annual Meeting of the Polymer Processing Society PPS-19*, Melbourne, 2003, CD, S2, 259, 1–5.

35. Deyrail, Y. and Cassagnau, P., Morphology development under shear flow of a cross-linked dispersed phase, in *Proceedings of the 19th Annual Meeting of the Polymer Processing Society PPS-19*, Melbourne, 2003, CD, S2, 162, 1–4.

36. Coran, A.Y. and Patel, R.P., Thermoplastic elastomers by blending and dynamic vulcanization, in *Polypropylene. Structure, Blends and Composites, Vol. 2, Copolymers and Blends*, Karger-Kocsis, J., Chapman & Hall, London, 1995, pp. 162–201.

37. Fortelný, I., Kovař, J., Sikora, A., et al., The structure of blends of polyethylene and polypropylene with EPDM elastomers, *Die Angew. Makromol. Chem.*, 132, 111–122, 1985.

38. Sundararaj, U. and Macosko, C.W., Drop breakup and coalescence in polymer blends: the effects of concentration and compatibilization, *Macromolecules*, 28, 2647–2657, 1995.

39. Scott, C.E. and Macosko, C.W., Processing and morphology of polystyrene/ethylene-propylene rubber reactive and nonreactive blends, *Polym. Eng. Sci.*, 35, 1938–1948, 1995.

40. Kumar, C.R., George, K.E., and Thomas, S, Morphology and mechanical properties of thermoplastic elastomers from nylon-nitrile rubber blends, *J. Appl. Pol. Sci.*, 61, 2383–2396, 1996.

41. Fritz, H.G., Bölz, U., and Cai, D., Innovative TPV two-phase polymers: formulation, morphology formation, property profiles and processing characteristics, *Polym. Eng. Sci.*, 39, 1087–1099, 1999.

42. Radusch, H.-J. and Pham, T., Morphologiebildung in dynamisch vulkanisierten PP/EPDM-Blends, *Kautschuk. Gummi. Kunststoffe*, 49(4), 249–257, 1996.

43. Radusch, H.-J., Lämmer, E., and Pham, T., Morphology formation in dynamic vulcanized PP-elastomer blends, in *Proceedings of the European Meeting of the Polymer Processing Society PPS-1995*, Stuttgart, 1995, pp. 3–6.

44. Pham, T., Radusch, H.-J., and Winkelmann, T., Thermoplastic elastomers by reactive compounding, in *Proceedings of the International Conference `Polymers in the Third Millenium'*, Montpellier, France, Sept. 2001, p. 123.

45. Rätzsch, M., Radusch, H.-J., et al., *Thermoplastische Elastomere guter Einfärbbarkeit und hoher Festigkeit und Elastizität sowie daraus hergestellte hochschlagzähe Polymerblends*, DE-Pat. 197 48 976.1 (LP 9729), 06.11.1997.

46. Corley, B., *Reaktive Kompatibilisierung Dynamischer Vulkanisate*, Ph.D. thesis, University Halle-Wittenberg, Halle, Germany, 1999.

47. Bhowmick, A.K., Chiba, T., and Inoue, T., Reactive processing of rubber-plastics blends: role of chemical compatibilizer, *J. Appl. Polym. Sci.*, 50, 2055–2064, 1993.

48. Coran, A.Y. and Patel, R.P., Rubber-thermoplastic-compositions. VIII. Nitrile rubber-polyolefin blends with technological compatibilization, *Rubber Chem. Tech.*, 56, 1045–1060, 1983.

49. Coran, A.Y. and Patel, R.P., Thermoplastic elastomers based on dynamically vulcanized elastomer-thermoplastic blends, in *Thermoplastic Elastomers,* 2nd Ed., Holden, G. L., Legge, N.R., Quirk, R., et al., Eds., Hanser Publ., Munich, 1996, p. 154–190.

50. Corley, B. and Radusch, H.-J., Intensification of Interfacial Interaction in dynamic vulcanization, *J. Macromol. Sci. Physics B*, 37(2), 265–273, 1998.

51. Oderkerk, J. and Groeninckx, G., Morphology development by reactive compatibilisation and dynamic vulcanization of nylon 6/EPDM blends with high rubber fraction, *Polymer*, 43, 2219–2228, 2002.

52. Oderkerk, J., de Schaetzen G., Goderis, B., et al., Micromechanical deformation and recovery processes of nylon 6/rubber thermoplastic vulcanizates as studied by atomic force microscopy and transmission electron microscopy, *Macromolecules*, 35, 6623–6629, 2002.

53. Oderkerk, J., Groeninckx, G., and Soliman, M., Investigation of the deformation and recovery behavior of nylon-6/rubber thermoplastic vulcanizates on the molecular level by infrared-strain recovery measurements, Macromolecules, 35, 3946–3954, 2002.
54. Radusch, H.-J. and Scharnowski, D., Recycling of dynamic vulcanizates, in *1st Intern. Conf. on Polymer Modification, Destruction and Stabilization MODEST*, Palermo, Sept. 2000.
55. Radusch, H.-J., Pham, T., and Scharnowski, D., Reprocessing of dynamic vulcanizates, in *Proceedings of the Regional Meeting of the Polymer Processing Society PPS-2001*, Antalya/Turkiye, Oct. 2001, p. 295.

10 Nanostructuring of *In Situ* Formed ABC Triblock Copolymers for Rubber Toughening of Thermoplastics

Christian Koulic, Robert Jérôme, and Johannes G.P. Goossens

CONTENTS

10.1 DEFORMATION MECHANISMS IN POLYMERS

Polymeric materials do not always possess mechanical properties that can comply with engineering purposes. Most thermoplastics have a low impact resistance that may, however, be improved (1–4), particularly by the addition of rubber particles (5–7). This strategy is by far more effective than the addition of a plasticizer that is detrimental to the glass transition temperature (T_g), modulus, and yield strength. The rubber particles are responsible for stress concentration that results in multiple crazing (8,9) or multiple shear deformation, depending on the matrix (10–13). The rubbery phase contributes only marginally to the toughness of the blend (9). The major disadvantage of the rubber modification of thermoplastics is the loss in stiffness (modulus and yield stress). This is the reason why much attention has been paid to optimizing parameters such as the rubber phase volume fraction, rubber particle size, distance between particles, and adhesion between the rubbery particles and the matrix. Wu (for semicrystalline polymers) (10,11) and Meijer and coworkers (for amorphous polymers) (15–17) proposed that a transition occurs from crazing to shear yielding when the matrix ligaments between the rubber particles or the characteristic thickness in layered structures are below a critical size. Therefore, the local deformation mechanism depends on whether the polymeric material is made locally thin, which means that the interparticle distance (ID) in rubber modified thermoplastics is of crucial importance. Meijer and coworkers (15–17) were able to correlate the aforementioned experimental structure-property relationship to the intrinsic behavior of the polymer matrices to be toughened. It was proposed that toughness could be improved by only acting on the intrinsic postyield behavior of the polymer, i.e., strain softening and hardening. The intrinsic deformation behavior has to be determined by methods not sensitive to localized phenomena (i.e., necking and crazing), e.g., by uniaxial compression testing (14). The intrinsic stress-strain curve shows an initial elastic region followed by yielding, strain softening, and strain hardening. The intrinsic difference between polystyrene (PS), known as a brittle material, and polycarbonate (PC), known as a ductile material, can be found in the postyield behavior: first, a drop in true stress after yielding (strain softening that induces strain localization) and, second, the increased slope of strain hardening at large strains (that stabilizes the localized plastic zone). A moderate strain softening and strong strain hardening should yield ultimate toughness because the moderate localization due to strain softening could thus be easily stabilized by strong strain hardening.

Up to now, the major drawback of the most efficient systems (PS/rubber blends) is the high amount of rubber (greater than 50 wt% of 200 nm sized preformed particles) necessary to meet the required interparticle distance of approximately 30 nm for PS (16). This high rubber content not only results in a drastic drop in modulus and yield stress, but the percolation threshold of the particles can also lead to phase inversion. The rubber content could be reduced to some extent by decreasing the particle size to 80 nm (17).

Obviously, viewing the results of attempting to control the scale and homogeneity of the rubber inclusions in the matrix (small ID with a low rubber content, i.e., less than 10 wt%) should improve our understanding of the toughening mechanism and lead to the production of useful new materials. The control of specific morphologies of immiscible polymer blends is the prerequisite for success in this strategy.

10.2 BLOCK COPOLYMERS AS IMPACT MODIFIERS

Pioneering work by Thomas and coworkers (18–22) and Hashimoto and coworkers (23–26) on block copolymers and their blends with homopolymers (blends of the AB/A type) stressed the importance of these copolymers in making polymeric nanostructured matrices. The effect of relevant parameters, such as the volume fraction of the constitutive A and B blocks and blend composition, was systematically studied. The phase structure of the blends is primarily governed by the length of the homopolymer chains (N_{Ah}) compared to the miscible block of the copolymer (N_{Ac}). Three regimes have been accordingly identified (27,28): (i) uniform solubilization ("wet-brush") when $N_{Ah} < N_{Ac}$. This case is similar to micellization in a solvent selective for one block; (ii) localized solubilization ("dry-brush") when N_{Ah} is approximately N_{Ac}. In this case, the homopolymer chains are solubilized in the copolymer domains (or vice versa) without affecting the conformation of the interfacial area; and (iii) macrophase separation when $N_{Ah} > N_{Ac}$. The copolymer phase separates from the homopolymer matrix.

It is thus possible to exercise full control over the phase morphology of AB/A copolymer/homopolymer blends. Morphological transitions from and to any of the states reported in the phase diagram of the copolymer are observed when the homopolymer is the minor phase dissolved in one phase of the copolymer. This was illustrated for a system consisting of a polystyrene-b-polyisoprene (PS-b-PIP) copolymer and a low molecular weight PS (hPS) (22b).

For the homopolymer matrix to be nanostructured, the copolymer must be the dispersed (minority) phase. Then, micelles of all sorts (spheres, cylinders, wormlike, vesicles) are formed in the matrix with or without a long range order, depending on the block copolymer concentration.

10.2.1 Nanostructuring with Block Copolymers Starting from the Liquid State

In the same vein, Bates and coworkers (29,30) demonstrated the potential of block copolymers to form nanostructured thermosets by blending an amphiphilic diblock copolymer, poly(ethylene oxide)-b-poly(ethylene-co-propylene) (PEO-PEP), with a polymerizable epoxy resin, a selective solvent for PEO. Upon increasing the amount of the epoxy resin precursor, the diblock phase microstructure evolved from lamellae, to gyroids, to cylinders, to body centered cubic packed spheres and ultimately to disordered micelles.

Upon the addition of a resin hardener and curing of the epoxy, the nanophase morphology was stabilized and macrophase separation of the epoxy from the block copolymer was avoided even though the PEO brush was expelled. Thus, a transition from wet-brush to dry-brush occurred upon curing.

To follow, these authors directed their research to dilute disordered blends (31–33) and illustrated the morphological features that can be formed upon dilution. Depending on the copolymer composition, spherical micelles or vesicular micelles are observed. In this example, one block of the amphiphilic copolymer [polybutadiene-b-poly(epoxidized isoprene) (PB-b-PIx)] is reacted with the epoxy precursor in order to prevent the brush from being expelled from the matrix upon curing. In addition to the fracture toughness (G_c), the glass transition temperature and storage modulus were measured (33a). G_c was improved, particularly in case of vesicles, and no significant effect on T_g was observed. Compared to spherical micelles, vesicles resulted in a much larger volume fraction of inclusions at constant copolymer content (approximately 2.5 wt%). This explains the loss in storage modulus, which is observed in this case. The large effect on the fracture toughness (± 170% improvement) was attributed to debonding of the vesicles and the subsequent localized matrix deformation. More recently (33b), the same group studied the effect of the formation of wormlike micelles in a bisphenol A/phenol Novolac matrix. The presence of such micelles led to a drastic increase in fracture resistance accompanied by an unexpected increase in glass transition temperature. This puzzling effect is still to be explained.

Similar morphological observations were reported by Kosonen et al. (34) also for a Novolac phenolic resin. The additive in this case was a polyisoprene-b-poly(2-vinylpyridine) copolymer (PIP-b-P2VP). The nitrogen atom–containing P2VP block interacts by H bonding with the phenol group of the resin. Spherical, cylindrical, and lamellar nanostructures were formed by blending the diblock and Novolac, followed by cross-linking of the continuous phase. These nanostructures are basically controlled by (i) the proper choice of the molecular weight of the P2VP block versus the molecular weight of Novolac before curing, (ii) the relative length of the P2VP block with respect to PIP in the diblock, and (iii) the weight fraction of PIP in the final blend.

Ritzenthaler et al. (35,36) extended this strategy to the synthesis of a nanostructured epoxy resin by using ABC triblock copolymers that are known to have a larger panel of phase morphologies than the diblocks. A high molar mass polystyrene-b-polybutadiene-b-polymethylmethacrylate (SBM) copolymer led to transparent, nanostructured thermosets whenever the polymethylmethacrylate (PMMA) block of the copolymer remained soluble within the thermoset polymer for the whole reaction. The authors further investigated the effect of blend composition and copolymer architecture on the final morphology of the thermoset. At low copolymer content, micelles are formed as a result of the immiscibility of both the PS and PB blocks with the epoxy monomer and the favorable interaction of the PMMA block with these monomers. At higher block copolymer contents (from 10 to 50 wt%), either a more complex "spheres on spheres" (raspberry) structure was observed, with the PB nodules localized around PS spheres, or a core-shell (onion) morphology consisting of PS spheres surrounded by PB shells was formed provided that the PB content was high enough. Addition of SBM was found to increase the epoxy network toughness from 0.65 to 2 MPa m$^{1/2}$ for a 50:50 blend for which phase inversion did not occur in contrast to what happens in the case of liquid rubber additives. At low copolymer content (10 wt%), the toughness increases linearly with the PB fraction of the triblock.

Rebizant et al. (37) reported recently on a similar strategy in order to induce the formation of a nanostructured epoxy. They basically merged the two above-mentioned systems since they used a reactive ABC(D) triblock copolymer: polystyrene-b-polybutadiene-b-polymethylmethacrylate-b-polyglycidylmethacrylate (SBMG), the fourth block being very short and acting as a reactive end group. By doing so, they demonstrated the effectiveness of the reaction induced modification of the epoxy thermoset compared to that with nonreactive SBM copolymers, which can be expelled from the network during its formation.

In a search for the conditions needed to produce "ultimate toughening," Van der Sanden (1), Van Melick (14), and Meijer and coworkers (17,38–40) attempted to validate their theoretical prediction that a homogeneous distribution of easily cavitating rubber particles of approximately 30 nm would induce a transition from crazing to shear yielding (in case of glassy matrices). The first attempt was based on chemically induced phase separation (CIPS) (17,38–40) and control of the phase morphology by low mobility and initial compatibility of the constitutive ingredients. Macroscopic demixing was avoided by the cross-linking of the dispersed phase. However, no significant improvement in macroscopic toughness was observed. The second envisioned strategy was similar to the ones previously discussed, i.e., self-assembly of a diblock copolymer in a monomer (styrene) prior to polymerization and formation of the matrix phase to be toughened in order to prepare nanosized core-shell particles. A major problem with this approach is the occurrence of macrophase separation, resulting in large domains of pure block copolymer (41). The macrophase separation could be circumvented by controlling the extent of miscibility of PS and poly(acrylate)-*b*-polyolefin diblock copolymers in a blend in which PS was chemically modified by copolymerization with *p*-(hexafluoro-2-hydroxy isopropyl) styrene (HFS) (42). Hydrogen bonding between the hydroxyl groups and the carbonyl groups of polybutylacrylate (PBA) enhanced the miscibility and led to randomly distributed polyolefin particles surrounded by a homogeneous PBA/PS matrix. The intrinsic deformation behavior was investigated by compression tests, whereas the microscopic mode of deformation was studied by time resolved small angle x-ray scattering, and it was shown that the macroscopic strain at break depends to a large extent on the diblock copolymer content and the degree of demixing between the rubber shell and PS matrix. Furthermore, it was illustrated that depending on the degree of demixing, the microscopic deformation mode changes from crazing to cavitation induced shear yielding. Although the glassy matrix was nanostructured, the improvement of the mechanical properties was limited. Further studies on pure triblock copolymer systems, such as poly(ethylene-butylene)-b-polybutylacrylate-b-polymethylmethacrylate (PEB-b-PBA-b-PMMA) and polycaprolactone-b-polybutadiene-b-polystyrene (PCL-b-PB-b-PS) stressed the importance of the role of cavitation (41,43).

All the strategies proposed until now for the production of nanostructured matrices are not suited to large-scale production because of the mandatory process in the liquid phase. Indeed, the final material is either prepared by solvent casting slow enough for the thermodynamic equilibrium to be reached or by polymerization (curing) of monomers. In the latter case, it is quite a problem to predict the final phase morphology because the interfacial brush regime may change during the polymerization (curing). Indeed, the brush is originally wet in the presence of the liquid monomer (or precursor), and its fate depends on the length of the chains formed upon polymerization (curing) and the type of reaction. Change in the brush swelling can modify the interfacial curvature and thus the phase morphology.

Production of nanostructured matrices by reactive melt-blending would be an ideal although challenging strategy. This is tested in the present study since it was recently shown possible to prepare well-defined block copolymers by interfacial melt-coupling (44).

10.2.2 PRODUCTION OF BLOCK COPOLYMERS BY REACTIVE MELT-BLENDING

For a long time, polyurethane, polyesters, and other segmented block copolymers have been synthesized by stepwise polymerization of two comonomers and a telechelic polymer, thus creating a premade block. In the same vein, block and graft copolymers can also be formed by coupling premade polymers end capped by mutually reactive groups (44). Conducted in the melt, coupling occurs at the interface between the immiscible constituents. Usually, only a limited amount of the chains are mutually reactive, the purpose being to form a small weight fraction of block or graft copolymers in a binary blend in order, e.g., to increase the interfacial adhesion between laminated or coextruded films or to compatibilize immiscible polymer blends. However, it is also possible to consider the quantitative coupling of all the chains, which requires a higher degree of reactivity of the groups involved and a rather low molecular weight of the reactive precursors (44,45).

Orr et al. (46) extensively studied the kinetics of the coupling reaction of end functional polymers for a series of pairs of reactive groups attached at the ends of the same PS chain and ranked them on the basis of increasing reactivity, i.e., acid with amine < hydroxyl with anhydride or acid < amine with epoxy < acid with oxazoline < acid with epoxy < aromatic amine with anhydride < aliphatic amine with anhydride, in good agreement with data for small size analogues in solution. Although these data were collected under homogeneous conditions, they can be safely extrapolated to interfacial coupling reactions. In the specific case of the primary amine–anhydride pair, quasiquantitative conversion can be reached within 2 min, which is a representative processing time.

The key role of the kinetics of the interfacial reaction has also been emphasized. For instance, the extremely fast cyclic anhydride–aliphatic primary amine reaction in the case of reactive PS and PIP chains of 25 kg/mol led to a nanophase morphology. The interfacial coupling of the same reactive chains was also carried out under static conditions. The same self-assembled structures were observed at the vicinity of the interface, which indicates that the interfacial area is renewed and made available for further reaction not only by mechanical mixing but also by the roughening of the interface as a result of the progressing reaction.

Inoue and coworkers (47–51) came up with the same conclusion for the solvent-free preparation of block copolymers by reactive blending of polysulfone and polyamide. The best results were also reported for primary amine–cyclic anhydride coupling reactions.

Yin et al. (52–58) further investigated the combined effect of the kinetics of the interfacial reaction and the molecular weight of the reactive precursors on the yield of the PS/PMMA coupling reaction. For the conversion of the reactive precursors into diblocks to be quantitative, it is mandatory that the molecular weight of the constitutive chains be rather low. For the interfacial reaction to progress, the copolymer chains have to leave the interface as soon as they are formed in order to make the interface available for further reactions. This condition is fulfilled in a limited range of molecular weights.

Clearly, the increase in molecular weight of the reactive precursors leads to a drastic change in phase morphology. From the micellar morphology observed for low molecular weight chains — almost quantitative conversion as measured by size exclusion chromatography (SEC) — the morphology shifts to quite a fine dispersion of PMMA particles in the PS matrix in case of higher molecular weight precursors (compatibilization of the PMMA phases dispersed within the PS matrix).

10.3 OBJECTIVES

The previous sections gave a general overview of the trends in modifying polymers, mostly in terms of toughness, with a block copolymer as the nanostructuring agent. The original and innovative strategies used to structure polymeric matrices always include a preparation step in the liquid phase, which is a drawback for any potential implementation on a larger scale.

The work presented in this chapter aims at merging the advantages of the two above-mentioned strategies and accordingly relieving all the drawbacks. The purpose is to produce thermodynamically stable nanostructured blends by a block copolymer formed by an interfacial reaction between two end functional precursors during the melt-blending. The model system developed in this study is based on a polyamide 12 (PA12) matrix, nanostructured by the self-assembly of an *in situ* formed polystyrene-b-polyisoprene-b-polyamide12 (PS-b-PIP-b-PA12) triblock copolymer. A linear ABC copolymer is a valuable nanostructuring agent because it allows a much broader and unique range of morphologies as compared to diblocks. This strategy aims at being universal, thus, being potentially applicable to other (including glassy) matrices.

The following sections first discuss the effect of the structure of the reactive diblock on the morphology. The relative volume fraction of each sequence will be systematically modified and thoroughly discussed in terms of the evolution of the interfacial curvature. This is followed by discussions on the transition from nanovesicles to "cucumberlike" core-shell nanoobjects by the addition of homopolystyrene (hPS) of well chosen molecular weight that will artificially modify the PS total volume fraction. The effect of the relative molecular weights of hPS and the PS sequence of the copolymer, thus, influencing the nature of the interfacial brush, will also be investigated. Subsequently, the mechanical properties are discussed and compared with some reference materials, such as PA12 modified with ethylene-propylene grafted with maleic anhydride. Finally, the strategy will be generalized for a broader scope of polymers.

TABLE 10.1
Details of the Reactive Diblock Copolymers and Their Nonreactive Counterparts

CODE	Type	M_n (10^{-3} g/mol)[a]	M_w/M_n	PS (wt%)	f (%)
R-1	PS-b-PIP-anh	16 (PS)[a]–18 (PIP)[b]	1.05	47	93
R-2	PS-b-PIP-anh	6 (PS)[a]–30 (PIP)[b]	1.04	17	96
R-3	PS-b-PIP-anh	22 (PS)–14 (PIP)[b]	1.03	61	92
R-4	PS-b-PIP-anh	26 (PS)–10 (PIP)[b]	1.02	72	91
R-5	PS-b-PIP-anh	29 (PS)–7 (PIP)[b]	1.04	80	93
R-6	PS-b-PIP-anh	31 (PS)[a]–7 (PIP)[b]	1.03	82	92
NR-1	PS-b-PIP	16 (PS)[a]–18 (PIP)[b]	1.05	47	—
NR-2	PS-b-PIP	6 (PS)[a]–30 (PIP)[b]	1.04	17	—
NR-3	PS-b-PIP	22 (PS)–14 (PIP)[b]	1.03	61	—
NR-4	PS-b-PIP	26 (PS)–10 (PIP)[b]	1.02	72	—
NR-5	PS-b-PIP	29 (PS)–7 (PIP)[b]	1.04	80	—
NR-6	PS-b-PIP	31 (PS)[a]–7 (PIP)[b]	1.03	82	—
R-7	PIP-b-PS-anh	15 (PS)[a]–16 (PIP)[b]	1.05	48	94
NR-7	PIP-b-PS	15 (PS)[a]–16 (PIP)[b]	1.05	48	—

[a] SEC with PS standards.
[b] 400 MHz ^{1}H NMR.

10.4 NANOSTRUCTURING OF A THERMOPLASTIC MATRIX BY REACTIVE BLENDING WITH DIBLOCK COPOLYMERS

Previous studies (52,59–61) emphasized that the kinetics of interfacial reaction of immiscible polymers in the melt had a decisive effect on the phase morphology. When all the chains are reactive and the mutually reactive groups are used in stoichiometric amounts, the reaction can go to completion for chains of relatively low molecular weight and of high mutual reactivity, i.e., high reaction rate. Because the coupling of PS-b-PIP diblock to PA12 must be as complete as possible, the highly reactive primary amine (PA12)/anhydride (PS-b-PIP) pair was considered in this study. The synthetic aspects of preparation of the block copolymers and the reactive end groups and the characterization thereof are discussed in detail elsewhere (62–65). Details of all used reactive block copolymers and their counterparts are listed in Tables 10.1 and 10.2.

10.4.1 NANOSTRUCTURING USING A SYMMETRIC REACTIVE BLOCK COPOLYMER

The symmetric PS-b-PIP-anh diblock (entry R-1 in Table 10.1, 47 wt% PS) was first melt reacted with 80 wt% PA12. For comparison, Figure 10.1a shows a transmission electron microscopy (TEM) micrograph of the nonreactive blending of PA12 with the NR-1 diblock. As expected for nonmiscible polymer blends, a coarse and

TABLE 10.2
Homopolystyrene (hPS) Used in This Study

Code	Type	M_n (10^{-3} g/mol)[a]	M_w/M_n[a]
S-1	PS	7	1.03
S-2	PS	16	1.02
S-3	PS	35	1.02

[a] SEC with PS standards.

heterogeneous macrophase separation is observed. It is, however, interesting to note that the lamellar mesophases typical of symmetric diblocks can be observed within the PS-b-PIP microdomains. The thickness of the constitutive PS lamellae is approximately 7 nm. When the nonreactive diblock is substituted by the end reactive counterpart, a drastic drop in the particle size is observed that is expected as a consequence of the formation of a triblock with a block identical to the matrix chains (Figure 10.1b). At high magnifications and in the direction perpendicular to the extrusion, the nanophases formed by the *in situ* generated triblock exhibit a general core-shell morphology, but a more careful analysis shows that the shell has a three layer substructure (see Figure 10.2). Two dark outer layers are clearly observed that have to be assigned to PIP selectively stained by OsO₄. PS is thought to form the intermediate layer because of its thickness (approximately 8 nm), which is consistent with the size of the lamellar morphology of the PS-b-PIP diblock itself. According to this picture, PA12 should be the core forming component leading to vesicular type nanostructures. Indeed, the *in situ* formed PS-b-PIP-b-PA12 triblock copolymer with constitutive blocks of comparable molecular weight ($M_{n,PA12}$ approximately 20

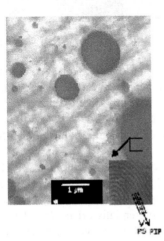
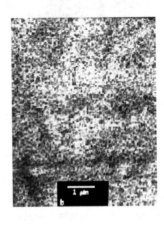

FIGURE 10.1 TEM micrographs for (left) the PA12/PS-b-PIP (NR-1, Table 10.1) blend and (right) the PA12/PS-b-PIP-anh (R-1, Table 10.1) blend (80:20).

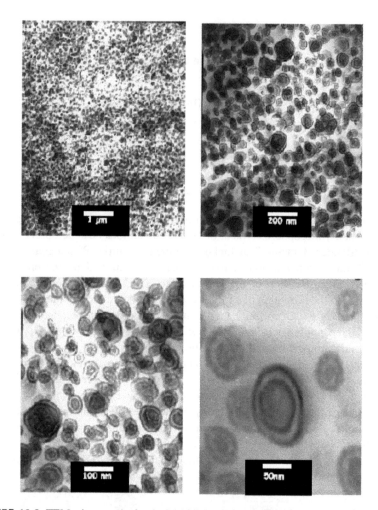

FIGURE 10.2 TEM micrographs for the PA12/PS-b-PIP-anh (R-1, Table 10.1) blend (80:20).

kg/mol) tends to self-assemble into a lamellar mesophase. However, a continuous lamellar mesophase cannot persist in the dilute regime under shear. The more stable lamellar morphology consists of bilayer vesicles with a large diameter compared to the wall thickness, thus with a low curvature (20,61,66). The vesicle wall is nothing but a double layer of PS-b-PIP-b-PA12 triblock copolymer lamellae as illustrated by the cartoon shown in Figure 10.3.

As a result of the presence of the nanostructures, the crystallinity of PA12 could potentially be perturbed. However, differential scanning calorimetry (DSC) analysis revealed that there was no change either in the melting temperature or in the degree of crystallinity. The nanostructures could be preferentially oriented by the flow field parallel to the extrusion direction. This possibility has been assessed by TEM analysis of the material parallel to the extrudate and is shown in Figure 10.4. Anisotropic nanophases are observed that look like vesicular cylinders. This type of

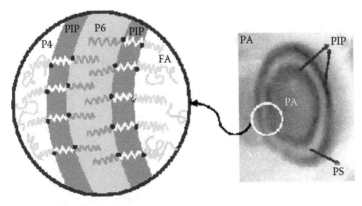

FIGURE 10.3 Schematic of the internal structure of the vesicular shell.

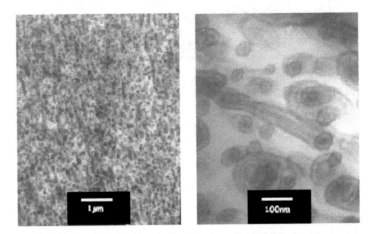

FIGURE 10.4 TEM micrographs of the PA12/PS-b-PIP-anh (R-1, Table 10.1) blend (80:20) observed in the direction of the extrudate at two different magnifications.

transverse orientation was already observed by Harada et al. (67) in the case of extrusion of a highly concentrated solution of lamellae forming block copolymer.

10.4.2 EFFECT OF THE FUNCTIONALITY AND THE LOCATION OF THE END GROUP ON THE NANOSTRUCTURING

Changing the final composition of the blend is a tool for tuning the architecture of the nanostructures in the ABC/C blends. In this respect, a direct way would consist in restricting the progress of the interfacial reaction such that AB diblock coexists with the ABC triblock. In order to test this strategy, the reactive diblock was premixed with a known amount of the nonreactive counterpart. Therefore, PA12 (80 wt%) was melt blended with a 50:50 (wt/wt) mixture of reactive PS-b-PIP-anh and the nonreactive counterpart (R-1/NR-1, Table 10.1). Figure 10.5 shows the onionlike morphology (68,69) that results from the selective location and self-organization of

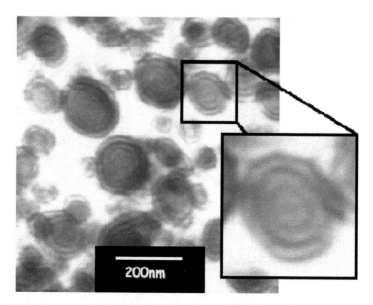

FIGURE 10.5 TEM micrograph for the PA12/[PS-b-PIP-anh (R-1, Table 10.1)/PS-b-PIP (NR-1, Table 10.1)] blend (80:10:10) with an onionlike morphology.

the (nonreactive) diblock copolymer within the interior of the former PS-b-PIP-b-PA12 vesicles in the PA12 matrix. This observation illustrates the possible tuning of the vesicular morphology of an ABC/C binary blend by the addition of a third component (AB).

According to the previous experiments, an ABC triblock with approximately the same volume fraction for each constitutive block easily forms vesicular nanoobjects when diluted in a C matrix under shear because of the low curvature imposed by the lamellae. In these experiments, the reactive group of the symmetric PS-b-PIP diblock was attached at the end of the PIP block. Because the composition of the triblock does not change if the reactive group is attached as an end group to PS rather than to PIP, this modification is not expected to change the type of nanostructures in the matrix. In order to confirm this prediction, a symmetric anh-PS-b-PIP diblock was synthesized by polymerizing first isoprene and then styrene followed by the same reaction for end group attachment. The reactive blending of 80 wt% PA12 with this reactive symmetric diblock (entry R-7, Table 10.1) led again to ellipsoidal nanoobjects with PIP localized in the shell (Figure 10.6). This observation is consistent with a shell consisting of two layers of triblock lamellae (PA12-PS-PIP/PIP-PS-PA12) with PIP in the central position. The transition in contrast from PIP (black) to PS (grey) to PA12 (white) is however less suitable for the direct observation of the internal structure of the shell than the previous transition from PA12 to PIP and finally to PS, with two easily detectable dark PIP sublayers. Nevertheless, the nanoobjects have the same size and shape whatever the triblock (ABC in Figure 10.2 vs. BAC in Figure 10.6). Moreover, the lamellar morphology in the bulk (not shown) is also the same, and these lamellae are now the key constituents of the nanoobjects formed with the same curvature under the same

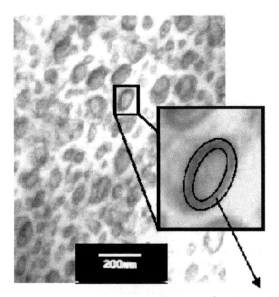

PIP intermediate layer of the double layer

FIGURE 10.6 TEM micrograph of the PA12/PIP-b-PS-anh (R-7, Table 10.1) blend (80:20) with an inverse vesicular morphology.

conditions. Although the nanostructuring remains that of the vesicular type, it is possible that a change in the vesicle substructure can affect the macroscopic properties of the blends.

10.4.3 NANOSTRUCTURING USING AN ASYMMETRIC REACTIVE BLOCK COPOLYMER

The effect of the PS volume fraction of the reactive PS-b-PIP-anh diblock copolymer on PA12 nanostructuring has also been studied by reactive blending of an asymmetric diblock precursor with the polyamide, while keeping the total molecular weight of the diblock constant (35 to 40 kg/mol). A PIP rich diblock copolymer (R-2, Table 10.1; 17 wt% PS) was first blended with PA12. Upon staining with OsO_4, dispersion of black nanodomains is observed, which are basically PS inclusion–containing PIP nodules. This expected substructuring of the PIP nanophases is hardly detectable by TEM because of the low PS content (Figure 10.7). This technical problem should be alleviated by inverting the composition of the diblock (R-6, Table 10.1, 82 wt% PS). The internal structure of the dispersed nanophases is then clearly observed by TEM at high magnification (Figure 10.8). The substantial increase in the PS volume fraction prevents vesicles from being formed and confirms that a lamellar forming system is the key condition for the development of vesicular nanostructures. When there is a phase transition from lamellae to cylinders (or spheres) for the neat AB diblock precursor, there is a transition from a vesicular to a core and shell-like morphology. Indeed, a PIP shell, which isolates PS from the PA12 matrix, is clearly observed after staining. Substructuring of the core is, however, observed that is

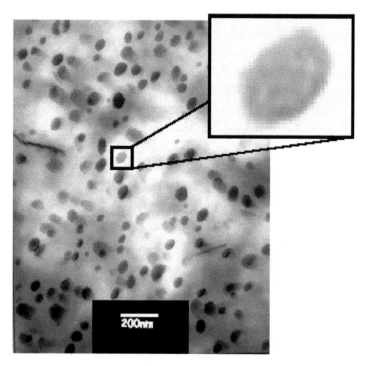

FIGURE 10.7 TEM micrograph of the PA12/PS-b-PIP-anh (R-2, Table 10.1) blend (80:20).

reminiscent of a slice of cucumber and that we will designate as a *cucumberlike* morphology afterward. TEM observation in the transversal direction shows exactly the same cucumberlike nanostructures as in Figure 10.8, which are therefore isotropic. One explanation for the dispersion of PIP containing nanophases in the core of the core-shell objects might be accumulation and phase separation of some unreacted diblock copolymer. In order to confirm this assumption, PA12 has been reactive blended with a 50:50 (wt/wt) mixture of the reactive (R-6) and nonreactive (NR-6) asymmetric diblock (Figure 10.9).

The nonreactive diblock is obviously accommodated in the core of the core-shell domains, whose size has accordingly increased. Because of the bimodal size distribution of the dispersed domains, it is not clear yet whether the two populations are distinct particles that coexist or identical particles far from equilibrium. In the former case, the small particles might consist essentially of triblock copolymers, whereas the larger particles might result from the phase separation of the unreacted diblock copolymers restricted in size by the triblock copolymer that forms a stabilizing envelope.

10.4.4 TRANSITION FROM NANOVESICLES TO CUCUMBER-LIKE CORE-SHELL NANOOBJECTS

In Section 10.4.1, a polystyrene-b-polyisoprene-b-polyamide 12 (PS-b-PIP-b-PA12) copolymer with a symmetric composition was formed by reactive melt-blending of

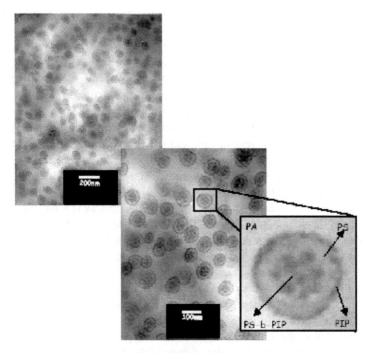

FIGURE 10.8 TEM micrographs of the PA12/PS-b-PIP-anh (R-6, Table 10.1) blend (80:20).

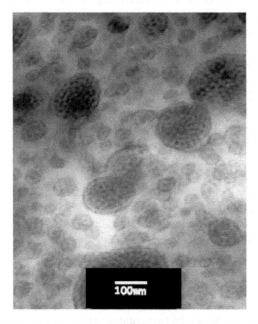

FIGURE 10.9 TEM micrograph of the PA12/[PS-b-PIP-anh (R-6, Table 10.1)/PS-b-PIP (NR-6, Table 10.1)] blend (80:10:10).

20 wt% of an anhydride end capped polystyrene-b-polyisoprene (PS-b-PIP-anh) with an amino end functional polyamide. Within a few minutes at 220°C, the ABC triblock copolymer is formed *in situ* and self-assembles into a liposomelike nanostructure. However, a core-shell nanostructure with a cucumber-like core organization was observed when the volume fraction of the PS block is increased, all the other conditions being the same. In this part, we focus on the transition between these two nanostructures by using only one reactive diblock (the symmetric one) and by increasing the PS volume fraction by the addition of homopolystyrene. The effect of the relative molecular weight of hPS and the PS block on the final phase morphology was also considered. The transition from vesicular to core-shell nanostructures can be triggered by using the symmetrical reactive diblock that generates vesicles and by adapting the composition of the final ternary blend to the composition at which the core-shell nanostructure is observed. This strategy is easily implemented by increasing the PS content in the 20 wt% of copolymer blended with PA12. Substitution of an R-1/hPS blend (35:65 wt/wt) for neat R-1 in the blend with PA12 mimics the composition of blends with core-shell nanostructures (see Figure 10.8). However, the key condition is that hPS spontaneously accumulates in the PS nanodomains of the triblock, which means that the molecular weight of this hPS is of a prime importance. It is only in the wet-brush regime that the hPS chains can swell the interfacial PS brush of the copolymer, i.e., whenever the hPS molecular weight is lower than that of the PS block of the triblock copolymer. This requirement was confirmed by considering the behavior of three hPSs of increasing molecular weight. First, the reactive diblock (R-1) was preblended with an hPS whose molecular weight is half that of the PS block of the reactive copolymer (S-1) (see Tables 10.1 and 10.2). Figure 10.10 shows that the cucumber-like core-shell morphology is observed in contrast to the vesicles expected for the neat symmetric copolymer (Figure 10.2). The actual PS content in the blend is now 16.4 wt%, which corresponds to 82 wt% in the R-1/S-1 blend added at a rate of 20 wt%. This is actually the composition of the asymmetric reactive diblock that promotes the cucumber-like core-shell morphology (Figure 10.8). The bulk morphology of this blend was observed by TEM as shown in Figure 10.11A. This phase morphology appears to be borderline between spheres (bcc) and hexagonally perforated lamellae (hpl), which is in line with a total PS volume fraction of 82 wt%, but is in contrast to the lamellar phase morphology observed for the neat copolymer with 47 wt% PS (Figure 10.11B) (70). The key observation is that PA12 can be obtained with the same nanostructure by reactive melt-blending with a constant amount of an additive (20 wt% in this work) that consists of either a mutually reactive diblock (e.g., PS-PIP-anh) of a well-defined composition or a blend of a mutually reactive diblock of the same molecular weight but different composition and a parent homopolymer (i.e., hPS) such that the composition of the additive remains unchanged. This flexibility allows for the characteristic sizes of the nanoobjects to be changed. The comparison of the core-shell particles illustrates that the thickness of the PIP rubbery shell is approximately two times larger in case of the three-component PA12/(R-1 + S-1) (80:20) blend (with R-1:S-1 = 35:65) compared to the two-component PA12/R-6 (80:20) reactive blend, as a result of which, $M_{n,PIP}$ = 18 kg/mol in the former blend and $M_{n,PIP}$ = 7 kg/mol in the latter one. This methodology is thus effective in tuning, e.g., the shell thickness

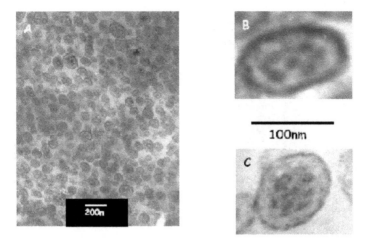

FIGURE 10.10 (A) TEM micrograph for the PA12/(R-1/S-1) (Tables 10.1 and 10.2) [80:20(35:65)] blend with a cucumber-like core-shell morphology. (B) This morphology is compared at higher magnification to (C) the counterpart shown in Figure 10.8. Emphasis has to be placed on the change in the shell thickness (dark shell of PIP).

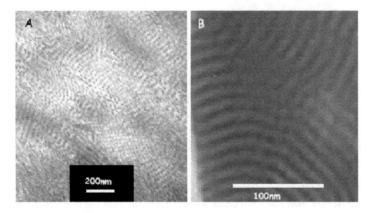

FIGURE 10.11 (A) TEM micrograph of the bulk morphology for the R-1/S-1 (Tables 10.1 and 10.2) (35:65) blend and (B) TEM micrograph of the bulk morphology for the R-1 copolymer (Table 10.1).

of thermoplastic core–rubber shell particles, which may be of great interest in helping us to understand the rubber toughening of thermoplastics by this type of core-shell additive. When the molecular weight of hPS is increased up to that of the PS block (S-2, Table 10.2), the phase morphology changes as illustrated by Figure 10.12A. The TEM micrograph shows that core-shell structures persist but with a modified internal substructure and a bimodal particle size distribution. In all cases, the PS cores are stabilized in the PA12 matrix by an interfacial monolayer of the *in situ* formed PS-b-PIP-b-PA12 copolymer. PS-b-PIP diblocks seem now to form cylindrical micelles in the PS domains as a result of the dry brush regime of the interface as reported by Kinning et al. (20,21,66) and more recently by Chen et al. (71). These

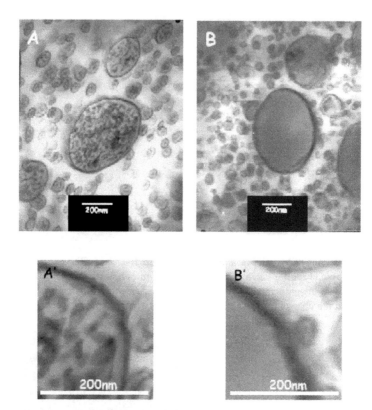

FIGURE 10.12 (A) TEM micrograph of the PA12/R-1/S-2 (Tables 10.1 and 10.2) [80:20(35:65)] blend and (B) TEM micrograph of the PA12/R-1/S-3 (Tables 10.1 and 10.2) [80:20(35:65)] blend. Comparison of the shell thickness for the two core-shell morphologies: (A′) PA12/(R-1/S-2) (Tables 10.1 and 10.2) [80:20(35:65)] blend and (B′) PA12/(R-1/S-3) (Tables 10.1 and 10.2) [80:20(35:65)] blend.

authors observed the same phenomenon for AB/hA blends in which the homopolyA (hA) is the matrix and the molecular weight ratio, M_{nA}/M_{nhA}, is close to one.

Finally, when the molecular weight of hPS is much higher than that of the PS block (S-3, Table 10.2), the same morphology as in the previous case persists (comparison of Figures 10.12A and 10.12B), except that the diblock cannot form micelles anymore in the PS microdomains. Any diblock excess tends then to accumulate at the PS/PA12 interface, which is thicker by approximately a factor of two compared to the previous blend in which only the triblock would be at the interface. From the magnification of the shell layer (Figures 10.12A′ and 10.12B′), it appears that the apparent thickness is uneven and that TEM is not reliable enough to measure the shell thickness.

10.4.5 MECHANICAL PROPERTIES

The macroscopic mechanical behavior of semicrystalline polymers has always been a challenge in materials research because it is a major criterion for the end-use of these materials. In addition to the elastic modulus, toughness is of a considerable importance. The mechanism commonly accepted for the toughening of polyamide by finely dispersed rubber particles relies on the localized yielding of the matrix material induced by cavitation or debonding of the rubber. Percolation of the yielded zones within the whole sample results in toughening. Because yielding is thermally activated, an increase in temperature is expected to generate yielding zones more easily and therefore to improve ductility. This explains the observation of a brittle-to-tough transition temperature (BTTT), which depends on the interparticle distance. In this part, the materials prepared in the previous sections are evaluated to determine whether they have the potential to match the above-mentioned properties. The nanostructures may fulfill most of the conditions for imparting to PA12 properly balanced properties, i.e., low interparticle distance at a low rubber content. The stiffness and toughness of the most relevant samples are compared to polyamide toughened by an ethylene-propylene copolymer grafted by maleic anhydride (ERP-g-MA). The determination of the local deformation mechanisms for the various morphologies is beyond the scope of this chapter and is the subject of further study.

10.4.5.1 PA12/EPR-g-MA Blends

Since the EPR-g-MA copolymer from Exxon Chemicals is nowadays the best toughening agent for polyamide available, it was melt blended with PA12, and the basic mechanical properties of the blend were measured. One reason is that the toughening of PA12 has been much less investigated than the more common PA6 and PA6,6. Figure 10.13 shows the impact properties of PA12 and PA12 blended with increasing amounts of EPR-g-MA.

PA12 is brittle at room temperature with an impact strength of 4 kJ/m^2. This changes when PA12 is melt blended with the reactive rubber. The impact strength

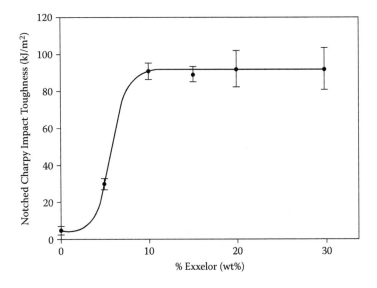

FIGURE 10.13 Dependence of the impact resistance on the EPR-g-MA content.

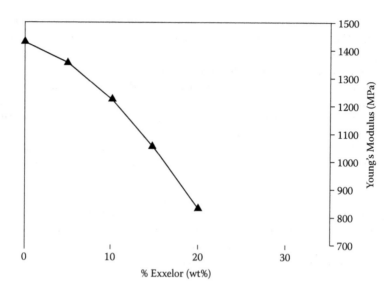

FIGURE 10.14 Dependence of the tensile modulus on the EPR-g-MA content.

substantially increases with the EPR-g-MA content. A brittle-to-tough transition is observed at approximately 7 to 8 wt% of EPR-g-MA with a maximum toughness for a 10 wt% rubber content [approximately 100 nm as observed by scanning electron microscopy (SEM) on surface fractured in liquid nitrogen followed by xylene extraction of the rubber nodules]. Consistently, a high extent of plastic deformation can be observed for the fractured surfaces in case of ductile fracture. Another important piece of information can be found in Figure 10.14, which shows a drastic drop in rigidity parallel to the brittle-to-tough transition related to a rubber content as high as 10 wt% required for maximum toughness.

10.4.5.2 PA12/PS-b-PIP-anh Blends: Vesicular Nanostructures

The mechanical properties of nanovesicles containing PA12 were measured. The compositionally symmetric PS-b-PIP-anh precursor (20 wt%) was melt blended with PA12, which corresponds to 10 wt% PIP 1,4-cis, thus, equal to the amount of EPR-g-MA used for observing maximum toughness. Figure 10.15 compares the fracture toughness for PA12 (0% rubber), for EPR-g-MA–containing reference (Excellor 10% rubber), and for the visicle-containing PA12 (vesicles 10% rubber).

Although the fracture toughness of PA12 is significantly increased by the addition of 20 wt% of the symmetric PS-b-PIP-anh, this increase cannot be compared with the beneficial effect of 10 wt% EPR-g-MA. Moreover, the effect on the elastic modulus is much more detrimental, which is a disappointing observation (Figure 10.16). Even though the nanovesicles formed by the triblock copolymer in PA12 improve the impact resistance, they do not match the "best seller" modifier EPR-g-MA.

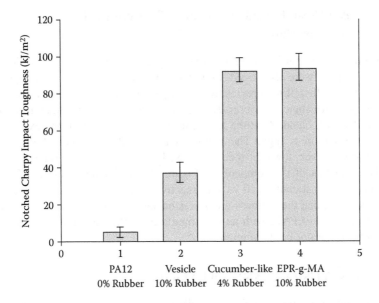

FIGURE 10.15 Comparison of fracture toughness for PA12 blended with different impact modifiers.

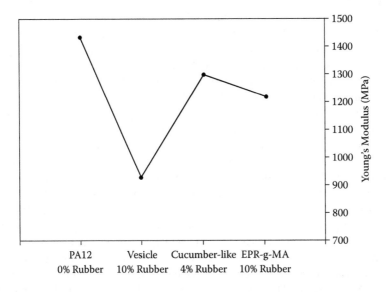

FIGURE 10.16 Comparison of the tensile moduli for PA12 blended with different impact modifiers.

10.4.5.3 PA12/PS-b-PIP-anh: Cucumber-Like Core-Shell Nanostructures

The mechanical properties of PA12 containing the so-called cucumber-like core-shell nanoobjects were also tested. Figure 10.15 compares the impact toughness of this system with that of the previously investigated samples. A tremendous increase in the impact resistance is observed, which is as high as the toughness of the PA12/EPR-g-MA blend. Anytime a ductile fracture is observed, the extent of the plastic deformation is high. The major advantage of the modified PA12 under consideration is a very low rubber content, i.e., 4 wt%. As a result, the elastic modulus is higher than that of the competitors (Figure 10.16), including the PA12/EPR-g-MA blend that must contain 10 wt% of reactive rubber to maximize its toughness. It must be noted that the stiffness-toughness balance has not yet been optimized for the reactive blend of PA12 with an asymmetric PS-b-PIP-anh diblock, accordingly, 4 wt% PIP might not be the minimum.

10.5 CONCLUSIONS AND OUTLOOK

In this chapter, new strategies for controlling the (nano)morphology of multiphase polymeric materials based on PA12 are discussed. A fine control is mandatory in order to fulfill the severe requirements needed in order to control the mechanical properties of the blend in terms of the balance between stiffness and toughness. Other strategies for structuring polymeric matrices always included a preparation step in the liquid phase, a drawback for implementation on a larger scale, and resulted in nonequilibrium, nonstable morphologies, which cannot be reprocessed.

This work was devoted to the study of the controlled nanostructuring of polymeric matrices by the self-assembly of a linear ABC triblock copolymer formed *in situ* during the processing of the blend. The production of a nanostructured polyamide matrix using reactive diblock copolymers was studied as a function of the relative weight fraction of the PS sequence within the block copolymer. As a consequence, the architecture of the final PS-b-PIP-b-PA12 triblock is modified, which in turn influences the self-assembly process of the copolymer within the matrix. If the weight fraction of each sequence was almost equal (compositionally symmetric), nanovesicles with a trilayer envelope (with a diameter less than 100 nm) were formed that resulted from the shear induced collapse of lamellar micelles. The molecular architectural requirements for inducing such nanostructures were emphasized.

Compositional symmetry is the key for the formation of nanovesicles. This was emphasized by locating the reactive phthalic anhydride at the end of the PS chain, leading to the production of inverse-nanovesicles. When the weight fraction of the PS sequence is increased, a transition in morphology is observed. Core-shell objects are formed with a cucumber-like core. It is not clear yet whether the core subinclusions are unreacted PS-b-PIP copolymer that could not be accommodated in the PS-b-PIP-b-PA12 domains. The key point is, however, the production of tunable core-shell particles with very low total rubber weight content.

Fine-tuning of the cucumber-like core-shell particles is of interest in the potential modification of the dimensional parameters of such objects. However, as result of the interrelation of the different parameters controlling morphology in polymer blends, no system allows the modification of one single parameter keeping all the rest constant. Even for a system that produces nanovesicles, a transition from vesicles to core-shell was possible by adding homopolystyrene of well chosen molecular weight. In order to induce a wet-brush, its molecular weight ought to be lower than that of the copolymer PS sequence. This allows the production of cucumber-like core-shell objects with a greater shell thickness at constant rubber weight fraction, total diameter, and therefore constant interparticle distance.

The evaluation of the effect of the nanostructures contained in PA12 on the mechanical properties of the final blend as done in this chapter focused on impact strength and Young's modulus, i.e., the toughness-stiffness balance, that could be compared to reference materials. The presence of the cucumber-like core-shell objects led to a drastic increase in toughness that compared favorably to that imparted to PA12 by the addition of a commercial impact modifier. Moreover, the very low rubber content of such materials allowed the modulus to be kept at a fairly high level, leading to a very good compromise between toughness and stiffness.

Most of this work has been carried out with a polyamide matrix because of intrinsic amino end functionality and a toughness-stiffness balance that could be improved by a rubber containing additive, i.e., the triblock formed *in situ*. Having in mind a general toughening strategy, it is of prime importance to demonstrate that the nanostructuring of PA12 can be extended to other (reactive) thermoplastics. For this purpose, a glassy matrix, i.e., a styrene and acrylonitrile copolymer (SAN) (65), was selected because a toughened version is available — acrylonitrile butadiene styrene (ABS) resins — that might serve as a reference. In contrast to most polycondensates, including polyamides, polymers prepared from vinyl monomers are not spontaneously end capped by a functional group. In case of SAN-NH_2, synthesis was carried out by controlled radical copolymerization based on the atom transfer radical polymerization (ATRP) mechanism (62). This matrix was reactive blended with 20 wt% of the asymmetric PS-b-PIP-anh copolymer. The typical cucumber-like core-shell was observed with the same characteristic features whatever the matrix, which confirms that the molecular architecture and composition of the ABC triblock have a decisive effect on the nanostructures that are formed in a variety of C matrices ranging from semicrystalline PA12 to glassy SAN. This shows that resitively blending of a diblock precursor with a matrix is quite a general strategy to obtain a nanodispersed rubbery component with well-defined shape and size. Other results on a blend of a preformed ABC copolymer with a C matrix (64), where C is PMMA, showed the universality of the approach, but further studies on both amorphous and semicrystalline polymers are required. Finally, studies are needed to understand the local deformation mechanisms, such as crazing and shear yielding, to correlate to the macroscopic properties, among which the occurrence of cavitation is of prime interest.

ACKNOWLEDGMENTS

Robert Jérôme and Christian Koulic are very much indebted to the Belgian Science Policy for financial support under Interuniversity Attraction Poles Programme (PAI V/03 — *Supramolecular Chemistry and Supramolecular Catalysis*).

REFERENCES

1. Van der Sanden, M.C.M., *Ultimate Toughness of Amorphous Polymers*, Ph.D. thesis, Eindhoven University of Technology, The Netherlands, 1994.
2. Bucknall, C., *Toughened Plastics*, Applied Sciences, London, 1977.
3. Utracki, L., *Polymer Alloys and Blends, Thermodynamics and Rheology*, Hanser Publishers, Munich, Germany, 1989.
4. Paul, D. and Newman, S., Eds., *Polymer Blends*, Academic Press, New York, 1978.
5. Kinloch, A. and Young, R., *Fracture Behavior of Polymers*, Elsevier Applied Sciences, London, 1985.
6. Gebizlioglu, O.S., Beckham, H.W., Argon, A.S., Cohen, R.E., and Brown, H.R., A new mechanism of toughening glassy polymers. 1. Experimental procedures, *Macromolecules*, 23, 3968–3974, 1990.
7. Argon, A.S., Cohen, R.E., Gebizlioglu, O.S., Beckham, H.W., Brown, H.R., and Kramer, E.J., A new mechanism of toughening glassy polymers. 2. Theoretical approach, *Macromolecules*, 23, 3975–3982, 1990.
8. Coumans, W.J., Heikens, D., and Sjoerdsma, S.D., Dilatometric investigation of deformation mechanisms in polystyrene-polyethylene block copolymer blends: correlation between Poisson ratio and adhesion, *Polymer*, 21, 103–108, 1980.
9. Newman, S. and Strella, S., Stress-strain behavior of rubber-reinforced glassy polymers, *J. Appl. Sci.*, 9, 2297–2310, 1965.
10. Wu, S., Phase structure and adhesion in polymer blends: a criterion for rubber toughening, *Polymer*, 26, 1855–1863, 1985.
11. Wu, S., A generalized criterion for rubber toughening: the critical matrix ligament thickness, *J. Appl. Polym. Sci.*, 35, 549–561, 1988.
12. Borggreve, R.J.M., Gaymans, R.J., and Schuijer, J., Impact behaviour of nylon-rubber blends. 5. Influence of the mechanical properties of the elastomer, *Polymer*, 30, 71–77, 1989.
13. Borggreve, R.J.M., Gaymans, R.J., and Eichenwald, H.M., Impact behaviour of nylon-rubber blends. 6. Influence of structure on voiding processes; toughening mechanism, *Polymer*, 30, 78–83, 1989.
14. Van Melick, H.G.H., *Deformation and Failure of Polymer Glasses*, Ph.D. thesis, Eindhoven University of Technology, The Netherlands, 2001.
15. Meijer, H.E.H., Govaert, L.E., and Smit, R., A multi-level finite elements method for modeling rubber-toughened amorphous polymers, in *Toughening of Plastics*, ACS, Boston, 50, 1999, pp. –.
16. Van der Sanden, M.C.M., Meijer, H.E.H., and Lemstra, P.J., Deformation and toughness of polymeric systems. 1. The concept of a critical thickness, *Polymer*, 34, 2148–2154, 1993.
17. Jansen, B.J.P., Rastogi, S., Meijer, H.E.H., and Lemstra, P.J., Rubber-modified glassy amorphous polymers prepared via chemically induced phase separation. 1. Morphology development and mechanical properties, *Macromolecules*, 34, 3998–4006, 2001.
18. Thomas, E., Kinning, D., Alward, D., and Henkee, C., Ordered packing arrangements of spherical micelles of diblock copolymers in two and three dimensions, *Macromolecules*, 20, 2934–2939, 1987.
19. Herman, D., Kinning, D., Thomas, E., and Fetters, L. A compositional study of the morphology of 18-armed poly(styrene-isoprene) star block copolymers, *Macromolecules*, 20, 2940–2942, 1987.

20. Kinning, D. and Thomas, E., Morphological studies of micelle formation in block copolymer/homopolymer blends, *J. Chem. Phys.*, 10, 5806–5825, 1990.

21. Kinning, D., Thomas, E., and Fetters, L., Morphological studies of micelle formation in block copolymer/homopolymer blends: comparison with theory, *Macromolecules*, 24, 3893–3900, 1991.

22. (a) Winey, K., Thomas, E., and Fetters, L., Swelling of lamellar diblock copolymer by homopolymer: influences of homopolymer concentration and molecular weight, *Macromolecules*, 24, 6182–6188, 1991. (b) Winey, K., Thomas, E., and Fetters, L., Ordered morphologies in binary blends of diblock copolymer and homopolymer and characterization of their intermaterial dividing surfaces, *J. Chem. Phys.*, 95, 9367–9375, 1991.

23. Han, C., Baek, D., Kim, J., Kimishima, K., and Hashimoto, T., Viscoelastic behavior, phase equilibria, and microdomain morphology in mixtures of a block copolymer and a homopolymer, *Macromolecules*, 25, 3052–3067, 1992.

24. (a) Koizumi, S., Hasegawa, H., and Hashimoto, T., Ordered structure of block polymer homopolymer mixture. 4. Vesicle formation and macrophase separation, *Makromol. Chem. Macromol. Symp.*, 62, 75, 1992. (b) Hashimoto, T., Yamasaki, K., Koizumi, S., and Hasegawa, H., Ordered structure in blends of block copolymers. 1. Miscibility criterion for lamellar block copolymers, *Macromolecules*, 26, 2895–2904, 1993.

25. Koizumi, S., Hasegawa, H., and Hashimoto, T., Ordered structure in blends of block copolymers. 3. Self-assembly in blends of sphere- or cylinder-forming copolymers, *Macromolecules*, 27, 4371–4381, 1994.

26. Koizumi, S., Hasegawa, H., and Hashimoto, T., Spatial distribution of homopolymers in block copolymer microdomains as observed by a combined SANS and SAXS method, *Macromolecules*, 27, 7893–7906, 1994.

27. Leibler, L., Theory of microphase separation in block copolymers, Macromolecules, 13, 1602–1617, 1980.

28. Leibler, L., Emulsifying effects of block copolymers in incompatible polymer blends, *Makromol. Chem., Macromol. Symp.*, 16, 1–17, 1988.

29. Hillmyer, M., Lipic, P.M., Hajduk, D., Almdal, K., and Bates, F., Self-assembly and polymerization of epoxy resin-amphiphilic block copolymer nanocomposites, *J. Am. Chem. Soc.*, 119, 2749–2750, 1997.

30. Lipic, P.M., Bates, F., and Hillmyer, M., Nanostructured thermosets from self-assembled amphiphilic block copolymer/epoxy resin mixtures, *J. Am. Chem. Soc.*, 120, 8963–8970, 1998.

31. Grubbs, R., Broz, M., Dean, J., and Bates, F., Selectively epoxidized polyisoprene-polybutadiene block copolymers, *Macromolecules*, 33, 2308–2310, 2000.

32. Grubbs, R., Dean, J., Broz, M., and Bates, F., Reactive block copolymers for modification of thermosetting epoxy, *Macromolecules*, 33, 9522–9534, 2000.

33. (a) Dean, J., Lipic, P., Grubbs, R., Cook, R., and Bates, F., Micellar structures and mechanical properties of block copolymer-modified epoxies, *J. Polym. Sci. Polym. Phys.*, 39, 2996–3010, 2001. (b) Dean, J., Verghese, N., Pham, H., and Bates, F., Nanostructure toughened epoxy resins, *Macromolecules*, 36, 9267–9270, 2003.

34. Kosonen, H., Ruokolainen, J., Nyholm, P., and Ikkala, O., Self-organized thermosets: blends of hexamethyltetramine cured novolac with poly(2-vinylpyridine)-block-poly(isoprene), *Macromolecules*, 34, 3046–3049, 2001.

35. Ritzenthaler, S., Court, F., David, L., Girard-Reydet, E., Leibler, L., and Pascault, J.P., ABC triblock copolymers/epoxy-diamine blends. 1. Keys to achieve nanostructured thermosets, *Macromolecules*, 35, 6245–6254, 2002.

36. Ritzenthaler, S., Court, F., David, L., Girard-Reydet, E., Leibler, L., and Pascault, J.P., ABC triblock copolymers/epoxy-diamine blends. 2. Parameters controlling the morphologies and properties, *Macromolecules*, 36, 118–126, 2003.

37. Rebizant, V., Abetz, V., Tournilhac, F., Court, F., and Leibler, L., Reactive tetrablock copolymers containing glycidyl methacrylate. Synthesis and morphology control in epoxy-amine networks, *Macromolecules*, 36, 9889–9896, 2003.

38. Jansen, B.J.P., Rastogi, S., Meijer, H.E.H., and Lemstra, P.J., Rubber-modified glassy polymers prepared via chemically induced phase separation. 2. Mode of microscopic deformation studied by in-situ small angle x-ray scattering during tensile deformation, *Macromolecules*, 34, 4007–4018, 2001.

39. Jansen, B.J.P., Rastogi, S., Meijer, H.E.H., and Lemstra, P.J., Rubber-modified glassy polymers prepared via chemically induced phase separation. 3. Influence of the strain rate on the microscopic deformation mechanism, *Macromolecules*, 32, 6283–6289, 1999.

40. Jansen, B.J.P., Rastogi, S., Meijer, H.E.H., and Lemstra, P.J., Rubber-modified glassy polymers prepared via chemically induced phase separation. 4. Comparison of properties of semi- and full-IPNs, and copolymers of acrylate-aliphatic epoxy systems, *Macromolecules*, 32, 6290–6297, 1999.

41. Van Casteren, I.A., *Control of Microstructures To Induce Ductility in Brittle Amorphous Polymers*, Ph.D. thesis, Eindhoven University of Technology, The Netherlands, 2003.

42. Van Casteren, I.A., Van Trier, R.A.M., Goossens, J.G.P., Lemstra, P.J., and Meijer, H.E.H., The influence of hydrogen bonding on the preparation and mechanical properties of PS-diblock copolymer blends, *J. Polym. Sci., Polym. Phys.*, 42, 2137–2160, 2004.

43. Van Asselen, O.L.J., Van Casteren, I.A., Goossens, J.G.P., and Meijer, H.E.H., Deformation behavior of triblock copolymers based on polystyrene: an FT-IR spectroscopy study, *Macromolecular Symposia*, 205, 85–94, 2004.

44. Macosko, C., Jeon, H., and Schulze, J., Block copolymers and compatibilization: reactively formed, in *Encyclopedia of Materials: Science and Technology*, Elsevier Science, London, 2001.

45. Jérôme, R. and Pagnoulle, C., Key role of structural features of compatibilizing polymer additives, in *Reactive Blending in Reactive Polymer Blending*, Baker, W. Scott, C., and Hu, G., Eds., 2001, pp. 22–112.

46. Orr, C., Cernohous, J., Guegan, P., Hirao, A., Jeon, H., and Macosko, C., Homogeneous reactive coupling of terminally functional polymers, *Polymer*, 42, 8171–8178, 2001.

47. Charoensirisomboon, P., Saito, H., Inoue, T., Weber, M., and Koch, E. Crystallization in polyamide 6/polysulfone blends: effect of polysulfone particle size, *Macromolecules*, 31, 4963–4969, 1998.

48. Ibuki, J., Charoensirisomboon, P., Ougizawa, T., Inoue, T., Koch, E., and Weber, M. Reactive blending of polysulfone with polyamide: a potential for solvent-free preparation of the block copolymer, *Polymer*, 40, 647–653, 1999.

49. Charoensirisomboon, P., Chiba, T., Solomko, S., Inoue, T., and Weber, M. Reactive blending of polysulfone with polyamide: a difference in interfacial behavior between in situ formed block and graft copolymers, *Polymer*, 40, 6803–6810, 1999.

50. Charoensirisomboon, P., Inoue, T., and Weber, M. Interfacial behavior of block copolymers in situ-formed in reactive blending of dissimilar polymers, *Polymer*, 41, 4483–4490, 2000.

51. Charoensirisomboon, P., Inoue, T., and Weber, M., Pull-out of copolymer in situ-formed during reactive blending: effect of the copolymer architecture, *Polymer*, 41, 6907–6912, 2000.
52. Yin, Z., Koulic, C., Pagnoulle, C., and Jérôme, R. Reactive blending of functional PS and PMMA: interfacial behavior of in-situ formed graft copolymers, *Macromolecules*, 34, 5132–5139, 2001.
53. Yin, Z., Koulic, C., Pagnoulle, C., and Jérôme, R., Controlled synthesis of anthracene-labeled ω-amine polystyrene to be used as a probe for interfacial reaction with mutually reactive PMMA, *Macromol. Chem. Phys.*, 203, 2021–2028, 2002.
54. Yin, Z., Koulic, C., Jeon, H.K., Pagnoulle, C., Macosko, C., and Jérôme, R., Effect of molecular weight of the reactive precursors in melt reactive blending, Macromolecules, 35, 8917–8919, 2002.
55. Yin, Z., Koulic, C., Pagnoulle, C., and Jérôme, R., Probing of the reaction progress at a PMMA/PS interface by using anthracene-labeled reactive PS chains, Langmuir, 19, 453–457, 2003.
56. Yin, Z., Koulic, C., Pagnoulle, C., and Jérôme, R., Dependence of the morphology development on the kinetics of reactive melt blending of immiscible polymers, *Can. J. Chem. Eng.*, 80, 1044–1050, 2002.
57. Yin, Z., Koulic, C., Pagnoulle, C., and Jérôme, R., Dependence of the interfacial reaction and morphology development on the functionality of the reactive precursors in reactive blending, Macromol. Symp., 198, 197–208, 2003.
58. Yin, Z., Effect of Molecular Characteristics of the Reactive Precursors on Reactive Polymer Blending, Ph.D. thesis, University of Liège, Belgium, 2002.
59. Orr, C., Adedeji, A., Hirao, A., Bates, F., and Macosko, C., Flow induced reactive self-assembly, *Macromolecules*, 30, 1243–1246, 1997.
60. Pagnoulle, C., *Reactive Compatibilization of SAN/EPR Blends*, Ph.D. thesis, University of Liège, Belgium, 2000.
61. Pagnoulle, C., Koning, C., Leemans, L., and Jérôme, R., Reactive compatibilization of SAN/EPR blends. 1. Dependence of the phase morphology development on the reaction kinetics, Macromolecules, 33, 6275–6283, 2000.
62. Koulic, C., *Reactive Blending as a Tool towards Nanostructured Polymers*, Ph.D. thesis, University of Liège, Belgium, 2004.
63. Koulic, C., Yin, Z., Pagnoulle, C., and Jérôme, R., Vesicular nanostructures prepared by reactive melt blending, *Angew. Chem. Int. Ed.*, 41, 2154–2156, 2002.
64. Koulic, C. and Jérôme, R., Nanostructured PMMA: from lamellar sheets to double-layered vesicles, Macromolecules, 37, 888–893, 2004.
65. Koulic, C. and Jérôme, R., Nanostructured polyamide by reactive blending. 1. Effect of the reactive diblock composition, Macromolecules, 37, 3459–3469, 2004.
66. Kinning, D.J., Winey, K., and Thomas, E.L., Structural transitions from spherical to nonspherical micelles in blends of poly(styrene-butadiene) diblock copolymer and polystyrene homopolymers, Macromolecules, 21, 3502–3506, 1988.
67. Harada, T., Bates, F., and Lodge, T., Transverse orientation of lamellae and cylinders by solution extrusion of a pentablock copolymer, Macromolecules, 36, 5440–5442, 2003.
68. Schlienger, M., Contribution à l'étude des copolymères triséquencés poly(styréne-b-isoprene-b-methacrylate de méthyle). Préparation d'alliages de polyméres, Ph.D. thesis, Mulhouse, France, 1976.
69. Riess, G., Schlienger, M., and Marti, S., New morphologies in rubber-modified polymers, *J. Macromol. Sci.-Phys.*, B17, 355, 1989.

70. Hamley, I.W., *The Physics of Block Copolymers*, Oxford University Press, Oxford, 1998.
71. Chen, H., Lin, S., Huang, Y., Chiu, F., Liou, W., and Lin, J., Crystallization in the vesicle walls templated by dry-brush block copolymer/homopolymer blend, Macromolecules, 35, 9434–9440, 2002.

11 Phase Morphology of Nanostructured Thermosetting Multiphase Blends

Jean-Pierre Pascault and Roberto J.J. Williams

CONTENTS

11.1 INTRODUCTION

Thermosetting polymers are polymer networks formed by the chemical reaction of monomers, at least one of which has a functionality higher than two. From the point of view of the polymer growth mechanism, two entirely different processes, step and chain polymerization, are distinguishable. Typical examples are phenolics (step growth polymerization with elimination of a low molar mass product), epoxies (step or chain growth polymerization depending on the comonomer used), and unsaturated polyesters (chain growth polymerization). In the course of the reaction, two principal structural transformations may occur: gelation and vitrification (1). Most commercial formulations include several other components in addition to monomers like catalysts, accelerators, pigments, fillers, rubbers and thermoplastic polymers. The initial formulation may be a homogeneous single phase system or a multiphase system. In the case of an initial homogeneous solution, the polymerization may lead to a multiphase structure because of topological, mechanistic, or thermodynamic reasons

(2). A typical example of thermodynamically driven phase separation is the generation of rubber modified thermosets or thermoset/thermoplastic polymer blends (3,4).

All physical properties or phenomena have a characteristic length, and when the size of an object is similar to a characteristic length, new and surprising properties can emerge. Multiphase structures may have a broad range of characteristic sizes. In the search of advanced materials with novel properties, materials strategies have targeted reducing the dimensions of the constituent components in multiphase polymers. In this chapter, we analyze different ways to synthesize multiphase thermosetting polymers where at least one of the phases exhibits a characteristic dimension in the range of 1 to 100 nm. These nanostructured materials have attracted interest in recent years because of their applications in areas such as low dielectric constant films for the microelectronics industry, (5–7) transparent materials with high refractive indices for optical applications (8–10), and toughening of polymer networks (11,12).

Thermosetting polymers have some advantages compared to thermoplastic ones. The low viscosity of thermoset precursors facilitates processing. On the other hand, network formation (e.g., chemistry, reaction rate, increase of viscosity) can compete with thermodynamics for controlling the morphology of the final material.

There are different possible ways to generate nanostructured thermosets:

a. In the course of polymerization
b. By self-assembly before polymerization and fixation by cross-linking
c. By the use of functionalized nanoparticles
d. After network formation

These depend on whether phase separation takes place (a) during network formation, (b) in the initial solution as a result of thermodynamic reasons, (c) because of the addition of preformed nanoparticles, or (d) after network formation by varying the temperature or by a chemical modification. A combination of these possibilities may also take place. For example, an initially nanostructured system may also undergo a phase separation in the course of polymerization. This is why nanostructure formation during a reaction is discussed first.

In the following sections, these different methods for synthesizing nanostructured thermosetting multiphase blends are analyzed. Particular emphasis is placed on the discussion of factors that control the morphologies generated.

11.2 NANOSTRUCTURE FORMATION IN THE COURSE OF POLYMERIZATION

Here, initial homogeneous solutions generate nanosize domains during the polymerization reaction. Two different cases may be considered: (a) phase separation taking place in unmodified thermoset formulations driven by topological and/or mechanistic factors, and (b) phase separation under controlled conditions in solutions of a modifier in thermoset precursors driven by thermodynamic factors.

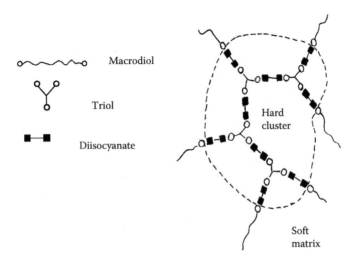

FIGURE 11.1 Schematic representation of a polyurethane network composed of a long diol, a triol, and a diisocyanate.

11.2.1 PHASE SEPARATION DRIVEN BY TOPOLOGICAL AND MECHANISTIC FACTORS

A typical example is the polymerization of a solution of a long (soft) diol, a short (hard) triol, and a short (hard) diisocyanate to give a polyurethane network containing hard topological clusters that act as cross-linking units (Figure 11.1). The distribution of cluster sizes is fully controlled by the initial composition of the system, the order of addition of different monomers, and the reactivities of the functional groups. For example, if every one of the monomers is present in the initial solution and the triol reacts with the diisocyanate at a higher rate than the diol, the clusters will grow big initially until depletion of the triol fixes their final dimensions. If the volume fraction of hard clusters surpasses a particular threshold value, percolation of the hard structure in the soft matrix takes place.

Another typical case of generation of nanostructures in unmodified thermosets is the free radical cross-linking polymerization of multivinyl monomers. In this case, compact cross-linked structures (microgel particles) are formed in the early stages of the reaction. Because of topological reasons, intramolecular cyclization reactions prevail over intermolecular cross-linking, leading to the early formation of nanosize domains of high cross-link density. This also results in a decrease in reactivity of pendant double bonds resulting from steric hindrance (13).

11.2.2 PHASE SEPARATION DRIVEN BY THERMODYNAMIC FACTORS

A suitable modifier (e.g., rubber, thermoplastic polymer, liquid crystal) may be dissolved in thermoset precursors, leading to a phase separation process in the course of polymerization. Thermodynamic factors that drive the polymerization induced phase separation (PIPS) have been extensively discussed in the literature

(1–4,14–16). They are: (a) the increase in the average size of the species of the thermosetting polymer and the corresponding decrease in the entropy of mixing (prevailing factor in the pregel stage); (b) the variation of the interaction parameter between the modifier and thermoset species as a result of the modification of the chemical structure produced by the reaction; and (c) the elastic contribution to free energy in the postgel stage.

PIPS is the usual technique for the synthesis of rubber modified thermosets, thermoset/thermoplastic blends and polymer dispersed liquid crystals (PDLCs). In most cases, the characteristic dimensions of phase segregated domains are in the micrometer range. Two questions are pertinent at this point: Why is it convenient to reduce the characteristic dimensions to the nanometer range in some particular systems? How can it be achieved? We will first try to provide an answer to both questions by analyzing some illustrative examples.

The characteristic size of dispersed domains produced in the course of a polymerization induced phase separation is determined by the competition between the coarsening rate (depending on the viscosity of the medium) and the polymerization rate (that fixes the time available for phase separation). If it is desired to reduce the size of dispersed domains from the usual micrometer range to the nanometer range, either the coarsening rate should be decreased or the polymerization rate increased.

Different actions may be taken to decrease the coarsening rate: (a) reducing the initial size of the dispersed domains by the use of an emulsifying agent, (b) increasing the viscosity of the medium at the time of phase separation by adding a third component or by producing phase separation close to gelation or vitrification, and (c) selecting conditions to produce phase separation in the postgel stage. Chemical structures and specific interactions between the modifier and the thermosetting polymer are particularly important in the determination of their miscibility. The more miscible they are, the higher the cloud-point conversion will be and the smaller the size of the generated dispersed domains.

The most obvious action to increase the polymerization rate is to increase the temperature. However, this also leads to a decrease in viscosity (and a corresponding increase in the coarsening rate) as well as to a shift of the miscibility region (the direction of this shift depends on whether the system exhibits an upper or a lower critical solution–temperature behavior). Complex trends have been found for the dependence of the average size of the dispersed domains versus temperature in particular systems (3,17). The polymerization rate may be also increased by adding specific catalysts or increasing the initiator amount used in a chainwise polymerization. It also may be increased by changing the type of initiation process, e.g. replacing thermal initiation by UV (ultraviolet light) cure. The industrial practice usually requires very fast polymerization processes like the UV cure. In this case, we may find the opposite situation: a way of obtaining a dispersed phase of required dimensions in extremely short times. A case where this can be achieved by using hydrogen bonded supramolecular polymers is discussed in this section (18).

Solutions of poly(methyl methacrylate) (PMMA) and diglycidyl ether of bisphenol A (DGEBA) are examples where the formation of nanosize domains is the result of the high initial miscibility between both components. Both components are completely miscible in the whole composition range. Depending on the curing agent

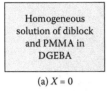
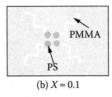
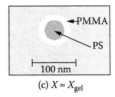

(a) $X = 0$ (b) $X \approx 0.1$ (c) $X \approx X_{gel}$

FIGURE 11.2 Schematic representation of different phase separation processes taking place in a solution of a poly(S-*b*-MMA) copolymer and PMMA in DGEBA that is homopolymerized in the presence of a tertiary amine: (a) initial homogeneous solution; (b) phase separation of PS blocks, at low conversions, leading to the formation of micelles stabilized by PMMA hairs; and (c) phase separation of PMMA at conversions close to gelation.

selected to polymerize DGEBA, phase separation may take place at advanced conversions or it may not occur at all. The use of a stoichiometric amount of 4,4-diaminodiphenyl sulfone (DDS) produced phase separation at conversions located in the pregel region, leading to a dispersion of PMMA domains with sizes in the micrometer range (19). However, when 4,4-methylenebis (3-chloro 2,6-diethylaniline) (MCDEA) was used as to cross-link DGEBA, no phase separation was detected in the whole conversion range (20). When DGEBA was homopolymerized in the presence of a tertiary amine, phase separation of a PMMA-rich phase took place close to gelation, leading to dispersed domains with sizes in the nanometer range (21). The average size of PMMA-rich domains could be increased by adding a small amount of a block copolymer P(S-*b*-MMA) to the initial solution (Figure 11.2a). As in this case, because thermoset precursors were poor solvents for homoPS (homopolystyrene), phase separation of the PS blocks occurred at low conversions, leading to PS micelles stabilized by miscible PMMA hairs (Figure 11.2b). Close to gelation, PMMA was phase separated with the micelles acting as nucleation centers, thus increasing the size of dispersed PMMA domains to the range of 100 to 200 nm (Figure 11.2c) (21). This is a rather unusual case where a block copolymer is added to increase the final size of the dispersed domains.

The use of block copolymers to decrease the dimensions of dispersed domains, can be illustrated by results obtained in a poly(phenylene ether) (PPE)/epoxy (DGEBA-MCDEA) system. In a first example (22), the selected diblock copolymer was a symmetrical poly(S-*b*-MMA) that was chosen because (a) homoPS interacts very favourably with PPE (this is one of a few examples of completely soluble polymers) and (b) homoPMMA was shown to remain miscible with the DGEBA-MCDEA polyepoxide system during the whole curing process regardless of the concentration and the temperature (20). But as phase separation of the PPE occurred before phase separation of the PS block of the diblock, a complex two scale structure with "splatlike" PPE-rich macrodomains coexisting with filamentlike and spherical microdomains is observed (Figure 11.3). This unusual morphology can be explained by the fact that the diblock is located at the interphase where it stops the coalescence of PPE-rich particles and drives the interfacial tension negative after a delay, as drawn schematically in Figure 11.4. Even if the emulsifier role of block copolymers is more complex in thermoplastic/thermoset (TP/TS) blends than in TP/TP blends, it can be used to control the interface structure.

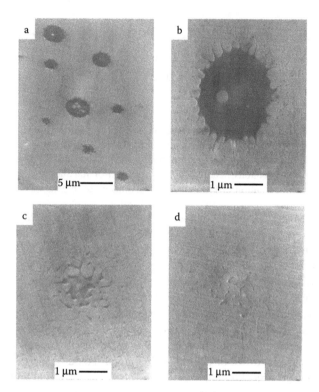

FIGURE 11.3 TEM micrographs of the ternary blend containing 10 wt% PPE plus 2 wt% poly(S-*b*-MMMA) after curing at 135°C at two magnifications: (a) x 3000 and (b to d) x 13000. (Reprinted with permission from Girard-Reydet et al., *Macromolecules*, 34, 5349–5353, 2001; copyright 2004 American Chemical Society.)

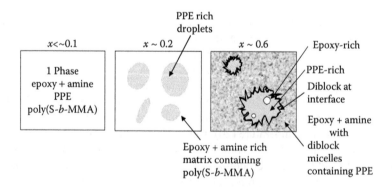

FIGURE 11.4 Schematic of the different stages of the demixing of the ternary blend containing 10 wt% PPE plus 2 wt% poly(S-*b*-MMMA), during reaction at 135°C. (Reprinted with permission from Girard-Reydet et al, *Macromolecules*, 34, 5349–5353, 2001; copyright 2004 American Chemical Society.)

In a second example (23), the selected block copolymer was the commercially available poly(S-*b*-EB-*b*-S); (EB: ethylene-*co*-buthylene). The central elastomeric block was grafted with maleic anhydride (MA), which was in turn capped with MCDEA (23). This block copolymer has the following properties: The central block is insoluble in the initial solution, and, as explained previously, PS blocks are soluble with PPE in all proportions. When PPE becomes phase separated, the block copolymer acts as an emulsifying agent, reducing the average size of the dispersed domains to the nanometer range. This is due to the dissolution of the PS blocks in the PPE domains, placing the EB blocks at the interface with the grafted MCDEA units extending towards the epoxy matrix. Dynamic mechanical spectra showed the appearance of a new relaxation located between β and α relaxations of the epoxy matrix, assigned to the elastomeric interphase. The critical stress intensity factor increased from 0.6 to 1 MPa m$^{1/2}$ (23).

Another approach to decrease the size of dispersed domains by one order of magnitude in the same PPE/epoxy system was to perform the cure at temperatures close to the glass transition temperature (T_g) of the initial solution (24). Under these conditions, the coarsening rate was decreased much more than the polymerization rate, leading, for some particular compositions, to a dispersed phase in the nanometer range. During the additional postcuring steps, necessary to reach a maximum epoxy conversion, the original morphologies were maintained.

Nanocomposites produced by cross-linking silsesquioxane precursors in the presence of sacrificial organic macromolecules is becoming an area of much interest because of the applications of the resulting nanoporous materials in such fields as the microelectronics industry, gas separation membranes, and catalysts. Silsesquioxane precursors may be synthesized by the hydrolysis and condensation of organotrialkoxysilanes, generating a distribution of multifunctional oligomers containing unreacted SiOH groups. Their condensation, at relatively high temperatures, leads to a cross-linked network. An appropriate organic polymer may be dissolved in the silsesquioxane precursor, leading to an initial homogeneous solution. Phase separation of the organic polymer occurs in the course of the cross-linking reaction, leading to dispersed organic domains with characteristic sizes determined by the cloud-point conversion. In turn, this largely depends on the chemical affinity between both components. Increasing the chemical compatibility will delay the cloud-point conversion, thus decreasing the characteristic size of dispersed domains. In a subsequent step, the organic phase is eliminated by thermal degradation leading to the desired nanoporous material (25).

One of the applications of these nanoporous films is for spin-on ultralow dielectric constant (ULK) materials (5–7). Current chip manufacturing processes require the films to be hydrophobic to avoid moisture adsorption and with closed cell pores (the average pore size being 15 nm or less) to optimize mechanical properties and avoid environmental contamination (26). Closed cell pores may be obtained by burning out the sacrificial polymer at temperatures where the cross-linked matrix exhibits enough rigidity to avoid pore collapse. But how can nanodomains of a characteristic size as low as 15 nm be generated? This question may be transposed into this other question: How can the chemical affinity between the sacrificial macromolecule and the silsesquioxane precursor be controlled?

A particular answer to the last question is given in a recent paper by Huang et al. (26). They used two different commercial methylsilsesquioxane precursors with similar distribution of molar masses but containing different amounts of unreacted SiOH groups. As a sacrificial molecule, they used poly(methyl methacrylate-*co*-(dimethylamino)ethyl methacrylate), poly(MMA-*co*-DMAEMA). The initial miscibility between both components was promoted by strong hydrogen bonding interactions between SiOH groups and the tertiary amino group in poly(MMA-*co*-DMAEMA). The lower the initial silanol content in the silsesquioxane precursor, the greater the immiscibility of both components. This led to lower cloud-point conversions during polymerization at 180°C and larger sizes of the nanodomains generated (Figure 11.5). The lower polycondensation rate of the silsesquioxane precursor containing less silanol groups might also contribute to the coarsening of dispersed domains.

The inverse problem of those analyzed in this section is related to cases where the polymerization is so fast that what is necessary is to find ways to increase the phase separation rate. This is the case for the UV cure of acrylate films exposed to radiation for very short periods (e.g., typically less than 1 sec). Phase separation of small molecules, like liquid crystals, can proceed at a fast rate under these conditions. UV cure is the usual process to synthesize polymer dispersed liquid crystals with dispersed phase sizes in the micrometer range (27). However, the problem is much more complex for conventional polymers because of their reduced mobility. A possible solution relies on the use of supramolecular polymers (28).

Supramolecular polymers are a new class of polymers in which repeating units are connected by noncovalent interactions. An example of hydrogen bonded supramolecular polymers developed on the basis of the self-complementary quadruple hydrogen bonding ureido-pyrimidinone group is shown in Figure 11.6 (18). These polymers may be dissolved in convenient monoacrylate-diacrylate solvents. Phase separation of the supramolecular polymer can efficiently take place during the UV cure, as schematically shown in Figure 11.7.

What is interesting is the fact that the relatively low average degree of polymerization of the supramolecular polymer in solution enables a fast phase separation. But association by hydrogen bonds can take place easily in the segregated domains, regenerating the properties of the bulk polymer. The size of the droplets containing supramolecular polymer decreased when increasing the amount of diacrylate, resulting in dispersed domain sizes varying from the micrometer to the nanometer range (18). The diacrylate addition decreased the length scale of the phase separation by shortening the vitrification time. Mechanical properties of the films containing the supramolecular polymer were comparable to films containing high molar mass polymers.

11.3 SELF-ASSEMBLY BEFORE POLYMERIZATION AND FIXATION BY CROSS-LINKING

The focus of this section is on systems that self-assemble before polymerization, leading to ordered or disordered nanostructures that may be fixed by the cross-linking

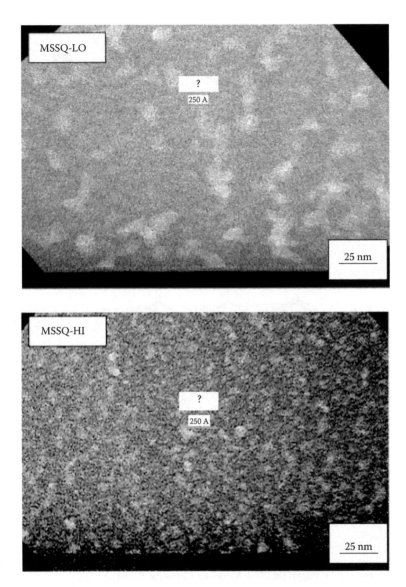

FIGURE 11.5 Bright field TEM images of nanoporous films obtained with silsesquioxane precursors containing (top) low and (bottom) high concentrations of residual SiOH groups and an initial 35 wt% of methacrylate copolymer. (Reprinted with permission from Huang et al., *Macromolecules*, 36, 7661–7671, 2003; copyright 2004 American Chemical Society.)

reaction. Examples of these systems are: (a) blends of amphiphilic block copolymers with reactive solvents and (b) liquid crystalline thermosets.

Amphiphilic block copolymers are made up of two or more blocks of different chemical natures. At least one of these blocks is immiscible with thermoset precursors while at least another one is initially miscible but may phase separate in the course of polymerization. As an example, Figure 11.8 shows schematics of experi-

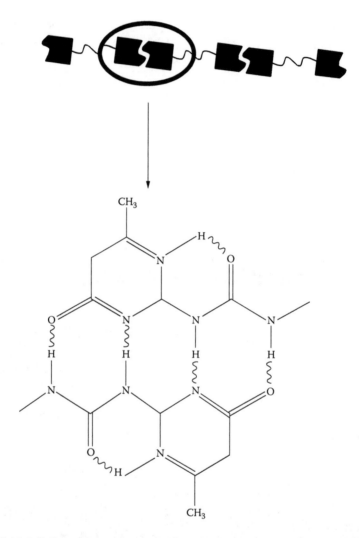

FIGURE 11.6 Self-complementary quadruple hydrogen bonding ureido-pyrimidinone group in a supramolecular polymer. (Reprinted with permission from Keizer et al., *Macromolecules*, 36, 5602–5606, 2003; copyright 2004 American Chemical Society.)

mental phase diagrams obtained for blends of a low molar mass poly(ethylene oxide)-*b*-(ethylene-*alt*-propylene), (PEO-PEP) diblock copolymer and an epoxy monomer based on diglycidyl ether of bisphenol A (DGEBA) (29). PEO is miscible with DGEBA in all proportions while PEP is immiscible. This gives place to different morphologies that depend on the amount and composition of the block copolymer (L: lamellar, G: bicontinuous cubic gyroid, C: hexagonally packed cylindrical, and S: body-centered cubic packed spherical). In the diluted solution range, the block copolymer may be present either as a dispersion of spherical micelles (a PEP core with PEO brushes solvated with the epoxy monomer) or as onion-like multilayer vesicles with both internal and external PEO layers solvated with the epoxy mono-

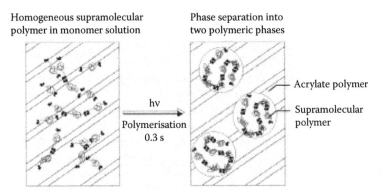

FIGURE 11.7 Schematic representation of PIPS using hydrogen bonded supramolecular polymers. *Left:* Low-molecular-mass supramolecular polymer in monomer solution. *Right:* High-molecular-mass supramolecular polymer. (Reprinted with permission from Keizer et al., *Macromolecules*, 36, 5602–5606, 2003; Copyright 2004 American Chemical Society.)

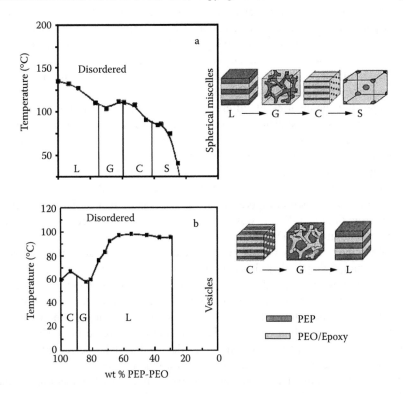

FIGURE 11.8 Phase diagrams of blends of a low molar mass poly(ethylene oxide)-*b*-(ethylene-*alt*-propylene), (PEO-PEP) diblock copolymer and an epoxy monomer based on diglycidyl ether of bisphenol A (DGEBA): (a) PEO-PEP copolymer with M_n = 2700 g/mol and a volume fraction of PEO, f_{PEO} = 0.51, and (b) PEO-PEP copolymer with M_n = 2100 g/mol and a volume fraction of PEO, f_{PEO} = 0.26. (Dean et al., *J. Polym. Sci., B: Polym. Phys.*, 39, 2996–3010, 2001; copyright 2004 John Wiley & Sons, Inc. With permission.)

mer. All the ordered structures exhibited characteristic dimensions in the nanometer range, and the resulting blends were transparent (29–31). The same happened with the dispersion of spherical micelles. However, vesicles may attain larger sizes, leading to cloudy dispersions (29).

An extensive literature exists on the rheology of block copolymers in selective solvents [e.g., refs. (32,33)]. A change of the solvent selectivity has a great influence on the order-disorder transition and can be used to control the rheological behaviour and the processing of solutions of TS precursors (34). When the block copolymer was blended with a stoichiometric mixture of a diepoxide and a diamine,, i.e., methylenedianiline (MDA),, the order-disorder transition temperatures (shown in the phase diagram of Figure 11.8a) were shifted to higher values (31). Therefore, the epoxy-amine polymerization was carried out in an already nanostructured system. The problem now is how to keep the initial nanostructure intact in the course of the epoxy-amine polymerization. If PEO remains completely soluble in the polymer network, there would be no significant perturbation capable of disintegrating the initial nanostructure. However, PEO is segregated from a DGEBA-MDA system during polymerization. Fortunately, for the selected molar mass of PEO segments, phase separation was delayed to conversions close to gelation or to the postgel stage (31). Under these conditions, coarsening of PEO-rich domains was not produced and the original nanostructure did not disintegrate. Figure 11.9 shows a TEM image of the final nanostructure for a particular initial composition leading to a hexagonally-packed cylindrical morphology (31).

Composition boundaries of different ordered morphologies should vary with conversion because of the changes in the chemical structure and in the average size of epoxy-amine species. Besides, phase separation produces an increase of the PEO concentration in the segregated phase, meaning that equilibrium morphologies can be shifted to regions of higher PEO concentrations after the beginning of phase separation. This was in fact the case for the PEO-PEP/DGEBA-MDA system with the phase diagram shown in Figure 11.8a, where for particular initial compositions, a change from G to L or S to C was captured by *in situ* small angle x-ray scattering (SAXS) measurements (31).

Self-assembly of block copolymers in the diluted concentration region leads to disordered morphologies that also may be fixed by the curing reaction. The shape of these morphologies depends on the volume fraction of the block miscible with the thermoset precursors (e.g., the epoxy-amine species). When the volume fraction of the miscible block increases, the resulting structures change from vesicles to wormlike micelles and, finally, to spherical micelles (35). Figure 11.10 shows TEM images of spherical micelles, wormlike micelles, and vesicles for cured blends of small amounts of block copolymers in epoxy formulations (36).

Vesicles can be effective in toughening epoxy at relatively low loadings because of their particular structures (29). They are closed objects consisting of a thin (approximately 10 nm) bilayer membrane formed by the block copolymer that encapsulates the thermoset precursors. This means that a small amount of the block copolymer can produce a large volume fraction of dispersed domains. It was found that the average size of vesicles increased during storage at room temperature, giving proof to the fact that they do not constitute an equilibrium morphology (35).

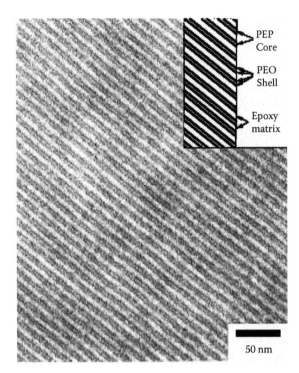

PEP
Core

PEO
Shell

Epoxy
matrix

50 nm

FIGURE 11.9 TEM image of the cylindrical nanostructure fixed in a blend of a PEO-PEP diblock copolymer in a cured epoxy. The morphology consists of cylindrical PEP cores surrounded by a PEO shell, enclosed by the epoxy matrix. (Reprinted with permission from Lipic et al., *J. Am. Chem. Soc.*, 120, 8963–8970, 1998; copyright 2004 American Chemical Society.)

Wormlike micelles are composed of long, thin tubes of the immiscible block stabilized by a corona of the block miscible in the thermoset precursors. An impressive 70-fold increase in the fracture resistance of epoxy formulations used in electronic applications was found when wormlike micelles were generated by addition of 5 wt% of an appropriate PEO-PEP block copolymer (36). When the PEO volume fraction in the block copolymer was increased, spherical micelles were obtained at the same concentration in the blend. This produced a 40-fold increase of the fracture resistance (36).

Several other studies of blends of block copolymers and thermoset precursors, where self-assembled nanostructures were preserved in the cured material, have been reported (37–45). The use of an amphiphilic block copolymer consisting of a crystallizable immiscible block — polyethylene (PE) — and a typical miscible block (PEO) in a blend with an epoxy formulation led to the formation of segregated crystalline domains (45). For block copolymer contents less than 30 wt%, PE was present as nanocrystals in the core of spherical micelles dispersed in the cured epoxy. Crystallization inside micelles proceeded by homogeneous nucleation and required large undercoolings. Higher concentrations of block copolymer led to coarsening of the crystal structure; for copolymer concentrations higher than 50 wt%, percolation

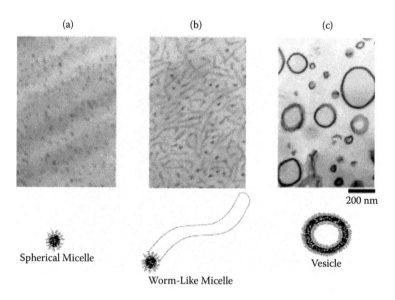

FIGURE 11.10 TEM images of (a) spherical micelles, (b) worm-like micelles, and (c) vesicles generated in cured blends of block copolymers and epoxy formulations. (Reprinted with permission from Dean et al., *Macromolecules,* 36, 9267–9270, 2003, copyright 2004 American Chemical Society.)

of the crystalline structure was observed, with spherulite sizes ranging from 10 to 30 μm in diameter (45).

The use of ABC triblock and ABCD tetrablock copolymers leads to a variety of nanostructures in the cured thermosets (46,47). A triblock copolymer composed of 22 wt% polystyrene, 9 wt% polybutadiene (PB), and 69 wt% poly(methyl methacrylate), with M_n = 122 kg/mol, was blended in a proportion of 50 wt% with two different epoxy formulations: one based on DGEBA-MCDEA and the other one on DGEBA-DDS.[46] Before the reaction, PS and PB blocks were immiscible with both thermoset precursors while the PMMA was completely miscible (although the pure PS homopolymer was initially soluble in both epoxy formulations, the PS-PB block was immiscible).

Figure 11.11a shows a TEM image of the self-assembled nanostructure before reaction observed for both epoxy formulations. Bright spherical particles, with diameters close to 60 nm, are the PS blocks. PB blocks are the dark spherical nodules with diameters in the range of 24 nm. These "raspberrylike" particles are dispersed in the matrix of epoxy precursors. PMMA blocks have one end covalently attached to the dark nodules and the other end immersed in the epoxy matrix because of its complete miscibility with both epoxy precursors. Partial phase separation of PMMA in the course of polymerization leads to two different types of morphologies in both systems (Figures 11.11b and 11.11c). For the DGEBA-MCDEA system, the initial nanostructure is fixed by the cross-linking reaction because segregation of PMMA was only partial, as revealed by the decrease of the matrix T_g at complete conversion and the fact that it occurred at advanced conversions (possibly after gelation). PMMA

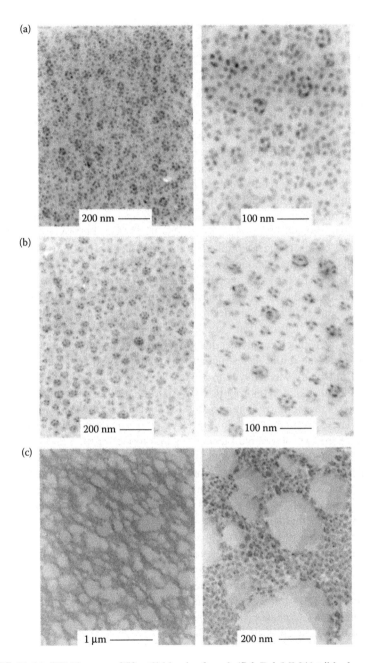

FIGURE 11.11 TEM images of 50 wt% blends of a poly(S-*b*-B-*b*-MMA) triblock copolymer in an epoxy formulation (DGEBA-MCDEA or DGEBA-DDS): (a) before the reaction to produce the DGEBA-MCDEA formulation (the picture is quite similar for the other epoxy system); (b) after the reaction for the DGEBA-MCDEA system; and (c) after the reaction for the DGEBA-DDS system. (Reprinted with permission from Ritzenthaler et al., *Macromolecules*, 35, 6245–6254, 2002; copyright 2004 American Chemical Society.)

segregation was confirmed by the presence of a shoulder in the tan δ relaxation peak. For the DGEBA-DDS system, phase separation of PMMA blocks took place at relatively low conversions (at a conversion of about 0.3 compared to a gel conversion of 0.6). This led to a macroscopic phase separation produced by the flocculation of the initial nanodomains.

Therefore, it may be concluded that the preservation of the initial nanostructure requires that the miscible block phase separates at high conversions or does not phase separate at all. This constitutes a limitation for the possible number of pairs of miscible blocks and thermoset precursors that separate at high conversions or remain miscible at full conversion.

A possible way to circumvent this limitation is to add a small amount of another block capable of reacting with the thermoset precursors. This can limit the extent of macroscopic phase separation, resulting in a nanostructured thermoset. This possibility was recently explored for a PS-PB-PMMA-PGMA [where PGMA: poly(glycidyl methacrylate)] tetrablock copolymer blended with the DGEBA-DDS formulation (47). A small amount (3 or 8 wt%) of a reactive PGMA block was added to the PS-PB-PMMA triblock copolymer. This block participates in the cross-linking reaction and avoids a macroscopic phase separation. Figure 11.12 shows the initial, intermediate, and final morphologies for the blend of the tetrablock copolymer with the DGEBA-DDS formulation. Initial morphologies were analogous to those observed for the triblock copolymer devoid of PGMA. The addition of 8 wt% PGMA block preserved the initial nanostructure throughout the polymerization. When the fraction of this block was reduced to 3 wt%, some coarsening of the structure took place. Coalescence of neighboring raspberry particles led to the formation of worm-like particles, consisting of homogeneous PS domains surrounded by a continuous PB layer. These morphologies might produce a significant increase in the fracture resistance of the nanostructured blend compared to that of the neat epoxy.

Liquid crystalline thermosets constitute another example of systems that exhibit an initial nano- or microstructure formation that may be fixed by cross-linking. Thermosets based on mesogenic monomers that retain their mesomorphic character after polymerization have been reported since the late 1980s (48,49). In some cases, low molar mass nematogenic monomers led to networks with a smectic molecular organization instead of the nematic phase of the precursors (50). The higher organization level was ascribed to the regular distance between cross-link sites. For networks with high cross-link densities, the mesophase could not be transformed into an isotropic phase by increasing temperature up to the decomposition of the organic material (50). In recent years, materials for optoelectronic applications that form supercooled LC-phases (liquid crystal glasses), have been obtained by photo-polymerization of reactive mesogens (51–53).

11.4 USE OF FUNCTIONALIZED NANOPARTICLES

Another approach to synthesize nanostructured thermosets is to dissolve function-alized nanoparticles (macromonomers) in a reactive solvent. If conditions are selected to avoid their phase separation initially or in the course of polymerization, nanoparticles keep their initial dimensions and give place to a nanostructured ther-

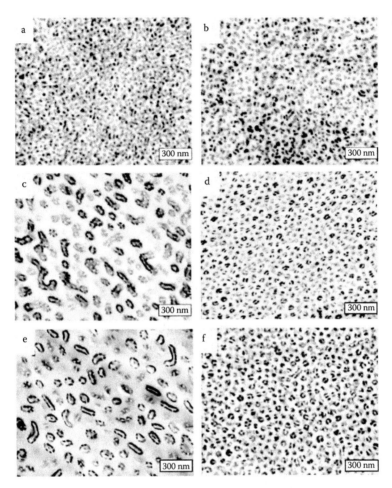

FIGURE 11.12 TEM images of 30 wt% blends of poly(S-*b*-B-*b*-MMA-*b*-GMA) tetrablock copolymers in a DGEBA-DDS formulation: (a) and (b) before the reaction; (c) and (d) after the cure at 135°C; and (e) and (f) after the postcure at 220°C. (a), (c), and (e) 3 wt% of the polyGMA block. (b), (d), and (f) 8 wt% of the polyGMA block. (Reprinted with permission from Rebizant et al., *Macromolecules*, 36, 9889–9896, 2003; copyright 2004 American Chemical Society.)

moset. Nanoparticles may contain only one reactive group, in which case they become part of pendant chains, or they can be functionalized with multiple reactive groups, thus behaving as cross-linking units of the polymer network. In this section, some examples of the use of functionalized nanoparticles as modifiers of a polymer network are discussed.

Polyhedral oligomeric silsesquioxanes (POSS), $(RSiO_{1.5})_n$, with $n = 6, 8, 10,...$, have nanosized cage structures that can be incorporated into linear or thermosetting polymers to improve their thermal and oxidation resistance and to reduce their flammability (54–59). Schematic structures of mono- and multifunctional POSS are shown in Figure 11.13 to Figure 11.16. Silsesquioxane cubes are the most typical

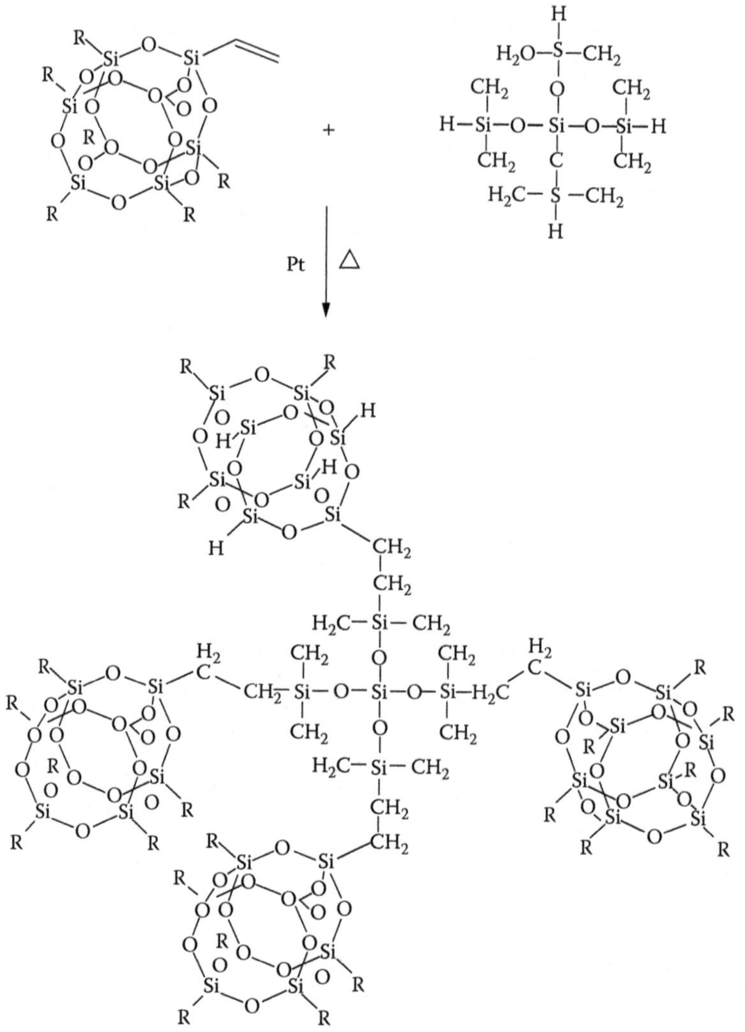

FIGURE 11.13 Chemical reactions leading to a cluster of four nonfunctionalized POSS cages as a model of a silica filler (R are isobutyl groups).

species of this family. They can be thought of as the smallest silica particles possible, providing an exactly defined hard particle (1.2 nm in diameter), to which eight organic groups may be appended. POSS modified polymers constitute the simplest models of silica reinforced composites.

Small amounts (5 to 10 wt%) of monofunctional POSS containing one epoxy group per cube were covalently bonded to epoxy-amine networks (60,61). The introduction of these bulky monomers led to a slight increase in the glass transition temperature and retarded the physical aging process in the glassy state. However, the use of 50 wt% of a monofunctional-epoxy POSS led to a macroscopic phase separation in POSS-rich and epoxy-rich regions (62). Another study reported the

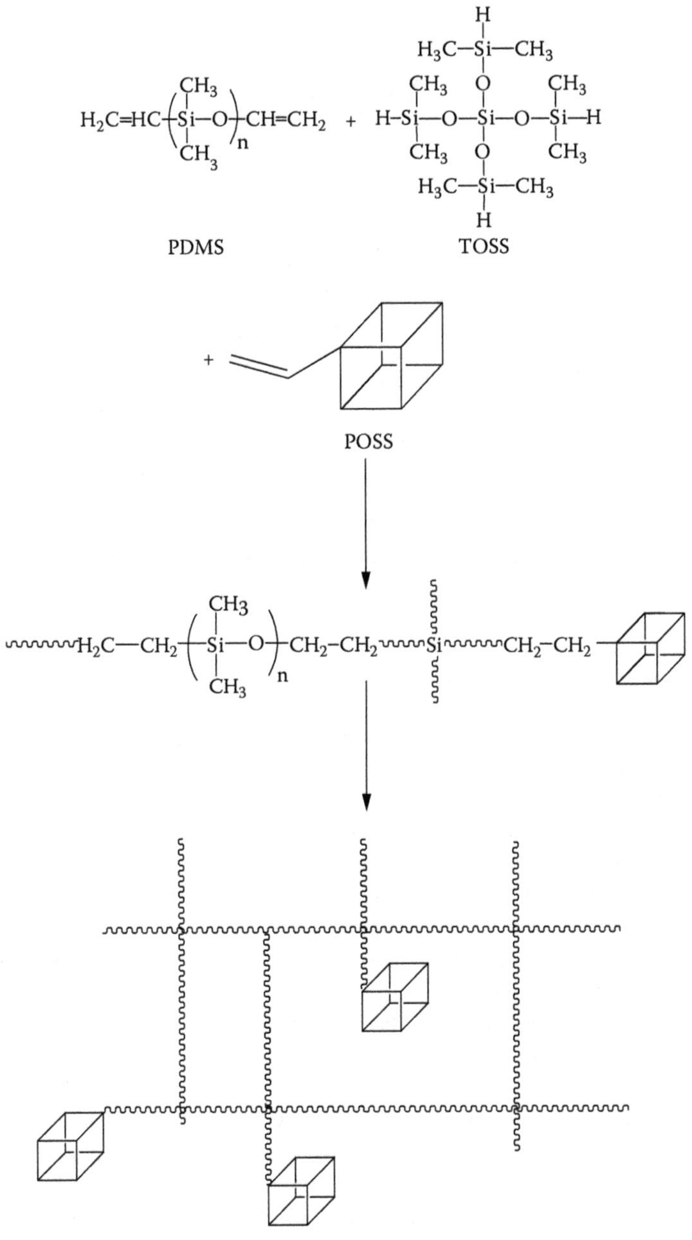

FIGURE 11.14 Chemical incorporation of vinyl-POSS cages into PDMS networks.

formation of aggregates composed of approximately three to four POSS molecules when using 20 wt% of a monofunctional norbornenylethyl-POSS to modify a poly(dicyclopentadiene) network (63). Avoiding the formation of POSS aggregates, or controlling their size, is the main problem related to the use of monofunctional POSS to generate nanostructured thermosets. This may be performed by selecting

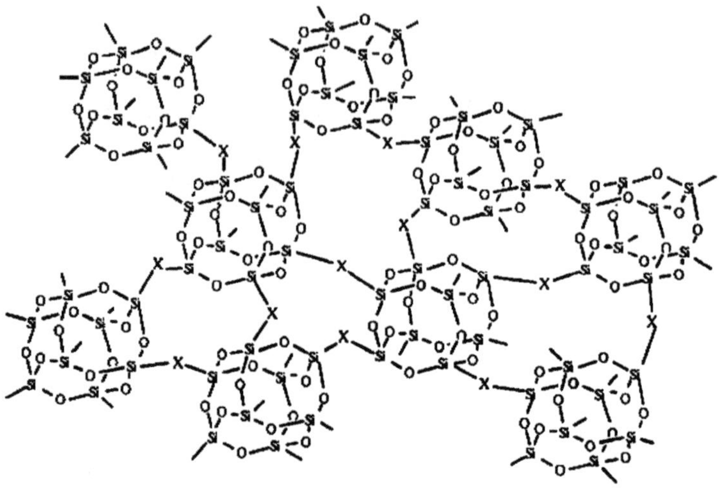

X = -CH₂-CH₂- (1 + 2 ⟶ A)

X = -O-Si(CH₃)₂-CH₂-CH₂- (1 + 4 ⟶ B, 3 + 2 ⟶ C)

X = -O-Si(CH₃)₂-CH₂-CH₂-Si(CH₃)₂-O- (3 + 4 ⟶ D)

FIGURE 11.15 Polymer network produced by the polycondensation of silsesquioxane cubes bearing complementary functionalities.

a POSS where the (seven) nonreactive organic groups are miscible with the polymer network or by performing the polymerization at a rate fast enough to have kinetic control over the phase separation.

Because of the tailorability of POSS molecules, they can also be designed to probe the molecular basis of reinforcement. Typical elastomers based on cross-linked polydimethylsiloxane (PDMS) are usually reinforced with fumed silica. Dynamic mechanical properties of the reinforced elastomer depend on the filler size (approximately 30 nm), the polymer-filler interface, and the state of aggregation. In order to study these effects, a model silica filler was synthesized according to the scheme depicted in Figure 11.13 (64). Both the individual vinyl-POSS cage and the cluster of four POSS cages were physically blended into silanol terminated PDMS before end linking. Cross-linking of silanol terminated PDMS was produced with tetraethoxysilane (TEOS), in the presence of a catalyst, and were subsequently blended with different amounts of single POSS cages and four POSS clusters; the final products were transparent films (64). As no chemical bonds were produced between the filler and the polymer network, a deleterious effect of filler addition on mechanical properties was found, which can be explained by the aggregation of the nanoparticles into larger domains. It was concluded that a necessary condition to produce reinforcement was to produce a uniform dispersion of the filler, a fact that could be accomplished by covalently bonding the filler to the polymer.

Figure 11.14 shows the chemical reactions involved in the formation of covalent bonds between the vinyl-POSS cages and the PDMS network (64). A substantial reinforcement of the polymer network was found in this case, particularly for POSS

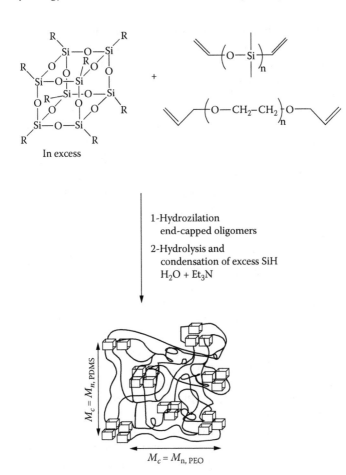

FIGURE 11.16 Scheme of the synthesis of POSS reinforced amphiphilic membranes.

concentrations equal to or higher than 10 wt% (64). However, because of the stoichiometry, only 3 wt% POSS could be covalently attached to the polymer network. Therefore, the covalent bonding of the filler to the polymer was not the direct source of reinforcement. Bonding produced an increase in compatibility. Without bonding, a macrophase separation of POSS occurred that led to a deterioration in the mechanical properties. For POSS amounts equal to or higher than 10 wt%, a very large dynamic strain softening was observed, giving an indication of the percolation of filler particles. Precisely, this was the origin of the significant increase in mechanical properties. But filler percolation would not have been possible without accomplishing the following requirements: (a) good compatibility of filler and polymer to obtain a good dispersion and (b) enough mobility of filler particles. Both conditions were achieved by introducing single covalent bonds between POSS particles and the polymer network.

The use of multifunctional POSS as cross-linking units in formulations of thermoset precursors has been reported for a variety of chemistries and curing procedures

(25,65–74). In general, these cube nanocomposites were found to be very homogeneous with no phase segregation even at a nanometer scale (71,74). One approach to synthesize these hybrid materials is to use a stoichiometric formulation with half of the cubes bearing eight functionalities of type A and the other half bearing eight B groups, where A reacts with B, leading to the network depicted in Figure 11.15. Examples that illustrate this possibility are the cross-linking of $[HSiO_{1.5}]_8$ or $[(HSiMe_2O)SiO_{1.5}]_8$ with $[(CH_2=CH)SiO_{1.5}]_8$ or $[(CH_2=CHSiMe_2O)SiO_{1.5}]_8$, in toluene, using platinum divinyltetramethyldisiloxane as catalyst (67). Because of steric reasons, it was not possible to attain complete conversion. Maximum conversions ranged from 43 to 81%, increasing with the length of the tether bridging the two cubes. After solvent evaporation, highly porous polymer networks were obtained.

Conducting the synthesis in an appropriate way may lead to the aggregation of cubes in nanosized domains. Figure 11.16 shows a scheme of the synthesis of POSS reinforced amphiphilic membranes (72). In a first step, an excess of $[HSiO_{1.5}]_8$ was reacted with both a diallyl-telechelic polyethylene oxide and a vinyl-telechelic polydimethylsiloxane, using platinum divinyltetramethyldisiloxane as catalyst. Upon completion of the hydrosilation reaction, the residual condensation of the excess SiH functionalities was performed by adding water and triethyl amine as catalyst. This step led to the formation of cube aggregates, as depicted in Figure 11.16. Because of the hydrophilic nature of PEO and the hydrophobic nature of PDMS and POSS, amphiphilic materials were obtained, i.e., swelling both in hydrocarbons and water as well as exhibiting rapid and reversible hydrophobic to hydrophilic surface rearrangements. Mechanical properties of these membranes, containing 6 to 13 wt% POSS, were comparable to PDMS networks reinforced by substantial amounts of fumed silica (72).

Another type of multifunctional macromonomers that may be introduced into thermoset formulations are cross-linked polymer microparticles (CPMs) or microgels (75–81). Preformed CPM may be dissolved in appropriate comonomers to generate a nanostructured thermoset by polymerization. They can be conveniently used in acrylate formulations cross-linked by photopolymerization (79). Because of the compact structure of preformed particles, adding a high CPM concentration to the initial formulation does not produce a significant increase in the solution viscosity (75,76,78). Moreover, CPM addition can lead to specific rheological behaviors such as shear-thinning or thixotropy.

The z-average radius of gyration, R_z, of CPM swollen by the mobile phase may be determined by size exclusion chromatography using an on-line multiangle laser light scattering detector. For typical CPM functionalized with acrylate or methacrylate groups, R_z varied in the range of 17 to 31 nm (79). When these CPM were dissolved in acrylic monomers and photopolymerized, the resulting films were transparent for CPM contents below 40 wt%. In this concentration range, CPM acted as individual cross-linking units. Figure 11.17 shows a TEM micrograph of a cured film containing 30 wt% CPM (79). The particles were in the 20 to 25 nm size range, which agreed with sizes of dry CPM particles determined by scanning electron microscopy (SEM).

However, at high CPM concentrations, cured films were opalescent, indicating some aggregation of the particles. This is probably the result of intramolecular

200 nm

FIGURE 11.17 TEM micrograph of a photocured acrylic film containing 30 wt% of cross-linked polymer microparticles (CPMs). (Valette et al., *Macromol. Mater. Eng.*, 288, 642–657, 2003; copyright 2004 John Wiley & Sons, Inc. With permission.)

reactions among CPM units favored by their large concentration in the mixture (chemical clustering). Dynamic mechanical spectra of films synthesized with large CPM concentrations showed a broad tan δ peak with a shoulder in the low temperature range, giving evidence of the existence of regions with different cross-link densities (79).

The solubility of CPM in the reactive solvent may be modulated by selecting the chemical structure of an acrylate monomer used as a stabilizing agent during the CPM synthesis. Good compatibility between the stabilizing agent and the reactive solvent leads to good CPM dispersion. Poor compatibility results in flocculation of a CPM-rich phase. A combination of stabilizers can be used to control the coarsening of the segregated phase (81).

Cross-linked elastomeric nanoparticles functionalized with carboxylic groups, with an average particle size close to 90 nm, were used to toughen epoxy-anhydride formulations (82). Covalent bonds produced by the reaction of –COOH groups with epoxy functionalities produced a good compatibility between the elastomeric filler and the epoxy polymer, avoiding a macroscopic phase separation. TEM images of fracture surfaces showed percolating structures of rubber particles (that kept their original dimensions) in the epoxy network. This could explain the significant increase in toughness produced by these elastomeric nanoparticles (82).

Functionalized hyperbranched polymers (HBPs) can be also used to increase the toughness of the polymer network (83–88). Key features of HBPs are the high degree of branching resulting from their structures and the very high end group functionality located at the surface. They usually exhibit low viscosities in solution due to their highly branched architecture that prevents chain entanglements. The multiple end groups can be conveniently functionalized to tailor the initial solubility in thermoset precursors and the conversion at which phase separation occurs in the course of polymerization. In turn, this determines the average size of HBP-rich domains in the cured material from the initial nanometer size to dimensions located in the micrometer range. A significant increase in the fracture resistance was observed when dispersed domains attained the micrometer range (89).

Another type of functionalized nanoparticles are colloidal dispersions of inorganic particles stabilized with organic molecules bearing reactive groups. Upon polymerization of these groups with an appropriate bifunctional comonomer, a nanocomposite consisting of a dispersion of nanosized inorganic particles covalently bonded to a polymer matrix may be obtained. A particular example is the synthesis of high refractive index thin films of ZnS/polythiourethane nanocomposites (8). ZnS nanoparticles end capped with thiophenol and 2-mercaptoethanol, with sizes in the range of 2 to 6 nm, could be synthesized as a stable colloidal dispersion in dimethylformamide. An isocyanate terminated polythiourethane was added to the colloidal dispersion of ZnS nanoparticles. The solution was concentrated under vacuum to a suitable viscosity and spin coated on a silicon wafer. Employing an adequate thermal cycle, the hydroxyl groups of 2-mercaptoethanol were reacted with isocyanate groups to generate a nanocomposite of ZnS particles covalently bonded through multiple polythiouretane linear bridges. Nanoparticles kept their original sizes throughout the process. The refractive index of the transparent films varied from 1.574 to 1.848 at 632.8 nm, increasing linearly with the wt% of ZnS particles.

11.5 NANOSTRUCTURE FORMATION AFTER NETWORK FORMATION

In some specific cases, it is possible to produce a nanostructure after the formation of the polymer network. This is, for example, the case for large molar mass mesogenic monomers that may be cross-linked in the isotropic state but phase separate into nematic domains with decreasing temperature. This self-organization, driven by a temperature decrease, becomes possible when the distance between cross-links is large enough to reorient the rigid parts of the polymer chains.

A glycidyl terminated oligoether based on 4,4-dihydroxy-methylstilbene (M_n = 3600 g/mol) exhibits a nematic-isotropic transition temperature, T_{NI} = 176°C (50). The glycidyl terminated oligoether was cross-linked using methylenedianiline either at 140°C, in the nematic state, or at 190°C, in the isotropic state. When cross-linked in the nematic state, the T_{NI} increased to 192°C. When the polymerization was performed in the isotropic state, nematic domains could be generated by decreasing the temperature. The transition temperature was T_{NI} = 157°C, revealing that nematic domains generated from the cured polymer were less ordered (or less pure) than similar domains formed before cross-linking (50).

Other examples where nanostructure formation is produced after the cure reaction, may be found in the field of amphiphilic networks. These materials contain covalently bonded immiscible hydrophilic and hydrophobic polymer chain segments, exhibiting a nanoseparated morphology (90,91). Macroscopic phase separation is hindered by the covalent bonds between both types of polymer chains. In order to prevent the occurrence of a macroscopic phase separation prior to or in the course of network formation, the hydrophilic monomer is replaced by a particular hydrophobic monomer that can regenerate the desired hydrophilic groups by a postcure hydrolysis. This reestablishes the immiscibility between both kinds of polymer segments and provokes a phase separation limited to the nanometer scale.

A particular example of this general strategy is the network formed by radical polymerization of poly(2-hydroxyethyl methacrylate) (HEMA), where HEMA is the hydrophilic monomer, and methacrylate-telechelic polyisobutylene (PIB), where PIB is the hydrophobic monomer (90–94). The desired network is synthesized using the hydrophobic monomer 2-(trimethylsilyloxy) ethyl methacrylate (SEMA) instead of HEMA. The polymer network synthesized by the free radical copolymerization of SEMA and the telechelic PIB, in a common solvent, could be nanostructured through a quantitative postcure hydrolysis of the trimethylsilyl protecting groups. Two T_g values were observed in the nanostructured material by differential scanning calorimetry. Depending on the initial composition, these values ranged from 96 to 111°C for PHEMA and −63 to −55°C for PIB. This provided evidence for the presence of a strong phase separated morphology, a fact that was also confirmed by SAXS and ¹H spin diffusion solid-state NMR (nuclear magnetic resonance). The sizes of PHEMA and PIB nanodomains ranged from 5 to 15 nm (Figure 11.18) (94).

Amphiphilic networks are capable of swelling both in water and in organic solvents. The swelling with aqueous cadmium chloride solution followed by exposure to H_2S resulted in CdS nanocrystals located in the hydrophilic domains (90).

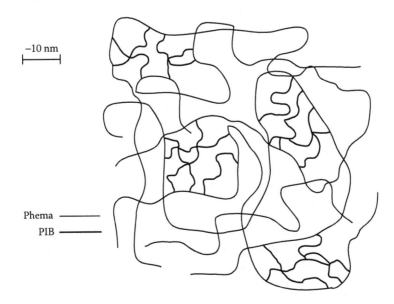

−10 nm

Phema ———
PIB ———

FIGURE 11.18 Schematic structure of an amphiphilic PHEMA-co-PIB network.

The size and connectivity of the dispersed inorganic crystals in a polymer matrix is a copy of the distribution of hydrophilic domains in the amphiphilic network.

11.6 CONCLUDING REMARKS

Different methods for generating nanostructures in thermosetting polymers are discussed in this chapter. A general classification was proposed based on whether phase segregation occurs before, during, or after polymerization, although combinations of these are also possible. Some illustrative examples were provided for every one of these cases.

The control of the size of the generated domains, at the nanometer level (1 to 100 nm) versus the micrometer level (0.1 to 10 μm), was discussed for the various types of phase separation processes. In some cases, nanostructure formation is the only possibility, as in the phase separation produced after network formation or in the formation of chemical clusters during polymerization. But in most cases, control of the selected formulation or the polymerization conditions — where the choice depends on the driving forces behind the phase separation — must be exercised to keep the generated structure in the nanometer level.

The initial miscibility of the components present in the system was identified as one of the key factors to controling the final dimensions of the generated domains. In the case of phase separation in the course of polymerization, a high initial miscibility leads to phase separation at conversions close to gelation or in the postgel stage, resulting in nanosize domains. In the case of structures that are self-assembled before polymerization as in the case of block copolymers, the miscible block must keep its miscibility to high conversions to avoid macroscopic phase separation. The same happens when functionalized nanoparticles are dissolved in thermoset precursors, as is the case of monofunctional POSS cages. In this case, the nonreactive organic branches attached to Si atoms must be miscible in the thermoset precursors to avoid their agglomeration prior to or during polymerization.

Most of the examples presented in this chapter are based on epoxy thermosetting polymers and most involve a step growth mechanism. Because of a quite different polymer growth process, the synthesis of nanostructured thermosetting blends polymerizing through a chain growth mechanism is, in many cases, a real challenge (unsaturated polyesters, for example).

But other factors can also be considered to control the characteristic size of the morphologies generated. For example, small amounts of block copolymers may be added as emulsifying agents to limit the coarsening of a phase segregated in the course of the polymerization. In this case, one of the blocks must be miscible with the segregated phase while the other one must be miscible with the matrix. However, if the initial solution contains both homopolymers together with a small amount of the block copolymer, and if both homopolymers are phase segregated at different conversions in the course of polymerization, the situation changes completely. Now, the block copolymer exerts its emulsifying effect on the domains that are first segregated, reducing their dimensions. But when the second homopolymer is segregated, the blocks of the copolymer present in the surface of the first generated domains act as nucleating agents of the second phase, leading to the coalescence of initial domains and the generation of biphasic particles (21).

Morphologies produced by the self-assembly of block copolymers in the initial solution with thermoset precursors may be also preserved by adding another block (containing functional groups that react with the thermosetting polymer) to the original immiscible-miscible blocks. This fixes the initial morphology by the introduction of chemical bonds with the thermoset even if the miscible block becomes phase separated at relatively low conversions.

Another factor that may be controlled is the relative rate of phase separation with respect to polymerization. Fast photopolymerization reactions may be adapted to generate the desired size of phase separated domains. However, this process may not provide enough time for the phase separation of high molar mass polymers. In this case, supramolecular polymers that increase their polymerization degree after phase separation may be used to reach the desired size of the dispersed domains.

REFERENCES

1. Pascault, J.P., Sautereau, H., Verdu, J., et al., *Thermosetting Polymers*, Dekker, New York, 2002.
2. Dusek, K. and Pascault, J.P., Reaction-induced (nano)inhomogeneities in polymer networks, in *The Wiley Polymer Networks Group Review Series*, Vol. 1, te Nijenhuis, K. and Mijs, W.J., Eds., Wiley, New York, 1998, pp. 277–299.
3. Williams, R.J.J., Rozenberg, B.A., and Pascault, J.P., Reaction-induced phase separation in modified thermosetting polymers, *Adv. Polym. Sci.*, 128, 95–156, 1997.
4. Pascault, J.P. and Williams, R.J.J., Formulation and characterization of thermoset-thermoplastic blends, in *Polymer Blends, Vol. 1: Formulation*, Paul, D.R. and Bucknall C.B., Eds., Wiley, New York, 2000, pp. 379–415.
5. Hedrick, J.L., Miller, R.D., Hawker, C.J., et al., Templating nanoporosity in thin-film dielectric insulators, *Adv. Mater.*, 10, 1049–1053, 1998.
6. Miller, R.D., In search of low-*k* dielectrics, *Science*, 286, 421–423, 1999.
7. Maier, G., Low dielectric constant polymers for microelectronics, *Prog. Polym. Sci.*, 26, 3–65, 2001.
8. Lü, C., Cui, Z., Li, Z., et al., High refractive index thin films of ZnS/polythiourethane nanocomposites, *J. Mater. Chem.*, 13, 526–530, 2003.
9. Lü, C., Cui, Z., Guan, C., et al., Research on preparation, structure and properties of TiO₂/polythiourethane hybrid optical films with high refractive index, *Macromol. Mater. Eng.*, 288, 717–723, 2003.
10. Caseri, W., Nanocomposites of polymers and metals or semiconductors: historical background and optical properties, *Macromol. Rapid Commun.*, 21, 705–722, 2000.
11. Karger-Kocsis, J., Frölich, J., Gryshchuk, O., et al., Synthesis of reactive hyperbranched and star-like polyethers and their use for toughening of vinylester-urethane hybrid resins, *Polymer*, 45, 1185–1195, 2004.
12. Zhu, B., Katsoulis, D.E., Keryk, J.R., et al., Toughening of polysilsesquioxane network by simultaneous incorporation of short and long PDMS chain segments, *Macromolecules*, 37, 1455–1462, 2004.

13. Dusek, K., Network formation by chain cross-linking (co)polymerization, in *Developments in Polymerization*, Vol. 3, Haward, R.N., Ed., Applied Science Publishers, Barking, UK, 1982, pp. 143–206.

14. Williams, R.J.J., Borrajo, J., Adabbo, H.E., et al., A model for phase separation during a thermoset polymerization, in *Rubber-Modified Thermoset Resins, Adv. Chem. Ser. 208*, Riew, C.K. and Gillham, J.K., Eds., American Chemical Society, Washington DC, 1984, pp. 195–213.

15. Borrajo, J., Riccardi, C.C., Williams, R.J.J., et al., Thermodynamic analysis of the reaction-induced phase separation in epoxy-based polymer dispersed liquid crystals (PDLC), *Polymer*, 39, 845–853, 1998.

16. Zucchi, I.A., Galante, M.J., Borrajo, J., et al., A model system for the thermodynamic analysis of reaction-induced phase separation: solutions of polystyrene in bifunctional epoxy/amine monomers, *Macromol. Chem. Phys.*, 205, 676–683, 2004.

17. Verchère, D., Sautereau, H., Pascault, J.P., et al., Rubber-modified epoxies: analysis of the phase separation process, in *Toughened Plastics I: Science and Engineering, Adv. Chem. Ser. 233*, Riew, C.K. and Kinloch, A.J., Eds., American Chemical Society, Washington, D.C., 1993, pp. 335–363.

18. Keizer, H.M., Sijbesma, R.P., Jansen, J.F.G.A., et al., Polymerization induced phase separation using hydrogen-bonded supramolecular polymers, *Macromolecules*, 36, 5602–5606, 2003.

19. Galante, M.J., Borrajo, J., Williams, R.J.J., et al., Double phase separation induced by polymerization in ternary blends of epoxies with polystyrene and poly(methylmethacrylate). *Macromolecules*, 34, 2686–2694, 2001.

20. Ritzenthaler, S., Girard-Reydet, E., and Pascault, J.P., Influence of epoxy hardener on miscibility of blends of poly(methyl methacrylate) and epoxy networks, *Polymer*, 41, 6375–6386, 2000.

21. Girard-Reydet, E., Sévignon, A., Pascault, J.P., et al., Influence of the addition of polystyrene-b-poly(methyl methacrylate) block copolymer (PS-b-PMMA) on the morphologies generated by reaction-induced phase separation in PS/PMMA/epoxy blends, *Macromol. Chem. Phys.*, 203, 947–952, 2002.

22. Girard-Reydet, E., Pascault, J.P., and Brown, H.R., Splat: a nonequilibrium morphology on the way to a microemulsion, *Macromolecules*, 34, 5349–5353, 2001.

23. Girard-Reydet, E., Sautereau, H., and Pascault, J.P., Use of block copolymers to control the morphologies and properties of thermoplastic/thermoset blends, *Polymer*, 40, 1677–1687, 1999.

24. Jansen, B.J.P., Meijer, H.E.H., and Lemstra, P.J., Processing of (in)tractable polymers using reactive solvents. Part 5: morphology control during phase separation, *Polymer*, 40, 2917–2927, 1999.

25. Kickelbick, G., Concepts for the incorporation of inorganic building blocks into organic polymers on a nanoscale, *Prog. Polym. Sci.*, 28, 83–114, 2003.

26. Huang, Q.R., Kim, H.C., Huang, E., et al., Miscibility in organic/inorganic hybrid nanocomposites suitable for microelectronic applications: comparison of modulated differential scanning calorimetry and fluorescence spectroscopy, *Macromolecules*, 36, 7661–7671, 2003.

27. Mucha, M. Polymer as an important component of blends and composites with liquid crystals, *Prog. Polym. Sci.*, 28, 837–873, 2003.

28. Brunsveld, L., Folmer, B.J.B., Meijer, E.W., et al., Supramolecular polymers, *Chem. Rev.*, 101, 4071–4098, 2001.

29. Dean, J.M., Lipic, P.M., Grubbs, R.B., et al., Micellar structure and mechanical properties of block copolymer-modified epoxies, *J. Polym. Sci., B: Polym. Phys.*, 39, 2996–3010, 2001.
30. Hillmyer, M.A., Lipic, P.M., Hajduk, D.A., et al., Self-assembly and polymerization of epoxy resin-amphiphilic block copolymer nanocomposites, *J. Am. Chem. Soc.*, 119, 2749–2750, 1997.
31. Lipic, P.M., Bates, F.S., and Hillmayer, M.A., Nanostructured thermosets from self-assembled amphiphilic block copolymer/epoxy resin mixtures, *J. Am. Chem. Soc.*, 120, 8963–8970, 1998.
32. Alexandris, P. and Lindman, B., *Amphiphilic Block Copolymers: Self Assembly and Applications*, Elsevier, New York, 2000.
33. Lodge, T.P., Xu, X., Ryu, C.Y., et al., Structure and dynamics of concentrated solutions of asymmetric block copolymers in slightly selective solvents, *Macromolecules*, 29, 5955–5964, 1996.
34. Fine, T., Beaume, F., Bonnet, A., et al., Order-disorder transition in AC, BC, ABC block copolymers in a selective and reactive solvent, *40th IUPAC Inter. Symp. Macromol.*, 2004, preprint no. 2085.
35. Dean, J.M., Grubbs, R.B., Saad, W., et al., Mechanical properties of block copolymer vesicle and micelle modified epoxies, *J. Polym. Sci., B: Polym. Phys.*, 41, 2444–2456, 2003.
36. Dean, J.M., Verghese, N.E., Pham, H.Q., et al., Nanostructure toughened epoxy resins, *Macromolecules*, 36, 9267–9270, 2003.
37. Grubbs, R.B., Broz, M.E., Dean, J.M., et al., Selectively epoxidized polyisoprene-polybutadiene block copolymers, *Macromolecules*, 33, 2308–2310, 2000.
38. Grubbs, R.B., Dean, J.M., Broz, M.E., et al., Reactive block copolymers for modification of thermosetting epoxy, *Macromolecules*, 33, 9522–9534, 2000.
39. Grubbs, R.B., Dean, J.M., and Bates, F.S., Methacrylic block copolymers through metal-mediated living free radical polymerization for modification of thermosetting epoxy, *Macromolecules*, 34, 8593–8595, 2001.
40. Mijovic, S., Shen, M., Sy, J.W., et al., Dynamics and morphology in nanostructured thermoset network/block copolymer blends during network formation, *Macromolecules*, 33, 5235–5244, 2000.
41. Kosonen, H., Ruokolainen, J., Nyholm, P., et al., Self-organized thermosets: blends of hexamethyltetramine cured novolac with poly(2-vinylpyridine)-*block*-poly(isoprene), *Macromolecules*, 34, 3046–3049, 2001.
42. Kosonen, H., Ruokolainen, J., Nyholm, P., et al., Self-organized cross-linked phenolic thermosets: thermal and dynamic mechanical properties of novolac/block copolymer blends, *Polymer*, 42, 9481–9486, 2001.
43. Kosonen, H., Ruokolainen, J., Torkkeli, M., et al., Micro- and macrophase separation in phenolic resol resin/PEO-PPO-PEO block copolymer blends: effect of hydrogen-bonded PEO length, *Macromol. Chem. Phys.*, 203, 388–392, 2002.
44. Guo, Q., Thomann, R., Gronski, W., et al., Phase behavior, crystallization, and hierarchical nanostructures in self-organized thermoset blends of epoxy resin and amphiphilic poly(ethylene oxide)-block-poly(propylene oxide)-block-poly(ethylene oxide) triblock copolymers, *Macromolecules*, 35, 3133–3144, 2002.
45. Guo, Q., Thomann, R., Gronski, W., et al., Nanostructures, semicrystalline morphology effect on the crystallization kinetics in self-organized block copolymer/thermoset blends, *Macromolecules*, 36, 3635–3645, 2003.

46. Ritzenthaler, S., Court, F., David, L., et al., ABC triblock copolymers/epoxy-diamine blends. 1. Keys to achieve nanostructured thermosets, *Macromolecules*, 35, 6245–6254, 2002.

47. Rebizant, V., Abetz, V., Tournilhac, F., et al., Reactive tetrablock copolymers containing glycidyl methacrylate. Synthesis and morphology control in epoxy-amine networks, *Macromolecules*, 36, 9889–9896, 2003.

48. Müller, H.P., Gipp, R., and Heine, H., *Liquid-Crystalline Diglycidyl Compounds, the Preparation of These, and the Use of These in Curable Epoxide Mixtures*, U.S. Pat. 4,764,581, 1988.

49. Mikroyannidis, J.A., Synthesis, characterization and polymerization of new epoxy compounds containing azomethine linkages, *Makromol. Chem.*, 190, 1867–1879, 1989.

50. Barclay, G.G., Ober, C.K., Papathomas, K.I., et al., Liquid crystalline epoxy thermosets based on dihydroxymethylstilbene: synthesis and characterization, *J. Polym. Sci., A: Polym. Chem.*, 30, 1831–1843, 1992.

51. Jandke, M., Hanft, D., Strohriegl, P., et al., Polarized electroluminescence from photocrosslinkable nematic fluorine bisacrylates, *SPIE Proc.*, 4105, 338, 2001.

52. Pfeuffer, T., Hanft, D., and Strohriegl, P., Vitrifying star-shaped liquid crystals: synthesis and application in cholesteric polymer networks, *Liq. Crystals*, 29, 1555–1564, 2002.

53. Strohriegl, P. and Grazulevicius, J.V., Charge-transporting molecular glasses, *Adv. Mater.*, 14, 1439–1452, 2002.

54. Lichtenhan, J.D., Vu, N.Q., Carter, J.A., et al., Silsesquioxane-siloxane copolymers from polyhedral silsesquioxanes, *Macromolecules*, 26, 2141–2142, 1993.

55. Sellinger, A., Laine, R.M., Chu, V., et al., Palladium- and platinum-catalyzed coupling reactions of allyloxy aromatics with hydridosilanes and hydridosiloxanes: novel liquid crystalline/organosilane materials, *J. Polym. Sci., A: Polym. Chem.*, 32, 3069–3089, 1994.

56. Lichtenhan, J.D., Polyhedral oligomeric silsesquioxanes: building blocks for silsesquioxane-based polymers and hybrid materials, *Comments Inorg. Chem.*, 17, 115–130, 1995.

57. Haddad, T.S. and Lichtenhan, J.D., The incorporation of transition metals into polyhedral oligosilsesquioxane polymers, *J. Inorg. Organomet. Polym.*, 5, 237–246, 1995.

58. Feher, F.J., Weller, K.J., and Schwab, J.J., Reactions of hydroxysilsesquioxanes and chlorosilsesquioxanes with phosphoranes, *Organometallics*, 14, 2009–2017, 1995.

59. Gilman, J.W., Schlitzer, D.S., and Lichtenhan, J.D., Low earth orbit resistant siloxane copolymers, *J. Appl. Polym. Sci.*, 60, 591–596, 1996.

60. Lee, A. and Lichtenhan, J.D., Viscoelastic responses of polyhedral oligosilsesquioxane reinforced epoxy systems, *Macromolecules*, 31, 4970–4974, 1998.

61. Lee, A. and Lichtenhan, J.D., Thermal and viscoelastic property of epoxy-clay and hybrid inorganic-organic epoxy nanocomposites, *J. Appl. Polym. Sci.*, 73, 1993–2001, 1999.

62. Abad, M.J., Barral, L., Fasce, D.P., et al., Epoxy networks containing large mass fractions of a monofunctional polyhedral oligomeric silsesquioxane (POSS), *Macromolecules*, 36, 3128–3135, 2003.

63. Constable, G.S., Lesser, A.J., and Coughlin, E.B., Morphological and mechanical evaluation of hybrid organic-inorganic thermoset copolymers of dicyclopentadiene and mono- or tris(norbornenyl)-substituted polyhedral oligomeric silsesquioxanes, *Macromolecules*, 37, 1276–1282, 2004.

64. Pan, G., Mark, J.E., and Schaefer, D.W., Synthesis and characterization of fillers of controlled structure based on polyhedral oligomeric silsesquioxane cages and their use in reinforcing siloxane elastomers, *J. Polym. Sci. B: Polym. Phys.*, 41, 3314–3323, 2003.

65. Sellinger, A. and Laine, R.M., Silsesquioxanes as synthetic platforms. 3. Photocurable, liquid epoxides as inorganic/organic hybrid precursors, *Chem. Mater.*, 8, 1592–1593, 1996.

66. Sellinger, A. and Laine, R.M., Silsesquioxanes as synthetic platforms. Thermally curable and photocurable inorganic/organic hybrids, *Macromolecules*, 29, 2327–2330, 1996.

67. Zhang, C., Babonneau, F., Bonhomme, C., et al., Highly porous polyhedral silsesquioxane polymers. Synthesis and characterization. *J. Am. Chem. Soc.*, 120, 8380–8391, 1998.

68. Li, G.Z., Wang, L., Toghiani, H., et al., Viscoelastic and mechanical properties of epoxy/multifunctional polyhedral oligomeric silsesquioxane nanocomposites and epoxy/ladderlike polyphenylsilsesquioxane blends, *Macromolecules*, 34, 8686–8693, 2001.

69. Neumann, D., Fisher, M., Tran, L., et al., Synthesis and characterization of an isocyanate functionalized polyhedral oligosilsesquioxane and the subsequent formation of an organic-inorganic hybrid polyurethane, *J. Am. Chem. Soc.*, 124, 13998–13999, 2002.

70. Kim, G.M., Qin, H., Fang, X., et al., Hybrid epoxy-based thermosets based on polyhedral oligosilsesquioxane: cure behavior and toughening mechanisms, *J. Polym. Sci., B: Polym. Phys.*, 41, 3299–3313, 2003.

71. Choi, J., Yee, A.F., and Laine, R.M., Organic/inorganic hybrid composites from cubic silsesquioxanes. Epoxy resins of octa(dimethylsiloxyethylcyclohexylepoxide) silsesquioxane, *Macromolecules*, 36, 5666–5682, 2003.

72. Isayeva, I.S. and Kennedy, J.P., Synthesis and characterization of novel POSS-reinforced amphiphilic membranes, *Polym. Mat. Sci. Eng.*, 89, 645–646, 2003.

73. Mya, K.Y., Huang, J., Xiao, Y., et al., Improvement of thermal-mechanical properties using polyhedral oligomeric silsesquioxanes (POSS)-modified epoxy resins, *Polym. Mat. Sci. Eng.*, 89, 757–758, 2003.

74. Choi, J., Kim, S.G., and Laine, R.M., Organic/inorganic hybrid epoxy nanocomposites from aminophenylsilsesquioxanes, *Macromolecules*, 37, 99–109, 2004.

75. Boggs, L.J., Rivers, M., and Bike, S.G., Characterization and rheological investigation of polymer microgels used in automotive coatings, *J. Coat. Technol.*, 68, 63–74, 1996.

76. Funke, W., Okay, O., and Joss-Muller, B., Microgels: intramolecularly crosslinked macromolecules with a globular structure, *Adv. Polym. Sci.*, 136, 139–234, 1998.

77. Valette, L., Pascault, J.P., and Magny, B., (Meth)acrylic cross-linked polymer microparticles: synthesis by dispersion polymerization and particle characterization, *Macromol. Mater. Eng.*, 287, 31–40, 2002.

78. Valette, L., Pascault, J.P., and Magny, B., Rheological properties of (meth)acrylic cross-linked polymer microparticles. 1. Comparison with linear polymers in bulk and in non-reactive solvents, *Macromol. Mater. Eng.*, 287, 41–51, 2002.

79. Valette, L., Pascault, J.P., and Magny, B., Use of acrylic functionalized (meth)acrylic cross-linked polymer microparticles in photopolymerized acrylic films, *Macromol. Mater. Eng.*, 288, 642–657, 2003.

80. Valette, L., Pascault, J.P., and Magny, B., Use of hydroxyl functionalized (meth)acrylic cross-linked polymer microparticles as chain transfer agent in cationic photopolymerization of cycloaliphatic epoxy monomer,1, *Macromol. Mater. Eng.*, 288, 751–761, 2003.

81. Valette, L., Pascault, J.P., and Magny, B., Use of hydroxyl functionalized (meth)acrylic cross-linked polymer microparticles as chain transfer agent in cationic photopolymerization of cycloaliphatic epoxy monomer,2, *Macromol. Mater. Eng.*, 288, 762–770, 2003.

82. Huang, F., Liu, Y., Zhang, X., et al., Effect of elastomeric nanoparticles on toughness and heat resistance of epoxy resins, *Macromol. Rapid Commun.*, 23, 786–790, 2002.

83. Malmström, E., Johansson, M., and Hult, A., Hyperbranched aliphatic polyesters, *Macromolecules*, 28, 1698–1703, 1995.

84. Boogh, L., Pettersson, B., and Månson, J.A.E., Dendritic hyperbranched polymers as tougheners for epoxy resins, *Polymer*, 40, 2249–2261, 1999.

85. Wu, H., Xu, J., Liu, Y., et al., Investigation of readily processable thermoplastic-toughened thermosets. V. Epoxy resin toughened with hyperbranched polyester, *J. Appl. Polym. Sci.*, 72, 151–163, 1999.

86. Mezzenga, R., Plummer, C.J.G., Boogh, L., et al., Morphology build-up in dendritic hyperbranched polymer modified epoxy resins: modeling and characterization, *Polymer*, 42, 305–317, 2001.

87. Mezzenga, R. and Månson, J.A.E., Thermo-mechanical properties of hyperbranched polymer modified epoxies, *J. Mater. Sci.*, 36, 4883–4891, 2001.

88. Ratna, D., Varley, R., Singh Raman, R.K., et al., Studies on blends of epoxy-functionalized hyperbranched polymer and epoxy resin, *J. Mater. Sci.*, 38, 147–154, 2003.

89. Fröhlich, J., Kautz, H., Thomann, R., et al., Reactive core/shell type hyperbranched blockcopolyethers as new liquid rubbers for epoxy toughening, *Polymer*, 45, 2155–2164, 2004.

90. Scherble, J., Thomann, R., Iván B., et al., Formation of CdS nanoclusters in phase-separated poly(2-hydroxyethyl methacrylate)/polyisobutylene amphiphilic conetworks, *J. Polym. Sci., B: Polym. Phys.*, 39, 1429–1436, 2001.

91. Iván, B., Almdal, K., Mortensen, K., et al., Synthesis, characterization, and structural investigations of poly(ethyl acrylate)/polyisobutylene bicomponent conetwork, *Macromolecules*, 34, 1579–1585, 2001.

92. Iván, B., Kennedy, J.P., and Mackey, P.W., Amphiphilic networks IV. Synthesis and characterization of, and drug release from poly(2-hydroxyethylmethacrylate)/poly-isobutylene, *Polym. Prepr. (ACS)*, 31, 217–218, 1990.

93. Iván, B., Kennedy, J.P., and Mackey, P.M., *Amphiphilic Networks*, U.S. Patent 5,073,381, 1991.

94. Domján, A., Erdödi, G., Wilhelm, M., et al., Structural studies of nanophase-separated poly(2-hydroxyethylk methacrylate)/polyisobutylene amphiphilic conetworks by solid-state NMR and small-angle X-ray scattering, *Macromolecules*, 36, 9107–9114, 2003.

12 Relationship between Phase Morphology, Crystallization, and Semicrystalline Structure in Immiscible Polymer Blends

R.T. Tol, V.B.F. Mathot, H. Reynaers, and G. Groeninckx

CONTENTS

12.1 INTRODUCTION

Blending of immiscible polymers is an efficient approach to designing polymeric materials with specific and beneficial properties. In many of these blends, a semicrystalline component is involved. The properties of such partly semicrystalline blends are not only governed by their phase morphology and blend interfacial structure, but also strongly depend on the crystallization behavior of the semicrystalline component (1). This crystallization behavior is often complex and is strongly influenced by the blend phase morphology generated.

In blend systems where the crystallizable polymer forms the continuous phase (i.e., blends with a crystallizable matrix), both the nucleation behavior and the spherulite growth rate have been shown to be affected by the presence of a dispersed phase compared to the crystallization of the neat crystallizable component (1). Here, migration of heterogeneous nuclei from one blend phase to the other and the disturbance of the crystallization growth front due to the presence of dispersed blend phase are the most important phenomena that influence the crystallization behavior of the matrix. These phenomena, however, mainly affect the final semicrystalline morphology; the crystallization behavior of the matrix is usually only slightly affected by virtue of a slightly lowered crystallization temperature.

Strong changes in the crystallization behavior, however, do occur as soon as the crystallizable polymer no longer forms a continuous phase but is forming discrete droplets within another polymer matrix, which is either crystalline or amorphous. For such blends, multiple crystallization peaks are often observed, reflecting crystallization in the dispersed phase at different and higher supercoolings. In some cases, the crystallization temperature of the dispersed phase and the neat component can differ by more than 100 centigrade degrees. Often, the strong influence on the blend properties of such a confined crystallization in immiscible polymer blends is underestimated. The same holds for the complex relations between the processing conditions, the phase morphology development, and the crystallization behavior and concomitant semicrystalline structure development.

In this chapter, we present an overview of the relations between blend phase morphology and crystallization behavior and resulting semicrystalline structure in polymer blends with a crystallizable dispersed phase. These crystallization phenomena, however, are not specific for polymer blends and not even for polymers, but can generally occur in systems where the crystallizable component is confined to a small volume. Therefore, in Section 12.2, we present an overview of the literature related to crystallization of dispersions of liquids, metals, and polymers, which serves as a sort of background for confined crystallization phenomena in immiscible blend systems, which is discussed in Section 12.3. In Section 12.4, the effect of confined crystallization on the semicrystalline structure of dispersions and polymer blends is discussed. Finally, in Section 12.5, we present some conclusions and provide prospects for future work in this area of research.

12.2 CRYSTALLIZATION IN A CONFINED VOLUME: REVIEW OF THE CRYSTALLIZATION PHENOMENA IN POLYMER AND NONPOLYMER DISPERSIONS

12.2.1 CONFINED CRYSTALLIZATION PHENOMENA IN LIQUID, METAL, AND POLYMER DISPERSIONS

The first investigations concerning the crystallization in small, discrete droplets probably date from 1880. Van Riemsdyk (2) reported that small gold melt droplets solidify at much higher supercoolings than the bulk material. Similar results were obtained for many other dispersed liquid metals (3–8). The reason for this particular crystallization phenomenon was related to a change in nucleation behavior inside the dispersed droplets. By dividing a bulk sample into numerous small droplets, the effects of "catalytic" sites responsible for nucleation in relatively large bulk samples are greatly suppressed because the number of droplets is large in comparison with the number of catalytic sites. As such, the remaining droplets are forced to nucleate at rates governed by the molecular characteristics of the sample.

The first report on crystallization of *polymer* droplets dates from 1959 when Price (9) showed that micrometer sized polyethyleneoxide (PEO) droplets crystallized only at a supercooling of 65°C, compared to 20°C in bulk. Subsequently, a number of interesting studies were performed by various authors on nucleation in polymer dispersions (10–14). Turnbull and coworkers (10,11,15) prepared stable suspensions of polyethylene (PE), polypropylene (PP), and n-alkanes in a thermodynamically inert liquid, with micrometer sized droplets. Koutsky et al. (13) used a similar approach for various polymers but used a silicon oil as the suspending medium to obtain polymer droplets in the size range of between 1 to 100 μm. As such, this so-called droplet technique proves to be a very efficient method to study nucleation phenomena in polymers. In a number of these investigations, distinct regions of nucleation could be observed, as can clearly be seen in Figure 12.1 for PE.

The investigated PE droplets were 1 to 3 μm in size. A small fraction of the droplets was found to crystallize above 100°C. Most of the droplets, however, crystallized between 85 and 87°C. Similar results were obtained by Koutsky et al. (13) who additionally found that the maximum supercooling reached was typical for the specific polymer under investigation.

The explanation for these phenomena can be linked to nucleation problems in the polymer, which is similar to the situation for liquid metals; the crystallization of polymers is known to be initiated by various heterogeneities present in the molten state. When the sample is finely subdivided, heterogeneous nucleation of the crystallizable polymer in the droplets is restricted to the volume of the droplet, and each droplet will crystallize according to the number and type of heterogeneities in it.

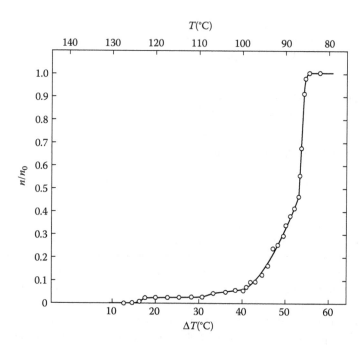

FIGURE 12.1 Fraction of solidified polyethylene as a function of supercooling. Cooling rate is 0.17 K/min. Region of no nucleation is down to 125°C. Region of heterogeneous nucleation is down to 100°C. Region of increasing nucleation with help from heterogeneous nuclei is perhaps from 100 to 85°C. Homogeneous nucleation occurs below 85°C (Wunderlich, B., *Macromolecular Physics, Vol. 2, Crystal Nucleation, Growth, Annealing*, 1976). (Reprinted with permission from Cormia et al., *J. Chem Phys.*, 37(6), 1333, 1962; copyright 1962 American Institute of Physics.)

When the droplet size is small enough, the number of droplets could exceed the number of heterogeneities present that normally promote crystallization as represented by the bulk crystallization temperature. In such a case, some fraction of the droplets can be nucleated by heterogeneities having a higher specific interfacial energy difference between the nucleus and the melt than the nuclei in bulk, which will crystallize at a lower temperature (16). Indications for the existence of such less active heterogeneous nuclei were obtained in early crystallization experiments on homopolymers. These experiments indicated an increase of the number of active nuclei at decreasing crystallization temperatures, confirming heterogeneous nucleation triggered by different types of heterogeneities active at increasing supercoolings (17–20). Finally, the droplets that do not contain any heterogeneities can undergo homogeneous nucleation at the largest obtainable supercooling. Here, polymer chains have to nucleate on their own. The multiple crystallization peaks can thus be considered to reflect the efficiency spectrum of the several nucleating heterogeneous nuclei available in the dispersed crystallizable polymer phase and possibly also crystallization triggered by homogeneous nucleation at the maximum supercooling. A relation between the temperature of the maximum supercooling realized and the melting point could be established for many metals and low molar mass substances (4,8,21), and also was found to be valid for all analyzed polymers (13): $T_{\text{hom}} = 0.8\ T_m^\circ$, where T_m° is the equilibrium melting temperature in Kelvin.

A quantitative approach describing the activity of the different types of heterogeneities as a function of the relative supercooling was derived by Frensch et al. (16). The free energy for formation of a critical primary nucleus at supercooling ΔT is proportional to the interfacial free energy difference $\Delta\gamma_i$ for a nucleus of type i:

$$\Delta F_i^* \propto \Delta\gamma_i\ /\ (\Delta T_i)^2 \tag{12.1}$$

This relation can be derived from the expression for the free energy of formation of a primary heterogeneous nucleus ("plateletlike" with chain folds) according to classical heterogeneous nucleation theory (17). Neglecting that the crystallization rate also depends on the temperature dependent mobility of the crystallizing chain segments, which would act to keep the ratio constant, we obtain:

$$\frac{\Delta\gamma_1}{\Delta\gamma_2} = \frac{T_1}{T_2}\left(\frac{\Delta T_1}{\Delta T_2}\right)^2 \tag{12.2}$$

where $\Delta\gamma$ is the specific interfacial free energy difference for the nucleus–foreign surface interface, which may be written as

$$\Delta\gamma_i = \gamma_{\text{cmi}} + \gamma_{\text{csi}} - \gamma_{\text{msi}} \tag{12.3}$$

where γ_{cmi} is the specific side surface (crystal-melt) free energy of the crystal nucleus, γ_{csi} is the specific crystal-substrate interfacial free energy, and γ_{msi} the specific melt-substrate surface free energy. For homogeneous nucleation, no foreign substrate is present, and nucleation must start by association of neighboring chains; therefore, $\Delta\gamma$ for homogeneous nucleation can be expressed as $2\gamma_{\text{cm}}$. Assuming $\Delta T_{\text{hom}} = T_m^\circ - T_{c,\text{hom}} = T_m^\circ/5$ (see above), we obtain by applying the relationship for homogeneous nucleation:

$$\frac{\Delta\gamma_i}{\gamma_{\text{cm}}} = 62.5\left(\frac{T_i}{T_m^\circ}\right)\left(\frac{\Delta T_i}{T_m^\circ}\right)^2 \tag{12.4}$$

In Figure 12.2, the relative interfacial energy difference $\Delta\gamma_i/\gamma_{\text{cm}}$ is plotted against $\dfrac{\Delta T_i}{T_m^\circ}$.

The relative degree of supercooling needed for heterogeneous nucleation is higher for nuclei with higher $\Delta\gamma_i/\gamma_{\text{cm}}$. The highest degree of supercooling is required for homogeneous nucleation, which is characterized by $\Delta\gamma_i/\gamma_{\text{cm}} = 2$.

These crystallization phenomena can be viewed as reflecting a quite basic crystallization mechanism where a crystallizable component is confined to a small volume. The crystallization in nanometer sized materials now is an intensively

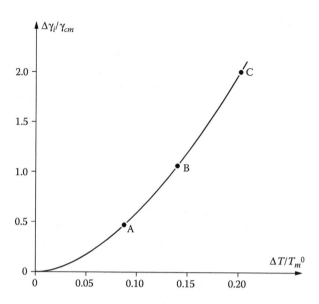

FIGURE 12.2 Relative interfacial energy difference $\Delta\gamma_i/\gamma_{cm}$ versus the relative degree of supercooling at which a heterogeneity nucleates the polymer. A and B: different types of heterogeneous nuclei; C: homogeneous nucleation. (Reprinted with permission from Frensch et al., *Multiphase Polymers: Blends Ionomers, ACS Symp. Series*, 395,101, 1989; copyright 1989 American Chemical Society.)

studied field; several studies have been performed on hydrocarbons in water emulsions, similar to the earlier experiments of Turnbull et al. (4–6), which have been used as model systems to describe the physical properties of emulsions (22–26). In these studies, it was found that the supercooling required to obtain crystallization in the droplets is significantly increased compared to the bulk material, which was attributed to a suppression of heterogeneous nucleation. Walstra and van Beresteyn (27) found an increased supercooling and slower crystallization rate that was inversely related to the droplet size for the crystallization of milk fat in the emulsified state. Charoenrein and Reid (28) combined the droplet emulsion technique and differential scanning calorimeter (DSC) to study heterogeneous and homogenous nucleation of ice in an aqueous system. Other interesting results on polymer dispersions were obtained by Barham et al. (29) on PE droplets and Taden and Landfester (30) for PEO droplets. Recently, Ibarretxe Uriguen et al. (31) prepared dispersions of polyolefins in water and found that these were suitable systems for realizing homogeneous crystallization, provided that they could be sufficiently stabilized against coalescence.

Several studies have been performed recently on the crystallization behavior of different types of crystallizable block copolymers with a microdomain morphology (32–42). Early observations of the effect of the microdomain volume of the crystallizable block on its crystallization behavior were reported by Lotz and Kovacs (43) and O'Malley et al. (44). They reported a two-step crystallization process for the PEO segments in polyethyleneoxide/polystyrene (PEO/PS) block copolymers, with one peak at the temperature of the homopolymer PEO and one at about −20°C. Molten block copolymers can spontaneously self-assemble into mesophases having spherical, cylindrical, or lamellar morphologies. The final solid-state structure adopted by a semicrystalline diblock copolymer depends on the interplay between this microphase separation and crystallization (33). Confined crystallization can take place in these microdomain volumes when the mesophase morphology of the block copolymer is retained after crystallization. Robitaille and Prudhomme (45) found pronounced supercooling of PEO in a PEO-PI-PEO (PI: polyisoprene) block copolymer. Loo et al. (32) studied homogeneous nucleation in a polyethylene/styrene-ethylene-butene (SEB63) diblock using time resolved small angle x-ray scattering (SAXS) and wide angle x-ray diffraction (WAXD) measurements and determined the microdomain volume of a polyethylene block to be 25 nm by transmission electron microscopy (TEM) measurements. Similar observations of homogeneous nucleation kinetics were found for poly(oxyethylene)-b-poly(oxybuty-lene)/poly(oxybutylene) blends by Xu et al. (34) and polyethylene-b-(vinylcyclo-hexane) diblock copolymers by Loo et al. (33). The crystallization behavior of different AB and ABC microphase separated block copolymers was analysed by Müller et al. (36,37). Polybutadiene(hydrogenated)-block-polyethylene oxide (PBh-b-PEO) copolymers with 12 nm PEO microdomains were investigated by Reiter et al. (39) using real-time atomic force microscopy (AFM).

12.2.2 HOMOGENEOUS VERSUS HETEROGENEOUS NUCLEATION KINETICS

The technique of dispersing a polymer proves to be a very effective method for studying nucleation in polymers. In many cases, a change in nucleation from heterogeneous to homogeneous nucleation can be expected upon decreasing the droplet size, resulting in much higher supercoolings than usual. In bulk samples, such supercoolings can only be reached after extensive purification in an attempt to remove all impurities, which is quite a task. Also, because the time needed for nucleation in small droplets is much longer than the time for space filling by crystal growth of the whole droplet, it is possible to study the nucleation behavior of polymers separately from that of crystal growth. For these reasons, interest in the crystallization kinetics of confined, supercooled polymers is steadily increasing.

It is, however, not easy to prove that homogeneous nucleation actually occurs. In a number of studies on droplet systems, the lowest crystallization temperature obtained is usually ascribed to homogeneous nucleation (36,44, 46–48), often referring to the empirical relation between T_m and the homogeneous nucleation temperature (see above). Obviously, this is necessary but not sufficient to prove homogeneous nucleation. The mechanism of homogeneous nucleation is expected to proceed in a totally different manner from that of heterogeneous nucleation. Instead of nucleation predetermined by the presence of heterogeneous nucleation sites, homogeneous nucleation is a random process of associations by chain segments. The process was nicely visualized by Reiter et al. (39) using real-time atomic force microscopy for a PBh-b-PEO copolymer. They found that each PEO phase, with a diameter of 12 nm crystallized independently in a random matter. Another approach was applied by Massa et al. (49) recently. By dewetting thin films of PEO on an unfavorable substrate, discrete ensembles of impurity free, micrometer size droplets of poly(ethylene oxide) are formed. By applying different heating and cooling experiments, it was shown that there was no correlation of crystallization with respect to different runs for these droplets, thus, indicating a homogeneous nucleation mechanism. Crystallization of partially "dirty" samples, in contrast, could be correlated, indicating a predetermined, heterogeneous mechanism.

The kinetics of such random, independent nucleation, a prerequisite for homogeneous nucleation, can be described by:

$$N/N_0 = \exp(-kt) \tag{12.5}$$

where N/N_0 represents the fraction of droplets not yet crystallized and k is a constant (10). This relation is valid for a monodispersed droplet distribution.

Indeed, in various papers, such random nucleation, characteristic for homogeneous nucleation, was found for polymer dispersions (13,49,50), microphase separated block copolymers (32–35,39,43), and also for nonpolymer emulsions (24,28). These results can be obtained by performing isothermal crystallization experiments and following droplet crystallizing as a function of time. Different techniques can be applied: optical microscopy (13,29,50), dilatometry (43), real-time SAXS and WAXD (32,33,35), real-time AFM (39), or DSC (34). A recent example is given in Figure 12.3 for PEO droplets of different sizes that were followed with an optical microscope.

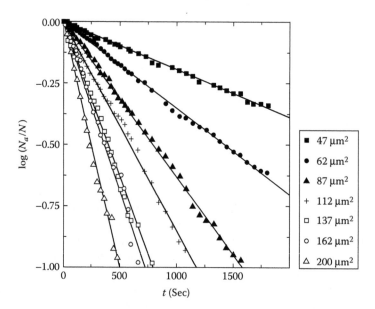

FIGURE 12.3 Plot of the logarithm of the uncrystallized fraction as a function of time for the isothermal homogeneous crystallization of droplets at $T = -5°C$. The data have been binned according to the base area of the droplets. The center of the bin range for the data sets is indicated in the legend (Massa and Dalnoki-Veress, *Phys. Rev. Lett.*, 92(25), 2555091, 2004.)

The constant k (the slope of the plotted data) is seen to decrease with decreasing droplet diameter, indicating a slower nucleation process for smaller droplets. This observation is expected from the classical critical size concept of nucleation, described by Turnbull and Fisher (51). In a number of cases, a strong temperature dependence of the crystallization rate was also found for the confined, supercooled polymer (32–34), which demonstrates that nucleation is the rate determining step in the crystallization of small droplets: Primary nucleation has a stronger temperature dependence than growth (i.e., secondary or tertiary nucleation) (17).

However, the presence of random kinetics is not a definitive proof of a homogeneous nucleation mechanism. Nucleation can be strongly affected by the nature of the interface, which could catalyze the solidification of the droplets. This was already recognized by Turnbull (6) for metal dispersions who showed that the solidification rate of supercooled liquid mercury droplets was strongly dependent upon the nature of the foreign substances on their surface. The effects of the substrate were also clearly identified by Barham et al. (29) for precipitated polyethylene droplets, resulting in different values for "homogeneous" nucleation that varied between 60 and 120°C, depending on the type of cover glass used for following the crystallization behavior with the optical microscope. Koutsky et al. (13) found for polymer droplets dispersed in oil that the crystallization was probably initiated at the oil-polymer melt interface. In hydrocarbon emulsions, the nature of the surfactant and aqueous phase was presumed to affect the crystallization rate (22–24).

If a real, volume dependent, homogeneous nucleation mechanism is operable, the kinetics will be proportional to the volume of the droplet, and Eq. 12.5 can be replaced by

$$N/N_0 = \exp(-Ivt) \tag{12.6}$$

where I is the nucleation rate expressed in number of nuclei per unit volume and time, and v is the average volume of the droplet (6,10). In the case of a heterogeneous surface dependent process, we obtain:

$$N/N_0 = \exp(-IAt) \tag{12.7}$$

where I is the nucleation rate expressed in number of nuclei per unit area and time, and A is the average area of the droplets (6).

In such analysis, however, the size polydispersity of the sample is important because of the difference in nucleation rate for different droplet sizes. Burns and Turnbull (11) found a slowing down of kinetics at longer crystallization times for emulsified isotactic polypropylene (iPP), which was explained by the size polydispersity of the sample (the biggest droplets crystallize first). Perepezko et al. (52) concluded that for the evaluation of the full crystallization kinetics it is essential to include the sample size distribution in the analysis. Upon the inclusion of the size polydispersity of the droplet sample, the isothermal solidification rate constants determined could distinguish between a volume dependent process and a heterogeneous surface area dependent mechanism. Recently, Massa et al. (50) have shown that for different PEO droplet systems, crystallization at high supercooling has kinetic constants that are proportional to the volumes of the domains, indicating volume dependent homogeneous processes instead of surface nucleated processes.

12.3 CONFINED CRYSTALLIZATION PHENOMENA IN IMMISCIBLE POLYMER BLENDS WITH DISPERSED MICRO- AND NANOMETER SIZED DROPLETS

The earliest reports on confined crystallization phenomena in immiscible blends with a crystallizable dispersed phase date back to 1980 (53). Some years later, it was Klemmer and Jungnickel (54) and Frensch et al. (16,55) who gave a systematic explanation of the crystallization of droplets in immiscible blends and introduced the term "fractionated crystallization" to describe the phenomena. This phenomenon was related to the observation of multiple crystallization peaks found at different supercoolings when the crystallizable polymer was dispersed inside a polymer matrix. These phenomena have been observed since then in a variety of immiscible blend systems (16,46–48,56–76). Recently, similar effects have been reported for thermosetting/thermoplastic blends or copolymers after phase separation that resulted in confined morphologies (41,77).

12.3.1 DROPLET SIZE AND DROPLET SIZE DISTRIBUTION

The blend morphology proves to be the most important parameter governing the crystallization behavior of immiscible blends with a crystallizable dispersed phase. Santana and Müller (46) observed a change in nucleation mechanism from predominantly heterogeneous to predominantly homogeneous for PP/PS blends as long as the size of the dispersed PP droplets is below a critical value (which was in the order of 1 to 2 μm). Results of Molinuevo et al. (67) on polycarbonate/polyethylene terephthalate (PC/PET) blends indicated that the finer the dispersion, the greater the inhibition of the crystallization of the PET droplets. A direct relation between the degree of fractionated crystallization and the number of droplets per unit volume is obtained when the total nucleation density developed during cooling remains constant (identical thermal histories, no nucleating agents, etc.), as indicated in Figure 12.4, calculated on the basis of the volume average droplet diameters.

We would expect that the phenomenon of fractionated crystallization could possibly be a useful tool to deduce whether a droplet-matrix morphology or a cocontinuous morphology has been obtained. A cocontinuous blend morphology consists of long, interpenetrating phase structures that lead to a network structure and is usually observed at relatively comparable concentrations of both blend components. In general, such a morphology is expected to give rise to crystallization around the bulk temperature because nucleation can trigger crystal growth within the total volume occupied by crystallizable polymer chains. Indeed, Everaert et al. (47) have shown that the onset composition for fractionated crystallization could be directly related to the center of the phase inversion region for a number of blend series. Also in papers by Ghysels et al. (48) for PP/TR (thermoplastic rubber) blends and by Pompe et al. (74) for reactively compatibilized polyamide 6/polypropylene (PA6/PP) blends, it was suggested that searching for the occurrence of fractionated

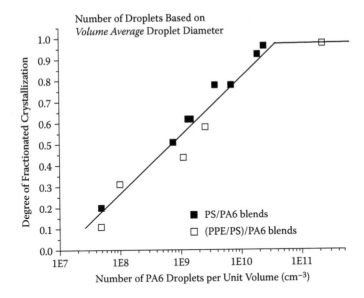

FIGURE 12.4 Degree of fractionated crystallization for PS/PA6 and (polyphenylene ether/polystyrene)/polyamide 6 [(PPE/PS)/PA6] blend compositions as a function of the number of PA6 droplets per unit volume, calculated using a volume average droplet diameter. Cooling rate is 10 K/min. (Tol et al., *Polymer*, 46, 369, 2005.)

crystallization can be an efficient tool to study phase morphology development (size evolution and type of morphology) in immiscible polymer blends. Tol et al. (75), however, showed that such an approach has to be applied with care. First of all, the crystallization behavior of the blend, using DSC, should always be compared with the morphology obtained after the application of thermal changes that mimic those applied by DSC. The thermal treatment can result in partial or full breakup of the cocontinuous morphology to a dispersed-matrix type of morphology (78). In addition, multiple crystallization can already take place inside the cocontinuous region, caused by very small droplet inclusions inside the cocontinuous domains.

It has been shown in a number of papers that the number of crystallization peaks caused by fractionated crystallization is independent of the applied cooling rate (16,75). Furthermore, the positions of the various crystallization peaks can be nicely compared for the same polymer in completely different blend systems as was observed by Everaert et al. (47) for polyoxymethylene (POM) and by Tol et al. (76) for PA6. However, fractionated crystallization of one type of polymer can be significantly influenced by the matrix polymer, as is discussed in Section 12.3.3. These observations seem to confirm the activity of groups of heterogeneous nuclei in the polymer. In a more quantitative approach by Frensch et al. (16), the areas under the DSC crystallization exotherms are related to a number of heterogeneities of a specific type. From the relative intensity of the different crystallization steps, calculations can be done with respect to the concentration of the respective heterogeneities if the mean size of the droplets is known. Considering a large number of small polymer droplets, each having a volume V_D, the fraction of droplets that contain exactly z heterogeneities of type i that can nucleate the polymer follows a Poisson distribution (79):

$$f_z^i = \left[(M^iV_D)^z / z!\right]\exp(-M^iV_D) \qquad (12.8)$$

where M^i is the concentration of heterogeneities of type i and M^iV_D is the mean number of heterogeneities per droplet with volume V_D.

The fraction of droplets, which contain at least one heterogeneity of type i, is given by:

$$f_{z>0}^i = 1 - \exp(-M^iV_D) \qquad (12.9)$$

or

$$M^i = -\left[\ln(1-f_{z>0}^i)\right] / V_D \qquad (12.10)$$

The fraction can be calculated from the relative partial area of each crystallization exotherm during cooling in the DSC. As such, the number of heterogeneities (M) of different types responsible for crystallization at different crystallization temperatures can be calculated. Here it is assumed that one nucleus is enough to crystallize the whole droplet. This approach has been applied by Everaert et al. (47) and Arnal et al. (68). The calculation of the heterogeneous nucleation densities could qualitatively explain the presence or absence of particular exotherms in the DSC cooling curves. The above equations, however, are only valid for monodispersed droplet distributions and also do not offer an explanation for the origin of the exotherms; they, thus, cannot discriminate between heterogeneous or homogeneous nuclei. As such, the usefulness of this approach is limited because of the droplet size dispersity, which can be expected to significantly alter the number and intensity of crystallization peaks (47,68).

Both Everaert et al. (47) and Tol et al. (75) found that the number of crystallization peaks was related to the width of the droplet size distribution. Broader droplet distributions displayed more crystallization peaks. The different peaks can be seen to correspond to the crystallization of groups of droplets of a certain droplet size interval containing a specific type of nuclei.

12.3.2 Heterogeneous Nucleation Density

A number of techniques have been applied that clearly demonstrate that the reason for the fractionated crystallization phenomena can be related to the lack of active heterogeneous nuclei inside the droplets. In a number of papers, Müller et al. (69–71) have shown the applicability of the self-nucleation technique to alter the nucleation density of the droplets. With this method, the nucleation density is increased enormously by heating up the material within the self-nucleation regimes where small

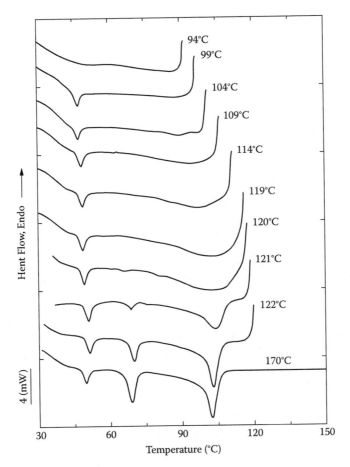

FIGURE 12.5 DSC cooling curves for (70:30) PS/LLDPE blends after self-nucleation at the indicated T_s temperatures. (Reprinted with permission from Arnal and Müller, *Macromol. Chem. Phys.*, 200, 2559, 1999; copryright 2000 Wiley-VCH.)

crystal fragments are still present in the melt (80). These tiny regions of a high degree of order, often stabilized by foreign substrates, may persist in the melt and will act as predetermined nuclei for crystallization upon cooling, causing the nucleation process to start at higher temperatures than would normally be the case. Mathot (81) showed that temperatures within the self-nucleation regimes prior to crystallization increased the crystallization temperature for various nylons (PA6, PA6.6, and PA4.6). Reported heterogeneous nucleation densities vary from approximately 10^6 nuclei/cm^3 after crystallization from the completely molten state to typically 10^{10} to 10^{12} nuclei/cm^3 after crystallization from the self-nucleation region (82). Figure 12.5 shows the effect of self-nucleation on a blend of polystyrene/linear low-density polyethylene (PS/LLDPE).

Decreasing the temperature below 122°C first diminishes the intermediate crystallization peak. The low temperature peak is not affected until 94°C, and its presence

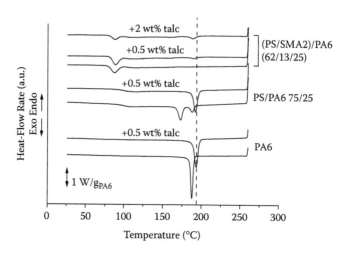

FIGURE 12.6 DSC cooling curves of PA6, a PS/PA6 (75:25) blend, and a (PS/SMA2)/PA6 [(62:13):25] (SMA2: styrene-maleic anhydride) blend with and without talc nucleating agent. (Tol et al., *Polymer*, in press, 2004.)

was not attributed to a lack of nuclei but was attributed to the presence of PE droplets with a much higher concentration of branched PE fractions, crystallizing at low temperature. Recently, Müller et al. (36) studied self-nucleation for different block copolymers with microdomain volumes of about 25 nm, confirming that the main reason for confined crystallization phenomena in these systems is related to the lack of nuclei in the microdomains.

An alternative method, that of fine talc powder as nucleating agent for PA6, was used by Tol et al. (76), which is shown in Figure 12.6. PA6 is effectively nucleated by the talc powder, resulting in an increase of the crystallization temperature (from 188 to 193°C). A PS/PA6 blend, having an average PA6 droplet size D_n of about 1.5 μm, is effectively nucleated, causing a complete suppression of fractionated crystallization. However, nucleation was not as effective for PA6 droplet sizes of about 150 nm. The talc experiments clearly show that the nucleation density within the droplets can be enhanced, though it is obvious that — in order to prevent homogeneous nucleation — there is a real need for nucleating agents that are small enough to be well incorporated in the small droplets.

12.3.3 EFFECT OF THE SECOND BLEND COMPONENT ON THE CRYSTALLIZATION BEHAVIOR OF THE DISPERSED COMPONENT IN IMMISCIBLE POLYMER BLENDS

It is clear, as shown above, that the nucleation density strongly governs the confined crystallization phenomena. In this section, it is shown that, besides the influence of the thermal history, nucleation behavior of the dispersed component in immiscible polymer blends can also be affected by the matrix blend component.

One important phenomenon is migration of impurities (83,84). During the melt-mixing process, heterogeneous impurities can migrate across the interface between both blend phases, with the driving force being the interfacial free energy of the impurity with respect to its melt phase. The possibility for migration of heterogeneities from one phase to the other during the mixing process depends on several factors (83), including the total amount of interfacial area, the time of mixing, and the interfacial tension of the blend. In general, the migration of heterogeneities during melt-mixing only slightly affects the crystallization temperature during cooling from the melt, but it can have a strong effect on the final semicrystalline structure. Because of the change in heterogeneous nucleation density caused by migration, it can be expected to have important consequences for the fractionated crystallization phenomena.

In Section 12.2, it was shown that the interface could have a nonnegligible influence on the confined crystallization phenomena for dispersions. Similar effects have been observed for polymer blends in which the crystallizable polymer was dispersed inside another polymer. Wenig et al. (85) demonstrated that the nucleation shifted from a predominantly thermal behavior to more of an athermal one with increasing content of PS in a blend with PP (matrix phase), which was explained by the effect of heterogeneous surface nucleation at PS interfaces. The type of interface and the wetting ability of the second component with the crystallizable phase are expected to affect interfacial nucleation; only an interface that wets well with the crystallizable polymer can cause heterogeneous nucleation (86,87). The wetting ability can be strongly influenced by the physical state (molten, rubbery, or vitrified) of the components. The effect of the physical state of the matrix phase has been shown in a number of investigations. Everaert et al. (47) demonstrated that vitrification of the (PPE/PS) matrix component prior to crystallization of POM droplets caused an increase in the nucleation density of the crystallizable POM droplets, resulting in a decrease in the fraction of droplets crystallizing at lower temperatures. A similar result was obtained by Tol et al. (75) for (PPE/PS)/PA6 blends. Figure 12.7 shows the effect of the physical state of the matrix phase on the crystallization of impurity-free droplets of PA6 crystallizing randomly at low tem-

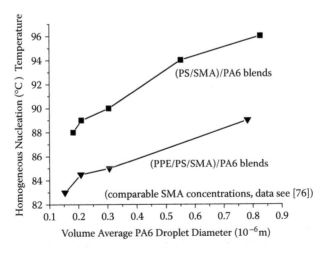

FIGURE 12.7 "Homogeneous nucleation" temperature as a function of PA6 droplet size for (PS/SMA2)/PA6 and (PPE/PS)/SMA2)/PA6 blends. (Tol et al., *Polymer*, 46, 2955, 2005.)

perature (88). It can be seen that the crystallization temperature for submicrometer PA6 droplets in (PS/SMA2)/PA6 (SMA2: styrene-maleic anhydride) blends is consequently 4 to 6 centigrade degrees higher for equal droplet sizes than in the case of (PPE/PS/SMA2)/PA6 blends. This can be attributed to the vitrification of the PS matrix phase, which has a glass transition temperature very close to the onset of the low temperature crystallization peak (100°C), whereas the vitrification of the (PPE/PS) phase occurs at a much higher temperature (150°C). So although the (PS/SMA2)/PA6 blend sample was crystallizing at random (88), PA6 droplet crystallization in these blends was thus most likely initiated at the PS-PA interface, in which case, it does not represent a true volume homogeneous nucleation.

The nucleation effects caused by the interface can be even more pronounced when the surrounding matrix polymer is crystallizable. A typical example is given in Figure 12.8 for an immiscible blend of polyvinylidene (PVDF) and polybutylene terephthalate (PBT), which are both crystallizable components (16). When the component with the lowest crystallization temperature forms the dispersed phase, i.e.,

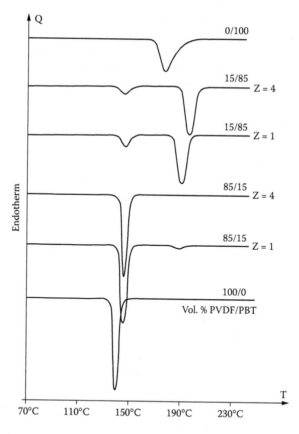

FIGURE 12.8 Coincident crystallization in PVDF/PBT immiscible blends. Z represents the number of extrusion cycles. (Reprinted with permission from Frensch and Jungnickel, *Colloid. Polym. Sci.*, 267,16,1989; copyright 1989 Springer-Verlag.)

PVDF/PBT (15:85), no fractionated crystallization of the PVDF droplets is found, which can be explained by the strong nucleating effect of the crystallizing PBT before the PVDF crystallization starts. However, when the component with the highest crystallization temperature forms the dispersed phase, only one crystallization peak is observed after four extrusion cycles ($Z = 4$). In this case, fractionated crystallization of PBT droplets occurs, which crystallize coincidently with the lower crystallizing PVDF matrix. The crystallizable PVDF matrix phase thus induces the crystallization of the PBT droplets at the crystallization temperature of the PVDF matrix. The crystallization temperature of the PVDF matrix phase displays a shift to higher temperatures upon blending with PBT, which was attributed to the nucleating efficiency of amorphous or crystallizing PBT domains.

Similar coincident crystallization effects were observed for a finely dispersed PA6 phase with PP by Moon et al. (63) in PP/PP-g-MA/PA6 blends, by Ikkala et al. (60) in PP/PA6 blends with different reactive compatibilizers, and by Psarski et al. (72) in polypropylene-acrylic acid (PP-AA/PA6) blends. In all these cases, PA6 was found to crystallize coincidently with PP bulk around 120°C.

12.3.4 EFFECT OF COMPATIBILIZATION ON THE CRYSTALLIZATION OF THE DISPERSED COMPONENT IN IMMISCIBLE POLYMER BLENDS

The strong relation between blend phase morphology and crystallization behavior of dispersed droplets for immiscible blends is clearly shown in the preceding sections. As such, an important contribution to the crystallization behavior of the dispersed droplets can be expected from compatibilization. The technique consists of a modification of the interfacial properties of a blend via addition of suitable graft or block copolymers, which are located at the interface between the phases of an immiscible blend and act as emulsifying agents. In this way, a blend with technologically desirable properties — characterized by a finely dispersed phase, good adhesion between the blend phases, and strong resistance to phase coalescence — is obtained. Compatibilization, either physical or reactive, forms an interesting approach for investigating the homogeneous nucleation and fractionated crystallization phenomena because of the possibility of decreasing the size of the droplets down to the submicrometer of even nanometer size range.

The number of fundamental studies investigating the effect of compatibilization on the droplet crystallization phenomena in immiscible polymer blends is limited (60,61,63,65,68,72–74). In most cases, fractionated crystallization is found to become much more pronounced upon addition of the compatibilizer, leading to much higher supercoolings and causing homogeneous nucleation or coincident crystallization (60,61,63,68,72,73,76). A clear example is given in Figure 12.9 for a PS/PA6 blend with PA6 as the dispersed phase, reactively compatibilized with SMA2 (76). As a result of the reactive compatibilizer, the PA6 droplet size decreases strongly, leading to a very pronounced shift in crystallization peaks to lower temperatures. A fraction of the droplets now crystallizes around 85 to 90°C, more than 100 centigrade degrees below the normal crystallization temperature of PA6, while the peak intensity

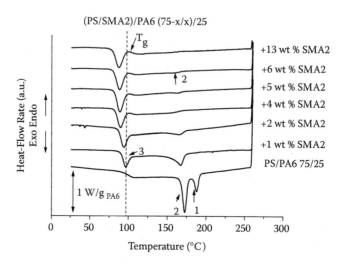

FIGURE 12.9 DSC cooling curves of (PS/SMA2)/PA6 (75:25) blends with different SMA2 concentrations. Cooling rate is 10 K/min. (Tol et al., *Polymer*, 46, 383, 2005.)

increases with increasing amount of compatibilizer (i.e., with decreasing droplet size). A significant effect is already found at a very low compatibilizer concentration (1 wt% SMA2).

Figure 12.10 provides a direct comparison of the crystallization behavior of uncompatibilized and reactively compatibilized PS/PA6 and (PPE/PS)/PA6 blends. The two lines represent the evolution of different crystallization peaks: Heterogeneous crystallization occurs between 10^7 to 10^{11} droplets/cm^3 in the uncompatibilized blends (left y axis) followed by the evolution of homogeneous crystallization at a larger number of droplets per unit volume (10^{11} to 10^{14} cm^{-3}) induced by reactive compatibilization (right y axis). Here, all the heterogeneous nuclei are exhausted. Using isothermal DSC, it was shown that the latter crystallization at the lowest crystallization temperatures (approximately 85°C) are random, a prerequisite for homogeneous nucleation (88). Such random nucleation kinetics was observed even at PA6 concentrations as high as 40 wt%, which is, to the best of our knowledge, the highest reported volume fraction crystallizable polymer in an immiscible blend displaying such behavior. This can be most likely attributed to the effect of the compatibilizer, which reduced the blend interfacial tension and droplet coalescence and strongly improved the level of dispersion for high volume fractions of one component. As reported earlier, however, these kinetics probably do not represent a true homogeneous nucleation process because nucleation was most likely initiated at the interface.

In a number of investigations, it was shown that in contrast to binary blends without a compatibilizer, the crystallization of the dispersed component in compatibilized blends cannot be solely explained by the size of the dispersion (60,61,63,76). The compatibilizer present at the interface of the blend components, or attached to the blend components after the reaction, can be expected to affect the crystallization behavior of the droplets. In fact, pronounced differences in crystallization were reported, depending on the miscibility of the compatibilizer with the blend compo-

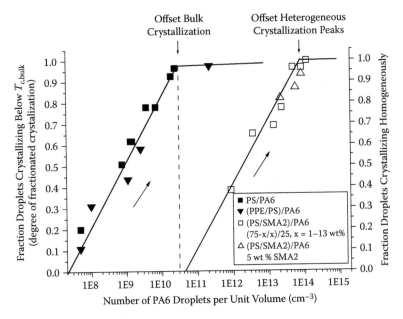

FIGURE 12.10 Fraction of droplets crystallizing heterogeneously below $T_{c,bulk}$ (left y axis) or "homogeneously" (right y axis) as a function of the number of PA6 droplets per unit volume for uncompatibilized and SMA2 compatibilized PS/PA6 and (PPE/PS)/PA6 blend compositions. (Tol et al., *Polymer*, 46, 383, 2005.)

nents. Ikkala et al. (60) proposed two classes of compatibilizers. The first class consists of compatibilizers that form a kind of immiscible interlayer between the two blend phases. Such a compatibilizer will prevent direct nucleating effects, such as migration and interface nucleation from one phase to the other. As such, only the size of the dispersion relative to the nucleation density of the dispersed phase and the nucleating activity of the compatibilizing agent itself play a role in the crystallization behavior. A second class consists of compatibilizers that have an analogues chemical structure to one or two of the blend components. The crystallization behavior in this case is much more complex and will also depend on the mutual nucleation ability, migration of impurities, etc. To study these effects, Ikkala et al. (60) investigated PP/PA6 blends compatibilized with different compatibilizers, with PP as the dispersed phase. It was found that the compatibilizers forming an *immiscible layer* (i.e., EBA-g-FA, SEBS-g-MAH, and ethylene-ethylacrylate-glycidyl-methacrylate (E-EA-GMA) compatibilizers in this case) prevent the strong nucleation of the PP droplets by the crystallizing PA6 matrix phase. The PP/PA6 blend with *miscible* polypropylene-graft-maleic anhydride (PP-g-MA) compatibilizer showed much less fractionated crystallization. The effect of the type of the compatibilizer on the mutual nucleation is also reported by Tang and Huang (62) to explain the decreased fractionated crystallization in PEO/PP/PP-g-MA blends, with PEO droplets. Here, the improved compatibility by the miscible compatibilizer PP-g-MA augmented their mutual nucleation [the nucleation of PEO droplets via the crystallizing of the (earlier crystallizing) PP matrix].

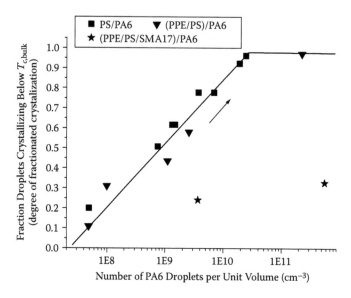

FIGURE 12.11 Degree of fractionated crystallization as a function of the number of PA6 droplets per unit volume for PS/PA6, (PPE/PS)/PA6, and (PPE/PS/SMA17) blend compositions. (Tol et al., *Polymer*, 46, 383, 2005.)

In blends where this mutual nucleation is less pronounced, the effects of the type of compatibilizer may be different. This is shown by Tol et al. (76) for the amorphous/crystalline PS/PA6 or (PPE/PS)/PA6 blend system. Figure 12.11 shows that the amount of fractionated crystallization is *decreased* for the *immiscible* compatibilizer SMA17 compared to the uncompatibilized case. As such, the presence of SMA17 at the interface seems to have caused an increase in nucleation density, probably explained by a change in impurity migration or nucleation effects caused by the compatibilizer at the interface.

For the *miscible* SMA2 compatibilizer the degree of fractionated crystallization strongly *increases,* as was shown in Figure 12.10. Here, similar degrees of homogeneous nucleation are reached for a comparable number of droplets per unit volume irrespective of the amount of SMA2 compatibilizer added, which indicates that the crystallization in this blend system is mainly dictated by the droplet size decrease. A similar observation was obtained by Moon et al. (63) for PP/PA6 blends with PA6 as the dispersed phase and the miscible PP-g-MA as a reactive compatibilizer. In this blend system, the effect of the miscible compatibilizer was also found to relate to the droplet size decrease of PA6, causing the dispersed PA6 phase to crystallize coincidently with the matrix at lower temperatures.

In most cases, the effect of the compatibilizer itself on the droplet crystallization was found to be negligible compared to the effects of the pronounced droplet size decrease for compatibilized blends. However, Tol et al. (76) did show a clear effect of the compatibilizer for high concentrations of reactive compatibilizer grafted to PA6 in PS/SMA2/PA6 blends. It depends on the mixing protocol applied to produce the blends; the effect was only observed for samples where the PA6 was first grafted with the compatibilizer in a separate mixing step. For other mixing protocols, the saturation of the interface with compatibilizer will mainly limit the amount of grafted polymer. Because of the reaction of the excess anhydride groups of the SMA2 compatibilizer with the PA6 amide groups, the number of PA6 chains available for crystallization is decreased drastically and the PA6 chains will be hindered during crystallization. This was found to increase the amount of fractionated crystallization for equal droplet size of PA6.

12.4 EFFECT OF CONFINED CRYSTALLIZATION ON THE SEMICRYSTALLINE STRUCTURE OF DISPERSIONS AND BLENDS

It is clear from the literature that confined crystallization can have a strong effect on crucial parameters for crystallization, like the degree of crystallinity and the semicrystalline morphology (i.e., lamellar thickness, long spacing, and crystal perfection). These parameters are of great importance for the thermal and mechanical properties of the final material.

12.4.1 Crystallinity and Semicrystalline Structure Ascertained by DSC, SAXS, and/or WAXD

Several authors report a drop in the degree of crystallinity of the dispersed phase in immiscible blends. A decrease of the PP crystallinity was observed for PP dispersed in PP/PS blends by Santana and Müller (46). Ghysels et al. (48) found that in PP/TR blends, the crystallization enthalpy was reduced from 98.5 J/g in pure PP to 61 J/g in a PP/TR (25:75) blend. Aref-Azar et al. (53) studied PS/HDPE blends in which the HDPE phase was finely dispersed and found a decrease in crystallinity of HDPE from 80% in the homopolymer to 55% in the PS/HDPE (95:5) blend. Sanchez et al. (73) also recently found lower crystallization and melting enthalpies for PA6, dispersed as droplets in ULDPE-graft-DEM (diethylmaleate)/PA6 blends. These effects were also found for the confined crystallization of microphase separated domains in crystallizable block copolymers. Block copolymers of PEO-PI-PEO having a minor amount of PEO showed a decrease of the crystallinity from 96% in the PEO homopolymer to 69% in the block copolymer and a decrease of the PEO melting point (45). Similar observations were made for PEO-PS and PEO-PS-PEO copolymers (43,44). Interesting results were also obtained for thermoplastic/thermosetting blends with a semicrystalline polymer as thermoplastic. For DDS cured DGEBA/PCL$_{5000}$ blends with 10 and 20 wt% PCL, small PCL droplets of micrometer size are formed as a result of reaction induced phase separation during curing (77). The crystallinity of the PCL in these blends, exhibiting crystallization at much lower temperatures compared to bulk PCL, was very low and close to zero.

A clear decrease as evidence by both DSC and WAXD crystallinity with decreasing droplet size was observed by Tol et al. (88) for PS/PA6 blends with PA6 droplet sizes between 1 to 10 μm. Nucleated blends causing complete suppression of the fractionated crystallization phenomena, however, still showed a decrease of crystallinity with droplet size, indicating the disturbing effect of the droplet size on the crystal morphology. Although in most of the cases, it is assumed that the decrease of crystallinity can be attributed to the formation of thinner and less perfect crystalline structures caused by the higher supercooling, time resolved SAXS and WAXD experiments of Everaert et al. (89) strongly indicate that the size of the droplets decreases the lateral dimensions of the crystallites.

A linear correlation between L_1, corresponding to the lateral dimensions of the crystallites, and the droplet diameter and degree of crystallinity is obtained, as shown in Figure 12.12. L_2, which is a measure of the lamellar thickness, is nonlinearly correlated with the degree of crystallinity, indicating that the decrease in $X_{c,WAXD}$ cannot be solely attributed to the formation of thinner lamella at higher supercoolings (89).

The lateral dimensions, however, cannot only be limited by the droplet size but are also likely affected by the high nucleation density attained in supercooled polymers, which can result in the formation of very small, disordered, blocky crystals because of increasingly nonadjacent reentry of chains leading to overcrowded surfaces. Barham et al. (29) stated that homogeneous nucleation of PE dispersed in droplets proceeds in a regime III nucleation mode in which the nucleation rate is so high that there is a lack of time and space for lateral growth, leading to the formation of small and highly imperfect crystalline structures. Hoffmann and Weeks (90) found extremely small folded chain crystals during homogeneous crystallization of polychlorotrifluoroethylene and attributed this to the catastrophic nucleation in homogeneously crystallizing materials. A similar effect was observed for low density homogeneous ethylene-1-octene copolymers with high degrees of comonomer (91). Here, the mobility of the crystallizable chain is restricted by the shift of T_c toward T_g, leading to a higher probability of formation of small nuclei at lower crystallization temperatures. As a consequence, decreased lateral dimensions results in small blocky crystallites. Eventually, with still increasing comonomer content, fringed micelles result in a fractal growth type of crystallization, where clusters of "loosely packed ethylene sequences" are formed.

12.4.2 Metastable Crystal Structures Formed at Lower Temperatures: Reorganization Phenomena

Because the equilibrium structure in most crystallizable polymers is not attained, except in case of fully extended chain crystals, crystallized polymers in general are semicrystalline. Because of their inherent metastability, they often show reorganization effects upon heating after crystallization. Such reorganization can include "perfectioning" and thickening of the crystallites, while at high temperatures also, the lateral dimensions can increase by additional crystallization. Confined crystallization, drastically decreasing the crystallization temperature of the polymer, can be expected to lead to a strong enhancement of reorganization phenomena in these systems.

Time resolved SAXS data by Mischenko et al. (65) indicated the presence of lamellar type density fluctuations, supporting the presence of an ordering process preceding the formation of real crystallites of PA6 in PA6/PE-AA and PA6/PP-AA

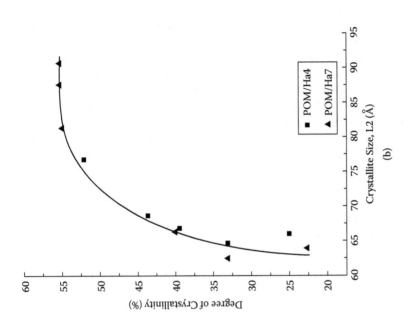

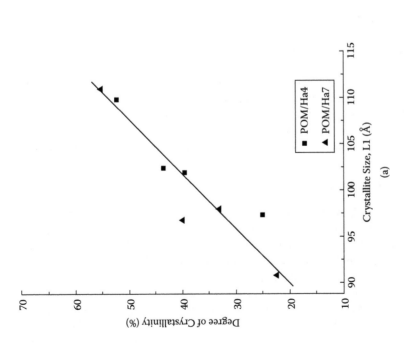

FIGURE 12.12 Correlation between the relative crystal sizes (a) L_1 and (b) L_2 calculated from WAXD and the total degree of crystallinity as measured from WAXD. (Reprinted with permission from Everaert et al., *Polymer*, 44(12), 3489, 2003; copyright 2003 Elsevier.)

blends with PA6 droplets. Reiter et al. (39) and Rottele et al. (92) observed strong improvements of crystalline order in heating via DSC for PEO crystals in 12 nm compartments in PB_h-b-PEO block copolymers. A broad melting range was observed that was related to the superposition of sharp melting transitions of the individual crystals, which take place at different temperatures, corresponding to a multitude of possible metastable states. Stabilization by reorganization was found to take place entirely in the crystalline state: Melting-recrystallization processes were inhibited in the small cells. Tol et al. (88,93) found extensive reorganization for PA6 crystals in the small (150 nm) PA6 droplets in (PS/SMA2)/PA6 blends. The final melting of PA6, measured at a heating rate of 10°C/min, took place more than 100 centigrade degrees above the onset of crystallization and the calculated melting area was much larger compared to the crystallization area. The observed discrepancy is obtained because of imperfect crystallization at low temperatures followed by reorganization of the formed crystallites and further crystallization during the subsequent heating run, resulting in a higher melting enthalpy.

Recently, Tol et al. (94) have used a fast scanning chip calorimeter to study reorganization of PA6 crystals formed at low temperature. The thin-film chip calorimeter is based on a thermal conductivity gauge, which consists of a Si_3N_x membrane with a thin-film thermopile and a resistive film–heater placed at the center of the membrane (95,96). Because of the strongly reduced sample mass, it is possible to obtain much higher cooling and heating rates. As such, it is possible to deduce the sample's thermal history from the heating scan because reorganization and relaxation processes can be avoided, which obscure the relationship between crystallization, the resulting morphology, and melting. This technique has already been successfully applied to study the melting and reorganization of PET (96).

Figure 12.13 shows melting curves of bulk PA6 and a (PS/SMA2)/PA6 blend with small PA6 droplets after isothermal crystallization at 85°C. For both samples, a double melting behavior is observed. The first melting peak shifts to higher temperatures with increasing heating rates and the second melting peak shifts to lower temperatures with increasing heating rates; the latter clearly points to an increased hindrance of reorganization during heating of the crystals formed at low temperature. This is also confirmed by the changing intensities of the two peaks.

12.5 CONCLUSIONS

In this chapter, we have shown that the crystallization behavior of a polymer confined to a small volume can be drastically affected compared to the bulk state. Because of the lack of nucleation sites in the confined polymer, crystallization takes place in different steps at higher supercoolings, possibly finally leading to homogeneous nucleation. It is, in fact, a quite general mechanism, with similar observations for liquid metals and low molar mass hydrocarbon liquids.

For polymer blends, a strong correlation between the blend morphology (droplet size and droplet size distribution) and the crystallization behavior of the confined, dispersed phase is found. The confined crystallization phenomena, also leading to so-called fractionated crystallization in polymer blends, can be prevented by addition

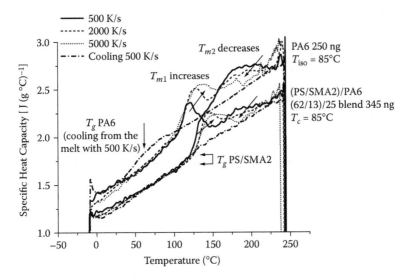

FIGURE 12.13 Specific heat capacities of PA6 and (PS/SMA2)/PA6 [(62:13):25] blend (PA6 droplet size of 150 nm), measured with a fast heating chip calorimeter at different heating rates, after crystallization from the melt at 85°C.

of a sufficient number of small nuclei, thus confirming that the phenomenon is basically related to nucleation problems inside the small droplets. Compatibilization, which is a technologically important technique in the processing of immiscible blends, has been shown to have various effects on the droplet crystallization depending on the compatibilization efficiency, concentration of the compatibilizer, type of compatibilizer, and mixing protocol. Further attention should be paid to these complex effects and should be extended to industrial formulations and processing conditions, for example, the effect of shear flow on the confined crystallization phenomena, which is still largely unknown.

Both for droplet emulsions and polymer blends, the interface has a significant influence on the crystallization behavior, often leading to nucleation initiated at the droplet–dispersing medium interface. These effects can be expected to become increasingly important upon decreasing the volume of the dispersed polymer to the nanometer range. Confined crystallization also strongly affects the final semicrystalline structure, often leading to a strong decrease in crystallinity and strong reorganization of crystals, formed at low temperature, during heating. Here, the use and further development of new, sophisticated techniques to monitor nanostructures in real-time and to allow controlled, fast scanning and cooling rates will be increasingly important.

ACKNOWLEDGMENTS

The authors are indebted to the Research Fund of the Katholieke Universiteit Leuven (GOA 98/06) and to the Fund for Scientific Research-Flanders, Belgium, for the financial support given to the Laboratory of Macromolecular Structural Chemistry.

REFERENCES

1. Groeninckx, G., Vanneste, M., and Everaert, V., Crystallization, morphological structure and melting of polymer blends, in *Polymer Blends Handbook*, Vol. 1, Utracki, L.A., Ed., Kluwer Academic Publishers, Dordrecht, 2002, p. 203.
2. Van Riemsdyk, A.D., *Ann. Chim. Phys.*, 20, 66, 1880.
3. Vonnegut, B., Variation with temperature of the nucleation of supercooled liquid tin and water drops, *J. Colloid Sci.*, 3, 563, 1948.
4. Turnbull, D. and Cech, R.E., Microscopic observation of the solidification of small metal droplets, *J. Appl. Phys.*, 21, 804, 1950.
5. Turnbull, D., Formation of crystal nuclei in liquid metals, *J. Appl. Phys.*, 21, 1022, 1950.
6. Turnbull, D., Kinetics of solidification of supercooled liquid mercury droplets, *J. Chem. Phys.*, 20, 411, 1952.
7. Perepezko, J.H. and Paik, J.S., Undercooling behavior of liquid metals, in *Rapidly Solidified Amorphous and Crystalline Alloys*, Kear, B.H., Giessen, B.C., and Cohen, M., Eds., North Holland Publ. Comp., New York, 1982, p..
8. Jackson, K.A., Nucleation from the melt, *Ind. Eng. Chem.*, 57, 28, 1965.
9. Price, F.P., *IUPAC Symposium on Macromolecules*, 1959, paper #1B2.
10. Cormia, R.L., Price, F.P., and Turnbull, D., Kinetics of crystal nucleation in polyethylene, *J. Chem Phys.*, 37(6), 1333, 1962.
11. Burns, J.R. and Turnbull, D., Kinetics of crystal nucleation in molten isotactic propylene, *J. Appl. Phys.*, 37(11), 4021, 1966.
12. Gornick, F., Ross, G.S., and Frolen, L.J., Crystal nucleation in polyethylene: the droplet experiment, *ACS Polym. Prepr., Div. Polym. Chem.*, 7, 82, 1966.
13. Koutsky, J.A., Walton, A.G., and Baer, E., Nucleation of polymer droplets. *J. Appl. Phys.*, 38(4), 1832, 1967.
14. Burns, J.R. and Turnbull, D., Nucleation of crystallization in molten isotactic polybutene-1, *J. Polym. Sci. Part A2*, 6, 775, 1968.
15. Turnbull, D. and Cormia, R.L., Kinetics of crystal nucleation of some n-alkanes, *J. Chem. Phys.*, 34, 820, 1961.
16. Frensch, H., Harnischfeger, P., and Jungnickel, B.J., Fractionated crystallization in incompatible polymer blends, *Multiphase Polymers: Blends and Ionomers, ACS Symp. Series*, 395,101, 1989.
17. Wunderlich, B., *Macromolecular Physics, Vol. 2, Crystal Nucleation, Growth, Annealing*, Academic Press, Inc., New York,1976.
18. Price, F.P., The development of crystallinity in polychlorotrifluoroethylene, *J. Amer. Chem. Soc.*, 74, 311,1952.
19. Sharples, A., The formation of nuclei in crystalline polymers, *Polymer*, 3, 250, 1962.
20. Boon, J., Challa, G., and van Krevelen, D.W., Crystallization kinetics of isotactic polystyrene. II. Influence of thermal history on number of nuclei, *J. Polym. Sci. Part A2*, 6, 1835, 1968.
21. Buckle, E.R., *Nature*, 188, 631, 1961.
22. McClements, D.J., Dickinson, E., and Povey, M.J.W., Crystallization in hydrocarbon-in-water emulsions containing a mixture of solid and liquid droplets, *Chem. Phys. Lett.*, 172, 6, 449, 1990.
23. Dickinson, E., Kruizenga, F-J., Povey, M.J.W., et al., Crystallization in oil-in-water emulsions containing liquid and solid droplets, *Colloids Surf. A*, 81, 273, 1993.

24. McClements, D.J., Dungan, S.R., German, J.B., et al., Droplet size and emulsifier type affect crystallization and melting of hydrocarbon-in-water emulsions, *J. Food. Sci.*, 58(5), 1148, 1993.

25. Herhold, A.B., Erta, D., Levine, A.J., et al., Impurity mediated nucleation in hexadecane-in-water emulsions, *Phys. Rev. E*, 59 (6), 6946, 1999.

26. Montenegro, R., Antonietti, M., Mastai, Y., et al., Crystallization in miniemulsion droplets, *J. Phys. Chem. B*, 107, 5088, 2003.

27. Walstra, P. and van Beresteyn, E.C.H., Crystallization of milk fat in the emulsified state, *Neth. Milk Dairy J.*, 29, 35, 1975.

28. Charoenrein, S. and Reid, D.S., The use of DSC to study the kinetics of heterogeneous and homogeneous nucleation of ice in aqueous systems, *Thermochimica Acta*, 156, 373, 1989.

29. Barham, P.J., Jarvis, D.A., and Keller, A., A new look at the crystallization of polyethylene. III. Crystallization from the melt at high supercoolings, *J. Polym. Sci. Phys. Ed.*, 20, 1733, 1982.

30. Taden, A. and Landfester, K., Crystallization of poly(ethylene oxide) confined in miniemulsion droplets, *Macromolecules*, 36(11), 4037, 2003.

31. Ibarretxe Uriguen, J., Bremer, L., Mathot, V.B.F., et al., Preparation of water-borne dispersions of polyolefins: new systems for the study of homogeneous nucleation of polymers, *Polymer*, 45, 5961, 2004.

32. Loo, Y.-L., Register, R.A., and Ryan, A.J., Polymer crystallization in 25-nm spheres, *Phys. Rev. Lett.*, 84(18), 4120, 2000.

33. Loo, Y.-L., Register, R.A., Ryan, A.J., et al., Polymer crystallization confined in one, two, or three dimensions, *Macromolecules*, 34, 8968, 2001.

34. Xu, J.-T., Fairclough, P.A., Mai, S.-M., et al., Isothermal crystallization kinetics and melting behavior of poly(oxyethylene)-b-poly(oxybutylene)/poly(oxybutylene) blends, *Macromolecules*, 35, 6937, 2002.

35. Lee, W., Chen, H.L., and Lin, T.L., Correlation between crystallization kinetics and microdomain morphology in block copolymer blends exhibiting confined crystallization, *J. Polym. Sci. Part B: Polym. Phys.*, 40(6), 519, 2002.

36. Müller, A.J., Balsamo, V., Arnal, M.L., et al., Homogeneous nucleation and fractionated crystallization in block copolymers, *Macromolecules*, 35(8), 3048, 2002.

37. Müller, A.J., Arnal, M.L., and Lopez-Carrasquero, F., Nucleation and crystallization of PS-b-PEO-b-PCl triblock copolymers, *Macromolecular Symposia*, 183, 199, 2002.

38. Sun, L., Zhu, L., Ge, Q., et al., Comparison of crystallization kinetics in various nanoconfined geometries, *Polymer*, 45, 2931, 2004.

39. Reiter, G., Castelein, G., Sommer, J.-U., et al., Direct visualization of random crystallization and melting in arrays of nanometer-size polymer crystals, *Phys. Rev. Lett.*, 87(22), article no. 226101, 2001.

40. Balsamo, V., Urdaneta, N., Perez, L., et al., Effect of polyethylene confinement and topology on its crystallisation with semicrystalline ABC triblock copolymers, *Eur. Polym. J.*, 40(6), 1033, 2004.

41. Guo, Q.P., Thomann, R., Gronski, W., et al., Nanostructures, semicrystalline morphology, and nanoscale confinement effect on the crystallization kinetics in self-organized block copolymer/thermoset blends, *Macromolecules*, 36(10), 3635, 2003.

42. Albuerne, J., Marquez, L., Müller, A.J., et al., Nucleation and crystallization in double crystalline poly(p-dioxanone)-b-poly(epsilon-caprolactone) diblock copolymers, *Macromolecules*, 36(5), 1633, 2003.

43. Lotz, B. and Kovacs, A.J., Phase transitions in block-copolymers of polystyrene and polyethylene oxide, *ACS Polym. Prepr., Div. Polym. Chem.*, 10(2), 820, 1969.

44. O'Malley, J.J., Crystal, R.G., and Erhardt, P.F., Synthesis and thermal transition properties of styrene-ethylene oxide block copolymers, *ACS Polym. Prepr., Div. Polym. Chem.*, 10(2), 796, 1969.

45. Robitaille, C. and Prudhomme, J., Thermal and mechanical properties of a poly(ethylene oxide-b-isoprene-b-ethylene oxide) block polymer complexed with NASCN, *Macromolecules*, 16, 665, 1983.

46. Santana, O.O. and Müller, A.J., Homogeneous nucleation of the dispersed crystallizable component of immiscible polymer blends, *Polymer Bulletin*, 32(4), 471, 1994.

47. Everaert, V., Groeninckx, G., and Aerts, L., Fractionated crystallization in immiscible POM/(PS/PPE) blends. Part 1. Effect of blend phase morphology and physical state of the amorphous matrix phase, *Polymer*, 41, 1409, 2000.

48. Ghysels, A., Groesbeek, N., and Yip, C.W., Multiple crystallization behaviour of polypropylene/thermoplastic rubber blends and its use in assessing blend morphology, *Polymer*, 23, 1913, 1982.

49. Massa, M.V., Carvalho, J.L., and Dalnoki-Veress, K., Direct visualization of homogeneous and heterogeneous crystallization in an ensemble of confined domains of poly(ethylene oxide), *Eur. Phys. J. E*, 12, 111, 2003.

50. Massa, M.V. and Dalnoki-Veress, K., Homogeneous crystallization of poly(ethylene oxide) confined to droplets: the dependence of the crystal nucleation rate on length-scale and temperature, *Phys. Rev. Lett.*, 92(25), 2555091, 2004.

51. Turnbull, D. and Fischer, J.C., Rate of nucleation in condensed systems, *J. Chem. Phys.*, 17, 71, 1949.

52. Perepezko, J.H., Höckel, P.G., and Paik, J.S., Initial crystallization kinetics in undercooled droplets, *Thermochimica Acta*, 388,129, 2002.

53. Aref-Azar, A., Hay, J.N., Marsden, B.J., et al., Crystallization characteristics of polymer blends. I. Polyethylene and polystyrene, *J. Polym. Sci. Phys. Ed.*, 18, 637, 1980.

54. Klemmer, N. and Jungnickel, B.J., Über eine kristallisationskinetische anomalie in mischungen aus polyethylen und polyoxymethylen, *Colloid Polym. Sci.*, 262, 381, 1984.

55. Frensch, H. and Jungnickel, B.J., Some novel crystallization kinetics peculiarities in finely dispersing polymer blends, *Colloid. Polym. Sci.*, 267,16,1989.

56. Pukánszky, B., Tüdös, F., Kalló, A., et al., Multiple morphology in polypropylene/ethylene-propylene diene terpolymer blends, *Polymer*, 30, 1399, 1989.

57. Frensch, H. and Jungnickel, B.J., Fractionated and self-seeded crystallization in incompatible polymer blends, *Plast. Rubber Comp. Process. Appl.*, 16, 5, 1991.

58. Baïtoul, M., Saint-Guirons, H., Xans, P., et al., Etude par analyse thermique differentielle de mélanges diphasiques de polyethylene basse densite et de polystyrene atactique, *Eur. Polym. J.*, 17, 1281, 1991.

59. Avella, M., Martuscelli, E., and Raimo, M., The fractionated crystallization phenomenon in poly(3-hydroxybutyrate) poly(ethyleneoxide) blends, *Polymer*, 34(15), 3234, 1993.

60. Ikkala, O.T., Holsti-Miettinen, R.M., and Seppälä, J., Effects of compatibilization on fractionated crystallization of PA6/PP blends, *J. Appl. Polym. Sci.*, 49, 1165, 1993.

61. Tang, T. and Huang, B., Fractionated crystallization in polyolefins-nylon 6 blends, *J. Appl. Polym. Sci.*, 53, 355, 1994.

62. Tang, T. and Huang, B., Compatibilization of polypropylene/poly(ethylene oxide) blends and crystallization behavior of the blends, *J. Polym. Sci. Part B: Polym. Phys.*, 32, 1991, 1994.

63. Moon, H.S., Ryoo, B.K., and Park, J.K., Concurrent crystallization in polypropylene/ nylon-6 blends using maleic anhydride grafted polypropylene as a compatibilizing agent, *J. Appl. Polym. Sci., Part B, Polym. Phys.*, 32, 1427, 1994.

64. Morales, R.A., Arnal, M.L., and Müller, A.J., The evaluation of the state of dispersion in immiscible blends where the minor phase exhibits fractionated crystallization, *Polymer Bulletin*, 35(3), 379, 1995.

65. Mischenko, N., Groeninckx, G., Reynaers, H., et al., A study of phase transitions in nylon 6/functionalized polyolefin blends by simultaneous SAXS and WAXS real-time experiments, *4th AIM Conference on Advanced Topics in Polymer Science*, Pisa, Italy, 1996.

66. Manaure, A.C., Morales, R.A., Sánchez, J.J., et al., Rheological and calorimetric evidences of the fractionated crystallization of ipp dispersed in ethylene/ α-olefin copolymers, *J. Appl. Polym. Sci.*, 66, 2481, 1997.

67. Molinuevo, C.H., Mendez, G.A., and Müller, A.J., Nucleation and crystallization of PET droplets dispersed in an amorphous PC matrix, *J. Appl. Polym. Sci.*, 70, 1725, 1998.

68. Arnal, M.L., Matos, M.E., Morales, R.A., et al., Evaluation of the fractionated crystallization of dispersed polyolefins in a polystyrene matrix, *Macromol. Chem. Phys.*, 199(10), 2275, 1998.

69. Arnal, M.L. and Müller, A.J., Fractionated crystallization of polyethylene and ethylene/α-olefin copolymers dispersed in immiscible polystyrene matrices, *Macromol. Chem. Phys.*, 200, 2559, 1999.

70. Arnal, M.L., Müller, A.J., Maiti, P., et al., Nucleation and crystallization of isotactic polypropylene droplets in an immiscible polystyrene matrix, *Macromol.Chem. Phys.*, 201(17), 2493, 2000.

71. Manaure, A.C. and Müller, A.J., Nucleation and crystallization of blends of polypropylene and ethylene/ α-olefin copolymers, *Macromol. Chem. Phys.*, 201(9), 958, 2000.

72. Psarski, M., Pracella, M., and Galeski, A., Crystal phase and crystallinity of polyamide 6/ functionalized polyolefin blends, *Polymer*, 41, 4923, 2000.

73. Sanchez, A., Rosales, C., Laredo, E., et al., Compatibility studies in binary blends of PA6 and ULDPE-graft-DEM, *Makromol. Chem. Phys.*, 202, 2461, 2001.

74. Pompe, G., Potschke, P., and Pionteck, J., Reactive melt blending of modified polyamide and polypropylene: Assessment of compatibilization by fractionated crystallization and blend morphology, *J. Appl. Polym. Sci.*, 86(13), 3445, 2002.

75. Tol, R.T., Mathot, V.B.F., and Groeninckx, G., Confined crystallization phenomena in immiscible polymer blends with dispersed micro- and nanometer sized PA6 droplets. Part 1. Uncompatibilized PS/PA6, (PPE/PS)/PA6 and PPE/PA6 blends, *Polymer*, 46, 369, 2005.

76. Tol, R.T., Mathot, V.B.F., and Groeninckx, G., Confined crystallization phenomena in immiscible polymer blends with dispersed micro– and nanometer sized PA6 Droplets. Part 2. Reactively compatibilized PS/PA6 and (PPE/PS)/PA6 blends, *Polymer*, 46, 383, 2005.

77. Vanden Poel, G., *Crystallizable Thermoplastic/Thermosetting Polymer Blends. Curing Kinetics, Phase Behaviour and Morphology, Crystallization and Semicrystalline Structure*, Ph.D. thesis, Katholieke Universiteit Leuven, Belgium, 2003.

78. Tol, R.T., Groeninckx, G., Vinckier, I., et al., Phase morphology and stability of co-continuous (PPE/PS)/PA6 and PS/PA6 blends: effect of rheology and reactive compatibilization, *Polymer*, 45(8), 2587, 2004.

79. Pound, G.M. and LaMer, V.K., Kinetics of crystalline nucleus formation in super-cooled liquid tin, *J. Am. Chem. Soc.*, 74, 2323, 1952.

80. Blundell, D.J., Keller, A., and Kovacs, A.J., A new self-nucleation phenomenon and its application to the growing of polymer crystals from solution, *J. Polym. Sci.*, B(4), 481, 1966.

81. Mathot, V.B.F., The crystallization and melting region, in *Calorimetry and Thermal Analysis of Polymers*, Mathot, V.B.F., Ed., Carl Hanser Verlag, Munich, 1994, p. 232.

82. Fillon, B., Wittmann, J.C., Lotz, B., et al., Self-nucleation and recrystallization of isotactic polypropylene (alpha-phase) investigated by differential scanning calorimetry, *J. Polym. Sci.Part B: Polym. Phys.*, 31,1383, 1993.

83. Bartczak, Z., Galeski, A., and Krasnikova, N.P., Primary nucleation and spherulite growth-rate in isotactic polypropylene polystyrene blends, *Polymer*, 28, 1627, 1987.

84. Bartczak, Z., Galeski, A., and Pracella, M., Spherulite nucleation in blends of isotactic polypropylene with high-density polyethylene, *Polymer*, 27, 537, 1986.

85. Wenig, W., Fiedel, H.W., and Scholl, A., Crystallization kinetics of isotactic polypropylene blended with atactic polystyrene, *Colloid Polym. Sci.*, 268, 528, 1990.

86. Turnbull, D, J., Kinetics of heterogeneous nucleation, *Chem. Phys.*, 18, 198, 1950.

87. Geil, P.H., in *Polymer Single Crystals*, Krieger, R.E., Ed., New York, 1973, p..

88. Tol, R.T., Mathot, V.B.F., and Groeninckx, G., Confined crystallization phenomena in immiscible polymer blends with dispersed micro- and nanometer sized PA6 droplets. Part 3. Crystallization kinetics and crystallinity of nanometer sized PA6 droplets crystallizing at high supercooling, *Polymer*, 46, 2955, 2005.

89. Everaert, V., Groeninckx, G., Koch, M.H.J., et al., Influence of fractionated crystallization on the semicrystalline structure of (POM/(PS/PPE)) blends. Static and time-resolved SAXS, WAXD and DSC studies, Polymer, 44(12), 3489, 2003.

90. Hoffman, J.D. and Weeks, J.J., Rate of spherulitic crystallization with chain folds in polychlorotrifluoroethylene, *J. Chem. Phys.*, 37, 1723, 1962.

91. Mathot, V.B.F., Scherrenberg, R.L., Pijpers, T.F.J., et al., Structure, crystallization and morphology of homogeneous ethylene-propylene, ethylene-1-butene and ethylene-1-octene copolymers with high comonomer contents, in *New Trends in Polyolefin Science and Technology*, Hosoda, S., Ed., Research Signpost, Trivandrum, India, 1996, p. 71.

92. Rottele, A., Thurn-Albrecht, T., Sommer, J.U., et al., Thermodynamics of formation, reorganization, and melting of confined nanometer-sized polymer crystals, *Macromolecules*, 36(4), 1257, 2003.

93. Tol, R.T., Mathot, V.B.F., Goderis, B., et al., Confined crystallization phenomena in immiscible polymer blends with dispersed micro– and nanometer sized PA6 droplets. Part 4. Polymorphous structure and (meta)-stability of PA6 crystals formed in different temperature regions, *Polymer,* 46, 2966, 2005.

94. Tol, R.T., Minakov, A.A., Adamovsky, S.A., et al., Melting and reorganization of confined, sub-micrometer PA6 droplets in a PS/SMA2 matrix via ultra fast calorimetry, to be submitted, 2004.

95. Adamovsky, S.A., Minakov, A.A., and Schick, C., Scanning microcalorimetry at high cooling rate, *Thermochim. Acta*, 403, 55, 2003.

96. Minakov, A.A., Mordvindsev, D.A., and Schick, C., Melting and reorganization of poly(ethylene terephthalate) on fast heating (1000 K/s), *Polymer*, 45(11), 3755, 2004.

13 Rheology-Morphology Relationships in Immiscible Polymer Blends

Peter Van Puyvelde and Paula Moldenaers

CONTENTS

13.1 INTRODUCTION

The blending of immiscible polymers is an economically attractive route to develop new materials that combine the desirable properties of more than one polymer without investing in new chemistry. The final properties of the blend are strongly influenced by the size distribution, orientation, and type of the generated microstructure, which in turn is determined by the fluid properties and the flow history. For dilute blends, consisting of Newtonian components, the rheological behavior and morphology development is relatively well understood. Most of the theories and experiments that contributed to the understanding of the morphology evolution in such blends under well-defined flow conditions are summarized in recent reviews by Ottino et al. (1) and Tucker and Moldenaers (2). In this chapter, the material covered in these reviews is extended by incorporating phenomena that pertain to morphology development in more realistic blend systems. Firstly, the structure-rheology relationship in compatibilized blends is addressed. This is clearly of major technological importance since compatibilizers are the main driving force for the future growth of the blend market (3). Secondly, since polymers are viscoelastic materials, effects of elasticity of the components on the morphology development are obviously of interest. Whereas the previously mentioned reviews focused mainly on the connection between theory and experiment — which is clearly easier for Newtonian components — the effect of component elasticity has not been covered extensively. Nevertheless, if we want to understand the morphology development in realistic systems, component elasticity should be taken into account. Finally, most publications dealing with the relationship between rheology and morphology are restricted to rather dilute systems because of the difficulties involved in studying — both theoretically as well as experimentally — droplet interactions that affect deformation and breakup phenomena at higher concentrations. However, recently, progress has been made in relating the rheological response to the underlying morphology in more concentrated systems, and thus the subject merits attention in this chapter.

Many mixtures of high molecular weight polymers are partially miscible, separating into two coexisting phases in some range of temperature and concentration. Most often, such blends display a lower critical solution temperature (LCST), becoming phase separated as the temperature is raised. Although the determination of phase diagrams under quiescent conditions is already known for a long time, investigations of the phase behaviour during flow are still rather scarce [e.g., ref. (4)]. Nevertheless, it has been shown for different blend combinations that both shear induced mixing as well as shear induced demixing can occur [e.g., refs. (5–7)]. This complex situation is avoided in this chapter by focusing on morphology development in totally immiscible systems without the additional concern of partial miscibility.

In order to connect the microstructural response to macroscopic observables — such as, for instance, stress — experiments should be performed in which the dynamics of the structure in the presence of flow is measured. Different approaches can be used, depending upon the degree of information that needs to be extracted. Basically these techniques can be classified into two broad categories. The first approach is the use of so-called direct imaging techniques such as transmission electron microscopy (TEM) and scanning electron microscopy (SEM). The use of these techniques has become quite popular in blend research in the past few decades [e.g., refs. (8–10)]. Although the obtained picture is a direct representation of the structure, sample preparation is often cumbersome and time-consuming. Moreover, these "postmortem" techniques have the undoubtable disadvantage that they can only be applied after the processing step when the sample has solidified. Hence, direct visualization of the morphology generation during flow is impossible. In this chapter, the interest is in understanding the structure development during flow; hence, we focus on studies in which so-called indirect techniques are used that follow *in situ* the structure development during flow. The emphasis here will be on studies that use various rheological and rheo-optical methods [see Fuller (11) for an over-view] to gain insight in the relationship between flow and structure in immiscible polymer blends Although many excellent theoretical studies have been published that model and/or simulate various morphological processes in blends [see, e.g., ref. (2) for a review], we restrict ourselves in this paper mainly to experimental studies.

13.2 RHEOLOGY AND MORPHOLOGY OF COMPATIBILIZED BLENDS

Recently, the effect of compatibilization on various morphological processes in immiscible polymer blends has been studied intensively [for an overview see ref. (12)]. Compatibilizers are either added before blending (physical compatibilization) or are generated by interfacial reactions (reactive compatibilization) during blending (13,14). In this part, we mainly focus on the effects of the compatibilizer on the structural evolution, rather than on the effectiveness of specific compatibilizers or the physical properties of the ultimate compatibilized blends. Hence, the distinction between physical and reactive compatibilization is not essential, although most of the systematic experimental work deals with physically compatibilized blends.

13.2.1 DEFORMATION AND BREAKUP

In two-phase polymer blends, the structural evolution is a competition between the hydrodynamic deforming stresses ($\eta_m \gamma$) and interfacial restoring stresses (Γ/R) in which η_m and γ are the viscosity and deformation rate of the matrix, respectively, and Γ and R are the interfacial tension and radius of the droplets, respectively. The ratio of these stresses is denoted as the capillary number Ca. For capillary numbers less than a critical value (Ca_{cr}), the droplet attains a steady shape and orientation, whereas above the critical capillary number, the droplet eventually breaks up. The dependence of this critical capillary number on the viscosity ratio p of the blend, i.e., the ratio of droplet and matrix viscosity, was measured by Grace (15) in both simple shear and planar elongational flow. This pioneering experimental work was performed by gradually increasing the strain rate on a single droplet until its breakup. Figure 13.1 summarizes the results of this study in simple shear flow, which indicate that no breakup is possible when the viscosity ratio of the blend exceeds a value of 4. Elongational flow (not shown in the figure) can break up droplets of any viscosity ratio. An extensive data set, including flows intermediate between simple shear and elongation, was provided by Bentley and Leal (16). Recently, it has been shown that the critical capillary number in simple shear flow decreases with increasing dispersed phase concentration (17). The data can still be described by a critical capillary number, but in its definition, the matrix viscosity needs to be replaced by the viscosity of the blend itself.

Droplet breakup mechanisms and the shapes of deformed Newtonian droplets have been widely studied [e.g., see refs. (18–20)]. The results are also schematically depicted in Figure 13.1. If Ca << 1, the steady shape is slightly ellipsoidal (depending on the viscosity ratio) with the long axis oriented at 45° to the flow direction. As the capillary number increases, the steady state deformation increases and the droplet

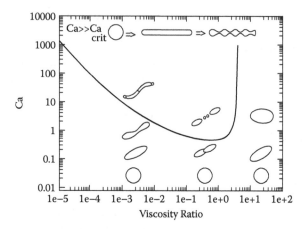

FIGURE 13.1 Schematic representation of the different modes of deformation and breakup in shear flow. The full line corresponds to an empirical fit for the critical capillary number using the data of Grace, *Chem. Eng. Commun.*, 14, 225–277, 1982.

major axis shifts toward the flow direction. When the capillary number slightly exceeds the critical value, the mode of breakup depends on the viscosity ratio (see Figure 13.1). For $p \ll 1$, the droplet assumes a sigmoidal shape from which small droplets are released (tip streaming). For p approximately 1, the central portion of the droplet "necks" until the droplet breaks into two daughter droplets, with small satellite droplets in between. For $Ca \gg Ca_{cr}$, the droplet deforms into a long slender fibril that eventually breaks up by capillary wave instability (21,22).

In the presence of a compatibilizer, the dynamics of deformation and breakup is altered. A compatibilizer can affect these morphological processes by reducing the interfacial tension [e.g., see refs. (23–25)], thereby lowering the hydrodynamic stress required to break droplets of a certain size. The presence of compatibilizers at the interphase entails, however, other phenomena as well. The bulk flow can induce concentration gradients of the compatibilizer along the drop surface. These gradients might lead to Marangoni stresses that, for instance, stabilize the droplets against further deformation. Gradients in block copolymer concentration were recently shown to be present in polymeric systems by visualizing a fluorescently labeled polystyrene-polymethylmethacrylate (PS-PMMA) block copolymer at the surface of a PMMA drop in a PS matrix (26). A quantitative study on single compatibilized droplets was performed by Hu et al. (27). These authors found that the compatibilizer stabilized droplets with $p > 1$, i.e., drops of a given Ca — with known droplet size and interfacial tension — deformed less when a compatibilizer was present, and that Ca_{cr} increased because of compatibilization. On the other hand, when $p < 1$, compatibilization led to destabilization and breakup either through an asymmetric breakup process at low concentrations of blockcopolymer or through tip streaming. In addition, it has been observed that the presence of compatibilizer could lead to the formation of long, sheetlike droplets, i.e., the transition from affine to fiberlike deformation was delayed by the presence of a compatibilizer (26,28). In some cases, compatibilized droplets even displayed widening (28), i.e., the dimension along the vorticity direction was larger than the original droplet diameter. This is interesting from an application point of view because this enlarged area opens routes for generating diffusion barrier materials with plateletlike microstructures.

There is relatively limited experimental work on nondilute compatibilized blends. Ramic et al. (29) concluded, using *in situ* microscopy, that small amounts of compatibilizer did not affect droplet breakup. However, this conclusion was not derived from a direct observation of deformation and breakup but was based on the measurement of steady state droplet sizes. Velankar et al. (30) derived the mean capillary number of compatibilized droplets directly from the linear viscoelastic properties. For p approximately 1, the steady shear capillary number was found to increase strongly with the addition of a compatibilizer; an observation that supports the picture of Marangoni stresses.

13.2.2 COALESCENCE IN COMPATIBILIZED BLENDS

When small droplets in an uncompatibilized polymer blend are sheared at low capillary number, their size increases because of coalescence. This process represents a limit to the ability to maintain fine microstructures. Coalescence in uncompatibilized blends has been reviewed by Chesters (31) and others (32,33). The critical event in coalescence is the collision between two droplets. Such an event leads to a flat or dimpled interface over which the droplets are separated by a thin film of matrix fluid. Hydrodynamic forces push the droplets together during some finite interaction time, causing the matrix fluid to drain from the film. If the film thickness reaches a critical value, van der Waals interactions become dominant and the film ruptures, leading to a coalescence event. Such droplet growth dynamics in uncompatibilized blends have been studied experimentally by several groups (29,34,35).

In compatibilized polymer blends, several experimental studies — most often based on postmortem analysis of the structure — clearly show that compatibilization causes a dramatic decrease in the rate of coalescence [e.g., see refs. (36–41)]. At least two physical descriptions have been postulated to explain coalescence suppression by compatibilizers (see Figure 13.2). The first one is based on the existence of Marangoni stresses; indeed, when the matrix fluid is swept out of the gap between

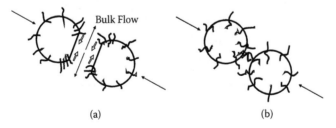

(a)   (b)

FIGURE 13.2 Schematic representation of the theoretical approaches to explain coalescence suppression in compatibilized polymer blends. (a) Bulk flow convects the compatibilizer out of the film region, and the stress due to the resulting gradient in interfacial tension (Marangoni stress) retards drainage of the film between the droplets. (b) Compression of the block copolymer chains in the matrix phase leads to repulsion between approaching droplets.

colliding droplets, it can introduce a concentration gradient of the compatibilizer on the droplet surface, which leads to an opposing Marangoni stress. The latter would slow down the film drainage and hence the coalescence efficiency. A second hypothesis is based on the compression of the compatibilizer block extending into the matrix phase when two droplets approach. This compression reduces the number of possible conformations that the moiety of the compatibilizer can achieve in the matrix phase, thus generating an elastic repulsion between the droplets (steric stabilization). Until now, the mechanism of coalescence suppression is not fully understood. Using microscopy during flow, Ramic et al. (29) and Hudson et al. (42) reported a severe reduction in coalescence efficiency even when only a very small amount of compatibilizer was present. Similar conclusions were drawn by Nandi et al. (43,44). Drainage times as well as detailed trajectories of two colliding compatibilized droplets were measured by Hu et al. (27) and Ha et al. (45). For the low viscosity ratio systems studied, the film drainage time increased monotonically as the amount of (reactive) compatibilizer was increased, consistent with a slowing down of the film drainage due to Marangoni stresses. For the high viscosity ratio systems, the effect of increasing the compatibilizer concentration was nonmonotonic and could not be explained merely by Marangoni stresses. Although Ha et al. (45) speculated about various possible mechanisms, no conclusive answer could be given.

Recently, it has been shown experimentally that interfacial viscoelasticity is present in compatibilized blends. Several research teams observed a slow relaxation process in small amplitude oscillatory shear experiments on compatibilized blends that was associated with interfacial elasticity (41, 46–50). Van Hemelrijck et al. (41) showed that the frequency at which this slow process occurs is independent of coalescence time but is strongly dependent on compatibilizer concentration (see Figure 13.3). Two distinct relaxation shoulders are present that are nicely predicted

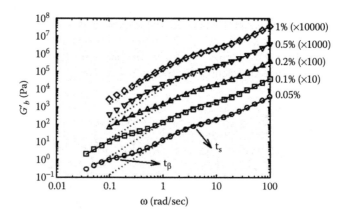

FIGURE 13.3 Vertically shifted G_b' for compatibilized blends polydimethylsiloxane-polyisoprene (PDMS-PI) with compatibilizer concentrations ranging from 0.05 to 1 wt% (concentrations expressed based on the amount of dispersed phase, preshear at 4.8 sec^{-1}). The slow relaxation phenomenon is clearly dependent on the compatibilizer concentration. (Van Hemelrijck et al., *J. Rheol.*, 48, 148–158, 2004. With permission.)

by the Palierne model (solid line). The dotted lines in Figure 13.3 correspond to the uncompatibilized blends, which only show a single relaxation shoulder, reflecting droplet form relaxation. It is also seen that the slow relaxation phenomenon is clearly dependent on the compatibilizer concentration. Different possible physical mechanisms were considered leading to a probable link between the interfacial elasticity and Marangoni stresses that decrease the coalescence efficiency in compatibilized blends.

13.2.3 RHEOLOGY OF COMPATIBILIZED BLENDS

From a rheological point of view, it would be desirable to incorporate the various microstructural processes into a single constitutive equation that is able to describe the full rheological response of a compatibilized blend. Although some experimental studies are available that relate the linear viscoelastic response of the blend to the occurrence of interfacial grafting reactions in reactively compatibilized blends [e.g., see refs. (50–52)], no general mechanistical picture is yet available that describes the influence of compatibilization on the rheological properties of the material.

Iza et al. (53) showed that the addition of a compatibilizer did not affect the steady shear properties but significantly modified the behavior during start-up of shear flow. The change in start-up dynamics was also reported by Van Hemelrijck et al. (54) for blends containing a small amount of compatibilizer. The characteristic overshoot of the first normal stress difference (N_1) was similar to that of uncompatibilized blends, but it shifted to shorter times, indicative of a faster breakup process. However, highly compatibilized systems no longer showed an overshoot in N_1, suggesting a lack of deformation of the compatibilized droplets. Their behaviour actually became rather suspensionlike. Recently, Velankar et al. (55) discussed the steady state rheological properties of compatibilized blends with a viscosity ratio ranging from 0.1 to 2.65. The steady shear viscosity of the compatibilized blends was found to be higher than for the corresponding uncompatibilized blends, independent of the viscosity ratio. Moreover, compatibilized blends were found to be less shear thinning than uncompatibilized ones. In addition, the relative interfacial normal stress, which reflects the contribution of droplets to the first normal stress difference, was higher for the compatibilized blends at low viscosity ratio, whereas at high viscosity ratios, hardly any effect of the compatibilizer was observed (see, e.g., Figure 13.4). Most of these observations are qualitatively consistent with the occurrence of Marangoni stresses that stabilize droplets against deformation in shear flow.

One of the key parameters during melt-processing is the viscosity of the blend. Since the 1970s, various reports on anomalously low viscosities for immiscible polymer blends have appeared [see reviews by Utracki (56,57)]. In some cases, the blend viscosity even becomes lower than that of the pure components. Zhao and Macosko (58) showed that systems that are more incompatible, i.e., having a larger Flory-Huggins interaction parameter, show a larger negative viscosity deviation. Interfacial slip has been proposed to account for the observed anomalies (59,60); in the interfacial region, polymer chains are less entangled than in the bulk phases and will hence form a low viscosity zone. Hence, when a shear stress is imposed on the material, the resulting shear rate in the interface region will become higher than in

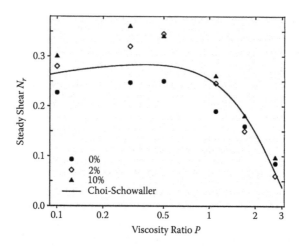

FIGURE 13.4 Relative interfacial normal stress N_r under steady shear at 1.2 sec^{-1} as a function of viscosity ratio p. The blends consist of 10% polyisobutylene (PIB) in PDMS, and the compatibilizer concentrations are based on the dispersed phase content. (Velankar et al., J. Rheol., 48, 725–744, 2004. With permission.)

the bulk phases leading to a slip layer. Recently, Zhao and Macosko (58) studied interfacial slip in specially made multilayer structures. Such a layered composite structure is a good model system to test interfacial slip because of its large interfacial area. To test for interfacial slip, the pressure drop across a slit die in an extruder was measured (see Figure 13.5). Addition of a premade diblock copolymer was able to suppress the interfacial slip, at least after a sufficiently long time to allow for block copolymers to diffuse to the interfaces. *In situ* formed graft copolymers, however, directly suppressed slip. Similar conclusions were drawn by Van Puyvelde et al. (61) who studied the effect of reactive compatibilization on the suppression of interfacial slip in multilayer structures using standard rheological techniques.

13.3 INFLUENCE OF COMPONENT ELASTICITY ON THE RHEOLOGY AND MORPHOLOGY OF POLYMER BLENDS

Commercial blends typically consist of high molecular weight polymers that are processed at high temperatures in the molten state. Such components typically exhibit significant elasticity, especially at the high shear rates encountered in processing equipment. Hence, we can expect significant deviations from the dynamics of blends consisting of Newtonian components.

13.3.1 EFFECT OF COMPONENT ELASTICITY ON DEFORMATION AND BREAKUP

At present, it is not known how the flow induced morphology of a polymer blend is related to the viscoelastic properties of the component polymers, such as the shear

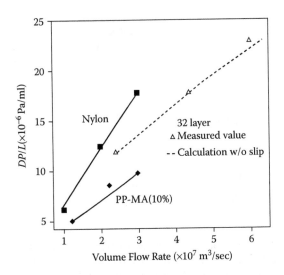

FIGURE 13.5 Effect of reactive compatibilization on the pressure drop (i.e., a measure for the viscosity) across a slit die in an extruder. The measured values of the pressure drop for a multilayer blend (32 layers) consisting of nylon and MA-modified polypropylene (PP) (open symbols) follow the prediction of no slippage between the layers whereas the values for an uncompatibilized multilayer structure (not shown) show a pressure drop well below the theoretical curve, pointing to slippage between the layers. (Zhao and Macosko, *J. Rheol.*, 46, 145–167, 2002. With permission.)

rate dependent viscosity and the normal stresses. In addition, the interpretation of experimental results is hampered by the lack of a theoretical framework. The latter is a result of the wide variety of constitutive equations that are available to describe the viscoelastic nature of materials, none of which is of general nature. In view of this, it seems that the morphological evolution associated with viscoelasticity is best explored in the simplest possible case, i.e., that of a single droplet in a well controlled flow field (the "infinite dilution" limit) where the elastic properties of the components are systematically changed, keeping all the other parameters fixed.

Although the available studies are in part conflicting, there seems to be a general consensus that droplet phase elasticity leads to less deformed, more stable droplets relative to comparable Newtonian droplets. Matrix elasticity tends to decrease the droplet stability [e.g., see refs. (62–76)]. However, in most of these studies, both the elasticities of droplet and matrix and the viscosity ratio were varied simultaneously, making it difficult to isolate the effect of elasticity. Recently, however, a few experimental studies have been published in which either the droplet elasticity or the matrix elasticity has been varied, thus keeping the viscosity ratio constant. Guido et al. (77,78) studied the deformation of a single droplet in a Boger fluid under steady state, slow shear flow by means of video enhanced microscopy. The main experimental result is that drop orientation in the flow direction is significantly enhanced as compared to the Newtonian case when the normal stress of the matrix is increased. Lerdwijitjarud et al. (79–81) studied the deformation of viscoelastic droplets in Newtonian matrices in simple shear flow, keeping the viscosity ratio fixed. The value

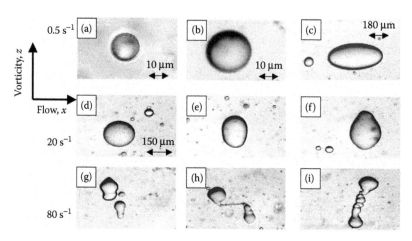

FIGURE 13.6 Video micrographs of a PIB based Boger fluid in a PDMS matrix. At low shear rates (A to C: 0.5 sec⁻¹), (A,B) a weak deformation is observed for small droplets and (C) a larger deformation is observed for the larger droplet. With increased flow rate (D to F: 20 sec⁻¹; G to I: 80 sec⁻¹), there is an increasing tendency for the droplets to align along the vorticity direction. (Migler, J. Rheol., 44, 277–290, 2000. With permission.)

of the steady state capillary number was found to increase with the first normal stress difference of the droplets, indicating again that elasticity stabilizes the droplets.

Although the tools to study the rheology-morphology relation in blends consisting of elastic components are available (e.g., rheology, various optical techniques) a thorough understanding of the effect of elasticity on the deformation and breakup — including the important determination of dimensionless parameters in addition to the viscosity ratio and the capillary number — is not available at present. Moreover, in some cases, remarkable effects have been observed along the vorticity axis. Levitt et al. (82) observed drop widening in the neutral direction of the shear flow when isolated droplets of polypropylene were sheared in a high elasticity polystyrene matrix. They proposed that the width of the flattened drops was dependent on the difference in storage modulus between the matrix and the droplet phase. In addition, Migler (83) and Hobbie and Migler (84) demonstrated that under conditions of high droplet elasticity, droplets can align in the vorticity direction rather than, as expected, in the flow direction (see, e.g., Figure 13.6). This "exotic" behavior was attributed to the high normal forces in the droplets and the presence of closed streamlines that form in the flow gradient plane and that cause the droplets to expand in the vorticity direction. These observations demonstrate that the incorporation of elasticity in blend components introduces some peculiar features that hamper — at least for the time being — the establishment of a complete understanding and a theoretical foundation of the deformation and breakup process.

13.3.2 EFFECT OF COMPONENT ELASTICTIY ON THE COALESCENCE

A number of theoretical studies have focused on understanding the mechanisms involved in coalescence using both numerical and scaling arguments applied at the level of individual Newtonian droplets in Newtonian matrices [see ref. (2) and references therein]. Even for this Newtonian reference case, only a few experimental investigations have examined the problem at the scale of single droplets (27,85–88), the degree of complexity commonly accepted as the starting point of a fundamental understanding of the problem. Instead, most of the experimental studies on coalescence involve postmortem types of analysis in which the coalescence mechanism is indirectly inferred from observations of changes in the droplet size distributions in rather concentrated blends resulting from varying flow conditions [e.g., see ref. (36)]. Hence, it is not very surprising that at present, systematical *in situ* studies on the effect of elasticity on the coalescence phenomena are lacking in the literature.

Elmendorp and van der Vegt (89) suggested that the molecular weight of the matrix might have a significant effect on coalescence even though its rheology remains Newtonian. Specifically, it was suggested that the film drainage process is facilitated in high molecular weight polymer liquids even if this film fluid does not show significant elasticity. This suggestion has been confirmed by the tedious experimental work by Park et al. (87). These authors carefully measured droplet trajectories during "glancing collisions" for systems containing matrix fluids with varying molecular weights but keeping the viscosity ratio constant. It was found that coalescence is facilitated as the molecular weight of the matrix fluid is increased. Although the molecular weights of the fluids used were not sufficient to obtain very significant elasticities, it might be anticipated that the conclusions drawn from this study are also valid for matrix materials exhibiting realistic elasticities. The easier coalescence was attributed to an enhanced film drainage process. This might seem somewhat counterintuitive since many possible alternative mechanisms would suggest that increasing the molecular weight might rather hinder the drainage process. For example, various studies report an increased viscosity with respect to the bulk viscosity when polymer fluids are confined between boundaries that are separated by a few molecular diameters [e.g., see refs. (90,91)]. In addition, we could anticipate that, when the polymeric film thins during coalescence, steric hindrance could start playing a role when the molecular weight is increased, hence, slowing down the drainage process. However, the results of Park et al. (87) seem to point toward an enhancement of the film drainage whenever the matrix molecular weight increases. These authors suggested the existence of a "slip layer" around polymer interfaces that might be responsible for the easier coalescence. Unfortunately, no experiments are available that measure the velocities in the draining film that could confirm the existence of a slip layer. However, the study is rather unique, and the ideas put forth should certainly be pursued in future experimental work.

13.4 CONCENTRATION EFFECTS ON THE RHEOLOGY-MORPHOLOGY RELATION

In contrast to the experiments on dilute or semiconcentrated blends conducted under well controlled conditions, experiments with dispersed phase concentrations of 20% and more are mostly carried out with industrial polymers in rather complex flow fields as encountered in mixers and extruders. Hence, it is difficult to draw conclusions from these studies. Rather than the usual droplet-matrix structure, concentrated blends may contain fibrillar [e.g., see refs. (92–95)] or cocontinuous morphologies [e.g., see refs. (96–99)]. In most studies, the short processing times and variable kinematic conditions make the generated structures transient rather than steady state. Here, we report on *in situ* morphology-rheology experiments on concentrated blend systems. The aim is to provide an overview of methods in which rheology or rheo-optical techniques can be used to deduce morphological information. Since more concentrated blends are less transparent, the majority of the reported studies use rheology to investigate the blend morphology.

13.4.1 RHEOLOGY OF CONCENTRATED BLENDS

In the concentration range around the phase inversion, the morphology can become very complex. The repeatedly discussed cocontinuous structures seem to have been produced most often by transient flow histories [e.g., see ref. (100)]. To what extent they occur during steady shear flow has not been established yet. In concentrated systems, fibrillar structures are found both under transient as well as under steady state conditions [e.g., see refs. (94,101)]. Measuring the steady state rheology of such systems is not straightforward because of the difficulty of achieving reproducible results (102). For dilute, uncompatibilized blends, 3000 to 4000 strain units have been reported to be adequate to reach a steady state morphology. However, Jansseune et al. (103) demonstrated that shearing for up to 80,000 strain units was required in concentrated systems. Jansseune et al. (103) studied in detail the steady state rheology of concentrated blends, varying the relevant parameters such as concentration, shear rate, and viscosity ratio. Using stress relaxation measurements, the contributions of the components and the interface could be separated. An example of such measurements is shown in Figure 13.7. Here, the interfacial contribution to the first normal stress difference is plotted as a function of concentration for a blend consisting of polyisobutylene (PIB) and polydimethylsiloxane (PDMS). On both ends of the composition range, a rise in interfacial contribution is found with increasing amount of dispersed phase. Toward the middle of the composition range, the interfacial contributions reach a plateau or even decrease toward a local minimum. Such a pattern was already observed by Ziegler and Wolf (95) and by Jeon et al. (104) for, respectively, the extent of shear thinning and the storage modulus as a function of concentration. The local minima in the interfacial contribution, as observed in Figure 13.7, can be explained by the formation of fibrils as such structures contribute to a lesser extent to the stress as compared to a droplet-matrix structure.

Although it is not completely clear yet whether cocontinuous morphologies are real stationary morphologies, much attention has been paid to these structures because in such structures both components can fully contribute to the properties of the blend. Cocontinuous morphologies are formed in a certain concentration range, with the phase inversion concentration regarded as its center. Many studies try to correlate this interval with component properties. Various models have been proposed that relate the viscosity ratios of the blend components to the phase inversion point [e.g., see refs. (93,105,106)], but none was found to have a general predictive value. Bourry and Favis (107) introduced an elasticity ratio as an important parameter

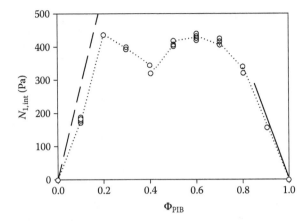

FIGURE 13.7 Interfacial contributions to the first normal stress difference in steady state conditions (PIB in PDMS, 5 sec^{-1}). The interfacial contribution to the stress has been determined by applying a specific rheological protocol: During stress relaxation experiments, the relaxation curve can be divided into two parts, an almost instantaneously relaxing part, reflecting the component contributions, and a more slowly relaxing part, resulting from the interface. Extrapolation of the latter part to the moment at which the flow is stopped yields the interfacial contribution during flow. (Adapted from Jansseune et al., *J. Rheol.*, 47, 829–845, 2003.)

to understand phase inversion. Their model accounts for the tendency of the more elastic phase to form the matrix. Bourry and Favis achieved much better agreement with their experimental data, particularly at high shear rates, as compared to the predictions merely based on viscous effects. An interesting study in this respect was performed by Steinmann et al. (108). These authors showed that for various blends, the ratio of elastic moduli G, evaluated at frequencies with the same loss modulus G, was a good indication of phase inversion. This requirement — in contrast to the more conventional view in which the moduli are evaluated at constant shear rate — was explained by the necessity of stresses needing to be continuous across the interphase. The work done in this field clearly indicates that further investigations concerning the influence of viscosity and component elasticity have to be done.

13.4.2 RELAXATION AFTER CESSATION OF FLOW

Systematic relaxation experiments on concentrated blends, either studied by rheology or by other *in situ* methods, are scarce in the literature. However, they are of technological interest since, at some point during processing, stress is removed and the structure might start to relax, resulting in a changing morphology. When starting from a fibrillar structure, the relaxation behavior can be quite complex. Elemans et al. (109) reported two relaxation mechanisms for highly concentrated blends. In the first one, neighbouring aligned strands coalesce with each other, reducing their interfacial area. Similar observations have been made for the annealing of fibrillar and cocontinuous structures in a variety of blend systems with dispersed phase content greater than 30% (110,111). In the second mechanism, the fibrils break under

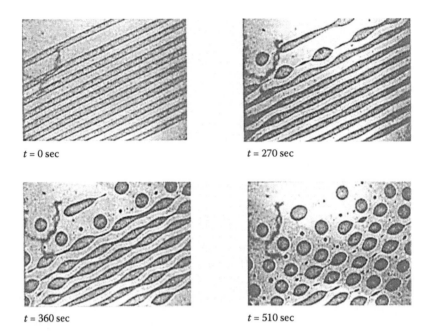

$t = 0$ sec $t = 270$ sec

$t = 360$ sec $t = 510$ sec

FIGURE 13.8 Process of the out-of-phase breakup of polyamide 6 (PA-6) threads in a matrix of PS. The initial thread diameter is 70 μm, and the viscosity ratio is 0.04. (Adapted from Knops et al., *AIChE J.*, 47, 1740–1745, 2001.)

the action of Rayleigh instabilities. However, the simultaneous breakup process of a liquid cylinder surrounded by other liquid fibrils is slightly different from the breakup process of an isolated liquid fibril of comparable diameter. In concentrated systems, the start of large amplitude distortions is delayed for a certain period because of the presence of neighbouring threads. Once this "incubation" period is complete, the growth rate of the distortions is the same for all the threads. Depending on the distance between the threads, the breakup can occur either by an in-phase mode when the fibrils are very close together or by an out-of-phase mode when the fibrils are somewhat further apart (see Figure 13.8).

It needs to be stressed that this out-of-phase mechanism requires an almost perfect alignment of the fibrils. This situation hardly occurs during real processing. Jansseune et al. (103) demonstrated that the relaxation in a real concentrated polymer blend is much faster than the idealized aligned situation would allow. They demonstrated that the quiescent coalescence process was capable of explaining their results.

13.5 CONCLUSIONS

In this chapter, an overview is given of some important elements allowing us to come to a general predictive framework for the morphology-rheology relationship in immiscible polymer blends. In the past, many researchers have focused on the study of the idealized case, which are a mixture of Newtonian components. In this chapter, it has been shown that in recent years research has emerged that addresses the effect of other aspects such as compatibilization, component elasticity, and concentration effects. Although not all the effects are understood yet, the research on their effect on the morphology development in blends will eventually lead to future improvements in both processing routes as well in product formulations. It is our strong belief that *in situ* and time resolved measurements — such as the ones cited throughout this paper — will form a fundamental cornerstone in these developments.

ACKNOWLEDGMENTS

Peter Van Puyvelde is indebted to the FWO-Vlaanderen for a postdoctoral fellowship. This research has been partially funded by a GOA project (03/06) from the Research Council K.U. Leuven and by a FWO-project (G.0623.04).

REFERENCES

1. Ottino, J.M., DeRoussel, P., Hansen, S., et al., Mixing and dispersion of viscous liquids and powered solids, *Adv. Chem. Eng.*, 25, 105–204, 1999.
2. Tucker, C.L. and Moldenaers, P., Microstructural evolution in polymer blends, *Ann. Rev. Fluid Mech.*, 34, 177–210, 2002.
3. http://bcc.ecnext.com/coms2/summary__002_000859_000000_000000_0002_1BCC, Business Communications Company, CT, USA, 2001.
4. Gerard, H., Cabral, J.T., and Higgins, J.S., Flow-induced enhancement of concentration fluctuations in polymer mixtures, *Phil. Trans. R. Soc. Lond., Math., Phys. Eng. Sci.*, 361, 767–778, 2003.
5. Nakatani, A.I., Kim, H., Takahashi, H., et al., Shear stabilization of critical fluctuations in bulk polymer blends studied by small angle neutron scattering, *J. Chem. Phys.*, 93, 795–802, 1990.
6. Fernandez, M.L., Higgins, J.S., Horst, R., et al., Flow instabilities in polymer blends under shear, *Polymer*, 36, 149–159, 1995.
7. Vlassopoulos, D., Rheology of critical LCST polymer blends: poly(styrene-co-maleic anhydride)/poly(methyl methacrylate), *Rheol. Acta*, 35, 556–566, 1996.
8. Flaris, V., Baker, W.E., and Lambla, M., A new technique for the compatibilization of polyethylene/polystyrene blends, *Polym. Net. Blends*, 6, 29–34, 1996.
9. Loyens, W. and Groeninckx, G., Phase morphology development in reactively compatibilized polyethylene terephtalate/elastomer blends, *Macromol. Chem. Phys.*, 203, 1702–1714, 2002.
10. Oommen, Z., Zachariah, S.R., Thomas, S., et al., Melt rheology and morphology of uncompatibilized and in situ compatibilized nylon-6/ethylene propylene rubber blends, *J. Appl. Polym. Sci.*, 92, 252–264, 2004.
11. Fuller, G.G., *Optical Rheometry of Complex Fluids*, Oxford University Press, New York, 1996.
12. Van Puyvelde P., Velankar, S., and Moldenaers, P., Rheology and morphology of compatibilized polymer blends, *Curr. Opinion Coll. Interf. Sci.*, 6, 457–463, 2001.
13. Lorenzo, M.L.D., Compatibilization criteria and procedures for binary blends: a review, *J. Polym. Eng.*, 17, 429–459, 1997.
14. Koning, C., Duin, M.V., Pagnoulle, C., et al., Strategies for compatibilization of polymer blends, *Prog. Polym. Sci.*, 23, 707–757, 1998.

15. Grace, H.P., Dispersion phenomena in high viscosity immiscible fluid systems and application of static mixers as dispersion devices in such systems, *Chem. Eng. Commun.*, 14, 225–277, 1982.

16. Bentley, B.J. and Leal, L.G., An experimental investigation of drop deformation and breakup in steady, two-dimensional linear flows, *J. Fluid Mech.*, 167, 241–283, 1986.

17. Jansen, K.M.B., Agterof, W.G.M., and Mellema, J., Droplet breakup in concentrated emulsions, *J. Rheol.*, 45, 227–236, 2001.

18. Rumscheidt, F.D. and Mason, S.G., Particle motions in sheared suspensions. XII. Deformation and burst of fluid drops in shear and hyperbolic flow, *J. Colloid Int. Sci.*, 16, 238–261, 1961.

19. Stone, H.A., Dynamics of drop deformation and breakup in viscous fluids, *Ann. Rev. Fluid Mech.*, 26, 65–102, 1994.

20. Guido, S. and Villone, M., Three-dimensional shape of a droplet under simple shear flow, *J. Rheol.*, 42, 395–415, 1999.

21. Tomotika, S., On the instability of a cylindrical thread of a viscous liquid surrounded by another viscous liquid, *Proc. R. Soc. Lond. Ser. A*, 150, 322–337, 1935.

22. Stone, H.A., Bentley, B.J., and Leal, L.G., An experimental study of transient effects in the breakup of viscous drops, *J. Fluid Mech.*, 173, 131–158, 1986.

23. Chapleau, N., Favis, B.D., and Carreau, P.J., Measuring the interfacial tension of polymers in the presence of an interfacial modifier: migrating the modifier to the interface, *J. Polym. Sci. Polym. Phys.*, 36, 1947–1958, 1998.

24. Elemans, P.H.M., Janssen, J.M.H., and Meijer, H.E.H., The measurement of interfacial tension in polymer-polymer systems: the breaking thread method, *J. Rheol.*, 34, 1311–1325, 1990.

25. Xing, P.X., Bousmina, M., Rodrigue, D., and Kamal, M.R., Critical experimental comparison between five techniques for the determination of interfacial tension in polymer blends: model system of polystyrene-polyamide, *Macromolecules*, 33, 8020–8034, 2000.

26. Jeon, H.K. and Macosko, C.W., Visualization of block copolymer distribution on a sheared drop, *Polymer*, 44, 5381–5386, 2003.

27. Hu, Y.T., Pine, D.J., and Leal, L.G., Drop deformation, breakup and coalescence with compatibilizer, *Phys. Fluids*, 12, 484–489, 2000.

28. Levitt, L. and Macosko, C.W., Shearing of polymer drops with interface modification, *Macromolecules*, 32, 6270–6277, 1999.

29. Ramic, A.J., Stehlin, J.C., Hudson, S.D., et al., Influence of block copolymer on droplet breakup and coalescence in model immiscible polymer blends, *Macromolecules*, 33, 371–374, 2000.

30. Velankar, S., Van Puyvelde, P., Mewis, J., et al., Effect of compatibilization on the breakup of polymeric drops in shear flow, *J. Rheol.*, 45, 1007–1019, 2001.

31. Chesters, A.K., The modeling of coalescence processes in fluid-liquid dispersions: a review of current understanding, *Trans. IchemE Ser. A.*, 69, 259–270, 1991.

32. Janssen, J.M.H., *Dynamics of Liquid-Liquid Mixing*, Ph.D. thesis, Eindhoven University of Technology, The Netherlands, 1993.

33. Yang, H., Park, C.C., Hu, Y.T., et al., The coalescence of two equal sized drops in a two-dimensional linear flow, *Phys. Fluids*, 13, 1087–1103, 2001.

34. Vinckier, I., Moldenaers, P., Terracciano, A.M., et al., Droplet size evolution during coalescence in semiconcentrated model blends, *AIChE J.*, 44, 951–958, 1998.

35. Rusu, D. and Peuvrel-Disdier, E., In-situ characterization by small angle light scattering of the shear-induced coalescence mechanisms in immiscible polymer blends, *J. Rheol.*, 43, 1391–1409, 1999.

36. Sundararaj, U. and Macosko, C.W., Drop breakup and coalescence in polymer blends: the effects of concentration and compatibilization, *Macromolecules*, 28, 2647–2657, 1995.
37. Lepers, J.C., Favis, B.D., and Lacroix, C., The influence of partial emulsification on coalescence suppression and interfacial tension reduction in PP/PET blends, *J. Polym. Sci. Polym. Phys.*, 37, 939–951, 1999.
38. Lyu, S.P., Bates, F.S., and Macosko, C.W., Coalescence in polymer blends during shearing, *AIChE J.*, 46, 229–238, 2000.
39. Lyu, S.P., Jones, T.D., Bates, F.S., et al., Role of block copolymers on suppression of droplet coalescence, *Macromolecules*, 35, 7845–7855, 2002.
40. Van Puyvelde, P., Velankar, S., and Moldenaers, P., Effect of Marangoni stresses on the deformation and coalescence in compatibilized immiscible polymer blends, *Polym. Eng. Sci.*, 42, 1956–1964, 2002.
41. Van Hemelrijck, E., Van Puyvelde, P., Velankar, S., et al., Interfacial elasticity and coalescence suppression in compatibilized polymer blends, *J. Rheol.*, 48, 143–158, 2004.
42. Hudson, S.D., Jamieson, A.M., and Burkhart, B.E., The effect of surfactant on the efficiency of shear-induced drop coalescence, *J. Colloid Int. Sci.*, 265, 409–421, 2003.
43. Nandi, A., Mehra, A., and Kakhar, D.V., Suppression of coalescence in surfactant stabilized emulsions by shear flow, *Phys. Rev. Lett.*, 83, 2461–2464, 1999.
44. Nandi, A., Khakhar, D.V., and Mehra, A., Coalescence in surfactant stabilized emulsions subjected to shear flow, *Langmuir*, 17, 2647–2655, 2001.
45. Ha, J.W., Yoon, Y., and Leal, L.G., The effect of compatibilizer on the coalescence of two drops in flow, *Phys. Fluids*, 15, 849–867, 2003.
46. Riemann, R.E., Cantow, H.J., and Friedrich, C., Rheological investigation of form relaxation and interface relaxation processes in polymer blends, *Polym. Bull.*, 36, 637–643, 1996.
47. Riemann, R.E., Cantow, H.J., and Friedrich, C., Interpretation of a new interface-governed relaxation process in compatibilized polymer blends, *Macromolecules*, 30, 5476–5484, 1997.
48. Jacobs, U., Fahrlander, M., Winterhalter, J., et al., Analysis of Palierne's emulsion in the case of viscoelastic interfacial properties, *J. Rheol.*, 43, 1495–1509, 1999.
49. Moan, M., Huitric, J., Mederic, P., et al., Rheological properties and reactive compatibilization of immiscible polymer blends, *J. Rheol.*, 44, 1227–1245, 2000.
50. Asthana, H. and Jayaraman, K., Rheology of reactively compatibilized polymer blends with varying extent of interfacial reaction, *Macromolecules*, 32, 3412–3419, 1999.
51. Ernst, B., Koenig, J.F., and Muller, R., Rheological characterization of interfacial crosslinking in blends of reactive copolymers, *Macromol. Symp.*, 158, 43–56, 2000.
52. Lacroix, C., Bousmina, M., Carreau, P.J., et al., Properties of PETG/EVA blends. 2. Study of reactive compatibilization by nmr spectroscopy and linear viscoelastic properties, *Polymer*, 37, 2949–2956, 1996.
53. Iza, M., Bousmina, M., and Jérôme, R., Rheology of compatibilized immiscible viscoelastic polymer blends, *Rheol. Acta*, 40, 10–22, 2001.
54. Van Hemelrijck, E., Van Puyvelde, P., and Moldenaers, P., Rheology and morphology of compatibilized and uncompatibilized polymer blends, *Proceedings of the Meeting of the European Polymer Processing Society*, Guimaraes, Portugal, 2002.
55. Velankar, S., Van Puyvelde, P., Mewis, J., et al., Steady-shear rheological properties of model compatibilized blends, *J. Rheol.*, 48, 725–744, 2004.
56. Utracki, L.A., Melt flow of polymer blends, *Polym. Eng. Sci.*, 23, 602–609, 1983.

57. Utracki, L.A. and Kamal, M.R., Melt rheology of polymer blends, *Polym. Eng. Sci.*, 22, 96–114, 1982.

58. Zhao, R. and Macosko, C.W., Slip at polymer-polymer interfaces: rheological measurements on coextruded multilayers, *J. Rheol.*, 46, 145–167, 2002.

59. Lin, C.C., A mathematical model for viscosity in capillary extrusion of two-component polymer blends, *Polym. J.*, 11, 185–192, 1979.

60. Shih, C.K., Rheological properties of incompatible blends of two elastomers, *Polym. Eng. Sci.*, 16, 742–746, 1976.

61. Van Puyvelde, P., Oommen, Z., Koets, P., et al., Effect of reactive slip in nylon-6/EPR blends, *Polym. Eng. Sci.*, 43, 71–78, 2003.

62. Vanoene, H., Modes of dispersion of viscoelastic fluids in flow, *J. Colloid Int. Sci.*, 40, 448–467, 1972.

63. Flumerfelt, R.W., Drop breakup in simple shear fields of viscoelastic fluids, *Ind. Eng. Chem. Fundam.*, 11, 312–318, 1972.

64. Bartram, E., Goldsmith, G.L., and Mason, S., Particle motions in non-Newtonian media, *Rheol. Acta*, 14, 776–782, 1975.

65. Chin, H.B. and Han, C.D., Studies on droplet deformation in extensional flows, *J. Rheol.*, 23, 557–590, 1979.

66. Elmendorp, J., *A Study on Polymer Blending Microrheology*, Ph.D. thesis, Technical University of Delft, The Netherlands,1986.

67. de Bruijn, R.A., *Deformation and Breakup of Drops in Simple Shear Flow*, Ph.D. thesis, Technical University Eindhoven, The Netherlands, 1989.

68. Milliken, W. and Leal, L.G., Deformation and breakup of viscoelastic drops in planar extension flows, *J. Non-Newt.Fluid Mech.*, 40, 335–379, 1991.

69. Varanasi, P.P., Ryan, M.E., and Stroeve, P., Experimental study on the breakup of model viscoelastic drops in uniform shear flow, *Ind. Eng. Chem. Res.*, 33, 1858–1866, 1994.

70. Sundararaj, U., Dori, Y., and Macosko, C.W., Sheet formation in immiscible polymer blends: model experiments on initial blend morphology, *Polymer*, 36, 1957–1968, 1995.

71. Mason, T.G. and Bibette, J., Shear rupturing of droplets in complex fluids, *Langmuir*, 13, 4600–4613, 1997.

72. Tretheway, D.C. and Leal, L.G., Deformation and relaxation of Newtonian drops in planar extensional flows of a Boger fluid, *J. Non-Newtonian Fluid Mech.*, 99, 81–108, 2001.

73. Mighri, F., Ajji, A., and Carreau, P.J., Influence of elastic properties on drop deformation in elongational flow, *J. Rheol.*, 41, 1183–1201, 1997.

74. Mighri, F., Carreau, P.J., and Ajji, A., Influence of elastic properties on drop deformation and breakup in shear flow, *J. Rheol.*, 42, 1477–1490, 1998.

75. Mighri F. and Huneault, M.A., Drop deformation and breakup mechanisms in viscoelastic model fluid systems and polymer blends, *Can. J. Chem. Eng.*, 80, 1028–1035, 2002.

76. Tretheway, D.C. and Leal, L.G., Surfactant and viscoelastic effects on drop deformation in 2-D extensional flow, *AIChE J.*, 45, 929–937,1999.

77. Guido, S., Simeone, M., and Greco, F., Deformation of a Newtonian drop in a viscoelastic matrix under steady shear flow: experimental validation of slow flow theory, *J. Non-Newtonian Fluid Mech.*, 114, 65–82, 2003.

78. Guido, S., Simeone, M., and Greco, F., Effects of matrix viscoelasticity on drop deformation in dilute polymer blends under slow shear flow, *Polymer*, 44, 467–471, 2003.

79. Lerdwijitjarud, W., Sirivat, A., and Larson, R.G., Influence of elasticity on dispersed-phase droplet size in immiscible polymer blends in simple shearing flow, *Polym. Eng. Sci.*, 42, 798–809, 2002.

80. Lerdwijitjarud, W., Larson, R.G., Sirivat, A., et al., Influence of a weak elasticity of dispersed phase on droplet behaviour in sheared polybutadiene/poly(dimethylsiloxane) blends, *J. Rheol.*, 47, 37–58, 2003.

81. Lerdwijitjarud, W., Sirivat, A., and Larson, R.G., Influence of dispersed-phase elasticity on steady-state deformation and breakup of droplets in simple shearing flow of immiscible polymer blends, *J. Rheol.*, 48, 843–862, 2004.

82. Levitt, L., Macosko, C.W., and Pearson, S.D., Influence of normal stress difference on polymer drop deformation, *Polym. Eng. Sci.*, 36, 1647–1655, 1996.

83. Migler, K.B., Droplet vorticity alignment in model polymer blends, *J. Rheol.*, 44, 277–290, 2000.

84. Hobbie, E.K. and Migler, K.B., Vorticity elongation in polymeric emulsions, *Phys. Rev. Lett.*, 82, 5393–5396, 1999.

85. Guido, S. and Simeone, M., Binary collisions of drops in simple shear flow by computer-assisted video optical microscopy, *J. Fluid Mech.*, 357, 1–20, 1998.

86. Yang, H., Park, C.C., Hu, Y.T., et al., Coalescence of two equal-sized drops in a two-dimensional linear flow, *Phys. Fluids*, 13, 1087–1106, 2001.

87. Park, C.C., Baldessari, F., and Leal, L.G., Study of molecular weight effects on coalescence: interface slip layer, *J. Rheol.*, 47, 911–942, 2003.

88. Yang, H., Zhang, H., Moldenaers, P., et al., Rheo-optical investigation of immiscible polymer blends, *Polymer*, 39, 5731–5737, 1998.

89. Elmendorp, J.J. and van der Vegt, A.K.., Study of blending microrheology. Part IV. Influence of coalescence on blend morphology origination, *Polym. Eng. Sci.*, 26, 1332–1338, 1986.

90. Reiter, G., Demirel, A.L., and Granick, S., From static to kinetic friction in confined liquid films, *Science*, 263, 1741–1744, 1994.

91. Shaffer, J.S., Dynamics of confined polymer melts: topology and entanglements, *Macromolecules*, 29, 1010–1013, 1996.

92. Chapleau, N. and Favis, B.D., Droplet/fibre transitions in immiscible polymer blends generated during melt processing, *J. Mat. Sci*, 30, 142–150., 1995.

93. Luciani, A. and Jarrin, J., Morphology development in immiscible polymer blends, *Polym. Eng. Sci.*, 36, 1619–1626, 1996.

94. Martin, P., Carreau, P.J., and Favis, B.D., Investigating the morphology/rheology interrelationships in immiscible polymer blends, *J. Rheol.*, 44, 569–583, 2000.

95. Ziegler, V. and Wolf, B.A., Viscosity and morphology of the two-phase system PDMS/PDMS-ran-MPS, *J. Rheol.*, 43, 1033–1045, 1999.

96. Favis, B.D. and Chalifoux, J.P., Influence of composition on the morphology of polypropylene/polycarbonate blends, *Polymer*, 29, 1761–1767, 1988.

97. Mekhilef, N., Favis, B.D., and Carreau, P.J., Morphological stability, interfacial tension, and dual phase continuity in polystyrene-polyethylene blends, *J. Polym. Sci. Part B: Polym. Phys.*, 36, 293–308, 1997.

98. Mekhilef, N. and Verhoogt, H., Phase inversion and dual-phase continuity in polymer blends: theoretical predictions and experimental results, *Polymer*, 37, 4069–4077, 1996.

99. Potschke, P. and Paul, D.R., Formation of cocontinuous structures in melt-mixed immiscible polymer blends, *J. Macromol. Sci.-Polym. Rev.*, 43, 87–141, 2003.

100. Tol, R.T., Groeninckx, G., Vinckier, I., et al., Phase morphology and stability of co-continuous (PPE/PS)/PA6 and PS/PA6 blends: effect of rheology and reactive compatibilization, *Polymer*, 45, 2587–2601, 2004.

101. Huitric, J., Mederic, P., Moan, M., et al., Influence of composition and morphology on rheological properties of polyethylene/polyamide blends, *Polymer*, 39, 4849–4856, 1998.

102. Astruc, M. and Navard, P., A flow-induced phase inversion in immiscible polymer blends containing liquid-crystalline polymer studies by in-situ optical microscopy, *J. Rheol.*, 44, 693–712, 2000.

103. Jansseune, T., Moldenaers, P., and Mewis, J., Morphology and rheology of concentrated biphasic blends in steady shear flow, *J. Rheol.*, 47, 829–845, 2003.

104. Jeon, H.S., Nakatani, A.I., Hobbie, E.K., et al., Phase inversion of polybutadiene/polyisoprene blends under quiescent and shear conditions, *Langmuir*, 17, 3087–3095, 2001.

105. Utracki, L.A., On the viscosity-concentration dependence in immiscible polymer blends, *J. Rheol.*, 35, 1615–1637, 1991.

106. Jordhamo, G.M., Manson, J.A., and Sperling, L.H., Phase continuity and inversion in polymer blends and simultaneous interpenetrating networks, *Polym. Eng. Sci.*, 26, 517–527, 1986.

107. Bourry, D. and Favis, D.B., Cocontinuity and phase inversion in HDPE/PS blends: Influence of interfacial modification and elasticity, *J. Polym. Sci. Part B: Polym. Phys.*, 36, 1889–1899, 1998.

108. Steinmann, S., Gronski, W., and Friedrich, C., Cocontinuous polymer blends: influence of viscosity and elasticity ratios of the consituent polymers on phase inversion, *Polymer*, 42, 6619–6629, 2001.

109. Elemans, P.H.M., van Wunnik, J.M., and van Dam, R.A., Development of morphology of immiscible polymer blends, *AIChE J.*, 43, 1649–1651, 1997.

110. Bouilloux, A., Ernst, B., Lobbrecht, A., et al., Rheological and morphological study of the phase inversion in reactive polymer blends, *Polymer*, 38, 4775–4783, 1997.

111. Willemse, R.C., Co-continuous morphologies in polymer blends, *Polymer*, 40, 2175–2178, 1999.

112. Knops, Y.M.M., Slot, J.J.M., Elemans, P.H.M., et al., Simultaneous breakup of multiple viscous threads surrounded by viscous liquid, *AIChE J.*, 47, 1740–1745, 2001.